AF214157

Aerodynamics

A Computational Introduction

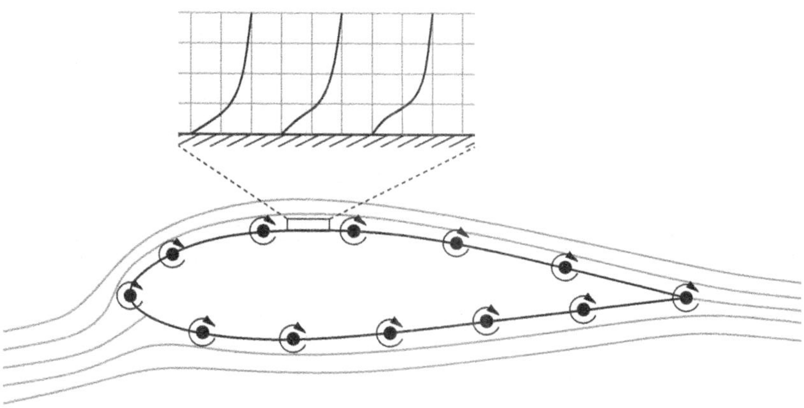

Krzysztof J. Fidkowski

University of Michigan

WILEY

Copyright © 2026 by John Wiley & Sons, Inc. All rights reserved, including rights for text and data mining and training of artificial intelligence technologies or similar technologies.

Published by John Wiley & Sons, Inc., Hoboken, New Jersey.

No part of this publication may be reproduced, stored in a retrieval system, or transmitted in any form or by any means, electronic, mechanical, photocopying, recording, scanning, or otherwise, except as permitted under Section 107 or 108 of the 1976 United States Copyright Act, without either the prior written permission of the Publisher, or authorization through payment of the appropriate per-copy fee to the Copyright Clearance Center, Inc., 222 Rosewood Drive, Danvers, MA 01923, (978) 750-8400, fax (978) 750-4470, or on the web at www.copyright.com. Requests to the Publisher for permission should be addressed to the Permissions Department, John Wiley & Sons, Inc., 111 River Street, Hoboken, NJ 07030, (201) 748-6011, fax (201) 748-6008, or online at http://www.wiley.com/go/permission.

The manufacturer's authorized representative according to the EU General Product Safety Regulation is Wiley-VCH GmbH, Boschstr. 12, 69469 Weinheim, Germany, e-mail: Product_Safety@wiley.com.

Trademarks: Wiley and the Wiley logo are trademarks or registered trademarks of John Wiley & Sons, Inc. and/or its affiliates in the United States and other countries and may not be used without written permission. All other trademarks are the property of their respective owners. John Wiley & Sons, Inc. is not associated with any product or vendor mentioned in this book.

Limit of Liability/Disclaimer of Warranty: While the publisher and the authors have used their best efforts in preparing this work, including a review of the content of the work, neither the publisher nor the authors make any representations or warranties with respect to the accuracy or completeness of the contents of this work and specifically disclaim all warranties, including without limitation any implied warranties of merchantability or fitness for a particular purpose. Certain AI systems have been used in the creation of this work. No warranty may be created or extended by sales representatives, written sales materials or promotional statements for this work. The fact that an organization, website, or product is referred to in this work as a citation and/or potential source of further information does not mean that the publisher and authors endorse the information or services the organization, website, or product may provide or recommendations it may make. This work is sold with the understanding that the publisher is not engaged in rendering professional services. The advice and strategies contained herein may not be suitable for your situation. You should consult with a specialist where appropriate. Further, readers should be aware that websites listed in this work may have changed or disappeared between when this work was written and when it is read. Neither the publisher nor authors shall be liable for any loss of profit or any other commercial damages, including but not limited to special, incidental, consequential, or other damages.

For general information on our other products and services or for technical support, please contact our Customer Care Department within the United States at (800) 762-2974, outside the United States at (317) 572-3993 or fax (317) 572-4002.

Wiley also publishes its books in a variety of electronic formats. Some content that appears in print may not be available in electronic formats. For more information about Wiley products, visit our web site at www.wiley.com.

Library of Congress Cataloging-in-Publication Data:

Names: Fidkowski, Krzysztof J. author
Title: Aerodynamics : a computational introduction / Krzysztof J. Fidkowski.
Description: Hoboken, New Jersey : John Wiley & Sons Inc, 2026. | Includes bibliographical references and index.
Identifiers: LCCN 2025027027 | ISBN 9781394319961 hardback | ISBN 9781394320004 adobe pdf | ISBN 9781394319985 epub
Subjects: LCSH: Aerodynamics | Aerospace engineering
Classification: LCC TL570 .F53 2026
LC record available at https://lccn.loc.gov/2025027027

Cover Design: Wiley
Cover Images: © skandar/stock.adobe.com, Courtesy of Krzysztof Fidkowski

This book is dedicated to anyone who is curious about flight and likes to tinker with a virtual wind tunnel, the computer.

Contents

Preface

Over 120 years after the Wright brothers' first powered flight, aerodynamics remains of interest among aerospace engineers and in popular culture. Aircraft have become ubiquitous as global air travel continues to grow, and many people have first-hand experience with flight. Yet heavier-than-air flight still carries a certain intrinsic mystique: how can 10 tons of thrust keep a 150 ton aircraft in the air? Such a question undoubtedly motivated the careers of many aerospace engineers, who, possibly even after earning multiple degrees in the field, may still be seeking an intuitive answer.

The goal of this book is to build intuition in aerodynamics, without getting bogged down in lengthy derivations or solution approaches that rely extensively and exclusively on mathematics. The motivation for this goal is that most of the physical concepts underlying aerodynamics are not that difficult to grasp, but that they can be obscured by mathematical complexities, particularly when applying the concepts to realistic aerodynamic objects, such as airfoils and wings. Here, computational techniques offer an advantage: often based directly on underlying physical concepts, they are sufficiently powerful to be applicable to realistic geometries in aerodynamics.

Many books on aerodynamics already exist, some focusing on theory, others on applications, and many recent ones touching on computation. When computational approaches are presented, they are often included as additional material after the theoretical exposition, for some of the topics. On the other hand, books dedicated to computational aerodynamics focus heavily on numerical methods with less emphasis on the physics. This book offers a different pedagogical approach: presenting computational techniques up front, without eschewing the theory and mathematics. In addition to the observation that basic elements of these numerical models are often easier for students to understand than mathematically heavy theory, an additional motivation is that most aerodynamics today is done computationally. Theory is still useful for back-of-the-envelope calculations, and for approximately checking results, but the derivations of the theoretical results and their conclusions are generally easier to grasp after having understood the key physical concepts. Computational experiments convey these physical concepts well, and this textbook is therefore replete with numerical demonstrations, codes, and comparisons to theory.

The target audience for this book is students in an undergraduate aerodynamics course. Professionals in the aerospace field may also benefit from this book's presentation of computational approaches. Prerequisites include calculus, knowledge of some physics and fluid dynamics, and a little experience in programming. Topics covered are the essentials of aerodynamics: basic fluid dynamics, kinematics, dynamics, potential flow, airfoil theory, boundary layers, wing theory, and compressible flow. Each chapter contains numerous problems and examples, with complete but succinct codes that students can implement and extend. Although presented exclusively in MATLAB, the codes are for the most part sufficiently simple that they can be implemented in other programming languages as well.

Krzysztof J. Fidkowski

1

Introduction

This chapter reviews basic fluid dynamics, mathematics, and numerical methods that are relevant to aerodynamics. The reader should have some prior experience with these topics, but if not, there are many great books that dive deeper into the material, including, on the fluids side, Landau and Lifshitz (1959), Acheson (1990), and Batchelor (2000).

1.1 Physics of Fluids

The working medium of our aerodynamics studies will be a fluid, and this section defines it and discusses its properties compared to other states of matter. It then presents a molecular interpretation of some thermodynamic properties and the link between molecular and continuum descriptions.

1.1.1 States of Matter

Matter makes up all things around us, including the objects that we handle, the water that we drink, and the air that we breathe. It is made up of molecules* and can take on different states, the most common of which are listed below, along with some of their defining characteristics.

* Molecules are small, on the order a of a few angstroms (10^{-10} m).

- *Solid*: In a solid, illustrated in Figure 1.1, molecules do not move much and are closely packed. A solid maintains its shape, which means that it does not require a vessel to contain it. Relevant to our studies is the property that a constant stress, which is a force per unit area, applied to a solid results in a finite strain, which is a relative displacement. There are exceptions, e.g. a solid that breaks under too much stress, but in general, pushing on a solid deforms it into a configuration that balances the applied stress, so that there are no further deformations. An example is a wing under loaded conditions: it deforms, but if functioning properly, it carries the required load.

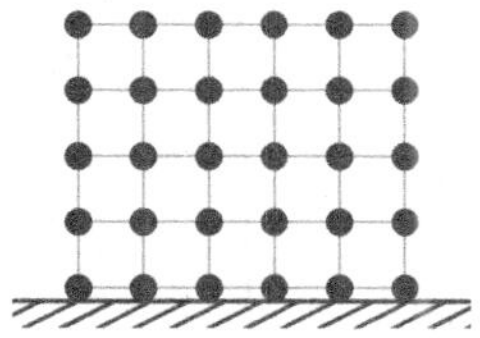

Figure 1.1 Solid molecules in a lattice.

- *Liquid*: In a liquid, shown in Figure 1.2, molecules are close together but can move past each other. This movement means that a liquid has no fixed shape but rather takes the shape of its container. The fact that molecules are close together means that the density, or mass per unit volume, of a liquid is still generally high, comparable to a solid. Important for our study is how a fluid responds to forces. In contrast to a solid, a constant stress applied to a fluid results in a strain *rate*, rather than a fixed amount of strain. That is, pushing on a fluid makes it deform continuously rather than reaching a new configuration in which the forces balance. Imagine moving an oar through water:

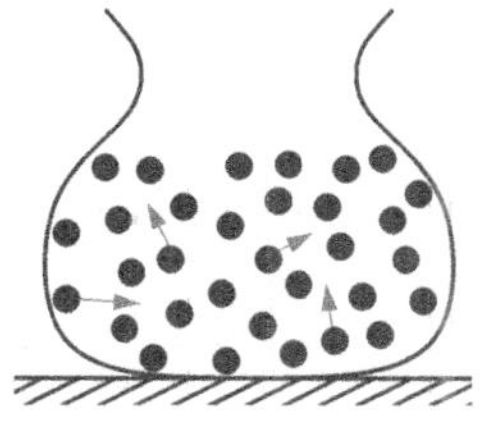

Figure 1.2 Liquid in a container.

Aerodynamics: A Computational Introduction, First Edition. Krzysztof J. Fidkowski.

© 2026 John Wiley & Sons, Inc. Published 2026 by John Wiley & Sons, Inc.

the oar keeps moving when you apply a force, and even though you feel resistance, it is only due to the motion. In contrast, applying a force to an oar in a solid, e.g. in frozen water, would not result in constant motion but rather (very) slight deformation of the ice (and likely the oar).

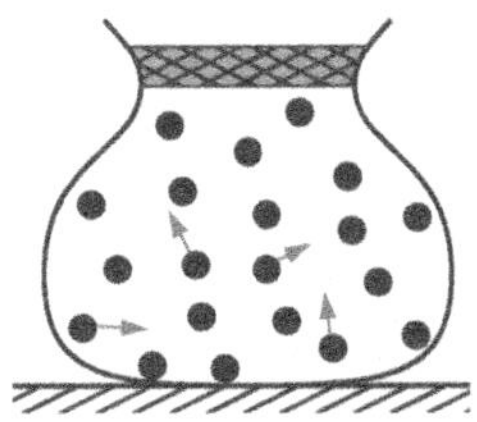

Figure 1.3 Gas in a capped container.

- *Gas*: A gas, shown in Figure 1.3, is similar to a liquid, in that molecules move past each other, so that a gas also does not have a fixed shape. In contrast to a fluid, the molecules in a gas are relatively far apart, which means that gases generally have lower densities compared to liquids and solids. A capped container is thus usually required to keep a gas from dispersing into its surroundings. In aerodynamics, we work mostly with air in its gaseous state, although other gases, such as the carbon dioxide atmosphere of Mars, can also be relevant. Due to the similarity between gases and liquids, particularly in the equations that govern their behavior, we will sometimes consider flows of liquids in certain examples and problems.
- *Plasma*: A plasma is similar to a gas but has the additional property that electrons are free to flow between molecules. Their study is important in numerous fields, including aerospace engineering, e.g. in space weather and electric propulsion. For our aerodynamics studies, however, we will not consider flying through plasmas.

There are other states, such as Bose–Einstein condensates (near absolute-zero atoms in the same quantum states), quark-gluon plasmas (high temperature free quarks and gluons), and more.

The term *fluid* arises in the study of both gases and liquids, and for our purposes, a fluid can refer to either state of matter. Indeed, the subject of fluid dynamics is broad and includes aerodynamics. There are differences,[†] but the fundamental equations are sufficiently similar that we can draw from examples involving both liquids and gases when studying aerodynamics. In addition, some of the language used in describing a fluid will be similar to that of a solid, e.g. stress and strain, but we repeat here an important difference between fluids and solids: a fluid deforms continuously when subjected to a shear stress, while a solid undergoes a fixed amount of deformation. In fluids, we therefore relate stress to the *rate* of deformation instead of to the *amount* of deformation.

† For example, different densities of water and air lead to ship designs in hydrodynamics that differ from aircraft designs in aerodynamics.

1.1.2 Local Thermodynamic Equilibrium

In a fluid, molecules move and collide with each other and with the molecules of the surroundings, for example, those of a container. The velocities of the molecules generally differ, as do their speeds. Given the large number of molecules in any appreciable amount of fluid,[*] it would be very hard to keep track of each molecule, and so a statistical description of a fluid is typically used. This means that quantities like averages and root-mean square deviations will be more relevant than individual molecular properties.

* There are 2.7×10^{22} molecules in a liter of air at standard conditions.

One relevant average quantity for a fluid is its *mean free path*, λ, which is the average distance a molecule travels before colliding with another molecule. How large is λ? That depends on the fluid, for example the size of the molecules, and the conditions, for example how many molecules occupy a given volume and their average speed. For air at *Standard Temperature and Pressure (STP)*, λ is about 68 nm.[†] This is a very small number compared to, say, the size of an aircraft, which we would measure in meters or tens of meters. Such a huge difference in length scales means that when we talk about flow over an aircraft, we can approximate the fluid as a *continuum*: any variations due to randomness of the molecular

† 1 nm = 1 nanometer = 10^{-9} m. 68 nm is about one one-thousandths the width of a human hair.

motion will average out to be imperceptible when considering the interaction of the aircraft with the surrounding air.

Higher up in the atmosphere, the air becomes rarer, and the mean free path increases. There are fewer and fewer molecules occupying a given volume, so collisions become less likely. On the edge of space, λ grows to about 10 cm and increases to about a meter in very low Earth orbit. This length is comparable to that of a satellite, as illustrated in Figure 1.4, which means that a continuum fluid approximation is probably not very good at these altitudes. The molecules do not collide often enough to be in equilibrium with an object, such as a satellite, that is interacting with them. Instead, analysis of the "flow" of molecules in this case must take into account the behavior of individual molecules as they hit and bounce off the object.

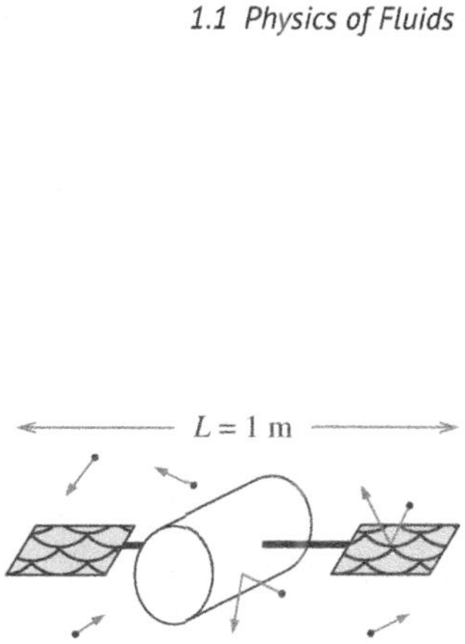

Figure 1.4 Air molecules surround a satellite in very low Earth orbit, but they are too far apart to be treated as a continuum fluid.

To quantify the distinction between a continuum and discrete molecules, we define the *Knudsen number* as the ratio of the mean free path to a characteristic length, L,

$$Kn \equiv \frac{\lambda}{L}. \tag{1.1}$$

The characteristic length, L, is some measure of the size of the object, such as the chord or span of an aircraft wing, the side length of a CubeSat, or the diameter of a pipe.‡ The two relevant regimes are then

‡The Knudsen number is not an intrinsic property of the fluid. Instead, it also depends on the object studied. We will see the same property with another important quantity, the Reynolds number, in Chapter 6.

1. $Kn \ll 1$: continuum: the fluid is in *Local Thermodynamic Equilibrium (LTE)*, which means that collisions between molecules are sufficiently numerous and occur sufficiently quickly such that molecular properties follow regular statistical distributions. The state of the fluid may change in time, but (1) the time scale for macroscopic changes is much longer than that associated with collisions between molecules; and (2) the length scale for macroscopic changes is much larger than the distance over which the collisions occur. An aircraft flying in the lower atmosphere is an example of a fluid–object combination in LTE.
2. $Kn \sim 1$ or $Kn > 1$: discrete molecules: in this case, few molecules are impacting the object, and we must treat the molecules individually in the analysis of the interaction between the fluid and the object. The length and time scales associated with the collisions between the molecules are comparable to those of the object and its motion. Hence, disturbances to molecules due to their interaction with the object do not propagate to other molecules sufficiently quickly or in a sufficiently small region to allow for a statistical description of the fluid. An example of a system not in LTE is a satellite in orbit.

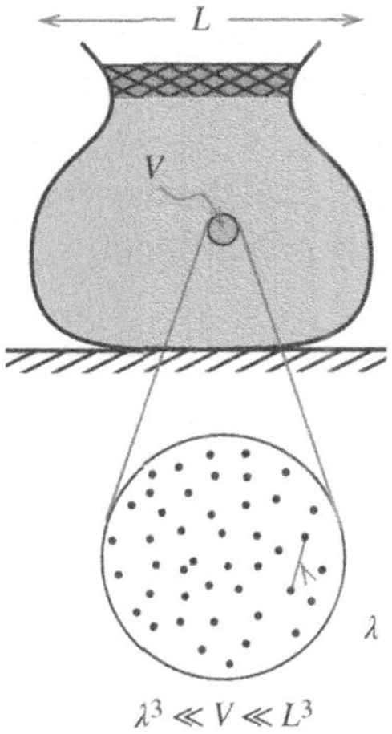

Figure 1.5 Volume, V, used to define a "point" quantity in a fluid.

Of interest in aerodynamics will be continuum fluids, which means that we assume LTE, $Kn \ll 1$. In a continuum, we can define point quantities as averages over a volume V that is small compared to L^3, but large compared to λ^3, as illustrated in Figure 1.5. For example, the *density* of a fluid is

$$\rho = \frac{\text{mass}}{\text{volume}} = \frac{1}{V}\sum_{i=1}^{N} m_i, \tag{1.2}$$

where N is the number of molecules in V, i indexes the molecules, and m_i is the mass of the ith molecule. We can also calculate the *momentum per unit volume* as

$$\rho\vec{v} = \frac{\text{momentum}}{\text{volume}} = \frac{1}{V}\sum_{i=1}^{N} m_i\vec{v}_i, \tag{1.3}$$

where $\vec{v}_i$ is the velocity of the ith molecule. The pointwise *velocity* is then given by the quotient

$$\vec{v} = \frac{\rho\vec{v}}{\rho} = \sum_{i=1}^{N} m_i\vec{v}_i \Big/ \sum_{i=1}^{N} m_i. \tag{1.4}$$

1.1.3 Fluid Properties

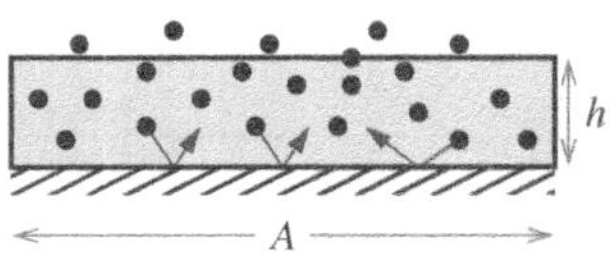

Figure 1.6 Pressure on a surface due to molecular collisions.

In addition to density and velocity, two more relevant properties of a fluid are *pressure*, p, and *temperature*, T. Pressure is the force per unit area exerted by the fluid on a surface. This force is due to the collisions of molecules with the surface: as a molecule bounces off the surface, it transfers some of its momentum, the normal component, to the surface. Figure 1.6 illustrates this for a surface area A exposed to the fluid.

Consider a thin slice of fluid of height h perpendicular to the surface. Suppose that the fluid is at "rest" – this just means that the average molecular velocity is zero. However, the molecules are still moving in random directions and colliding with each other.* Let u_{rms} be the root mean square speed in one direction. This is the average of the speeds along one direction of the fluid, here taken as perpendicular to the surface.

* In fact, molecules move very fast: for air at STP, molecules move almost 500 m/s, which is over 1100 mph!

We can obtain the force on the surface by dimensional analysis. Using SI units, force is measured in Newtons, $1\ \text{N} = 1\ \text{kg} \cdot \text{m/s}^2$. This is mass times speed squared divided by length. So one molecule of mass m_i exerts a force of $F_i \propto m_i u_{\text{rms}}^2/h$, where h is the relevant distance because we are looking at the normal direction momentum transfer.† The total force is the sum over all molecules, so that the pressure is

† Another way to think about this relationship is to take $h = \lambda$ so that the time in between a molecule's collisions with the surface scales as h/u_{rms}. Since $m_i u_{\text{rms}}$ is the imparted momentum, the force is the momentum transfer per unit time, which is $m_i u_{\text{rms}}^2/h$.

$$p = \frac{F}{A} \propto \sum_i \frac{m_i u_{\text{rms}}^2}{hA} = \sum_i \frac{m_i u_{\text{rms}}^2}{V} = \rho u_{\text{rms}}^2. \tag{1.5}$$

This scaling of pressure should make sense: if the density is higher, the molecules are more massive or more numerous, while a higher speed increases the momentum and frequency of collisions.

The temperature measures the kinetic energy of the random motions of the molecules,‡

‡ For temperature, we actually have to take into account the motion in all three directions in three dimensions, but assuming isotropic random motion, $|\vec{v}|_{\text{rms}}^2 = u_{\text{rms}}^2 + v_{\text{rms}}^2 + w_{\text{rms}}^2 = 3u_{\text{rms}}^2$, so the two rms speeds are directly proportional.

$$T \propto u_{\text{rms}}^2. \tag{1.6}$$

Combining this equation with the one for pressure, we see that pressure is proportional to the density and temperature, which we can write as

$$p = \rho R T. \tag{1.7}$$

This equation is the *ideal gas law*. The constant of proportionality is the specific gas constant, R, which has SI units of J/kg · K.§ This law assumes a perfect gas, which is one in which molecular motion is random, inter-molecular forces are ignored, and molecules interact only through elastic collisions. Air at STP can be well-approximated as a perfect gas.

§ For air, $R = 287$ J/kg · K. Note, this is the universal gas constant divided by the molar mass.

One more fluid property that will come up often is viscosity, μ, which relates the gradient in velocity to the *shear stress*, τ,

$$\tau = \mu \frac{\partial u}{\partial y}. \tag{1.8}$$

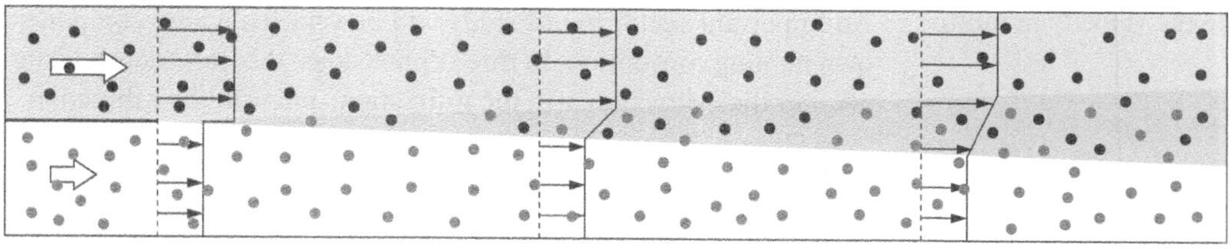

Figure 1.7 Two fluid streams of different velocities mixing after a divider. The upper molecules have a higher average horizontal velocity than the lower ones. They mix due to random molecular velocities in the vertical direction.

In this equation, the y direction is perpendicular to the surface on which the shear stress is measured. Imagine two streams of fluid moving at different average velocities, as illustrated in Figure 1.7. These two streams are initially separated by a divider. After the divider, the streams begin to mix due to the random velocities of the molecules. Some faster molecules move into the slower stream and vice versa. The resulting mixing of momentum is a diffusion process: it is fast at first but then slows down as the gradient gets smaller. Eventually, the shear stress goes to zero because $\frac{\partial u}{\partial y} \to 0$: in this state, the momentum exchange between the two fluid streams is zero, since on average, molecules crossing the dividing plane have the same momentum and are equally likely to move up as they are to move down.

By dimensional analysis of Equation 1.8, the SI units of μ are kg/m · s. We can form a quantity of this dimension from density, rms speed, and mean free path,¶

$$\mu \propto \rho u_{\text{rms}} \lambda. \tag{1.9}$$

Since density and mean free path are inversely proportional, the viscosity scales with u_{rms}, and hence the square root of the temperature. The higher the temperature, the higher the viscosity due to the faster motion of the molecules.‖ Also, we note that μ is the *dynamic viscosity*, whereas $\nu \equiv \mu/\rho$ is the *kinematic viscosity* with dimensions of length squared per time.

¶ The mean free path is the relevant length scale because momentum transfer between molecules occurs by collisions.

‖ This direct relationship between viscosity and temperature is a property of gases. Many liquids exhibit the opposite, inverse relationship because increasing temperature breaks molecular bonds and allows molecules to slide past each other more easily.

1.2 Mathematics Review

1.2.1 Conventions

In the study of aerodynamics, we will encounter various mathematical quantities, including spatial and temporal coordinates, non-dimensional groups such as the Reynolds number, scalar fields like pressure, vector fields like the velocity, and tensors such as stress and rate of strain. A notation convention helps to distinguish these quantities, and the one adopted in this text is as follows:

c : (regular font) a *scalar*

$\vec{v}$: (arrow) a *spatial vector*, e.g. $\vec{v} = v_1\hat{x}_1 + v_2\hat{x}_2 + v_3\hat{x}_3$, where $\hat{x}_i$ is the unit vector in the ith Cartesian direction.

$\mathbf{u}$: (bold) a *state vector*, e.g. $\mathbf{u} = [\rho, \rho u, \rho v, \rho E]^T$.

$\mathbf{A}$: (bold, typically uppercase) a *matrix* or *tensor*, e.g.

$$\mathbf{A} = \begin{bmatrix} A_{11} & A_{12} \\ A_{21} & A_{22} \end{bmatrix}. \tag{1.10}$$

An important vector in our study will be velocity, whose components have their own naming convention. In three dimensions, the Cartesian velocity components are u in the x direction, v in the y direction, and w in the z direction.*

1.2.2 Units and Dimensions

The *dimension* of a physical quantity refers to the fundamental property that it represents, e.g. length (L), mass (M), time (T), or some combination. For example, speed has dimension of length divided by time, and density is mass divided by length cubed. The *units* of measurement refer to the "yardstick" used in measuring the dimension, e.g. meters versus feet for length, or kilograms versus pounds for mass.*

One use of dimensions is to quickly check equations for *dimensional consistency*: terms of different dimensions cannot be compared or added together. Similarly, equations should be checked for *vector consistency*, for example to avoid adding or comparing a vector and a scalar.

Another use of dimensions is to obtain *non-dimensional groups* that can simplify or generalize equations. For example, in viscous flows, the Reynolds number, a ratio of convective to viscous transport scales, will often appear, and in compressible flow, we will rely heavily on the Mach number, which is a ratio of flow speed to the speed of sound. The *Buckingham Pi theorem* states that from n variables in k physical dimensions, we can form $p = n - k$ non-dimensional groups.†

Finally, *dimensional analysis* can be used to determine the dimensions of a quantity given an equation or as an approximate derivation technique for expressing one quantity in terms of others. For example, given a wing of area $A\,[\mathrm{m}^2]$ flying at speed $v\,[\mathrm{m/s}]$ in a fluid of density $\rho\,[\mathrm{kg/m}^3]$, the force on the wing can be estimated by considering its dimensions: Newtons $= \mathrm{kg}\cdot\mathrm{m/s}^2$ can be formed from $\rho v^2 A$. So we should expect a force on the wing (e.g. lift or drag) to scale with $\rho v^2 A$. Note that this analysis does not give an exact answer, since even if the force scaled exactly with $\rho v^2 A$, the scaling factor is not known from dimensional analysis alone.

Example 1.1 Vector consistency

Consider the expression

$$\nabla\cdot\vec{v}\times\nabla p, \tag{1.11}$$

where $\vec{v}$ is a vector field and p is a scalar field. Is this expression valid? The answer depends on how we group the operators. Treating the dot and cross product as of equal rank in the order of operations, one would proceed from left to right. Evaluating the divergence $\nabla\cdot\vec{v}$ yields a scalar, which then cannot be crossed with the vector ∇p. In this view, the expression is invalid. However, if we carry out the cross product first, then $\vec{v}\times\nabla p$ is a vector, we can take its divergence, and the expression is valid. Parentheses help clarify the intent of keeping the expression valid, and in this case, we would write

$$\nabla\cdot\left(\vec{v}\times\nabla p\right). \tag{1.12}$$

Conversely, the expression $\left(\nabla\cdot\vec{v}\right)\times\nabla p$ would clearly be invalid.

* We can write the velocity in multiple ways:

$$\vec{v} = u\hat{i} + v\hat{j} + w\hat{k} = u\hat{x} + v\hat{y} + w\hat{z} = \begin{bmatrix} u \\ v \\ w \end{bmatrix} = [u \;\; v \;\; w]^T,$$

where $\hat{i}=\hat{x}$, $\hat{j}=\hat{y}$, and $\hat{k}=\hat{z}$ are unit vectors in the three Cartesian directions.

* In this text, we will adopt SI units (m = meters, s = seconds, kg = kilograms), and sometimes, we will use these units when discussing the dimensions of a quantity.

† For example, given $n = 5$ variables: density $\rho\,[\mathrm{kg/m}^3]$, speed $v\,[\mathrm{m/s}]$, dynamic viscosity $\mu\,[\mathrm{kg/m\cdot s}]$, length, $\ell\,[\mathrm{m}]$, and acceleration, $g\,[\mathrm{m/s}^2]$, in $k=3$ dimensions [m, kg, s], we can form $p=2$ non-dimensional groups. These are not unique, and one choice is

$$\mathrm{Re} \equiv \frac{\rho v \ell}{\mu}, \qquad r \equiv \frac{v^2}{\ell g}.$$

1.2.3 Linear Algebra

Matrices and vectors are useful for setting up and solving linear equations. Such equations will arise when we study numerical methods for predicting flow over an airfoil or a wing. The unknowns in these cases will be parameters in a model of the flow, and the equations will be the physical constraints that the flow must satisfy.

This section briefly reviews a few linear algebra concepts relevant to this text. First, the superscript T following a vector or matrix denotes the *transpose* operation, which means flipping rows into columns and vice versa. The dot, or inner, product between two column vectors* $\mathbf{u}, \mathbf{v}$ of length n is

$$\mathbf{u} \cdot \mathbf{v} = \mathbf{u}^T\mathbf{v} = \mathbf{v}^T\mathbf{u} = \sum_{i=1}^{n} u_i v_i = \text{a scalar.} \tag{1.13}$$

Two vectors are orthogonal if their dot product is zero. We can multiply a column vector $\mathbf{u} \in \mathbb{R}^m$ from the left by a matrix $\mathbf{A} \in \mathbb{R}^{n\times m}$. This operation produces a vector $\mathbf{v} = \mathbf{A}\mathbf{u} \in \mathbb{R}^n$, the entries of which are dot products between $\mathbf{u}$ and the rows of $\mathbf{A}$,

$$\begin{bmatrix} v_1 \\ v_2 \\ \vdots \\ v_n \end{bmatrix} = \begin{bmatrix} A_{11} & A_{12} & \dots & A_{1m} \\ A_{21} & A_{22} & \dots & A_{2m} \\ \vdots & \vdots & \ddots & \vdots \\ A_{n1} & A_{n2} & \dots & A_{nm} \end{bmatrix} \begin{bmatrix} u_1 \\ u_2 \\ \vdots \\ u_m \end{bmatrix}. \tag{1.14}$$

Note that such a matrix–vector product requires m multiplications and additions per entry of $\mathbf{v}$, for a total of $\mathcal{O}(mn)$ operations.†

In a similar way, we can multiply matrices, for example $\mathbf{A} \in \mathbb{R}^{n\times m}$ and $\mathbf{B} \in \mathbb{R}^{m\times p}$ (note the inner dimension m has to match), to obtain a matrix $\mathbf{C} = \mathbf{AB} \in \mathbb{R}^{n\times p}$ with entries $C_{ik} = A_{ij}B_{jk}$.‡ Note that a matrix–matrix product requires p matrix–vector multiplications, so the number of operations is $\mathcal{O}(mnp)$. Multiplying a column vector $\mathbf{v} \in \mathbb{R}^n$ on the right by a row vector $\mathbf{u} \in \mathbb{R}^m$ produces, via an outer product, a matrix $\mathbf{vu} \in \mathbb{R}^{n\times m}$.

The *determinant* of a square matrix, $\det(\mathbf{A})$, is a scalar computed from a systematic combination of products of the matrix entries.§ Properties of the determinant include:

1. $\det(\mathbf{I}) = 1$, where $\mathbf{I}$ is the identity matrix
2. $\det(\mathbf{AB}) = \det(\mathbf{A})\det(\mathbf{B})$
3. $\det(\mathbf{A}^T) = \det(\mathbf{A})$
4. $\det(\mathbf{A}^{-1}) = 1/\det(\mathbf{A})$

The *inverse* of a square matrix, $\mathbf{A}^{-1}$, if it exists, is a matrix that multiplies $\mathbf{A}$ to give the identity matrix, $\mathbf{A}\mathbf{A}^{-1} = \mathbf{A}^{-1}\mathbf{A} = \mathbf{I}$. A *singular* matrix is one for which the inverse does not exist, and the determinant in this case is zero. We will sometimes use the colon (:) notation to denote parts of matrices, e.g. $\mathbf{A}(i, :)$ denotes the ith row of matrix $\mathbf{A}$.

The *rank* of a matrix, $\text{rank}(\mathbf{A})$ is the number of linearly-independent columns, which is the same as the number of linearly-independent rows. An $n \times n$ matrix must have *full rank* of n to be invertible. The *null space* of $\mathbf{A}$ is the set of all vectors $\mathbf{v}$ for which $\mathbf{Av} = \mathbf{0}$. The dimension of the null space is $(n - \text{rank}(\mathbf{A}))$, so a full-rank matrix has an empty null space.

* We repeat the convention of using boldface for general vectors that are arrays of numbers, such as unknowns in a system of equations, and reserving the arrow for spatial vectors.

† The notation $\mathcal{O}(\cdot)$ is an approximate estimate of the size of a number, e.g. cost in operations. For operations, we treat multiplications and additions similarly and ignore leading multiplicative factors.

‡ Summation is implied on the index j. More on this in Section 1.2.4.

§ For example, for a 2×2 matrix,

$$\det\left(\begin{bmatrix} a & b \\ c & d \end{bmatrix}\right) = ad - bc.$$

For larger matrices, the determinant can be computed by combining determinants of 2×2 submatrices (minors).

Example 1.2 A 3 × 3 matrix
Consider the following matrix

$$\mathbf{A} = \begin{bmatrix} 1 & 2 & 3 \\ 3 & 2 & 1 \\ -2 & 0 & 2 \end{bmatrix}. \quad (1.15)$$

We can show that **A** is not invertible by calculating its determinant,

$$\det(\mathbf{A}) = 1(2*2-0) - 2(3*2-(-2)*1) + 3(0-(-2)*2) = 0.$$

Since the determinant is zero, the matrix is not invertible. To identify the null space of **A**, we look for vectors $\mathbf{v} = [a, b, c]^T$ for which

$$\mathbf{A}\mathbf{v} = \mathbf{0} \quad \Rightarrow \quad \begin{bmatrix} 1 & 2 & 3 \\ 3 & 2 & 1 \\ -2 & 0 & 2 \end{bmatrix} \begin{bmatrix} a \\ b \\ c \end{bmatrix} = \mathbf{0}. \quad (1.16)$$

Carrying out the matrix–vector multiplication, the system for a, b, c is

$$a + 2b + 3c = 0, \quad (1.17)$$
$$3a + 2b + c = 0, \quad (1.18)$$
$$-2a + 2c = 0. \quad (1.19)$$

From the third equation, $a = c$, and using this equation to replace a with c in either the first or second equation shows that $b = -2c$.¶ So the null space consists of multiples of the vector

¶ The two equations are the same when $a = c$, so there is no unique solution for a, b, c.

$$\mathbf{v} = \begin{bmatrix} a \\ b \\ c \end{bmatrix} = \begin{bmatrix} 1 \\ -2 \\ 1 \end{bmatrix}. \quad (1.20)$$

Null space vectors are not unique: an arbitrary multiple of a null-space vector is also in the null space.

1.2.4 Index Notation

Index notation is a shorthand for concisely expressing operations on vectors and tensors in Cartesian coordinates, which are now labeled (x_1, x_2, x_3). The components of a three-dimensional vector $\vec{v}$ in index notation are written as v_i, where $i = 1,2,3$ indexes the dimension. One shorthand is the use of *implied summation* for all repeated indices. For example, a *dot product* becomes*

* The repeated index in the last expression implies summation. This is a shorthand to avoid writing the summation symbol.

$$\vec{v} \cdot \vec{w} = \sum_i v_i w_i = v_i w_i. \quad (1.21)$$

∂_i is used to denote differentiation with respect to x_i. So the *gradient* of a scalar field p is†

† We sometimes refer to $\partial_i p$ as the gradient. A free index, in this case i, indicates that the quantity is a vector, requiring $\hat{x}_i$ to make it complete.

$$\nabla p = \frac{\partial p}{\partial x_1}\hat{x}_1 + \frac{\partial p}{\partial x_2}\hat{x}_2 + \frac{\partial p}{\partial x_3}\hat{x}_3 = \partial_i p \hat{x}_i. \quad (1.22)$$

The *divergence* of a vector field can be written as

$$\nabla\cdot\vec{v}=\frac{\partial v_1}{\partial x_1}+\frac{\partial v_2}{\partial x_2}+\frac{\partial v_3}{\partial x_3}=\partial_i v_i. \tag{1.23}$$

The implied summation on the repeated index i means that the answer is a scalar. Two important symbols used often in index notation are the *Kronecker delta*, δ_{ij}, and the *alternating symbol*,‡ ϵ_{ijk},

‡ A quantity with more than one index, is a matrix for two indices, or a higher-dimensional equivalent for three or more indices.

$$\delta_{ij}=\begin{cases}1 & \text{if } i=j\\ 0 & \text{otherwise}\end{cases}, \tag{1.24}$$

$$\epsilon_{ijk}=\begin{cases}1 & \text{if } i,\ j,k=1,2,3 \text{ or } 2,3,1 \text{ or } 3,1,2\\ -1 & \text{if } i,\ j,k=3,2,1 \text{ or } 2,1,3 \text{ or } 1,3,2\\ 0 & \text{otherwise}\end{cases}. \tag{1.25}$$

Using ϵ_{ijk}, the ith component of the *cross product* of $\vec{v}$ and $\vec{w}$ can be written as $\epsilon_{ijk}v_jw_k$.§ Note that in this expression, summation is performed on j and k. Also, the ith component of the *curl* of a vector is $\epsilon_{ijk}\partial_j v_k$. When a derivative ($\partial_j$) operator is involved, the order of operations is important – the derivative does not commute. A useful relation for proving many vector identities is

§ Recall that the cross product of two vectors $\vec{v}$ and $\vec{w}$ is a vector quantity that can be calculated as the determinant of the following matrix:

$$\vec{v}\times\vec{w}=\begin{vmatrix}\hat{i} & \hat{j} & \hat{k}\\ v_1 & v_2 & v_3\\ w_1 & w_2 & w_3\end{vmatrix}.$$

The curl is similar but involves the differential operator ∇,

$$\nabla\times\vec{v}=\begin{vmatrix}\hat{i} & \hat{j} & \hat{k}\\ \frac{\partial}{\partial x_1} & \frac{\partial}{\partial x_2} & \frac{\partial}{\partial x_3}\\ v_1 & v_2 & v_3\end{vmatrix}.$$

$$\epsilon_{ijk}\epsilon_{klm}=\delta_{il}\delta_{jm}-\delta_{im}\delta_{jl}, \tag{1.26}$$

where the only implied summation is on k on the left-hand side.

Example 1.3 Proving a vector identity using index notation
Show that

$$\nabla\times(\nabla\times\vec{v})=\nabla(\nabla\cdot\vec{v})-\nabla^2\vec{v}, \tag{1.27}$$

for any 3D vector field $\vec{v}$. Note that this is a vector equation: both the left- and right-hand sides are vector quantities. We use index notation and i as the free index on both sides.¶ Starting with the left-hand side, we convert the double curl to index notation and then re-arrange,‖

¶ This free index has to be the same for both sides. Other indices, local to one side, can be different.

$$\begin{aligned}\epsilon_{ijk}\partial_j\epsilon_{klm}\partial_l v_m&=\epsilon_{ijk}\epsilon_{klm}\partial_j\partial_l v_m\\&=\left(\delta_{il}\delta_{jm}-\delta_{im}\delta_{jl}\right)\partial_j\partial_l v_m\\&=\partial_m\partial_i v_m-\partial_j\partial_j v_i.\end{aligned} \tag{1.28}$$

‖ The second step uses the identity from Equation 1.26, and the final step uses the result that $\delta_{im}v_m=v_i$ (summation implied).

The result is the ith component of $\nabla(\nabla\cdot\vec{v})-\nabla^2\vec{v}$, which is the right-hand side, so we have verified the identity.

Example 1.4 Clarifying order of operations with index notation
Index notation can also be used to clarify otherwise ambiguous vector expressions, without the need for parentheses. Consider again the expression from Example 1.1, given in Equation 1.11. Using index notation, the valid form of the

expression is the divergence of a cross product of two vectors, and so we write it as**

$$\nabla \cdot (\vec{v} \times \nabla p) = \partial_i \epsilon_{ijk} v_j \partial_k p. \tag{1.29}$$

** An additional nicety of index notation is that the order in which the terms appear generally does not matter, since the indices dictate how the terms are combined. So we could switch v_j and $\partial_k p$ in this expression. An exception is when derivatives are involved: the assumption is that ∂_i acts on a quantity to its right.

The repeated index i carries out the divergence, whereas the repeated indices j and k are used with the alternating symbol to form the cross product. There are no free indices in the expression, so the result is a scalar.

Note, the invalid form, where we evaluate $\nabla \cdot \vec{v}$ first, simply cannot be written with index notation, as there is no free index left on $\nabla \cdot \vec{v}$ to use in a cross product.

1.2.5 Cylindrical Coordinates

Index notation does not apply directly to non-Cartesian coordinates, such as cylindrical coordinates, which are useful in aerodynamics. In cylindrical coordinates, a vector is decomposed into radial (r), azimuthal (θ), and axial (z) components, as

$$\vec{v} = v_r \hat{r} + v_\theta \hat{\theta} + v_z \hat{z}, \tag{1.30}$$

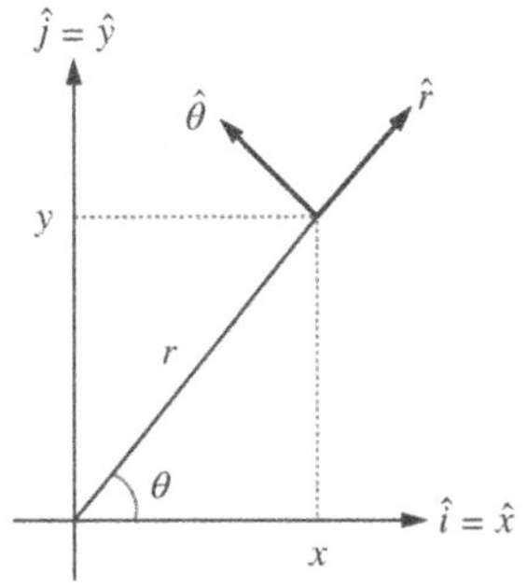

Figure 1.8 Cartesian and cylindrical basis vectors. While $\hat{i}$ and $\hat{j}$ remain constant, $\hat{r}$ and $\hat{\theta}$ depend on position. $\hat{z} = \hat{k}$ points out of the page.

where $\hat{r}$, $\hat{\theta}$, and $\hat{z}$ are unit vectors. Note that the directions of $\hat{r}$ and $\hat{\theta}$ depend on position, as shown in Figure 1.8.

Differential operators take on different forms in cylindrical coordinates. Some useful expressions are

$$\nabla p = \frac{\partial p}{\partial r}\hat{r} + \frac{1}{r}\frac{\partial p}{\partial \theta}\hat{\theta} + \frac{\partial p}{\partial z}\hat{z}, \tag{1.31}$$

$$\nabla \cdot \vec{v} = \frac{1}{r}\frac{\partial (r v_r)}{\partial r} + \frac{1}{r}\frac{\partial v_\theta}{\partial \theta} + \frac{\partial v_z}{\partial z}, \tag{1.32}$$

$$\nabla \times \vec{v} = \left[\frac{1}{r}\frac{\partial v_z}{\partial \theta} - \frac{\partial v_\theta}{\partial z}\right]\hat{r} + \left[\frac{\partial v_r}{\partial z} - \frac{\partial v_z}{\partial r}\right]\hat{\theta} + \left[\frac{1}{r}\frac{\partial (r v_\theta)}{\partial r} - \frac{1}{r}\frac{\partial v_r}{\partial \theta}\right]\hat{z}. \tag{1.33}$$

In two spatial dimensions, cylindrical coordinates become *polar coordinates*, and we use the names interchangeably.

Example 1.5 Transformation of cylindrical basis vectors
The cylindrical basis vectors $\hat{r}$ and $\hat{\theta}$ depend on position, and at any point in space, we can write them in terms of the Cartesian basis vectors $\hat{i}$ and $\hat{j}$.* From Figure 1.8, the components of the unit vector $\hat{r}$ along the x and y directions are $\cos\theta$ and $\sin\theta$, respectively, so we can write it as

* The exception is the origin, where $\hat{r}$ and $\hat{\theta}$ are not defined.

$$\hat{r} = \cos\theta \hat{i} + \sin\theta \hat{j}. \tag{1.34}$$

Similarly, the components of the unit vector $\hat{\theta}$ are $-\sin\hat{\theta}$ and $\cos\hat{\theta}$, so that

$$\hat{\theta} = -\sin\theta \hat{i} + \cos\theta \hat{j}. \tag{1.35}$$

These transformations enable conversion of vector quantities from cylindrical to Cartesian bases, and they provide a way to calculate derivatives of $\hat{r}$ and $\hat{\theta}$. For example, to evaluate the derivative $\frac{\partial \hat{\theta}}{\partial \theta}$, we can write†

$$\frac{\partial \hat{\theta}}{\partial \theta} = \frac{\partial}{\partial \theta}(-\sin\theta \hat{i} + \cos\theta \hat{j}) = -\cos\theta \hat{i} - \sin\theta \hat{j} = -\hat{r}. \tag{1.36}$$

† This evaluation relies on the fact that $\hat{i}$ and $\hat{j}$ are constant, so that their derivative with respect to θ is zero.

1.2.6 Scalar and Vector Fields

A macroscopic description of a fluid consists of a characterization of its properties, such as density, pressure, temperature, and velocity, in space and time. Sometimes, usually in idealized conditions, such a characterization is possible with analytical formulas in terms of spatial and temporal coordinates, also known as *field coordinates*. More often, the variation of a property will not be a simple analytical function, and numerical descriptions, through point values, averages over small control volumes, or functions inside such volumes, must be used instead. However, the goal of such methods remains the same: to describe how a property varies in space and/or time.

A *scalar field* is a function defined over all points in space and possibly time. Examples include pressure, temperature, and density of a fluid.* Given a representation of a scalar field, such as an analytical expression, of interest is often its visualization. We may wish to see the distribution of pressure over an airfoil, or the temperature inside a combustion chamber.

* In the continuum assumption, these quantities can be evaluated at any point, with the understanding that at the molecular level, a point is actually a very small control volume with many molecules, as shown in Figure 1.5.

In one dimension, scalar fields are fairly easy to visualize, especially if given an analytical form. A standard visualization consists of a plot like the one shown in Figure 1.9, where $p(x)$ is the scalar function and x is the single spatial coordinate.

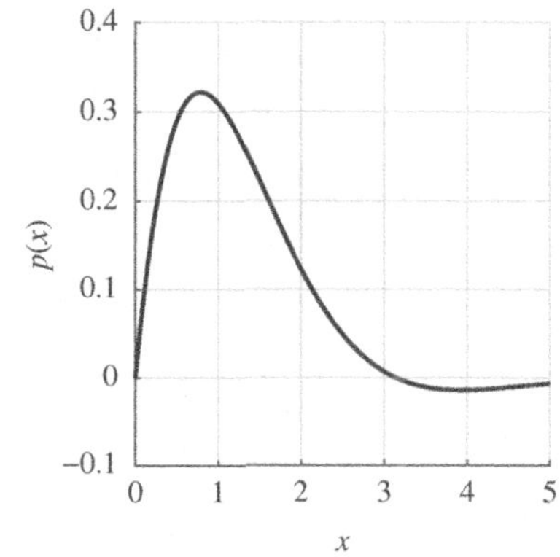

Figure 1.9 Plot of a one-dimensional scalar field $p(x)$.

Example 1.6 Visualizing a scalar field in one dimension
This example shows how to write a MATLAB code to make a one-dimensional plot of the scalar function

$$p(x) = \sin(x)e^{-x}, \quad x \in [0,5]. \tag{1.37}$$

The code in Listing 1.1 makes a vector of x locations and computes p for each of the entries in x. Note the use of `.*` instead of `*` to perform component-wise multiplication. It then plots the vector of p values against the vector of x values, connecting the points linearly in between.†

† The actual function is smooth and curved in between the sampled points, but if we use a large number of points, the piecewise-linear representation in the plot will be barely noticeable.

Listing 1.1 Code for visualizing a scalar field in one dimension.

```
function scalarviz1D
% Makes a plot of a 1D scalar function

x = linspace(0,5,101);   % vector of points
p = sin(x).*exp(-x);     % p values
figure(1); clf;
plot(x, p, 'k-', 'linewidth', 2);
box off; grid on; axis([0,5,-.1,.4]);
set(gca, 'xtick', linspace(0,5,6));
set(gcf, 'position', 400*[1,1,1,1]);
writeplot('$x$', '$p(x)$', '../figs/scalarviz1D.pdf', 4);
```

This code makes use of the function `writeplot.m`, given in Listing 1.2, which formats and exports a figure for output, and which will be used throughout this text.

Listing 1.2 Code for formatting and exporting a figure.

```
function writeplot(xlab, ylab, fname, f);
set(gca, 'fontsize', 12+f);
set(gca, 'ticklabelinterpreter', 'latex');
xlabel(xlab, 'interpreter', 'latex', 'fontsize', 14+2*f);
ylabel(ylab, 'interpreter', 'latex', 'fontsize', 14+2*f);
exportgraphics(gcf, ['../figs/', fname]);
```

In two dimensions, a function $p(x, y)$ is a little harder to visualize. One option is a surface plot, like the one shown in Figure 1.10, but such plots can be difficult to read quantitatively due to the perspective. Often, a better choice is a *contour plot*, like the one in Figure 1.11, where a contour line indicates where the scalar is constant. The areas between contours can be colored or shaded, or the contours can be labeled with numerical values if a quantitative visualization is desired.

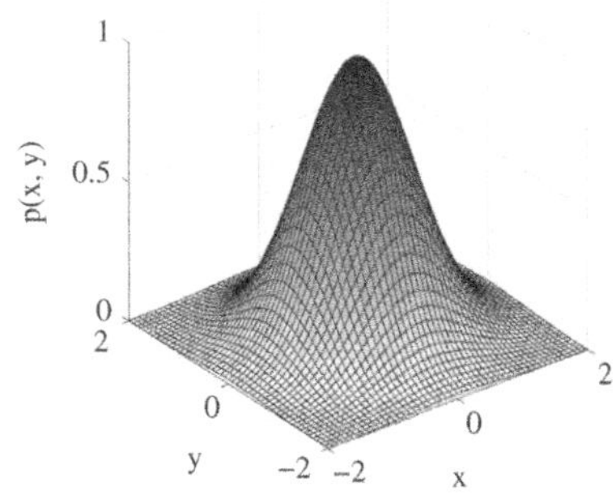

Figure 1.10 Surface plot of a two-dimensional scalar field.

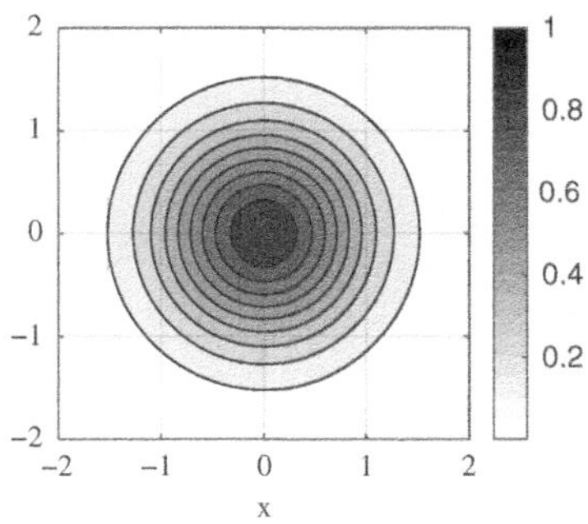

Figure 1.11 Contour plot of a two-dimensional scalar field.

Example 1.7 Visualizing a scalar field in two dimensions
Consider the two-dimensional function

$$p(x, y) = e^{-x^2-y^2}, \quad x, y \in [-2, 2]^2. \tag{1.38}$$

Using the `ndgrid` function, the code in Listing 1.3 makes two matrices of (x, y) coordinates, which form a lattice of points, and computes p for each point. The resulting variable `P` is a matrix of the p values at each point. The code uses this matrix together with the x and y coordinate matrices to make both a surface plot and a contour plot.

Listing 1.3 Code for visualizing a scalar field in two dimensions.

```
function scalarviz2D
% Makes surface and contour plots in 2D

s = linspace(-2,2,51);   % 1D points for plotting
[X,Y] = ndgrid(s,s);     % 2D matrices of points
P = exp(-X.^2 - Y.^2);   % p(x,y) at points
figure(1); clf;

% surface plot
surf(X,Y,P); colormap(flipud(gray(20)));
set(gca, 'fontsize', 14, 'ticklabelinterpreter', 'latex');
xlabel('x', 'fontsize', 16, 'interpreter', 'latex');
ylabel('y', 'fontsize', 16, 'interpreter', 'latex');
zlabel('p(x,y)', 'fontsize', 16, 'interpreter', 'latex');
set(gcf, 'position', 400*[1,1,1,1]);
exportgraphics(gcf, '../figs/scalarviz2D_surf.pdf');

% contour plot
contourf(X,Y,P); axis equal; colorbar('eastoutside');
set(gca, 'xtick', linspace(-2,2,5)); grid on;
set(gca, 'fontsize', 14, 'ticklabelinterpreter', 'latex');
xlabel('x', 'fontsize', 16, 'interpreter', 'latex');
ylabel('y', 'fontsize', 16, 'interpreter', 'latex');
exportgraphics(gcf, '../figs/scalarviz2D_cont.pdf');
```

‡The directional derivative is a derivative along a specified direction from a point. It is given by the dot product between the gradient and a unit vector along the direction.

Since a scalar p is constant on its contour, also known as an *isoline*, the *directional derivative* along the contour or isoline is zero.‡ That is,

$$\nabla p \cdot \vec{t} = 0,$$

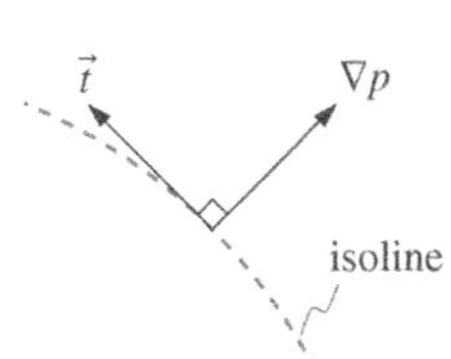

Figure 1.12 At any point, the gradient vector of a two-dimensional scalar field is perpendicular to the isoline through that point.

where $\vec{t}$ is the tangent vector and $\nabla p \cdot \vec{t} = \frac{\partial p}{\partial x} t_x + \frac{\partial p}{\partial y} t_y$ is the directional derivative. This means that *isolines are perpendicular to the gradient vector*, as illustrated in Figure 1.12.

In three dimensions, a function $p(x, y, z)$ can be depicted through *iso-surfaces*, which are surfaces of constant p. Similarly to the two-dimensional result, at each point in space, the gradient vector ∇p is perpendicular to the iso-surface that passes through that point. Without creative "cut-away" plots, only one iso-surface can be shown at the same time. A more quantitative alternative is to plot two-dimensional contours on select cut planes. Finally, unsteady fields, $p(x, y, z, t)$, can be depicted through animations or snapshots in time.

Vector fields are also functions of space, and possibly time, but as opposed to scalar fields, they are vector-valued. An example vector field is the velocity, $\vec{v}(x, y, z)$. This notation indicates that at each coordinate (x, y, z) in three-dimensional space, we have one velocity vector, $\vec{v}$, that contains three components. Visualizing vector fields is more difficult compared to scalar fields because of the additional components. One option is to draw arrows at select points in space, in the direction of the vector and with length proportional to the magnitude. Alternatively, contour plots can be made of individual components or the vector magnitude. Figure 1.13 illustrates the latter option.

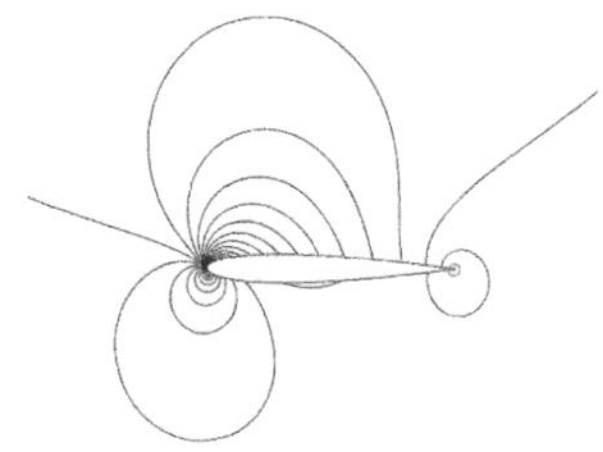

Figure 1.13 Velocity magnitude contours for flow over an airfoil, obtained from computational fluid dynamics data.

Example 1.8 The gradient vector

Consider analytical temperature and velocity fields in two dimensions, given by

$$T(r, \theta) = 10 + \frac{1}{r} \sin \theta, \quad \vec{v}(r, \theta) = re^{-r}\hat{\theta}. \tag{1.39}$$

The direction of maximum change of T is given by its gradient, which is, remembering to use the cylindrical coordinate formula from Equation 1.31,

$$\nabla T = \frac{\partial T}{\partial r}\hat{r} + \frac{1}{r}\frac{\partial T}{\partial \theta}\hat{\theta} = \frac{-1}{r^2} \sin \theta \, \hat{r} + \frac{1}{r^2} \cos \theta \, \hat{\theta}. \tag{1.40}$$

Evaluating this expression at several points, we have

$$\begin{aligned} &\text{At } (r, \theta) = (2{,}0): \quad -\frac{1}{4}(0)\hat{r} + \frac{1}{4}(1)\hat{\theta} = \frac{1}{4}\hat{\theta}, \\ &\text{At } (r, \theta) = (2, \frac{\pi}{4}): \quad -\frac{1}{4}\frac{\sqrt{2}}{2}\hat{r} + \frac{1}{4}\frac{\sqrt{2}}{2}\hat{\theta} = \frac{\sqrt{2}}{8}(\hat{\theta} - \hat{r}), \\ &\text{At } (r, \theta) = (2, \frac{\pi}{2}): \quad -\frac{1}{4}(1)\hat{r} + \frac{1}{4}(0)\hat{\theta} = -\frac{1}{4}\hat{r}. \end{aligned} \tag{1.41}$$

We can also calculate the direction of maximum change of the velocity magnitude. The velocity magnitude is $|\vec{v}| = re^{-r}$, and taking the gradient of this quantity, we obtain

$$\nabla|\vec{v}| = \frac{\partial|\vec{v}|}{\partial r}\hat{r} + \frac{1}{r}\frac{\partial|\vec{v}|}{\partial \theta}\hat{\theta} = (e^{-r} - re^{-r})\,\hat{r}. \tag{1.42}$$

This gradient vector points along the radial direction, $\hat{r}$, and its magnitude is independent of the angle θ. Obtaining a directional derivative of either the temperature or the velocity magnitude involves dotting the gradient vector with a unit vector in the desired direction.

1.2.7 Stokes's and Divergence Theorems

In our study of aerodynamics, we will encounter quantities that can be defined *locally*, at points in space, in terms of the properties of the flow. These quantities may include derivatives, but assuming continuous properties, such derivatives are well-defined. We will also encounter quantities that depend on control volumes or curves drawn in space. For example, the amount of mass flowing into or out of a control volume depends in general on the shape and location of the control volume. In certain cases, we will be able to relate the local quantities to the control-volume-dependent ones, through two useful mathematical theorems. These are

Stokes's theorem and the divergence or Gauss's theorem, both described in this section.

Stokes's theorem relates a line integral of the tangential component of a vector around a closed contour to a surface integral of the curl of that vector.

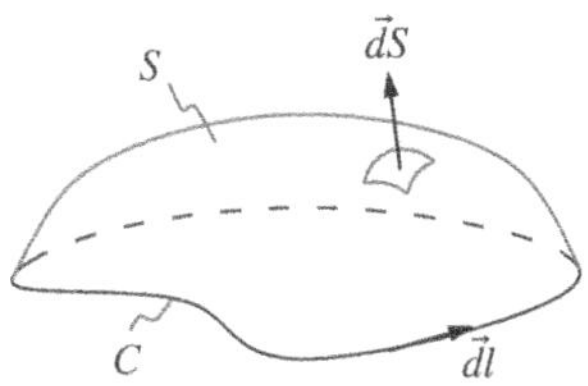

Figure 1.14 Surface, *S*, and enclosing contour, *C*, for Stokes's theorem. *S* can be thought of as a soap bubble attached to the hoop *C*.

Theorem 1.2.1 (Stokes's Theorem.) *Given a contour C enclosing a surface S, and a differentiable vector field $\vec{v}$,*

$$\oint_C \vec{v} \cdot \vec{dl} = \int_S (\nabla \times \vec{v}) \cdot \vec{dS}, \tag{1.43}$$

where $\vec{dl}$ is a differential line element tangent to C, and $\vec{dS}$ is a normal-pointing differential surface element on S. The direction of the contour integral is given by the right-hand rule implied by the direction of $\vec{dS}$, as illustrated in Figure 1.14.

We will use Stokes's theorem to relate vorticity, which is the curl of the velocity, to circulation, which is a contour integral of its tangential component. The relationship will be important for quantifying the amount of lift produced by certain flows.

The second theorem, which is the divergence or Gauss's theorem, relates the normal component of a vector field integrated over the surface of a control volume to the divergence of the vector integrated over the interior volume.

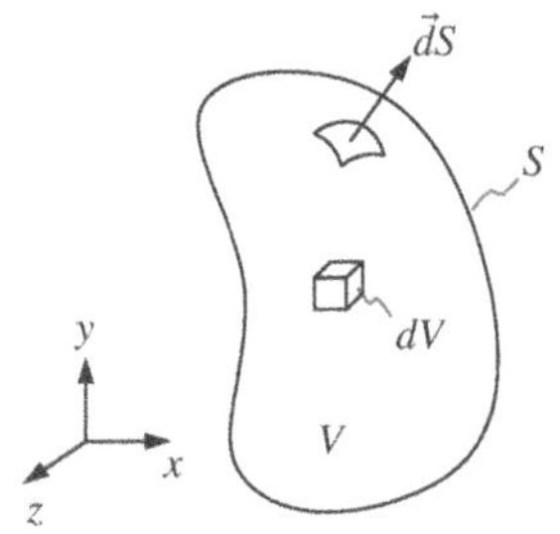

Figure 1.15 Volume and surface used in the divergence theorem in three dimensions.

Theorem 1.2.2 (Divergence/Gauss's Theorem.) *Given a surface S enclosing a volume V and a vector field $\vec{F}$*

$$\oint_S \vec{F} \cdot \vec{dS} = \int_V \nabla \cdot \vec{F}\, dV, \tag{1.44}$$

where dV is a differential volume element in V, and $\vec{dS}$ is an outward-pointing differential surface element on S, as illustrated in Figure 1.15.

The two-dimensional version of Equation 1.44 reads

$$\oint_C \vec{F} \cdot \vec{n}\, dl = \int_A \nabla \cdot \vec{F}\, dA, \tag{1.45}$$

where the curve C encloses an area A, dl is a differential line element on C, $\vec{n}$ is an outward-pointing normal on C, and dA is a differential area element on A, as illustrated in Figure 1.16.

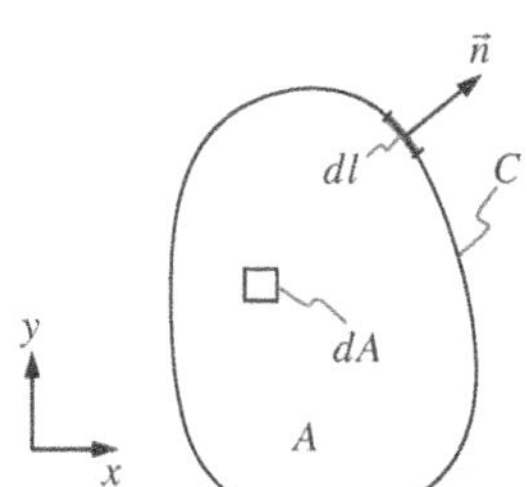

Figure 1.16 Area and contour used in the divergence theorem in two dimensions.

We will use the divergence theorem in the study of conservation laws, specifically in relating integral statements of conservation, such as mass entering and leaving a control volume, to differential forms that have to hold pointwise. The differential forms will yield differential equations that lend themselves to a systematic solution procedure for general flow problems.

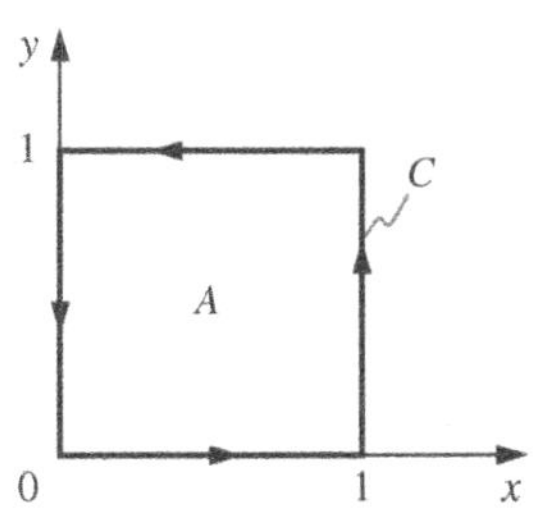

Figure 1.17 Contour and enclosed area used in Example 1.9 and Example 1.10.

Example 1.9 Use of Stokes's theorem

Given a vector field $\vec{v} = x\hat{i} + y\hat{j}$, and the contour shown in Figure 1.17, we would like to calculate the contour integral

$$\oint_C \vec{v} \cdot \vec{dl}. \tag{1.46}$$

One approach is to calculate this integral directly by breaking it up into four parts,

$$\oint_C \vec{v}\cdot d\vec{l} = \int_0^1 x\,dx \quad \text{(bottom edge, } \vec{v}\cdot d\vec{l} = x\,dx\text{)}$$

$$+\int_0^1 y\,dy \quad \text{(right edge, } \vec{v}\cdot d\vec{l} = y\,dy\text{)}$$

$$+\int_0^1 -x\,dx \quad \text{(top edge, } \vec{v}\cdot d\vec{l} = -x\,dx\text{)}$$

$$+\int_0^1 -y\,dy \quad \text{(left edge, } \vec{v}\cdot d\vec{l} = -y\,dy\text{)}$$

$$= \frac{1}{2} + \frac{1}{2} - \frac{1}{2} - \frac{1}{2} = 0.$$

Alternatively, we can use Stokes's theorem, with the enclosed area A as the surface S, and $d\vec{S} = dA\,\hat{k}$. Evaluating the curl of $\vec{v}$, we have,

$$\nabla\times\vec{v} = \begin{vmatrix} \hat{\imath} & \hat{\jmath} & \hat{k} \\ \dfrac{\partial}{\partial x} & \dfrac{\partial}{\partial y} & \dfrac{\partial}{\partial z} \\ x & y & 0 \end{vmatrix} = 0\hat{\imath} + 0\hat{\jmath} + 0\hat{k}. \tag{1.47}$$

So the right-hand side of Equation 1.43 is 0, and this means that the left-hand side, which is the quantity that we are after, is also 0.

Example 1.10 Use of the divergence theorem

Consider again the contour in Figure 1.17, but suppose that we are interested in

$$\oint_C \vec{v}\cdot\vec{n}\,dl. \tag{1.48}$$

Taking the direct approach first, we split the integral up into four parts and obtain

$$\oint_C \vec{v}\cdot\vec{n}\,dl = \int_0^1 -y\,dx \quad \text{(bottom edge, } \vec{v}\cdot\vec{n} = -y,\ y = 0\text{)}$$

$$+\int_0^1 x\,dy \quad \text{(right edge, } \vec{v}\cdot\vec{n} = x,\ x = 1\text{)}$$

$$+\int_0^1 y\,dx \quad \text{(top edge, } \vec{v}\cdot\vec{n} = y,\ y = 1\text{)}$$

$$+\int_0^1 -x\,dy \quad \text{(left edge, } \vec{v}\cdot\vec{n} = -x,\ x = 0\text{)}$$

$$= 0 + 1 + 1 + 0 = 2.$$

Alternatively, we can obtain the same answer by integrating the divergence of $\vec{v}$ and applying the divergence theorem, Equation 1.45.*

* The divergence of $\vec{v}$ is

$$\nabla \cdot \vec{v} = \frac{\partial x}{\partial x} + \frac{\partial y}{\partial y} = 2,$$

so applying the divergence theorem yields

$$\oint_C \vec{v} \cdot \vec{n}\, dl = \int_A \nabla \cdot \vec{v}\, dA = \int_A 2\, dA = 2.$$

1.3 Numerical Methods

Throughout this text, we will use numerical methods to solve equations, approximate flows, visualize results, and gain insight into flow physics. These numerical methods can be implemented in a variety of programming languages, and in this text, we will use MATLAB (The MathWorks Inc. 2024).

1.3.1 Integration of Ordinary Differential Equations

Ordinary differential equations (ODEs) involve derivatives with respect to one variable, such as time. A system of ODEs can be written generally as

$$\dot{\mathbf{u}} = \mathbf{f}(\mathbf{u}, t), \tag{1.49}$$

where $\dot{\mathbf{u}}$ is the time derivative of a state vector $\mathbf{u}$. The function $\mathbf{f}$ takes as input the instantaneous state vector $\mathbf{u}$ and possibly the time t, and it returns the rate of change (i.e. the slope) of $\mathbf{u}$ with respect to time.

When solving an ODE numerically, we can divide a time horizon of interest, $t \in [0, T]$, into N intervals of equal length $\Delta t \equiv T/N$.* Let $\mathbf{u}^n$, $0 \le n \le N$, denote the state at *time node n*, which is at time $t = t^n \equiv n\Delta t$. $\mathbf{u}^0$ is a known initial condition, and the goal in ODE integration is to obtain $\mathbf{u}^n$ for $n > 0$.†

* Unequal time steps can be used with some methods, and they can be more efficient when time accuracy requirements change during a simulation.

† Numerical methods are generally not going to give the exact answer due to approximations made in the numerical integration. For good numerical methods, the error will decrease quickly as $\Delta t \to 0$.

One of the simplest methods for advancing Equation 1.49 in time is *forward Euler*,

$$\mathbf{u}^{n+1} = \mathbf{u}^n + \Delta t\, \mathbf{f}(\mathbf{u}^n, t^n). \tag{1.50}$$

The state at t^{n+1} is given by extrapolating the state at t^n using the slope at t^n, which is just $\mathbf{f}(\mathbf{u}^n, t^n)$. This method is simple but only first-order accurate: the global error, e.g. in the solution at time $t = T$, is asymptotically proportional to Δt.

A simple improvement to Equation 1.50 is to account for a changing time slope of the state by using an average slope between t^n and t^{n+1}. This gives the two-stage *predictor–corrector* method,

$$\tilde{\mathbf{u}}^{n+1} = \mathbf{u}^n + \Delta t\, \mathbf{f}(\mathbf{u}^n, t^n), \tag{1.51}$$

$$\mathbf{u}^{n+1} = \mathbf{u}^n + \frac{1}{2}\Delta t\left[\mathbf{f}(\mathbf{u}^n, t^n) + \mathbf{f}(\tilde{\mathbf{u}}^{n+1}, t^{n+1})\right]. \tag{1.52}$$

The first equation, the predictor stage, is just forward Euler. The resulting state $\tilde{\mathbf{u}}^{n+1}$ is only used to provide a better slope in the actual update, the corrector stage, which uses the average of the slopes at n and $n + 1$.‡

‡ If we measure the cost of a method by the number of $\mathbf{f}$ evaluations, predictor–corrector is twice as expensive per time step compared to forward Euler. However, it is second-order accurate, which means that the global error asymptotically scales like $(\Delta t)^2$, a huge improvement compared to forward Euler.

Even higher-order time integration methods exist, with the most popular being the four-stage Runge–Kutta method (RK4),

$$\begin{aligned}
\mathbf{f}_0 &= \mathbf{f}(\mathbf{u}^n, t^n) \\
\mathbf{f}_1 &= \mathbf{f}\left(\mathbf{u}^n + \frac{1}{2}\Delta t\mathbf{f}_0, t^n + \frac{\Delta t}{2}\right) \\
\mathbf{f}_2 &= \mathbf{f}\left(\mathbf{u}^n + \frac{1}{2}\Delta t\mathbf{f}_1, t^n + \frac{\Delta t}{2}\right) \\
\mathbf{f}_3 &= \mathbf{f}(\mathbf{u}^n + \Delta t\mathbf{f}_2, t^n + \Delta t) \\
\mathbf{u}^{n+1} &= \mathbf{u}^n + \frac{\Delta t}{6}(\mathbf{f}_0 + 2\mathbf{f}_1 + 2\mathbf{f}_2 + \mathbf{f}_3)
\end{aligned} \tag{1.53}$$

Note the presentation in terms of $\mathbf{f}$ evaluations, which for efficiency is how the method would be coded. RK4 is fourth-order accurate, which means that the global error is proportional to $(\Delta t)^4$.§

Cheaper higher-order methods also exist, in the multi-step family, where states at earlier time nodes are used in the slope approximation and update. However, these methods require another method at the initial step(s), since states before $t = 0$ are not available. In addition, multi-stage methods like RK4 generally have better stability properties compared to multi-step methods.

The time-integration methods discussed here are *explicit*, which means that the formula for $\mathbf{u}^{n+1}$ can be written explicitly in terms of previous states and $\mathbf{f}$ evaluations that use their combinations. When the state changes rapidly, explicit methods cannot advance with too large of a time step because the solution will grow unbounded if all of the time scales in the ODE system are not resolved. When a system has very small time scales whose resolution does not significantly impact accuracy, the very small time step restriction can become onerous. Such a system is called *stiff*, and often, a more efficient approach to time integration becomes an *implicit* method, in which the update formula for $\mathbf{u}^{n+1}$ uses the unknown $\mathbf{u}^{n+1}$ in one or more evaluations of $\mathbf{f}$.¶

§ Although RK4 requires four $\mathbf{f}$ evaluations per time step, the cost is worth it when high accuracy is required: fourth order is vastly superior to first and second order.

¶ The presence of $\mathbf{u}^{n+1}$ inside $\mathbf{f}(\cdot)$ makes the update more difficult, especially when $\mathbf{f}(\cdot)$ is a nonlinear function, since then a nonlinear solver must be applied at each time step.

Example 1.11 Time integration for the ascent of a balloon

At time $t = 0$, a balloon is released from $(x, y) = (0,0)$ in a two-dimensional atmosphere. Its vertical velocity is constant at $v_y = 2\text{m/s}$, but its horizontal velocity varies according to

$$v_x = 2\sin(t/8)\cos(y/6) + \sin(x/3)e^{-y/50}. \quad (1.54)$$

We can simulate the path of the balloon using the forward Euler method. Since velocity is the time derivative of position, the two governing equations are‖

$$\dot{x} = 2\sin(t/8)\cos(y/6) + \sin(x/3)e^{-y/50}, \quad (1.55)$$

$$\dot{y} = 2. \quad (1.56)$$

Taking the state vector to be the balloon's position, $\mathbf{u} = [x, y]^T$, these two equations are in the form $\dot{\mathbf{u}} = \mathbf{f}(\mathbf{u}, t)$. Note that $\mathbf{f}$ depends on both the state (the position) and the time. To simulate the path, we set the position to the origin and then repeatedly call a function to compute $\mathbf{f}$ from the current state and time and use $\mathbf{f}$ to update the state. Listing 1.4 shows a code that implements the time integration and that produces the plot shown in Figure 1.18. Each solution has numerical error due to a finite time step, Δt, but we expect the error to decrease as the time step is reduced. The fact that the paths for the smaller two Δt are closer together compared to the largest Δt confirms this expectation.

‖ One can solve at least the y equation analytically, but then the resulting x equation is cumbersome. A numerical approach is therefore simpler in this case.

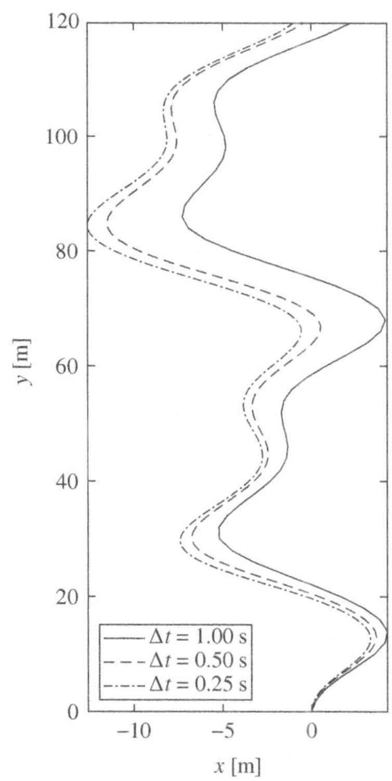

Figure 1.18 Balloon ascent paths using forward Euler and different time steps.

Listing 1.4 Code for simulating the ascent of a balloon.

```
function balloon
% simulates the ascent of a balloon in a known velocity field

u0 = [0;0]; % state = initial position [m]
T = 60;     % time horizon [s]
figure(1); clf;

% forward Euler time stepping
Nt = 60;    % number of time steps
ctype = {'k-', 'k--', 'k-.'};
```

```
for it = 1:3,
  dt = T/Nt; % time step
  U = zeros(2,Nt+1); U(:,1) = u0; % initial condition
  for n = 1:Nt,
    U(:,n+1) = U(:,n) + dt*getf(U(:,n),n*dt);
  end
  plot(U(1,:), U(2,:), ctype{it}, 'displayname', ...
       sprintf('$\\Delta t=%.2f$s', dt));
  hold on;
  Nt = Nt*2;
end

h = legend('fontsize', 14, 'location', 'southwest', ...
           'interpreter', 'latex');
set(gcf, 'position', 400*[1,1,1,2]);
writeplot('$x$ [m]', '$y$ [m]', '../figs/balloon.pdf', 4);

%-----------------------
function f = getf(u,t)
f = u; % this just makes f the same size as u
f(1) = 2*sin(t/8)*cos(u(2)/6) + sin(u(1)/3)*exp(-u(2)/50);
f(2) = 2;
```

1.3.2 Finite Differences for Partial Differential Equations

Partial differential equations (PDEs) involve derivatives with respect to multiple variables, e.g. multiple spatial coordinates and/or time. They require appropriate initial and boundary conditions in order to be well-posed. An initial condition is the state at time $t = 0$, and it is required when solving unsteady problems. Boundary conditions refer to specifying the state, or parts of the state, at the boundaries of the domain. Such information is generally required when solving either steady or unsteady problems. In some cases, the boundary information may be given in terms of derivatives of states, called *Neumann* boundary conditions, as opposed to the state information only, called *Dirichlet* boundary conditions. *Robin* boundary conditions involve both the state and its derivatives. On some boundaries, we may not need to specify any boundary conditions.*

*An example is the outflow of a supersonic gas from some domain. Information cannot propagate upstream in supersonic flow, so no boundary condition is required at such an outflow.

The *finite-difference method* provides a simple and systematic approach to numerically solving PDEs.† A grid of points is placed inside the spatial domain, usually with uniform spacing. Ideally, the domain will be a box or a shape that can be transformed into one or more boxes via a simple mapping. The derivatives in the PDE are then approximated at each point in the domain by using neighboring points. For example, a first derivative in one spatial dimension, $\frac{\partial u}{\partial x}$, can be approximated at node j in several ways,

† Other numerical techniques for PDEs exist, such as finite-volume and finite-element methods, but we will only consider the finite-difference method here.

$$\text{forward difference:} \quad \left.\frac{\partial u}{\partial x}\right|_j^F \approx \frac{u_{j+1} - u_j}{\Delta x}, \tag{1.57}$$

$$\text{backward difference:} \quad \left.\frac{\partial u}{\partial x}\right|_j^B \approx \frac{u_j - u_{j-1}}{\Delta x}, \tag{1.58}$$

$$\text{central difference:} \quad \left.\frac{\partial u}{\partial x}\right|_j^C \approx \frac{u_{j+1} - u_{j-1}}{2\Delta x}. \tag{1.59}$$

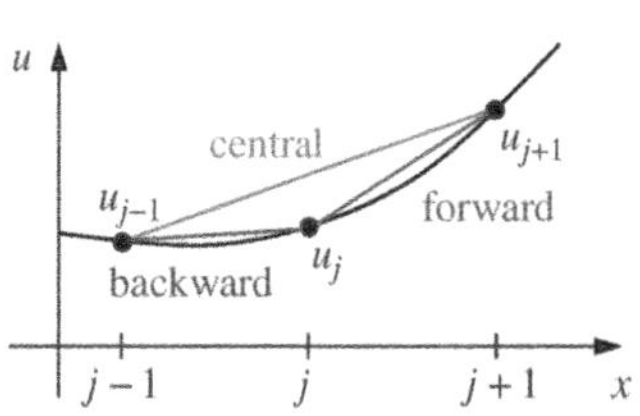

Figure 1.19 Stencils for approximating the first derivative in one dimension.

Figure 1.19 illustrates the u evaluations used in these three expressions. The set of points in the neighborhood of j required to approximate a derivative is known as the *stencil* of the finite-difference formula.

The forward and backward differences are first-order accurate, while the central difference is second-order accurate.‡ By applying the central-difference formula twice, we can obtain a discretization of the second derivative,

$$\left.\frac{\partial^2 u}{\partial x^2}\right|_j \approx \frac{u_{j-1} - 2u_j + u_{j+1}}{\Delta x^2}. \tag{1.60}$$

This is also known as a central difference, but for a second derivative. It is also second-order accurate.

‡ First-order accuracy means that the error in the approximation is $\mathcal{O}(\Delta x)$, while second-order accuracy means that the error is $\mathcal{O}(\Delta x^2)$.

Example 1.12 An unsteady diffusion problem
An example of a PDE is the prototypical unsteady diffusion problem for a one-dimensional scalar field $u(x, t)$,

$$\frac{\partial u}{\partial t} - \nu \frac{\partial^2 u}{\partial x^2} = 0, \tag{1.61}$$

where ν [m^2/s] is the diffusivity. Let the spatial domain be $[0, L]$, and the boundary conditions homogeneous Neumann (i.e. $\frac{\partial u}{\partial x} = 0$) at both ends. Discretizing the spatial derivative using a second-order central finite difference on a grid of spacing $\Delta x = L/N$ gives a *semi-discrete* ODE system for the nodal states u_j,

$$\frac{du_j}{dt} = \nu \frac{u_{j-1} - 2u_j + u_{j+1}}{\Delta x^2}. \tag{1.62}$$

We can discretize the boundary conditions with a first-order difference, which gives $u_0 - u_1 = 0$ and $u_N - u_{N-1} = 0$.§ For marching forward in time, the time step of an explicit method is restricted by the *heat number* $\mu \equiv \nu\Delta t/\Delta x^2$, which means that halving Δx requires a four-fold reduction in Δt. Forward Euler, used for simplicity in this example, requires $\mu < 0.5$ for stability. Listing 1.5 shows a code that implements a solver for this problem.

§ Higher-order one-sided stencils can be used to improve the boundary condition discretization, to match the second order of the interior.

Listing 1.5 Code for computing a forward Euler solution to unsteady diffusion.

```
function diffusion
% runs an unsteady diffusion simulation

L = 1.;                        % domain length
nu = 0.1;                      % diffusivity
N = 50;                        % number of intervals
dx = L/N;                      % interval length
x = linspace(0,L, N+1);        % node positions

u = exp(-100*(x-0.5*L).^2);  % initial condition
figure(1); clf;
plot(x,u,'k-', 'displayname', 't=0 ); hold on;

mu = 0.4;                      % heat number
dt = mu*dx^2/nu;               % time step
Nt = 50;                       % how often to plot
ctype = {'k--', 'k-.', 'k:'};

for n = 1:3*Nt
  up = u;
  for j = 2:N, up(j) = u(j) + mu*(u(j-1)-2*u(j)+u(j+1)); end;
  u = up; u(1) = u(2); u(N+1) = u(N);
  if (mod(n,Nt)==0),
    plot(x,u, ctype{n/Nt}, 'displayname', sprintf('t=%.2f', n*dt));
  end
end

% Write out the plot
```

Code notes:

- Physics and discretization parameters are hard-coded but could be made inputs.
- The time step is set based on the heat number, chosen less than 0.5.
- The initial condition is hard-coded as a Gaussian distribution.
- Inside the time-step loop, the solution is plotted at prescribed intervals.
- Discretizing the time derivative in Equation 1.62 and solving for u_j^{n+1} yield the update formula used in the loop over j.
- up is a temporary state used to prevent overwriting u as the update is computed.
- Neumann BCs are enforced simply by setting the boundary states to their adjacent values.

```
h = legend('fontsize', 14, 'location', 'northeast', 'interpreter',
    'latex');
set(gcf, 'position', 400*[1,1,1,.75]); box off;
set(gca, 'xtick', linspace(0,1,6), 'ytick', linspace(0,1,6));
writeplot('$x$', '$u(x)$', '../figs/diffusion.pdf', 4);
```

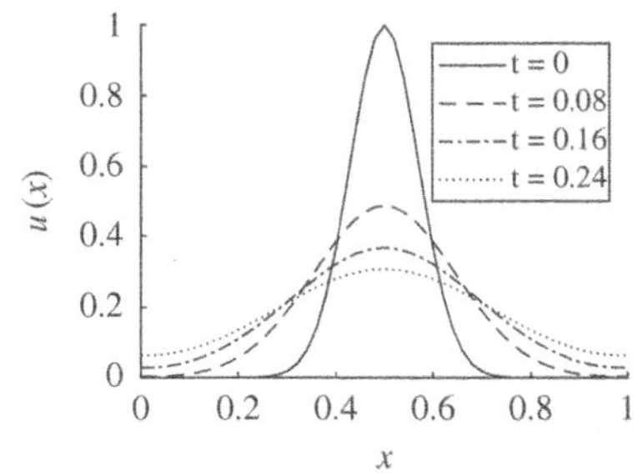

Figure 1.20 Solutions to unsteady diffusion at various times for the parameters in Listing 1.5.

Figure 1.20 shows snapshots of the solution after initialization with a Gaussian profile, at various times, measured in terms of the time, the units of which are determined by the units of L and ν. The discretization consists of 100 intervals in x. As the solution evolves, the state spreads out but does not translate, consistent with a diffusion process. The rate of spreading is fastest at $t = 0$, when the gradients are steepest. The discretization is conservative, and since the state does not leave through the boundaries, the integrated state (area under the profile) remains constant for all times.

1.3.3 Fourier-Series Representations of Functions

A function of a coordinate x can be represented as a linear combination of basis functions, such as sines or cosines. Instead of specifying the formula for the function, or point values at coordinates, we instead specify the amplitudes of the basis functions. For smooth functions, an accurate representation can often be obtained using a small number of basis functions, and this is an advantage of the Fourier-series approach. The representation can also be more convenient, as we only need to store a few amplitudes, and it can make certain math operations, such as integration, easier.

Consider a sine-series expansion. Suppose that we are given some function $f(x)$ over an interval $x \in [0, L]$. Suppose also that the function is 0 on the boundaries and varies smoothly in between. Our goal is to represent $f(x)$ as a sum of a few sine functions.* The representation is

* We use sines because these go to zero on the boundaries when the domain is scaled to be an integer multiple of π.

$$\begin{aligned} f(x) &\approx A_1 \sin\left(\pi \frac{x}{L}\right) + A_2 \sin\left(2\pi \frac{x}{L}\right) + \ldots + A_N \sin\left(N\pi \frac{x}{L}\right) \\ &= \sum_{m=1}^{N} A_m \sin\left(m\pi \frac{x}{L}\right). \end{aligned} \tag{1.63}$$

Here, $m \in [1, N]$, where N is the number of coefficients used in the representation. If we are given $f(x)$, how do we obtain these coefficients? A simple approach uses the orthogonality property of sine functions,†

† By the substitution $u = \pi x/L$, $du = \pi dx/L$, the integral in Equation 1.64 becomes one from 0 to L, and the right-hand side becomes $\delta_{mn} L/2$.

$$\int_0^{\pi} \sin(nu)\sin(mu)\, du = \frac{\pi}{2}\delta_{mn}, \tag{1.64}$$

where $\delta_{mn} = 1$ if $m = n$, 0 otherwise. To obtain the nth coefficient, A_n, we multiply Equation 1.63 by $\sin(n\pi x/L)$ and integrate from 0 to L,

$$\int_0^L \sin\left(n\pi \frac{x}{L}\right) f(x)\, dx = \int_0^L \sin\left(n\pi \frac{x}{L}\right) \sum_{m=1}^{N} A_m \sin\left(m\pi \frac{x}{L}\right)\, dx. \tag{1.65}$$

The right-hand side of this equation simplifies by applying Equation 1.64,

$$\sum_{m=1}^{N} A_m \int_0^L \sin\left(n\pi \frac{x}{L}\right) \sin\left(m\pi \frac{x}{L}\right) = \sum_{m=1}^{N} A_m \frac{L}{2}\delta_{mn} = \frac{L}{2}A_n. \tag{1.66}$$

Equating with the left-hand side gives an explicit formula for A_n;

$$A_n = \frac{2}{L}\int_0^L \sin\left(n\pi\frac{x}{L}\right) f(x)\, dx. \tag{1.67}$$

Example 1.13 Sine-series approximation of a function
Given the following function,

$$f(x) = x(x-5)(x^2 - 4x + 3), \quad x \in [0,5], \tag{1.68}$$

we construct sine-series approximations using various N. We could perform the integration in Equation 1.67 analytically, symbolically, or numerically. In the code in Listing 1.6, we use a numerical approach, trapezoidal integration, for simplicity.

Listing 1.6 Code for computing a Fourier representation of a polynomial function.

```
function sineseries
% approximates a polynomial with a Fourier sine series

L = 5;  % domain length
x = linspace(0,L,501);
f = x.*(x-L).*(x.^2 - 4*x + 3);
N = 6;  % max number of terms
figure(1); clf;
plot(x,f, 'k-', 'displayname', '$f(x)$'); hold on;
fn = zeros(size(x));
ctype = {'k--', 'k-.', 'k:'};
for n = 1:N,
  An = (2/L)*trapz(x, sin(n*pi*x/L).*f);
  fn = fn + An*sin(n*pi*x/L);
  if (mod(n,2)==0),
    plot(x, fn, ctype{n/2}, 'displayname', sprintf('$N=%d$', n));
  end;
end
h = legend('fontsize', 14, 'location', 'southwest', 'interpreter', '
    latex');
set(gcf, 'position', 400*[1,1,1,1]); box off; axis([0,5,-15,10]);
set(gca, 'xtick', linspace(0,5,6), 'ytick', linspace(-15,10,6));
writeplot('$x$', '$f(x)$', '../figs/sineseries.pdf', 4);
```

Code notes:

- The approximation fn is built up incrementally by adding a sine term one at a time.
- Plots are generated at even numbers of terms.
- Accuracy of the trapezoidal integration could be increased by using more points.

Figure 1.21 shows the plot produced by this code. With just two terms, the approximation is quite poor, but the accuracy quickly improves. By six terms, the approximation is close to on top of the function.

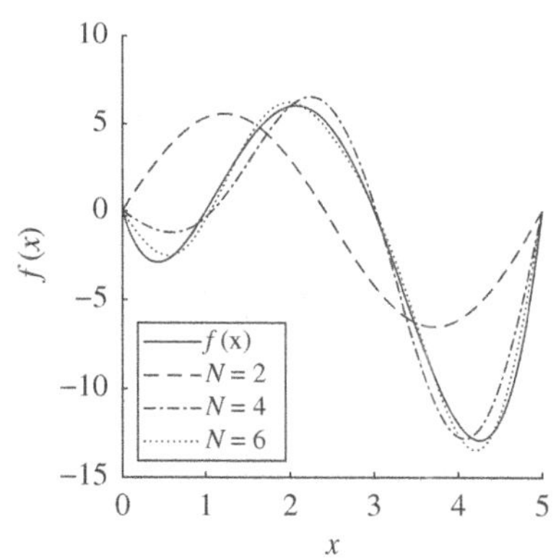

Figure 1.21 Fourier approximations of a function using different numbers of terms.

1.4 Summary and Concepts

This chapter reviewed basic elements of fluid dynamics, thermodynamics, mathematics, and numerical methods that are relevant to the study of aerodynamics by computational means. A molecular description of a fluid was used to build intuition for certain thermodynamic quantities, including pressure, temperature, and viscosity. The mathematics review focused on linear algebra, important for solving systems of equations, index notation for concisely describing complex vector and tensor equations, cylindrical coordinates, and two theorems relevant to control-volume analysis. Numerical methods were discussed for solving both ordinary and partial differential equations, and the Fourier series was reviewed for representing smooth functions.

The information presented is mostly sufficient for the content that follows, but it is by no means exhaustive or complete: fluid dynamics, thermodynamics, and numerical methods are all rich subjects with much more depth than touched on here. The cited references can serve as a starting point for further study in these topics.

The key concepts covered in this chapter are as follows:

- Liquids and gases qualify as fluids, and their distinction from solids is that a fluid deforms continuously when subjected to a shear stress.
- A fluid is made up of discrete molecules, and treating it as a continuum is possible when any small "point-like" control volume contains a large number of molecules. In such a condition, the fluid is in local thermodynamic equilibrium (LTE).
- The Knudsen number (Kn) measures the ratio of the mean free path to a characteristic length, and LTE occurs when $Kn \ll 1$.
- At a molecular level, temperature measures the kinetic energy of the random motions of the molecules, whereas pressure measures the momentum transferred by the molecules when they collide with a surface.
- The ideal gas law relates pressure, density, and temperature. We will assume that air is an ideal gas.
- The dynamic viscosity measures the shear stress caused by velocity gradients in a fluid. At a molecular level, it is proportional to the density, molecular speed, and the mean free path.
- Dimensional analysis is a powerful tool for deducing a formula for one quantity in terms of others based solely on the dimensions.
- Checking equations for vector and dimensional consistency can help catch mistakes when solving problems.
- For a linear system of equations to be solvable, the matrix determinant cannot be zero, which also means that the matrix has full rank and an empty null space.
- Index notation is a shorthand for expressing vector and tensor operations, including derivatives. It particularly helps eliminate confusion when working with tensors.
- A cylindrical basis is often convenient for representing vector fields in aerodynamics, but care must be taken when using it, since the basis varies with position.
- A directional derivative measures the rate of change of a field quantity in a specified direction.
- Stokes's and divergence theorems are useful in relating integral quantities over a control volume or its surface to local point quantities inside the control volume.
- Many numerical techniques are available for solving ordinary differential equations, which often appear when simulating the time evolution of a dynamical system, including a fluid.
- Partial differential equations can be solved on a grid of points using the finite-difference method, with accuracy levels that depend on the number of points used and the order of the finite differences.
- Fourier series provide a succinct representation of a smooth function using sinusoidal functions. We will use them in studying airfoils, wings, and slender bodies in supersonic flow.

1.5 Problems

1. Justify each answer to the below questions with an explanation.
   a Is $\hat{r} + \hat{\imath}$ a valid expression? If so, give the result in the Cartesian basis.
   b True or false: $\vec{v} \cdot \nabla p$ is a valid mathematical expression.
   c Is $\vec{v} \cdot \nabla \vec{w}$ a scalar, vector or tensor?
   d True or false: if $\oint_C \vec{v} \cdot \vec{n}\, dl = 0$, for some contour C, then $\nabla \cdot \vec{v} = 0$ everywhere inside C.
   e True or false: the expression $p/\rho + |\vec{v}|^2/2 + gh$ is dimensionally consistent.*
   f What is the solution to $\dot{\vec{x}} = \hat{\theta}$, with the initial condition $\vec{x}|_{t=0} = \vec{x}_0$?
   g True or false: the forward Euler method applied to the ordinary differential equation $\dot{u} = t$ yields the exact answer. Repeat for the predictor-corrector method.

* p is pressure, ρ is density, $\vec{v}$ is velocity, g is acceleration, and h is height.

2. Compute the following quantities and present your answers in both the Cartesian basis and the cylindrical basis.

$$\nabla r, \quad \nabla \theta, \quad \frac{\partial x}{\partial \theta}, \quad \frac{\partial \hat{r}}{\partial x}, \quad \frac{\partial \hat{r}}{\partial \theta}.$$

3. Use index notation to prove the following two identities for any three-dimensional scalar fields $p(\vec{x})$, $q(\vec{x})$, and vector fields $\vec{v}(\vec{x})$, $\vec{w}(\vec{x})$.
   A $\nabla \cdot (\nabla p \times \nabla q) = 0$
   B $\nabla \cdot (\vec{v} \times \vec{w}) = \vec{w} \cdot (\nabla \times \vec{v}) - \vec{v} \cdot (\nabla \times \vec{w})$

4. Consider the vector field $\vec{F} = x\hat{\imath} + (x + y)\hat{\jmath}$ and the half-disk control volume shown in Figure 1.22. Independently evaluate both sides of the divergence theorem equation for the given $\vec{F}$ and control volume, and show that the results match. Repeat for Stokes's theorem.

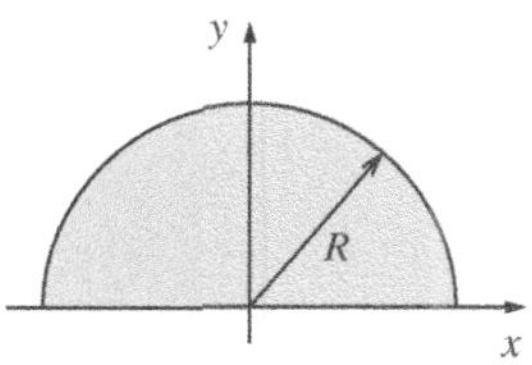

Figure 1.22 A half-disk control volume for verifying the divergence and Stokes's theorems.

5. A three-dimensional vector field is given by $\vec{F} = 3x\hat{\imath} + y\hat{\jmath} - 2z\hat{k}$. For a $2 \times 2 \times 2$ cube with a vertex at the origin, shown in Figure 1.23, calculate separately both sides of the divergence theorem equation and show that they are equal.

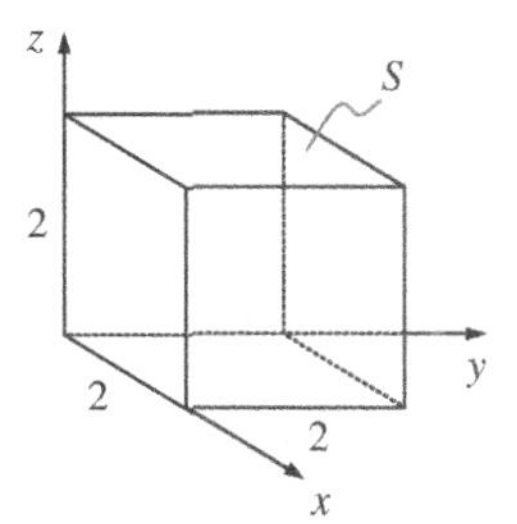

Figure 1.23 A cube control volume for divergence theorem calculations.

6. Assume that the temperature field near an airport can be approximated as linear in space and time,

$$T(x, y, z, t) = T_0 + ax + by + cz + ft, \tag{1.69}$$

where T_0, a, b, c, f are constants, temperature is measured in Kelvin, time is measured in seconds, and x, y, z are measured in meters. The control tower is located at the origin, $(x, y, z) = (0,0,0)$, and at $t = 0$ it reads a temperature of 300 K. Four aircraft flying at constant velocities near the airport report temperature readings at two times, as given in the table below.

Aircraft	velocity, $\vec{v}$, in m/s	T at $t = 0$ s	T at $t = 20$ s
A	$50\hat{\imath} + 40\hat{\jmath} + 5\hat{k}$	269.5 K	268.91 K
B	$-30\hat{\imath} + 80\hat{\jmath} - 4\hat{k}$	309.5 K	305.8 K
C	$75\hat{\imath} - 10\hat{\jmath} + 2\hat{k}$	293.0 K	294.94 K
D	$-40\hat{\imath} - 60\hat{\jmath} - 5\hat{k}$	292.5 K	294.21 K

Using this information, determine the unknowns T_0, a, b, c, f that characterize the temperature field.

7. Modify the code in Example 1.11 to use the predictor–corrector method and fourth-order Runge–Kutta. Study how the error in the final position decreases with the time step Δt for each method. Verify that you obtain the expected convergence rates for each method.

8. Extend the Fourier sine-series approximation code in Example 1.13 to calculate the L_2 error, E_{L2}, between the approximation, $f^N(x)$, and the true function, $f(x)$, defined by

$$E_{L2}^2 = \frac{1}{L}\int_0^L \left[f^N(x) - f(x)\right]^2 dx. \tag{1.70}$$

Study the convergence of E_{L2} versus N for two functions: a smooth one defined by $f_1 = \exp(100(x - L/2)^2)$, and a discontinuous one defined by $f_2(x) = 1$ for $L/3 \leq x \leq 2L/3$ and zero otherwise. Plot several approximations and discuss the differences in the convergence of E_{L2}.

2

Kinematics

Kinematics is the study of the motion of a fluid without regard to the forces present. It consists of describing how pieces of the fluid translate, rotate, and deform. In our discussion, a piece of fluid will be called a *fluid element*, and it will consist of a volume of fluid, δV, that is much smaller compared to a reference volume, L^3, but much larger than the volume associated with the mean free path, λ^3:

$$\lambda^3 \ll \delta V \ll L^3. \tag{2.1}$$

In two dimensions, we will often depict a fluid element by a square, as shown in Figure 2.1. In this chapter, we will describe where a fluid element moves, what properties it experiences, and how it rotates and changes shape. The relationship between this motion and the forces present is left to the study of dynamics in the next chapter.

2.1 The Material Derivative

A key concept in fluid kinematics is that of a time derivative. In a fluid, properties may vary spatially and in time, and partial derivatives can be used to describe how a property changes in one direction, or in time, while holding the other coordinates fixed. However, such a purely mathematical derivative is not always the most useful, in particular when describing the rate of change of a property of a moving fluid. In this section, we present a derivative calculation that takes into account the motion of a fluid.

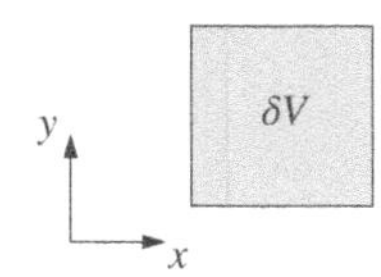

Figure 2.1 A square fluid element in two dimensions.

2.1.1 A Motivating Example

Consider an airplane* gliding to a landing at a constant speed V and descent angle θ, as shown in Figure 2.2.

Assuming an isothermal atmosphere of ideal-gas† air at temperature T, the hydrostatic balance‡ states that the pressure gradient in the vertical (y) direction is

* Not a fluid element, but a useful object for explaining the material derivative.

† Recall that this means $p = \rho RT$.

‡ See Example 3.9.

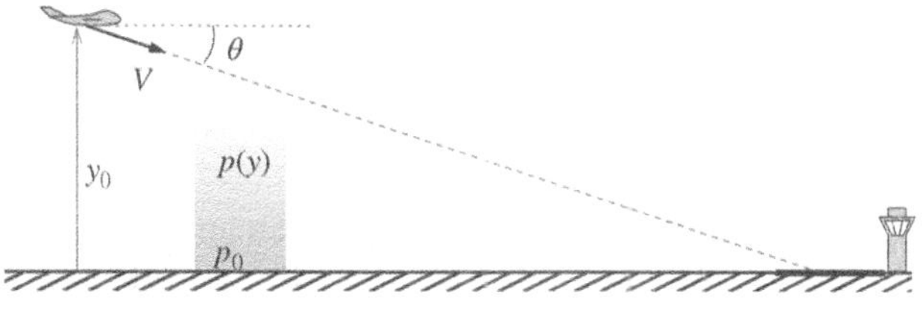

Figure 2.2 Airplane in a constant speed descent through an isothermal atmosphere.

Aerodynamics: A Computational Introduction, First Edition. Krzysztof J. Fidkowski.

© 2026 John Wiley & Sons, Inc. Published 2026 by John Wiley & Sons, Inc.

$$\frac{dp}{dy} = -\rho g = -\frac{pg}{RT}, \tag{2.2}$$

where g is the acceleration due to gravity. Integrating this equation yields an equation for how the pressure changes with altitude,

$$p(y) = p_0 e^{-gy/(RT)}, \tag{2.3}$$

where p_0 is the pressure at the ground ($y = 0$). We now ask, how will the atmospheric pressure measured by a barometer on the airplane vary as the plane descends? Given the last equation, we expect it to rise, more steeply closer to the ground due to the constant descent speed and exponential variation of p with respect to y. The rate of change of pressure experienced by the airplane is related to the material derivative that we will soon define.

Using some realistic numbers, we can simulate this situation too. Listing 2.1 presents a code that collects pressure data during the plane's descent and then makes a plot of the data versus time. Figure 2.3 shows the resulting pressure measured by the plane. We see qualitative agreement with expectations. The code also computes the rate of change of the pressure, using a simple forward difference, and Figure 2.4 shows this quantity. The trend should be expected, since both the pressure and its gradient increase in magnitude toward the ground.

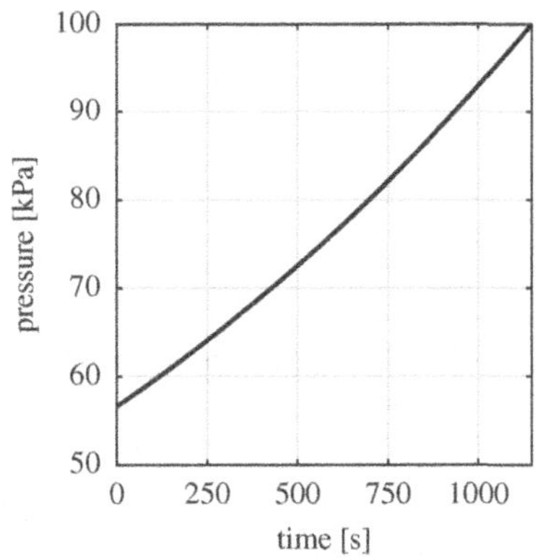

Figure 2.3 Pressure measured by a descending airplane.

Code notes:

- The `linspace` function creates a vector of points equally spaced in a specified interval.
- The altitude, time, and pressure variables are vectors, which is useful for making plots.

Listing 2.1 Code for simulating the pressure measured by a descending airplane.

```
function glider
% calculates pressure experienced by a descending airplane

y0 = 5000;    % initial height [m]
V  =   50;    % descent speed [m/s]
th =    5;    % descent angle [deg]
g  =  9.8;    % gravity acceleration [m/s^2]
R  =  287;    % air gas const [J/kg.K]
T  =  300;    % temperature [K]
p0 =  1e5;    % pressure on the ground [Pa]
N =   100;    % number of intervals for plotting

v = V*sind(th); % vertical descent speed
y = linspace(y0,0,N+1); % altitudes
t = linspace(0,y0/v,N+1); % times
p = p0*exp(-g*y/(R*T));    % pressures

% pressure versus time plot
figure(1); clf;
plot(t,p/1e3, 'k-', 'linewidth', 2); grid on;
set(gcf, 'position', [400, 400, 400, 400]);
set(gca,'xtick', linspace(0,1000,5));
writeplot('time [s]', 'pressure [kPa]', 'glidersim.pdf', 4);

% pressure rate of change versus time plot
p_t = diff(p)./diff(t);   % dp/dt approximation
tm = 0.5*(t(1:end-1) + t(2:end)); % time at midpoints
plot(tm, p_t, 'k-' ,'linewidth', 2, 'displayname', 'simulation');
hold on; grid on;
plot(t, p*g*v/(R*T), 'k--', 'displayname', 'exact');
h = legend('location', 'southeast'); set(h, 'fontsize', 14);
set(gca, 'xtick', linspace(0,1000,5));
writeplot('time [s]', '$dp/dt$ [Pa/s]', 'glidersimd.pdf', 4);
```

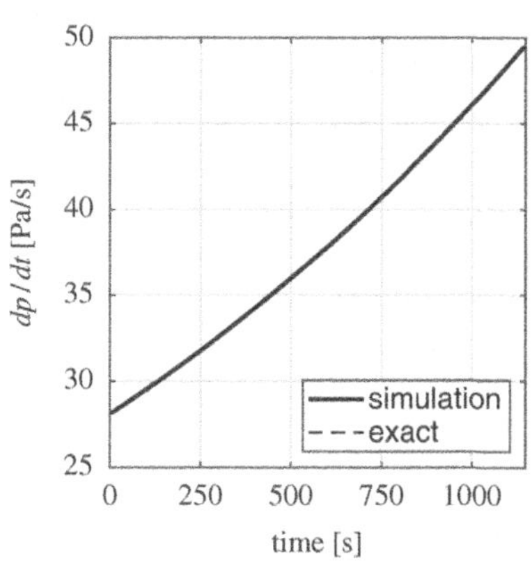

Figure 2.4 Pressure rate of change measured by a descending airplane.

Finally, note that this problem also lends itself to an analytical solution. If y_0 is the initial plane altitude, the altitude as a function of time is $y(t) = y_0 - vt$, where v is the magnitude of the vertical component of the descent velocity. Substituting for

y into the pressure equation, the pressure experienced by the plane as a function of time is

$$p(t)\Big|_{\text{airplane}} = p(y_0 - vt) = p_0 e^{-g(y_0 - vt)/(RT)}. \tag{2.4}$$

The time derivative of this quantity is

$$\frac{dp}{dt}\Big|_{\text{airplane}} = p_0 e^{-g(y_0 - vt)/(RT)} gv/(RT) = p(y)gv/(RT). \tag{2.5}$$

This time derivative is overlaid on the finite-difference result in Figure 2.4, and the agreement is very good.

2.1.2 Frames of Reference

A frame of reference is the coordinate system ascribed to some observer of the flow. When describing a flow field, a frame of reference is typically assumed, but it is not unique. For example, two observers could be viewing the same flow while moving relative to each other. In such a case, the fluid velocity measured by each observer will be different, even though the flow is the same. The frame of reference thus affects measurements and the velocity of a flow.

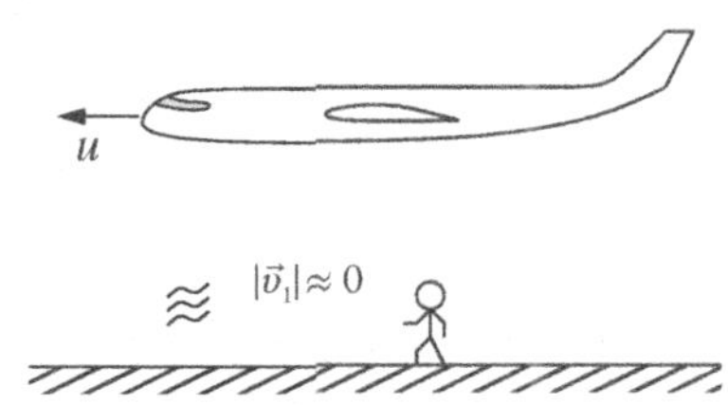

Figure 2.5 An observer on the ground sees a still atmosphere.

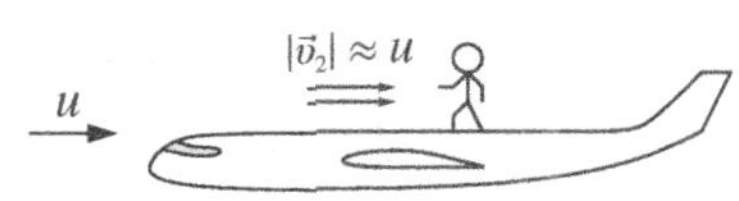

Figure 2.6 An observer on the aircraft feels an oncoming wind.

The fact that the flow is the same regardless of how it is observed leads to the principle of *Galilean invariance*. This property states that the mathematical equations used to describe the fluid must remain the same when written in a frame that is moving at a constant velocity $\vec{v}_0$ relative to some base frame (Pope 2000). A caveat here is that any velocity term appearing in the equations *does* change under this transformation. This is clear when considering an aircraft flying from the point of view of an observer on the ground versus one on the aircraft. The observer on the ground, shown in Figure 2.5, sees a mostly still atmosphere disturbed by the passage of the aircraft, whereas the observer on the aircraft, shown in Figure 2.6, sees an atmosphere moving quickly "at" the aircraft, i.e. in the direction opposite that of the aircraft's velocity. Galilean invariance states that both observers could describe the motion of the atmospheric air using the same form of the governing equations, but that their solutions for the air velocity would be offset by $\vec{v}_0$, the relative frame velocity.

We will distinguish between two types of frames of reference for describing the motion of a fluid element. These are the Lagrangian frame and the Eulerian frame, presented in an early work by Prandtl and Tietjens (1934).*

* Neither of these are uniquely defined: each fluid element has its own Lagrangian frame, while Eulerian-frame observers can be moving in arbitrary directions and velocities relative to the fluid.

- A *Lagrangian frame* is a frame that moves together with a fluid element. The observer is situated on a tagged fluid element as it translates, rotates, and deforms.
- An *Eulerian frame* is a frame that is stationary with respect to some observer of the fluid flow. The observer can take snapshots of the fluid flow at various times to yield field variables.

Consider water flowing in a river with variable depth, as illustrated in Figure 2.7. The water moves from left to right, and as it does so, it encounters a section of larger depth compared to the starting and ending values.

Suppose that a stationary observer by the river measures the depth distribution as a function of x. Assuming that the river bottom is fixed and that the water level is constant, this observer measures a constant $h(x)$ distribution. Thus, the depth field is steady in an Eulerian frame that is stationary with respect to the river bottom. However, now suppose that we look at the Lagrangian frame, which is one moving together with a fluid element in the water. The fluid element finds itself

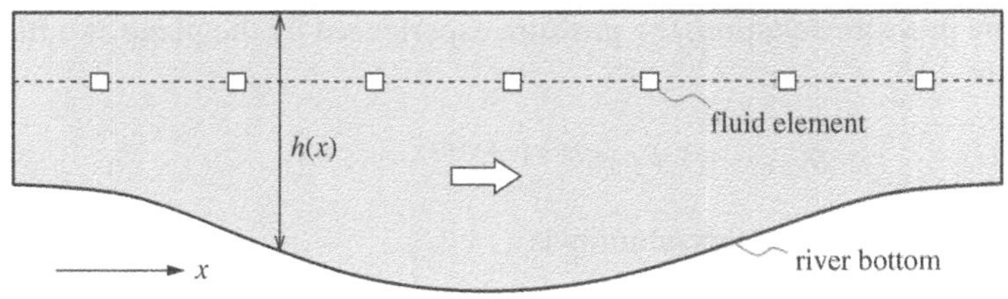

Figure 2.7 A situation in which the depth field, $h(x)$, is steady in an Eulerian frame, but unsteady in the Lagrangian frame.

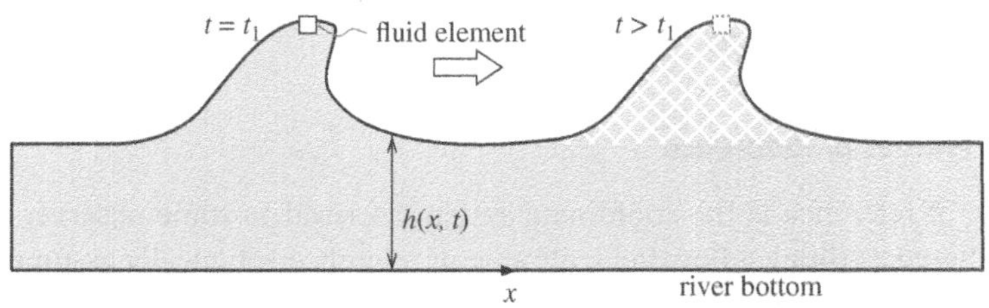

Figure 2.8 A situation in which the depth field, $h(x)$, is steady in the Lagrangian frame, but unsteady in an Eulerian frame.

in different depths at different times as it moves to the right, so the depth field appears unsteady in the Lagrangian frame of that fluid element.

Now consider a river with a flat bottom but with a traveling wave, as shown in Figure 2.8. In an Eulerian frame stationary relative to the river bottom, the flow no longer appears steady: an observer next to the river will measure, at one x location, different depths at different times as the wave passes by. However, in the Lagrangian frame, the depth now appears steady: a fluid element remains at the same depth as it rides the wave downriver.† The conclusion is that the frame of reference matters when characterizing a flow, in particular when determining if it is steady or not. The reference frame also affects the calculation of derivatives, as we will see in the next section.

† We assume that the fluid element remains at the crest of the wave, so that it is actually riding the wave.

2.1.3 Observed Rates of Change

Quantities of interest in aerodynamic flows often consist of local fluid properties, such as density, pressure, temperature, or velocity. For the following discussion, we use temperature, but the results extend to other local quantities.

Suppose that we would like to measure a time rate of change of temperature. If the temperature field changes in both space and time, the time derivative depends on the reference frame used. This applies to the partial derivative, $\frac{\partial T}{\partial t}$, since by definition, a partial derivative with respect to time is calculated at a fixed spatial position, and observers moving relative to each other will not agree on what fixed in space means.

Let us situate ourselves in one Eulerian frame, and suppose that in this frame we observe a temperature distribution of $T(\vec{x}, t)$, where $\vec{x}$ is the position measured in our frame, and t is time. We also observe a fluid velocity of

$$\vec{v}(\vec{x}, t) = u\hat{i} + v\hat{j} + w\hat{k}. \tag{2.6}$$

The Eulerian derivative is the rate of change measured at a fixed $\vec{x}$ in the Eulerian frame,

$$\text{Eulerian derivative of } T = \left.\frac{\partial T}{\partial t}\right|_{\vec{x}}. \tag{2.7}$$

The partial derivative indicates that we keep $\vec{x}$ fixed.*

There is one frame in which the time derivative is uniquely defined, and that is the Lagrangian frame. At any point in the flow, we can define a local Lagrangian frame as the one that moves with the fluid. The fluid velocity in this frame is zero, and a measure of the time derivative in this frame gives the rate of change experienced by a fluid element. This derivative is called the *material derivative*, also known as the *substantial* or *total* derivative.

* If another observer is moving relative to our frame, that observer's Eulerian derivative will in general differ from ours, since a fixed spatial coordinate in that frame is moving in our frame.

The material derivative of the temperature is the rate of change of temperature measured by a fluid element. Imagine that we drop a small and very light temperature probe into the fluid, and that this probe is carried along by the fluid. That is, the probe is at rest in the Lagrangian frame. The rate of change of temperature measured by this probe is the material derivative of the temperature.

We can also calculate the material derivative in any Eulerian frame. To do this, we just need to know the velocity of the fluid, in addition to the quantity that we are differentiating. The material derivative calculation in an Eulerian frame consists of two contributions:

material derivative = Eulerian derivative + convective derivative

The Eulerian derivative is the partial derivative defined in Equation 2.7. The convective derivative is the change in T that the fluid element sees due to its motion. If T were uniform in space, the convective derivative would be zero. However, when T is not uniform, then a moving fluid element measures a temperature change even if the temperature is steady in the Eulerian frame, as shown in Figure 2.9. For example, the fluid element can be moving from a colder to a warmer region in space. Mathematically, the convective derivative of temperature is written as

$$\vec{v} \cdot \nabla T = u\frac{\partial T}{\partial x} + v\frac{\partial T}{\partial y} + w\frac{\partial T}{\partial z}. \tag{2.8}$$

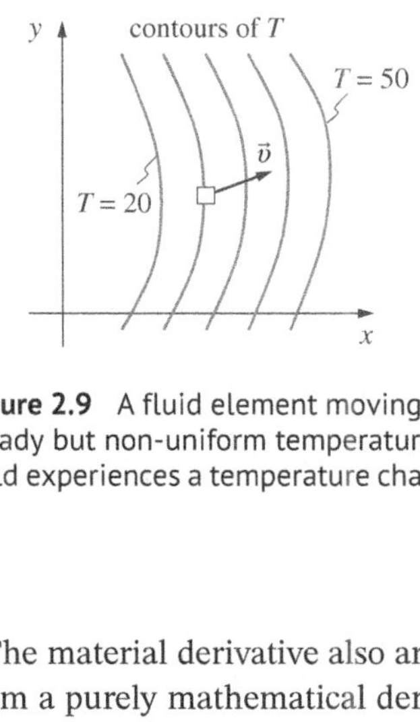

Figure 2.9 A fluid element moving in a steady but non-uniform temperature field experiences a temperature change.

This is the spatial derivative of T in the direction of motion, multiplied by the velocity magnitude. The directional derivative tells us how T is changing in space in the direction of motion, and the velocity magnitude tells us how fast the fluid is moving in that direction. Hence, their product is the rate of change of T due to the fluid's motion.

Combining the Eulerian and convective derivatives, we have the definition of the material derivative (of temperature)†

$$\frac{DT}{Dt} \equiv \frac{\partial T}{\partial t} + \vec{v} \cdot \nabla T. \tag{2.9}$$

The physical meaning of each term is

$\frac{DT}{Dt}$ = rate of change measured by a moving fluid element.

$\frac{\partial T}{\partial t}$ = rate of change seen by an observer at a fixed $\vec{x}$.

$\vec{v} \cdot \nabla T$ = change in T that the fluid element sees due to its motion.

† The material derivative also arises from a purely mathematical derivation. If the position of a fluid element is $\vec{x}(t)$, then the fluid element measures a temperature of $T(\vec{x}(t), t)$, and we can obtain the time rate of change of this quantity by using the chain rule, shown here in two spatial dimensions,

$$\begin{aligned}\frac{DT}{Dt} &= \frac{dT(\vec{x}(t), t)}{dt} \\ &= \frac{\partial T}{\partial t} + \frac{\partial T}{\partial x}\frac{dx}{dt} + \frac{\partial T}{\partial y}\frac{dy}{dt} \\ &= \frac{\partial T}{\partial t} + \vec{v} \cdot \nabla T.\end{aligned}$$

The material derivative applies not only to scalar fields but also to vector fields. Consider a vector field $\vec{F} = f\hat{\imath} + g\hat{\jmath} + h\hat{k}$, whose components f, g, and h are scalar fields. The material derivative of the vector is simply taken component-wise,

$$\frac{D\vec{F}}{Dt} = \frac{Df}{Dt}\hat{\imath} + \frac{Dg}{Dt}\hat{\jmath} + \frac{Dh}{Dt}\hat{k}. \tag{2.10}$$

A shorthand vector notation for this component-wise differentiation involves the gradient of $\vec{F}$,

$$\frac{D\vec{F}}{Dt} = \frac{\partial \vec{F}}{\partial t} + \vec{v} \cdot \nabla\vec{F}. \tag{2.11}$$

The last term, the convective derivative of a vector, can be a bit confusing in the form written. This is because $\nabla\vec{F}$ is a tensor quantity, and we would like to avoid using tensors when possible. To clarify the expression, we can group the terms in $\vec{v}\cdot\nabla\vec{F}$ in a more convenient form,

$$\vec{v}\cdot\nabla\vec{F} = (\vec{v}\cdot\nabla)\vec{F}, \tag{2.12}$$

where

$$\vec{v}\cdot\nabla = u\frac{\partial}{\partial x} + v\frac{\partial}{\partial y} + w\frac{\partial}{\partial z} \tag{2.13}$$

is a *scalar operator*. Therefore, the convective derivative is‡

$$\vec{v}\cdot\nabla\vec{F} = u\frac{\partial\vec{F}}{\partial x} + v\frac{\partial\vec{F}}{\partial y} + w\frac{\partial\vec{F}}{\partial z}. \tag{2.14}$$

‡ Alternatively, we can use index notation to write,

$$\frac{DF_i}{Dt} = \frac{\partial F_i}{\partial t} + v_j\partial_j F_i,$$

with summation implied on the repeated index j. In this form, there is no ambiguity about how the convective derivative is calculated.

Example 2.1 Material derivative in one spatial dimension
Consider a temperature field in one dimension given in one Eulerian frame by, in Kelvin (K),

$$T(x,t) = 300 + 0.5x + 0.1t, \tag{2.15}$$

and suppose that the fluid is moving with a velocity of $u = 2\text{m/s}$. The rate of change of temperature measured by a fluid element is

$$\frac{DT}{Dt} = \frac{\partial T}{\partial t} + u\frac{\partial T}{\partial x} = 0.1\frac{\text{K}}{\text{s}} + \left(2\frac{\text{m}}{\text{s}}\right)\left(0.5\frac{\text{K}}{\text{m}}\right) = 1.1\frac{\text{K}}{\text{s}}. \tag{2.16}$$

Note that this rate of change is higher than the Eulerian derivative of $\frac{\partial T}{\partial t} = 0.1\text{K/s}$ in this frame because the fluid is moving to higher x values, and the given temperature field increases with x.

Example 2.2 Material derivative in cylindrical coordinates
Given an unsteady temperature field and a steady velocity field in cylindrical coordinates,

$$T(t,r,\theta) = 10t + \frac{1}{r}\sin\theta, \tag{2.17}$$

$$\vec{v}(r,\theta) = re^{-r}\hat{\theta}, \tag{2.18}$$

we can calculate the material derivative of T as follows:§

$$\begin{aligned}\frac{DT}{Dt} &= \frac{\partial T}{\partial t} + \vec{v}\cdot\nabla T\\ &= 10 + \vec{v}\cdot\left(\frac{\partial T}{\partial r}\hat{r} + \frac{1}{r}\frac{\partial T}{\partial\theta}\hat{\theta}\right)\\ &= 10 + re^{-r}\hat{\theta}\cdot\left(\frac{-1}{r^2}\sin\theta\hat{r} + \frac{1}{r^2}\cos\theta\hat{\theta}\right)\\ &= 10 + \frac{e^{-r}}{r}\cos\theta.\end{aligned} \tag{2.19}$$

§ The second step uses the definition of the gradient in cylindrical coordinates, Equation 1.31, and the last step uses the fact that $\hat{\theta}\cdot\hat{r} = 0$ and $\hat{\theta}\cdot\hat{\theta} = 1$.

The material derivative of the velocity vector, $\frac{D\vec{v}}{Dt}$, is slightly more involved. It becomes easier to evaluate the scalar operator $\vec{v} \cdot \nabla$ instead of tackling the tensor $\nabla\vec{v}$,¶

$$\begin{aligned}
\frac{D\vec{v}}{Dt} &= \frac{\partial \vec{v}}{\partial t} + \vec{v} \cdot \nabla\vec{v} \\
&= \vec{v} \cdot \nabla\vec{v} \\
&= \underbrace{re^{-r}\hat{\theta}}_{\vec{v}} \cdot \underbrace{\left(\hat{r}\frac{\partial}{\partial r} + \frac{1}{r}\hat{\theta}\frac{\partial}{\partial \theta}\right)}_{\nabla} \underbrace{(re^{-r}\hat{\theta})}_{\vec{v}} \\
&= re^{-r}\frac{1}{r}\frac{\partial}{\partial \theta}\left(re^{-r}\hat{\theta}\right) \\
&= re^{-2r}\frac{\partial}{\partial \theta}\left(\hat{\theta}\right) \\
&= re^{-2r}(-\hat{r}) \\
&= -re^{-2r}\hat{r}.
\end{aligned} \tag{2.20}$$

¶ The second-to-last step uses $\frac{\partial\hat{\theta}}{\partial\theta} = -\hat{r}$, which was derived in Example 1.5 from the $\hat{\theta}$ and $\hat{r}$ expressions, repeated here:

$$\begin{aligned}
\frac{\partial\hat{\theta}}{\partial\theta} &= \frac{\partial}{\partial\theta}\left(-\sin\theta\hat{\imath} + \cos\theta\hat{\jmath}\right) \\
&= -\cos\theta\hat{\imath} - \sin\theta\hat{\jmath} \\
&= -\hat{r}.
\end{aligned}$$

2.2 Flow Lines

Visualization of the flow of a fluid is useful for gaining insight into where the flow is going and how it interacts with surrounding objects, such as wings and airfoils. One way to visualize a flow field is through *flow lines*, which show the motion of one or more fluid elements. This section discusses four basic flow lines:

1. Pathline: intrinsic to the Lagrangian view
2. Streamline: intrinsic to the Eulerian view
3. Streakline: easy to produce experimentally
4. Timeline: easy to produce experimentally

An early reference that analyzes flow lines is the work of Prandtl and Tietjens (1934).

2.2.1 Pathline

A *pathline* is a trace of the trajectory followed by an individual fluid element. It is illustrated in Figure 2.10 for two fluid element markers released at different times in an unsteady flow.

The shape of the pathline depends on the location from which the element marker is released and, if the flow is unsteady, the time at which the element marker is released. Mathematically, we can obtain an equation for tracing a pathline, $\vec{x}(t)$, if given the velocity field, $\vec{v}(x, t)$, and the location, $\vec{x}_0$, and time, t_0, of release. The equation is just the definition of the velocity as the time derivative of position,

$$\frac{d\vec{x}}{dt} = \vec{v}(\vec{x}, t) \qquad \text{subject to } \vec{x}(t = t_0) = \vec{x}_0. \tag{2.21}$$

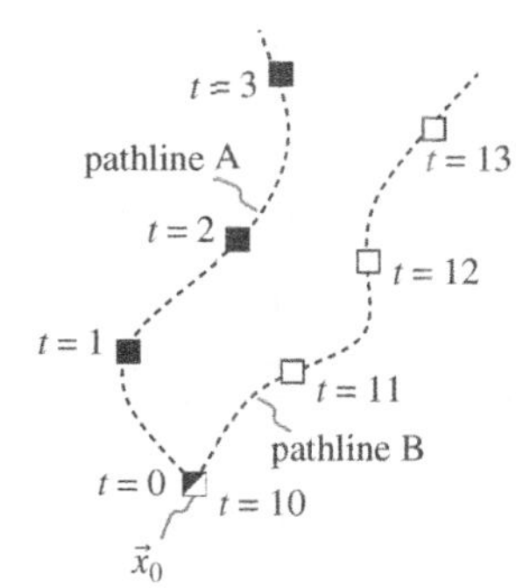

Figure 2.10 Two pathlines for fluid elements released at different times. Pathline A is the trajectory of a fluid element released at $t = 0$, while pathline B is the trajectory of an element released at $t = 10$. The flowfield is not steady in the Eulerian frame shown.

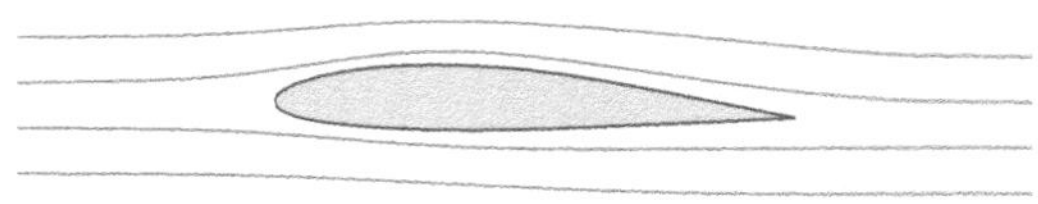

Figure 2.11 Streamlines for flow over an airfoil. At every point on the streamline, the velocity is tangent to the streamline.

2.2.2 Streamline

A *streamline* is a curve that is everywhere tangent to the velocity vector, drawn for one snapshot of the flow. Figure 2.11 shows an example of a set of streamlines. The flow is moving from left to right, although the streamlines by themselves do not indicate the direction of the flow. Arrows could be included on the streamlines to show the direction.

In an unsteady flow, the streamline pattern generally depends on the time at which the snapshot is taken. In a steady flow, streamlines coincide with pathlines.

Mathematically, streamlines are obtained by solving a differential equation similar to that of a pathline,

$$\frac{d\vec{x}}{ds} = \vec{v}(\vec{x}, t_0) \qquad \text{subject to } \vec{x}(s = s_0) = \vec{x}_0. \tag{2.22}$$

where t_0 is the time at which the snapshot is taken, $\vec{x}_0$ is the point through which the streamline is drawn, and s is a pseudo-time variable that parameterizes the positions along the streamline. Note that s does not appear on the right-hand side because the velocity field is evaluated at one fixed time, t_0. Integrating this equation with an arbitrary point $\vec{x}_0$ in space as the initial condition, and an arbitrary pseudo-time s_0 at release, gives the streamline that passes through that point.*

*A streamline can be drawn to extend in both directions from a chosen point, by using both $s > s_0$ and $s < s_0$.

2.2.3 Streakline

A *streakline* is a curve that is generated by a continuous release of markers from a single point. Figure 2.12 shows an example of a streakline for an unsteady flow. Fluid elements are marked at different times as they pass through $\vec{x}_0$. At some later time, a picture of the flow field is taken, and all of the marked fluid elements form a curve that is called the streakline.

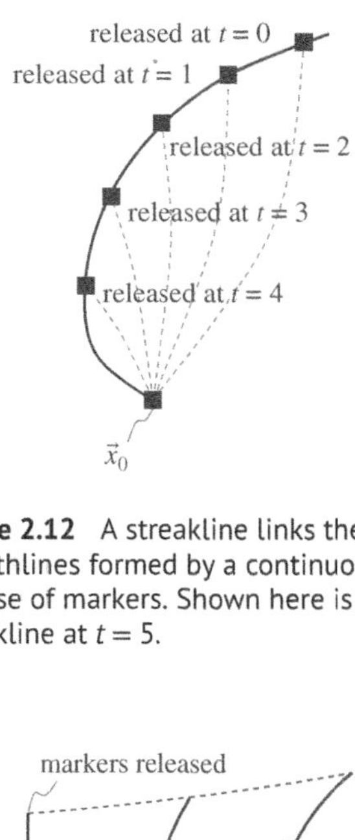

Figure 2.12 A streakline links the tips of pathlines formed by a continuous release of markers. Shown here is a streakline at $t = 5$.

Streaklines are easy to generate experimentally. For example, they are seen through visualization techniques that involve smoke or injected dyes, since the smoke or dye particles act like markers of fluid elements. A picture taken at some later time after the smoke or dye release started shows a streakline at that time. A picture of the smoke emanating from a smoldering candle is an example of a streakline.

Mathematically, the analysis of a streakline is a bit cumbersome because the points on a streakline are the endpoints of pathlines formed by markers released at different times. Examples 2.3 and 2.4 will show numerical and analytical techniques for drawing these lines. In a steady flow, streaklines, streamlines, and pathlines all coincide.

2.2.4 Timeline

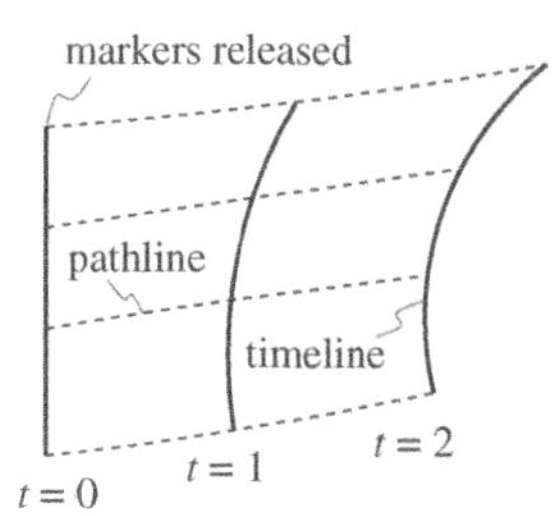

Figure 2.13 A timeline at some point in time is observed after releasing markers along a curve in space.

A *timeline*, also referred to as a material line, is a line or curve generated by markers released at the same time along a continuous line or curve in space. Figure 2.13 shows sample timelines at the time of release and at later times.

Timelines can be generated experimentally, for example by forming bubbles along a wire in a fluid, in short bursts, or releasing smoke or dye from many points along a line, instantaneously. They are useful for visualizing spatial differences in the velocity field, such as those that occur inside channels or close to boundaries of an object. Note, timelines generally do not coincide with pathlines, streamlines, or streaklines, all of which are generated using a single point of marker release, as opposed to a line.

2.2.5 Drawing Lines Numerically

Given a velocity vector field as a function of space and time, we can use the mathematical and physical definitions given in the previous section to draw flow lines numerically. Obtaining points on a pathline requires numerically integrating Equation 2.21, starting with the given point as the initial condition. A simple approach is to use the forward Euler method, illustrated in Figure 2.14, where $\vec{v}_n$ is the velocity field evaluated at position $\vec{x}_n$ and time t_n. In discrete form, Equation 2.21 becomes

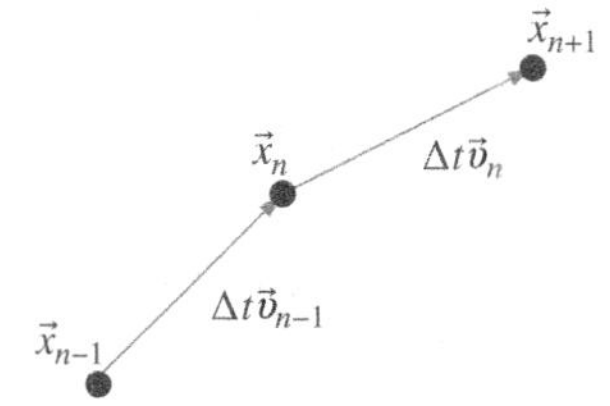

Figure 2.14 Forward Euler time integration to approximate points on a pathline.

$$\frac{\vec{x}_{n+1} - \vec{x}_n}{\Delta t} = \vec{v}(\vec{x}_n, t_n) \tag{2.23}$$

$$\Rightarrow \quad \vec{x}_{n+1} = \vec{x}_n + \Delta t\vec{v}\,(\vec{x}_n, t_n). \tag{2.24}$$

This is an update equation for the position given the velocity. Listing 2.2 shows a code that does this for a particular velocity field, given an initial position, initial time, and time horizon. It outputs the entire time history of positions obtained by the numerical integration.

Listing 2.2 Numerical simulation of a pathline using the forward Euler method.

```
function X = pathline(x0, t0, T)
% Draws a pathline from x0, t0, over a time horizon T.
% The output X contains positions at all time steps.

% forward Euler time stepping
Nt = 100;    % number of time steps
dt = T/Nt;   % time step
X = zeros(2,Nt+1); X(:,1) = x0; % initial condition
for n = 1:Nt,
  X(:,n+1) = X(:,n) + dt*getv(X(:,n),t0+n*dt);
  % X(:,n+1) = X(:,n) + dt*getv(X(:,n),t0); % for a streamline
end

%----------------------
function v = getv(x,t)
v = x; % for sizing
v(1) = 2*x(1) + t;
v(2) = x(2) - t;
```

Code notes:

- The number of time steps is hardcoded but could be made an input.
- The output matrix `X`, of size 2 by $N+1$, contains the pathline positions at all $N+1$ time nodes, the first being the initial condition.
- The velocity field is hardcoded in a function, `getv`.
- For the pathline, the time at which the velocity is calculated is the time of the nth time node, $t = t_0 + n\Delta t$.

We can use nearly the same code in order to obtain positions on a streamline. The only difference is that a streamline uses a fixed time, t_0, to call the velocity function. This constitutes a one-line difference in the code, as shown by the commented line in Listing 2.2.

From Figure 2.12, a streakline is a collection of pathline endpoints. To draw one, we call the pathline code multiple times, simulating the release of multiple markers. The markers start from the same point but are released at different times and propagate for different times, so that they all end up at the same physical time. Listing 2.3 implements this process. Note that it does not contain any time integration but instead relies on repeated calls to the pathline function. As such, it is more expensive, especially if the number of points in the streakline or the pathline discretization is large.

Code notes:

- $t_0 + T$ is the time at which the streakline is drawn.
- The index n loops over the marker release times.
- $t_0 + n\Delta t$ is the release time of the nth marker, and it propagates for the remaining time of $T - n\Delta t$.

Listing 2.3 Calculation of a streakline as a collection of pathline endpoints.

```
function X = streakline(x0, t0, T)
% Draws a streakline from x0, t0, over a time horizon T.
% The output X contains positions at all time steps.

Nt = 100;    % number of points
dt = T/Nt;   % time step
X = zeros(2,Nt+1);
for n = 0:Nt,
  P = pathline(x0, t0+n*dt, T-n*dt);
  X(:,n+1) = P(:,end);
end
```

A timeline can be obtained in a manner similar to a streakline, by calling the pathline code with the same initial and final times but different starting positions.

Example 2.3 Numerical flow lines

This example demonstrates how to draw flow lines numerically, using the codes developed in this section, for an unsteady two-dimensional velocity field,

$$\vec{v}(x, y, t) = (2x + t)\hat{\imath} + (y - t)\hat{\jmath}. \tag{2.25}$$

Our goal is to draw a pathline, a streamline, and a streakline corresponding to the release point $\vec{x}_0 = [1,2]$, an initial release time of $t_0 = 0$, and a time horizon of $T = 2$. Listing 2.4 calls the three flow line functions with these parameters and draws the flow lines.

Listing 2.4 Draws multiple flow lines on one figure.

```
function drawflowlines
% makes a plot of numerically-generated flow lines

x0 = [1;2]; % initial position
t0 = 0;     % initial time
T  = 2;     % time horizon

figure(1); clf;

X = pathline(x0, t0, T);
plot(X(1,:), X(2,:), 'k-', 'linewidth', 2);
hold on; grid on;

X = streamline(x0, t0, T);
plot(X(1,:), X(2,:), 'k--', 'linewidth', 2);

X = streakline(x0, t0, T);
plot(X(1,:), X(2,:), 'k-.', 'linewidth', 2);

axis([0,70,0,16]);
h = legend('pathline', 'streamline', 'streakline', ...
           'location', 'southeast', 'interpreter', 'latex');
set(h, 'fontsize', 14);
set(gcf, 'position', [400,400, 400, 400]);
writeplot('$x$', '$y$', '../figs/flowlinesnum.pdf', 4);
```

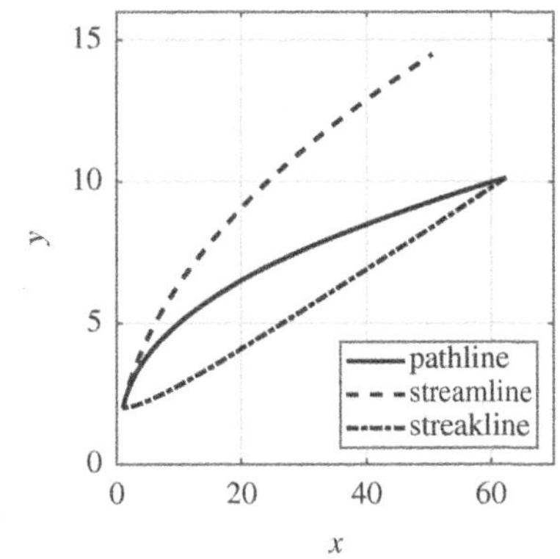

Figure 2.15 Numerically generated flow lines for an unsteady velocity field.

Figure 2.15 shows the resulting lines. Since the velocity field is unsteady, the lines are all different, although they share the release point. In addition, the streakline and pathline have the same two endpoints: the lower-left is the release location, and the upper-right is the first marker released on the streakline, which follows the trajectory shown by the pathline.

Example 2.4 Analytical flow-line calculations
For the two-dimensional unsteady velocity field given in Example 2.3, we can also find analytical equations for pathlines, streamlines, and streaklines.

1. We first look at the pathline of a marker released at $t_0 = 0$ into the fluid from $\vec{x}_0 = [1,2]$. The pathline is governed by Equation 2.21. Writing the two equations for the velocity components in Equation 2.25 gives

$$\frac{dx}{dt} = 2x + t, \quad \frac{dy}{dt} = y - t. \tag{2.26}$$

Since these equations are decoupled, we can treat them separately. The solution to the first one has the form* $x(t) = ce^{2t} + bt + a$, and substituting this form into the differential equation gives

$$x(t) = ce^{2t} - \frac{t}{2} - \frac{1}{4}. \tag{2.27}$$

Since $x(t = 0) = 1$, $c = 5/4$ and so

$$x(t) = \frac{5}{4}e^{2t} - \frac{t}{2} - \frac{1}{4}. \tag{2.28}$$

Following a similar procedure, the solution for y is

$$y(t) = e^t + t + 1. \tag{2.29}$$

Figure 2.16 shows a plot of the resulting pathline, which is obtained from the above equations for $0 < t < 2$. The plot agrees with the numerical result in Figure 2.15, which shows the result for the same time horizon, $T = 2$.

2. Next, we consider the streamline at $t = 0$ passing through $\vec{x}_0 = [1, 2]$. The differential equation for a streamline at a time $t = t_0$ is given by Equation 2.22. The snapshot time $t_0 = 0$ is given, and the pseudo-time parameter $s = 0$ is chosen to correspond to $\vec{x}_0$. For the given velocity, the two components of the streamline equation are

$$\frac{dx}{ds} = 2x + t_0, \quad \frac{dy}{ds} = y - t_0. \tag{2.30}$$

These are again decoupled and even simpler than for the pathline, since the independent variable s does not appear on the right-hand side. The solution to the first equation takes the form† $x(s) = ce^{2s} + b$, and substituting into the differential equation gives $b = -t_0/2$. Applying the initial condition $x(s = 0) = 1$ gives the solution

$$x(s) = e^{2s} - \frac{t_0}{2} = e^{2s}. \tag{2.31}$$

Following a similar procedure, the solution to the y equation is

$$y(s) = 2e^s + t_0 = 2e^s. \tag{2.32}$$

A plot of the resulting streamline, which is obtained from the above equations for $0 \le s \le 2$, is included in Figure 2.16, and it also agrees with the numerical result.

3. Finally, we look at the streakline at $t = 2$ formed by markers released continuously between $t = 0$ and $t = 2$ from $\vec{x}_0 = [1, 2]$. The position

*The solution for $x(t)$ must involve e^{2t} since differentiating with respect to t gives $2x = 2e^{2t}$. It must also have a linear function of t to offset the t on the right-hand side.

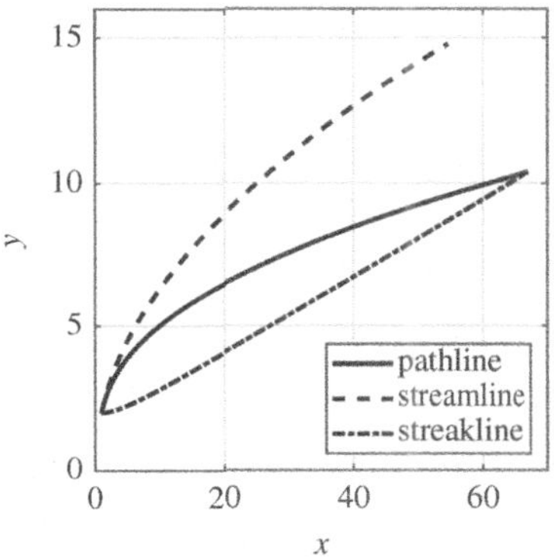

Figure 2.16 Flow lines obtained analytically in Example 2.4.

†The solution to the equation for x must involve e^{2s}, since differentiating with respect to s returns $2x$. There must also be a constant to cancel t_0.

at $t = t_0$ of a marker released at $t = t_0 - s$ is given by the pathline equation. We recall the general pathline solution from Equation 2.27 and Equation 2.29‡

‡ The y equation for the pathline was given as the particular solution, but before applying the initial condition, an arbitrary constant, c_2, is present in front of the exponential.

$$x(t) = c_1 e^{2t} - \frac{t}{2} - \frac{1}{4}, \quad y(t) = c_2 e^t + t + 1, \tag{2.33}$$

where c_1 and c_2 are determined by the release time of the marker. We are interested in a marker released at time $t = t_0 - s$ from $\vec{x}_0 = [1,\ 2]$. Using the general pathline solution, at $t = t_0 - s$, the marker must be at the release location, so

$$1 = c_1 e^{2(t_0 - s)} - \frac{t_0 - s}{2} - \frac{1}{4}, \tag{2.34}$$

$$2 = c_2 e^{t_0 - s} + t_0 - s + 1. \tag{2.35}$$

Solving for c_1 and c_2 yields

$$c_1 = e^{-2(t_0 - s)} \left(\frac{t_0 - s}{2} + \frac{5}{4} \right), \tag{2.36}$$

$$c_2 = e^{-(t_0 - s)} \left(s - t_0 + 1 \right). \tag{2.37}$$

The marker position at t_0 is then obtained from Equation 2.33 by substituting $t = t_0$ and the above formulas for c_1 and c_2, with the result

$$x(s) = e^{2s} \left(\frac{t_0 - s}{2} + \frac{5}{4} \right) - \frac{t_0}{2} - \frac{1}{4}, \tag{2.38}$$

$$y(s) = e^s \left(s - t_0 + 1 \right) + t_0 + 1.$$

A plot of the streakline, which is obtained from the above equations for $0 \le s \le t_0 = 2$, is given in Figure 2.16. We see agreement with the numerically generated flow lines in Figure 2.15.

2.3 Vorticity and Circulation

This section introduces two related quantities that are useful for describing the motion of a fluid. The first is vorticity, which is a local quantity that measures whether a fluid element is rotating. Of interest to us for much of the rest of the text will be flow in which the elements are not rotating, as this assumption simplifies the flow analysis. The second quantity is circulation, which is defined on a contour drawn in the flow and measures the net amount of vorticity contained in that contour.

We begin with *vorticity*, $\vec{\xi}$, which is a measure of how rotational a flow is. It is defined as the curl of the velocity,

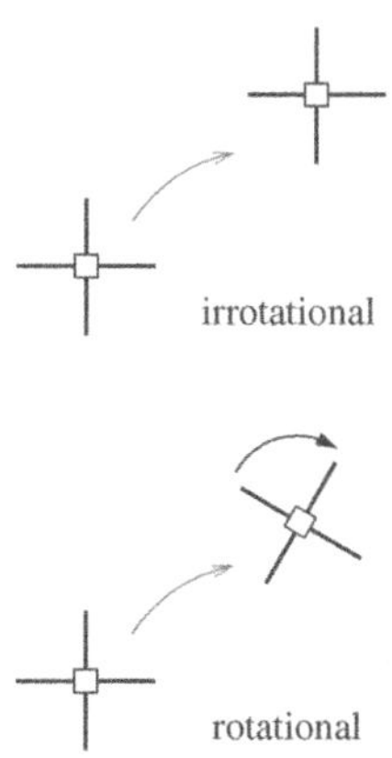

Figure 2.17 In a 2D irrotational flow, a paddlewheel consisting of two sticks rigidly joined together will not spin.

$$\vec{\xi} \equiv \nabla \times \vec{v}. \tag{2.39}$$

Note that $\vec{\xi}$ is a vector. When $\vec{\xi} = \vec{0}$, we say that the flow is *irrotational*, and when $\vec{\xi} \neq \vec{0}$, the flow is rotational. Intuitively, in two dimensions, a flow is rotational if a paddlewheel placed in the flow begins to spin, as illustrated in Figure 2.17.*

* A nuance with this thought experiment is that the paddlewheel testing rotation has to have its vertical and horizontal sections of equal size and joined rigidly. A light vertical or horizontal rod could rotate even in an irrotational flow, due to shear, to be discussed in Section 2.6.

An assumption that we will make often in aerodynamics is that the flow is irrotational, which greatly simplifies the solution process. Such an assumption may seem restrictive, but it is justified in many conditions. In particular, flow moving at constant velocity far in front of an aircraft is irrotational, as it has no spatial gradients, and we will see in the next chapter that zero vorticity is preserved as long as viscous effects are neglected. For most aerodynamic flows, viscosity affects only a small region of the flow close to walls, so the majority of the flow will generally be irrotational. Counterexamples exist, such as in large regions of separated flow for a stalled wing, but these are typically off-design conditions for an aerodynamic vehicle.

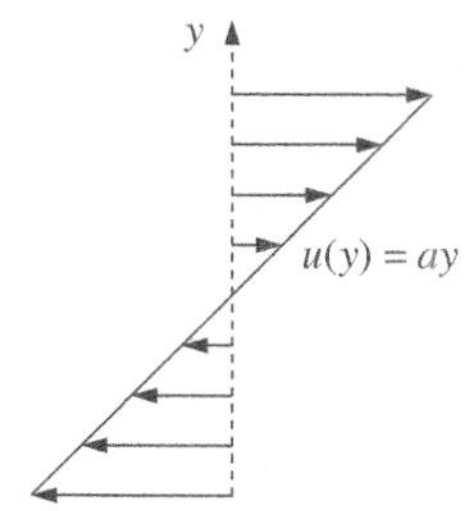

Figure 2.18 A unidirectional flow with vorticity.

Example 2.5 A flow with vorticity
Even a unidirectional flow, with only one component nonzero, can be rotational. Consider a velocity field for which $u = ay$, where a is a constant, and all other components zero,

$$\vec{v} = ay\hat{\imath} + 0\hat{\jmath} + 0\hat{k}. \tag{2.40}$$

Figure 2.18 illustrates the velocity vectors for this flow. The vorticity is

$$\vec{\xi} = \left(\frac{\partial w}{\partial y} - \frac{\partial v}{\partial z}\right)\hat{\imath} + \left(\frac{\partial u}{\partial z} - \frac{\partial w}{\partial x}\right)\hat{\jmath} - \left(\frac{\partial v}{\partial x} - \frac{\partial u}{\partial y}\right)\hat{k} = -a\hat{k}. \tag{2.41}$$

Since the vorticity is nonzero, the flow is rotational.

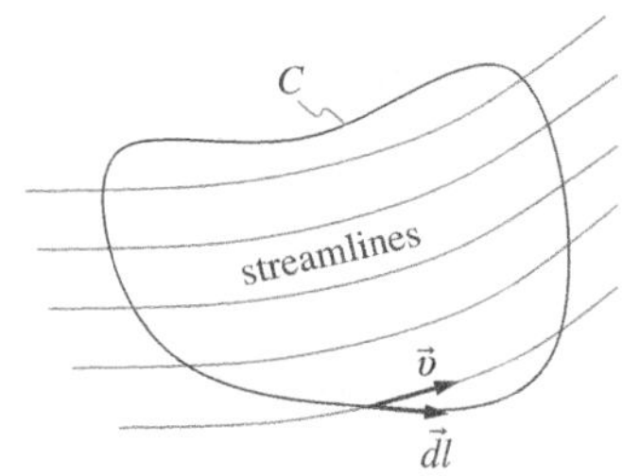

Figure 2.19 Contour used in the definition of circulation.

Related to vorticity is *circulation*, Γ, which is defined by a contour integral of velocity,

$$\Gamma \equiv -\int_C \vec{v} \cdot d\vec{l}, \tag{2.42}$$

where C is some closed contour and the integral is evaluated in a counterclockwise sense, as shown in Figure 2.19. Like vorticity, circulation also measures how much the fluid is rotating, but in an integral sense over the perimeter of a control volume, rather than in a local sense at one point. Given a velocity field and contour, we can calculate the circulation by integrating the tangential component of the velocity along the contour.

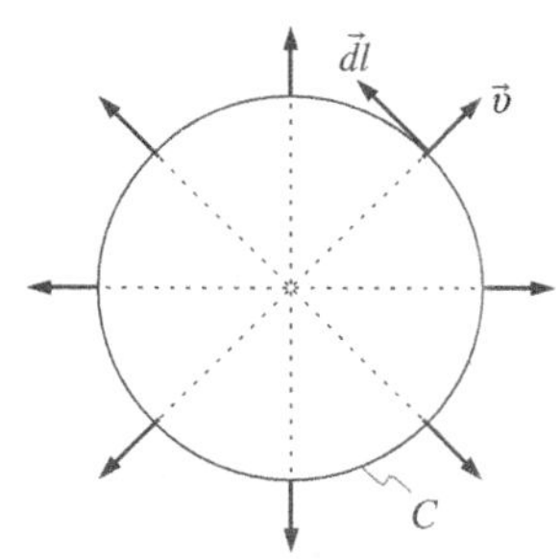

Figure 2.20 Circulation is zero for a circular contour in radial flow.

Example 2.6 Calculating circulation
Consider a circular contour centered at the origin, with a purely radial flow field. The circulation for this contour will be zero because the velocity has no component in the tangential direction anywhere on the contour, as illustrated in Figure 2.20.

$$\vec{v} \perp d\vec{l} \quad \Rightarrow \quad \vec{v} \cdot d\vec{l} = 0 \quad \Rightarrow \quad \Gamma = 0. \tag{2.43}$$

For the same circular contour in a purely tangential flow, the circulation will in general be nonzero, and it will depend on the velocity field and on the radius of the contour. Referring to Figure 2.21, if v_θ does not vary in θ, then

$$\vec{v} \cdot d\vec{l} = v_\theta R d\theta \quad \Rightarrow \quad \Gamma = -\int_C \vec{v} \cdot d\vec{l} = -2\pi R v_\theta. \tag{2.44}$$

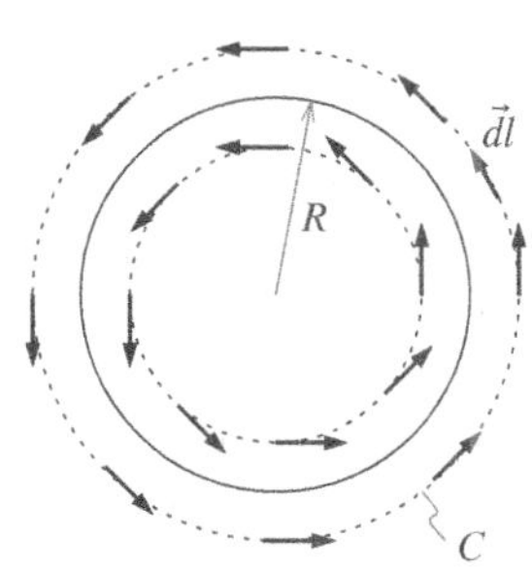

Figure 2.21 Circulation is nonzero for a circular contour in tangential flow.

Note that for positive v_θ, that is for flow moving counter-clockwise, Γ is negative. The convention of a minus sign in the definition of circulation ensures that, as we will soon see, positive lift corresponds to positive circulation for flow moving from left to right over an object such as an airfoil.

The local quantity of vorticity and the contour-specific quantity of circulation are related via Stokes's theorem. Applying Equation 1.43,

$$\int_C \vec{v} \cdot d\vec{l} = \int_S \underbrace{(\nabla \times \vec{v})}_{\vec{\xi}} \cdot d\vec{S}, \tag{2.45}$$

where for a two-dimensional contour in the x-y plane, $d\vec{S} = dS\,\hat{k}$, and dS is the differential element of surface area. Using the definition of circulation, we obtain

$$\Gamma = -\int_S \vec{\xi} \cdot d\vec{S}. \tag{2.46}$$

In words, the circulation for a closed contour C is equal to the negative of the integral of vorticity over the enclosed surface. Note that vorticity is a point quantity, while the value of circulation generally depends on the contour used for its calculation.

Example 2.7 Vorticity and Stokes's theorem
Given the following velocity field in cylindrical coordinates,

$$v_\theta(r) = \frac{Kr}{1 + \frac{r^2}{r_0^2}}, \qquad v_r = 0, \tag{2.47}$$

where K and r_0 are constants, we calculate the vorticity as a function of r. Using Equation 1.33 for the curl in cylindrical coordinates, the vorticity is†

$$\vec{\xi} = \nabla \times \vec{v} = \left[\frac{1}{r}\frac{\partial(rv_\theta)}{\partial r} - \frac{1}{r}\frac{\partial v_r}{\partial \theta}\right]\hat{z} = \frac{2K}{\left(1 + \frac{r^2}{r_0^2}\right)^2}\hat{z}. \tag{2.48}$$

This calculation shows that the vorticity for this flow is purely in the $+\hat{z}$ direction, out of the two-dimensional plane, and that it is independent of θ.

By explicit area integration, the integrated amount of vorticity,

$$\int_A \vec{\xi} \cdot \hat{z}\, dA,$$

contained in a circular area A of radius R around the origin, is‡

$$\int_A \vec{\xi} \cdot \hat{z}\, dA = \int_0^{2\pi}\int_0^R \frac{2K}{\left(1 + \frac{r^2}{r_0^2}\right)^2} r\,dr\,d\theta = \frac{2\pi K R^2}{1 + \frac{R^2}{r_0^2}}. \tag{2.49}$$

We can verify this answer by using Equation 2.46 and a circular contour of radius R. The circulation is given by

$$\begin{aligned} \Gamma = -\int_C \vec{v} \cdot d\vec{l} &= -\int_0^{2\pi} v_\theta(R) R\, d\theta \\ &= -2\pi R v_\theta(R) = -\frac{2\pi K R^2}{1 + \frac{R^2}{r_0^2}}. \end{aligned} \tag{2.50}$$

† Since only v_θ is nonzero, and since it only depends on r, the vorticity calculation simplifies to

$$\begin{aligned} \vec{\xi} &= \frac{1}{r}\frac{\partial}{\partial r}\left(\frac{Kr^2}{1 + \frac{r^2}{r_0^2}}\right)\hat{z} \\ &= \frac{1}{r}\frac{\partial}{\partial r}\left(\frac{K}{\frac{1}{r^2} + \frac{1}{r_0^2}}\right)\hat{z} \\ &= \frac{1}{r}\left(-\frac{K}{\left(\frac{1}{r^2} + \frac{1}{r_0^2}\right)^2}\frac{(-2)}{r^3}\right)\hat{z} \\ &= \frac{1}{r}\frac{2Kr}{\left(1 + \frac{r^2}{r_0^2}\right)^2}\hat{z} \\ &= \frac{2K}{\left(1 + \frac{r^2}{r_0^2}\right)^2}\hat{z}. \end{aligned}$$

‡ The derivation of this expression is as follows:

$$\begin{aligned} &\int_0^{2\pi}\int_0^R \frac{2K}{\left(1 + \frac{r^2}{r_0^2}\right)^2} r\,dr\,d\theta \\ &= 2\pi K r_0^2 \int_0^R \frac{1}{\left(1 + \frac{r^2}{r_0^2}\right)^2}\frac{2r}{r_0^2}\,dr \\ &= 2\pi K r_0^2 \left[-\frac{1}{1 + \frac{r^2}{r_0^2}}\right]_0^R \\ &= 2\pi K r_0^2 \left[-\frac{1}{1 + \frac{R^2}{r_0^2}} + 1\right] \\ &= 2\pi K r_0^2 \left[\frac{R^2/r_0^2}{1 + \frac{R^2}{r_0^2}}\right] \\ &= \frac{2\pi K R^2}{1 + \frac{R^2}{r_0^2}}. \end{aligned}$$

Stokes's theorem states that the circulation is equal to the negative of the enclosed vorticity, and this agrees with the explicit area calculation above.

2.4 Conservation of Mass

A physical law that is relevant to the study of kinematics is conservation of mass. This law constrains the velocity field because fluid elements have mass, and their motion transports mass through space. Under certain assumptions, such as steady flow, the fluid motion cannot be arbitrary if mass conservation is enforced.

The equation for mass conservation is based on the principle that mass is neither created nor destroyed. Consider a control volume V, with surface S, as shown in Figure 2.22. The law of mass conservation states that

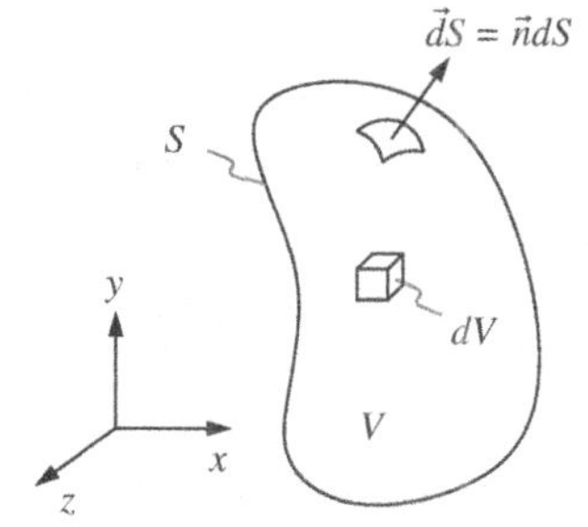

Figure 2.22 Control volume used for conservation of mass analysis.

Law 2.4.1 (Conservation of mass.) *The mass flow rate out of a control volume equals the time rate of decrease of mass in the control volume.*

The mass flowing out of the control volume is given by an integral over the surface S,

$$\text{mass flowing out of } V = \int_S \rho\vec{v}\cdot\vec{n}\,dS \quad \left[\frac{\text{mass}}{\text{time}}\right], \tag{2.51}$$

where the expression in the brackets denotes the dimensions of the integral. The integrand $\rho\vec{v}\cdot\vec{n}$ is the *mass flux* through the surface,

$$\rho\vec{v}\cdot\vec{n} = \text{mass flux} \quad \left[\frac{\text{mass}}{\text{time}\cdot\text{area}}\right]. \tag{2.52}$$

The rate of decrease of mass in the control volume is

$$\text{rate of mass decrease in } V = -\frac{d}{dt}\int_V \rho\,dV \quad \left[\frac{\text{mass}}{\text{time}}\right]. \tag{2.53}$$

Equating the two expressions, we obtain the *integral form of the mass conservation equation,**

$$\frac{d}{dt}\int_V \rho\,dV + \int_S \rho\vec{v}\cdot\vec{n}\,dS = 0. \tag{2.54}$$

* The mass conservation equation is also called the *continuity equation.*

We can apply the divergence theorem to convert the surface integral in Equation 2.54 to a volume integral,

$$\int_S \rho\vec{v}\cdot\vec{n}\,dS = \int_V \nabla\cdot(\rho\vec{v})\,dV. \tag{2.55}$$

In addition, the time derivative can be brought inside the volume integral in the first term of Equation 2.54,†

$$\frac{d}{dt}\int_V \rho\,dV = \int_V \frac{\partial\rho}{\partial t}\,dV. \tag{2.56}$$

† The enabling assumption is that the control volume is not changing in time; otherwise, a surface integral term would have to be included.

The resulting equation is

$$\int_V \left[\frac{\partial\rho}{\partial t} + \nabla\cdot(\rho\vec{v})\right] dV = 0. \tag{2.57}$$

Since this equation must hold for any control volume V, the integrand must be identically zero‡

$$\frac{\partial \rho}{\partial t} + \nabla \cdot (\rho \vec{v}) = 0. \tag{2.58}$$

‡ If this were not the case at some point in space, we could place the control volume V around that point and arrive at a contradiction.

This is the *differential form of the mass conservation equation.*§ The differential form can also be written using the material derivative by applying the product rule to the second term, giving

$$\frac{D\rho}{Dt} + \rho \nabla \cdot \vec{v} = 0. \tag{2.59}$$

§ The differential form is more useful from a solution point of view, although it is slightly more restrictive – for example, it is not valid if $\rho\vec{v}$ is not differentiable.

Example 2.8 Differential form of mass conservation
This example shows how the differential form of mass conservation can be used to solve for the density given velocity. Consider a three-dimensional velocity field given by

$$\vec{v} = x\hat{i} + y\hat{j} + z\hat{k}. \tag{2.60}$$

Suppose that at $t = 0$, the density is uniform and equal to 1 throughout the domain,

$$\rho(\vec{x}, t = 0) = 1. \tag{2.61}$$

We can determine the density at later times, $\rho(\vec{x}, t)$, by using the differential form of the mass conservation equation. Applying Equation 2.59, we first calculate the divergence of the velocity field,

$$\nabla \cdot \vec{v} = \frac{\partial u}{\partial x} + \frac{\partial v}{\partial y} + \frac{\partial w}{\partial z} = 3. \tag{2.62}$$

Substituting into Equation 2.59, we have

$$\frac{D\rho}{Dt} + 3\rho = 0. \tag{2.63}$$

We can integrate this equation in time¶ to obtain the density as observed by one fluid element,

$$\rho(t) = e^{-3t}, \tag{2.64}$$

¶ The time integration involving the material derivative can be carried out like a standard total derivative, d/dt, with the understanding that the result holds for one fluid element as it moves through the flow.

where we used the initial condition of $\rho = 1$ to determine the constant of integration. Since the density started out spatially uniform, and since the above expression has no spatial dependence, all fluid elements will have this density so that the density will remain spatially uniform. This means that the answer applies to the entire domain, and we can write

$$\rho(x, t) = \rho(t) = e^{-3t}. \tag{2.65}$$

Example 2.9 Displacement thickness of a boundary layer
When air, which has a small amount of viscosity, flows over a flat plate, a boundary layer forms.‖ Consider a situation in which air moving uniformly in the horizontal direction at speed U_∞ encounters a stationary horizontal flat plate. Suppose that after a certain distance x from the leading edge of the plate, the x-velocity is given by

‖ We will study boundary layers in Chapter 6, and for now, it is sufficient to think of a boundary layer as a thin region near the wall where the velocity changes rapidly.

$$u(x,y) = U_\infty \left(1 - e^{-y/\delta(x)}\right). \tag{2.66}$$

We assume that the y-velocity is negligible everywhere, and that the air is incompressible. The situation is illustrated in Figure 2.23.

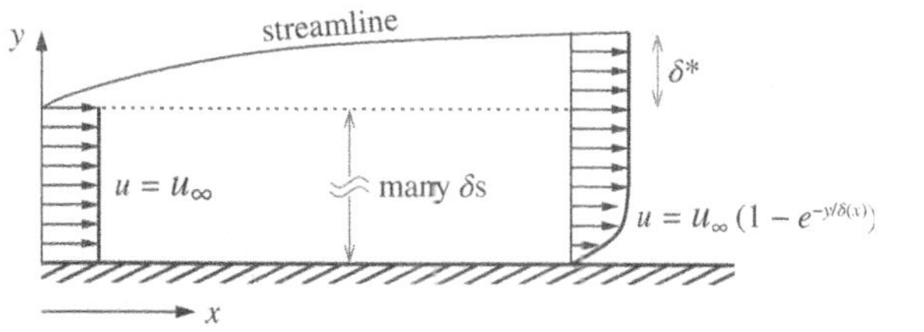

Figure 2.23 Use of conservation of mass to calculate the displacement thickness in a boundary layer.

In terms of $\delta(x)$, we can calculate $\delta^*(x)$, which is the displacement of a streamline far away from the flat plate, also known as the displacement thickness. By "far away," we mean a streamline at a y location for which $y \gg \delta(x)$, where $u \to U_\infty$.

To perform the calculation, we define a control volume bounded by the wall on the bottom, the streamline on the top, a vertical line at $x = 0$ on the left, and a vertical line at an arbitrary x on the right. No fluid crosses the streamline or the wall, so by conservation of mass, the mass of fluid flowing in through the left side of the control volume must equal the mass of fluid flowing out through the right side. Since the fluid is assumed incompressible, the same statement holds for the volume and area of the fluid.**

** This is a two-dimensional problem, so conservation of volume means conservation of area, since there is no variation in any property in the z direction.

We denote by y_1 the y-location of the dotted line in Figure 2.23. The rate at which fluid enters the control volume is

$$\text{area rate of fluid entering } V = y_1 U_\infty. \tag{2.67}$$

This calculation is straightforward because the flow is uniform at $x = 0$. For any other x, an integral is required to obtain the rate at which the fluid is leaving the control volume,

$$\begin{aligned}
\int_0^{y_1+\delta^*(x)} u(y)\,dy &= \int_0^{y_1+\delta^*(x)} U_\infty \left(1 - e^{-y/\delta(x)}\right) dy \\
&= U_\infty \left[y + \delta(x) e^{-y/\delta(x)}\right]_0^{y_1+\delta^*(x)} \\
&= U_\infty \left(y_1 + \delta^*(x) + \delta(x) e^{-\frac{y_1+\delta^*(x)}{\delta(x)}} - \delta(x)\right).
\end{aligned}$$

Since $y_1 \gg \delta(x)$ (the streamline is far away from the flat plate), the exponential in the above expression is negligible compared to the other terms. We therefore drop it to obtain

$$\text{area rate of fluid leaving } V = U_\infty \left(y_1 + \delta^*(x) - \delta(x)\right). \tag{2.68}$$

Equating this rate with the rate of area flowing in, Equation 2.67, we have, for the particular velocity profile defined in Equation 2.66, $\delta^*(x) = \delta(x)$. Different velocity profiles will yield different results, but as long as y appears normalized by $\delta(x)$, i.e. $y/\delta(x)$, $\delta^*(x)$ will be proportional to $\delta(x)$.

2.5 Velocity Potential and Stream Function

A velocity field is described most generally by using a vector $\vec{v}$ that consists of two or three scalar fields, which are the components in two or three dimensions. However, not all combinations of these scalar fields are allowed if the flow satisfies certain kinematic constraints, such as incompressibility or irrotationality. We show in this section that under these constraints, the characterization of the flow velocity can be reduced to one scalar field.* We consider two situations when such a reduction is possible:

1. Irrotational flow.
2. Incompressible flow in two dimensions.

Consider first irrotational flow. In this case, the velocity must satisfy, from Equation 2.39,

$$\nabla \times \vec{v} = 0. \tag{2.69}$$

This condition is automatically satisfied by a velocity vector field that is the gradient of some scalar potential,

$$\vec{v} = \nabla \phi, \tag{2.70}$$

because then†

$$\nabla \times \vec{v} = \nabla \times (\nabla \phi) = 0. \tag{2.71}$$

$\phi = \phi(x, y, z)$ is called the *velocity potential.*

Second, we consider incompressible flow, i.e. flow in which the density is constant. The requirement of conservation of mass reduces to

$$\frac{\partial \rho}{\partial t} + \nabla \cdot (\rho \vec{v}) = 0, \quad \rho = \text{const.} \quad \Rightarrow \quad \nabla \cdot \vec{v} = 0. \tag{2.72}$$

This requirement can be satisfied identically in two dimensions by choosing a velocity field that is the curl of a vector field solely in the $\hat{k}$ direction – i.e. one that can be described by a single scalar field.‡ In particular, if we write

$$\vec{v} = \nabla \times (\psi \hat{k}), \tag{2.73}$$

for some scalar field $\psi(x, y)$, then the flow satisfies $\nabla \cdot \vec{v} = 0$, since§

$$\nabla \cdot \vec{v} = \nabla \cdot \left(\nabla \times (\psi \hat{k})\right) = 0. \tag{2.74}$$

$\psi = \psi(x, y)$ is called the *stream function.* In Cartesian coordinates, Equation 2.73 is

$$u = \frac{\partial \psi}{\partial y}, \quad v = -\frac{\partial \psi}{\partial x}, \tag{2.75}$$

and in cylindrical coordinates, the velocity components are

$$v_r = \frac{1}{r}\frac{\partial \psi}{\partial \theta}, \quad v_\theta = -\frac{\partial \psi}{\partial r}. \tag{2.76}$$

A flow that is both incompressible and irrotational can be described by both a velocity potential and a stream function. In this case, additional constraints appear on ϕ and ψ. First, using the velocity potential, the requirement of incompressibility is

$$\vec{v} = \nabla \phi, \ \nabla \cdot \vec{v} = 0 \quad \Rightarrow \quad \nabla \cdot (\nabla \phi) = 0 \quad \Rightarrow \quad \nabla^2 \phi = 0. \tag{2.77}$$

* The description of a velocity vector field by a scalar field is similar to electromagnetics, where, for example a vector electric field can be calculated from a scalar electric potential.

† The last equality is an identity for any scalar field: the curl of the gradient of a scalar field is zero. We can show this by looking at the index notation representation of the ith component of the curl,

$$(\nabla \times \nabla \phi)_i = \epsilon_{ijk} \partial_j \partial_k \phi.$$

The contribution to the implied sum by any pair (j, k) is balanced by an opposite contribution of the pair (k, j) due to the sign switch in ϵ_{ijk} and the fact that derivatives commute.

‡ This scalar field must be a function of only x and y, not z.

§ The last equality is an identity: the divergence of a curl is always zero, and this can also be shown using index notation.

Therefore, to describe incompressible flow, the velocity potential must satisfy *Laplace's equation.*

Starting instead from the stream function, the requirement of irrotationality is

$$\begin{aligned}\vec{v} = \nabla \times (\psi\hat{k}),\ \nabla \times \vec{v} = 0 &\Rightarrow \nabla \times (\nabla \times (\psi\hat{k})) = 0\\ &\Rightarrow \nabla\left(\nabla \cdot (\psi\hat{k})\right) - \nabla^2(\psi\hat{k}) = 0\\ &\Rightarrow \nabla^2\psi = 0,\end{aligned} \tag{2.78}$$

where the last equality follows from the fact that $\nabla \cdot (\psi\hat{k}) = \partial\psi/\partial z = 0$ for $\psi = \psi(x, y)$. Hence, to describe irrotational flow, the stream function must also satisfy Laplace's equation.

In summary,

- In an irrotational flow, $\nabla \times \vec{v} = 0$, the velocity can be obtained from a velocity potential, $\vec{v} = \nabla\phi$.
- In a 2D incompressible flow, $\nabla \cdot \vec{v} = 0$, the velocity can be obtained from a stream function, $\vec{v} = \nabla \times (\psi\hat{k})$.
- For a 2D incompressible and irrotational flow, both a velocity potential and a stream function exist, and they both satisfy Laplace's equation.

Example 2.10 Converting between ϕ and $\vec{v}$

Given a velocity potential, obtaining the velocity vector is straightforward. Suppose that

$$\phi = x^2y - y^3 + 2yz^2. \tag{2.79}$$

Then

$$u = \frac{\partial\phi}{\partial x} = 2xy, \tag{2.80}$$

$$v = \frac{\partial\phi}{\partial y} = x^2 - 3y^2 + 2z^2, \tag{2.81}$$

$$w = \frac{\partial\phi}{\partial z} = 4yz. \tag{2.82}$$

Going the other way, from the velocity to the potential, can be a little more difficult. We can integrate the velocity field components u, v, and w with respect to x, y, and z, respectively, but with each integration, we must include to-be-determined functions of the other variables. For example, assuming that the above velocity field is given, but that we do not know the velocity potential, we can calculate it three times¶

$$\begin{aligned}\phi &= \int u\,dx = x^2y + f_1(y, z),\\ \phi &= \int v\,dy = x^2y - y^3 + 2yz^2 + f_2(x, z),\\ \phi &= \int w\,dz = 2yz^2 + f_3(x, y).\end{aligned} \tag{2.83}$$

¶ Integrating a partial derivative with respect to one variable introduces a function of the other variables as the arbitrary constant. Here, these functions are f_1, f_2, and f_3.

Comparing the expressions reveals that

$$f_1(y,z) = -y^3 + 2yz^2 + C,$$
$$f_2(x,z) = C,$$
$$f_3(x,y) = x^2y - y^3 + C,$$

where C is a constant, so that||

$$\phi = x^2y - y^3 + 2yz^2 + C.$$

|| Starting from the velocity field, we can only determine ϕ up to a constant, since adding a constant to ϕ does not have an effect on $\vec{v} = \nabla\phi$.

The fact that we were able to match up the expressions for ϕ is due to the flow being irrotational. For a rotational flow, no ϕ would exist, and we would not be able to match the expressions after integrating the velocity components. A similar procedure is used to convert between the stream function and a two-dimensional velocity field.

The above discussion shows that analysis of an incompressible and irrotational flow amounts to solving Laplace's equation, either Equation 2.77 or Equation 2.78. There are various analytical and numerical approaches for solving Laplace's equation. However, before attempting a solution, we must specify the appropriate boundary conditions. For an external flow, such as that over an airfoil, the boundary conditions are illustrated in Figure 2.24. These consist of no flow crossing the body surface and the velocity approaching uniform freestream far from the object. The solution technique therefore proceeds as follows:

- Solve Laplace's equation with the appropriate boundary conditions to obtain either $\psi(x,y)$ or $\phi(x,y)$. In some cases, an analytical solution is possible, e.g. for flow over a cylinder, as presented in Chapter 4. Most geometries, however, will require a numerical approach, such as a panel method, presented in Chapter 5.
- Obtain the velocity field from either $\vec{v} = \nabla\phi$ or $\vec{v} = \nabla \times (\psi\hat{k})$. The velocity can be computed at arbitrary points, either on the surface of the object or in the domain of the flow. If streamline visualization is desired, contours of the stream function, ψ, can be plotted without computing the velocity vector field.
- Calculate pressures from Bernoulli's equation, which is discussed in Chapter 3. This equation relates pressure to velocity magnitude, and hence by knowing the pressure and velocity in the freestream, we can calculate the pressure close to the object, including on its surface.
- Integrate pressures over the object of interest to obtain the force exerted by the fluid on the object.

The boundary condition for the stream function at a solid wall deserves additional analysis, as it is related to a physical interpretation of the stream function.

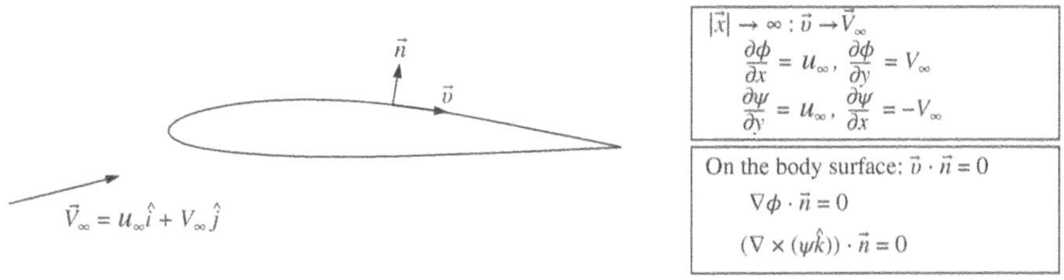

Figure 2.24 Boundary conditions on ϕ and ψ for an external flow.

Consider two points A and B connected by an arbitrary path, as illustrated in Figure 2.25. Recall that

$$\frac{\partial \psi}{\partial y} = u, \qquad \frac{\partial \psi}{\partial x} = -v. \tag{2.84}$$

Integrating the stream function along the path from A to B gives

$$\begin{aligned}
\psi(B) - \psi(A) &= \int_A^B \nabla\psi \cdot \vec{t}\, dl \\
&= \int_A^B \left(\frac{\partial \psi}{\partial x} t_x + \frac{\partial \psi}{\partial y} t_y\right) dl \\
&= \int_A^B \left(-\frac{\partial \psi}{\partial x} n_y + \frac{\partial \psi}{\partial y} n_x\right) dl \\
&= \int_A^B (v n_y + u n_x)\, dl \\
&= \int_A^B \vec{v} \cdot \vec{n}\, dl,
\end{aligned} \tag{2.85}$$

where $\vec{n} = [n_x, n_y]$ is the normal to the path, and $\vec{t} = [t_x, t_y] = [-n_y, n_x]$ is the tangent to the path. The above equation says that the flow rate through any path joining points A and B will be the same and equal to $\psi(B) - \psi(A)$. Note that the flow rate is independent of the path between A and B, which is a consequence of the fact that $\nabla \cdot \vec{v} = 0$.

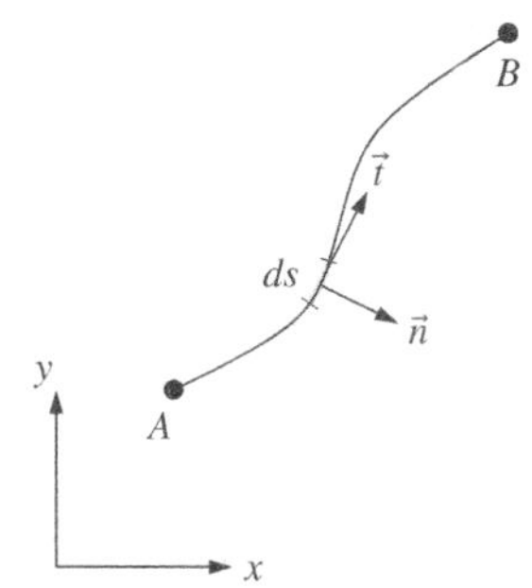

Figure 2.25 Path between two points used for a physical interpretation of the stream function, ψ. The difference in the stream function between two points is the 2D volumetric flow rate between the points.

If the path connecting A and B is a streamline, then $\vec{v} \cdot \vec{n} = 0$ everywhere on the path, so that $\psi(B) = \psi(A)$. This means that *the stream function is constant on a streamline of the flow*. In particular, since a wall or other solid surface is a streamline, ψ is constant on a wall, and a convention, albeit not a necessary one, is to set $\psi = 0$ there. Note that if there are multiple objects, the walls of each may have different stream function values, and in such a case, it would not be possible to set $\psi = 0$ on all of them. In addition, there is a relationship between streamlines and iso-potential lines, where an *iso-potential line* is defined as a line where ϕ is constant. The gradient $\nabla\phi$ is perpendicular to an iso-potential line, and $\vec{n}$ is perpendicular to a streamline. Evaluating the dot product

$$\nabla\phi \cdot \vec{n} = \vec{v} \cdot \vec{n} = 0 \tag{2.86}$$

shows that $\nabla\phi$ is perpendicular to $\vec{n}$, which means that *iso-potential lines and streamlines are everywhere orthogonal*. This orthogonality is illustrated in Figure 2.26 for flow over a cylinder.

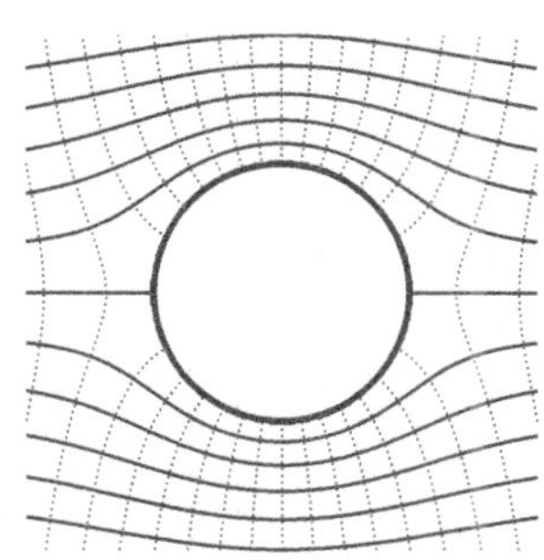

Figure 2.26 Streamlines (solid) and iso-potential lines (dotted) for incompressible and irrotational flow over a cylinder. Note that streamlines are everywhere perpendicular to iso-potential lines.

Example 2.11 From ψ to $\vec{v}$ and ϕ
The flow shown in Figure 2.26 can be described by the following stream function,

$$\psi(r, \theta) = br \sin\theta \left[1 - \frac{a^2}{r^2}\right] + c \ln \frac{r}{a}, \tag{2.87}$$

where a, b, and c are constants.** The constant a is the radius of the cylinder, and evaluating ψ at $r = a$ results in 0, showing that the cylinder surface is a streamline of the flow. The constant c is related to circulation, and for the flow shown in Figure 2.26, $c = 0$.

**We will derive this stream function for flow over a cylinder in Chapter 4.

To determine the velocity vector, we differentiate the stream function,

$$v_r = \frac{1}{r}\frac{\partial\psi}{\partial\theta} = b\cos\theta\left[1 - \frac{a^2}{r^2}\right], \tag{2.88}$$

$$v_\theta = -\frac{\partial\psi}{\partial r} = -b\sin\theta\left[1 + \frac{a^2}{r^2}\right] - \frac{c}{r}. \tag{2.89}$$

The velocity potential exists if the flow is irrotational, which is the case for this flow, because the vorticity evaluates to zero:

$$\begin{aligned}
\nabla\times\vec{v} &= \left[\frac{1}{r}\frac{\partial(rv_\theta)}{\partial r} - \frac{1}{r}\frac{\partial v_r}{\partial\theta}\right]\hat{z} \\
&= \left[\frac{1}{r}\frac{\partial}{\partial r}\left(-b\sin\theta\left(r + \frac{a^2}{r}\right) - c\right) - \frac{1}{r}(-\sin\theta)\left(1 - \frac{a^2}{r^2}\right)\right]\hat{z} \\
&= \left[-\frac{1}{r}b\sin\theta(1 - a^2/r^2) + \frac{1}{r}\sin\theta(1 - a^2/r^2)\right]\hat{z} \\
&= \vec{0}.
\end{aligned}$$

We calculate the potential by integrating the velocity field,

$$\phi(r,\theta) = \int v_r dr = b\cos\theta\left[r + \frac{a^2}{r}\right] + f(\theta), \tag{2.90}$$

$$\phi(r,\theta) = \int r v_\theta d\theta = b\cos\theta\left[r + \frac{a^2}{r}\right] - c\theta + g(r). \tag{2.91}$$

Comparing the above expressions, we see that we need $f(\theta) = -c\theta + C$, and $g(r) = C$, where C is a constant, in order for the expressions to match. Putting everything together,

$$\phi(r,\theta) = b\cos\theta\left[r + \frac{a^2}{r}\right] - c\theta + C. \tag{2.92}$$

In contrast to the stream function, the velocity potential varies along the surface of the cylinder, which is consistent with tangential flow along the surface. To make the streamlines and iso-potential lines shown in Figure 2.26, we can draw contours of ψ and ϕ. Listing 2.5 shows a code that does this, using the `contour` function.

Code notes:

- The stream function and velocity potential are defined everywhere in the domain, and the resulting flow described inside a closed object such as a cylinder can be considered fictitious.
- The last terms in the calculation of ψ and ϕ raise the value of these scalars inside the cylinder, so that no lines are shown there for the ranges chosen in the contours.
- The coordinate variables `x`, `y`, `r`, `theta`, and `psi` are all matrices.
- The `atan2` function computes the four-quadrant inverse tangent, by taking into account the signs of both x and y.
- The contour levels for `psi` and `phi` are chosen manually to show a sufficient collection of contours in the chosen domain.

Listing 2.5 Plotting streamlines and iso-potential lines as contours of ψ and ϕ.

```
function streamisopot
% plots streamlines and iso-potential lines for a cylinder

% coefficients defining the stream function
a = 1; b = 1; c = 0;

% Cartesian coordinates (for plotting contours)
[x,y] = ndgrid(linspace(-2*a,2*a,200));

% corresponding polar coordinates
r = sqrt(x.^2 + y.^2);
t = atan2(y,x);

figure(1); clf; hold on;
```

```
% streamlines = contours of the stream function
psi = b*r.*sin(t).*(1-a^2./r.^2) + c*log(r/a) + (sign(1-r)+1)*50;
contour(x,y,psi, [-1.5:.3:1.5], 'k-', 'linewidth', 1);

% iso-potential lines = contours of the evelocity potential
phi = b.*cos(t).*(r + a^2./r) - c*t + (sign(1-r)+1)*50;
contour(x,y,phi, [-2.7:.3:2.7], 'k ', 'linewidth', 1);

% draw the cylinder geometry
t = linspace(0,2*pi,100);
plot(a*cos(t),a*sin(t),'k-', 'linewidth', 3);

axis equal; axis off
exportgraphics(gcf, '../figs/streamisopot.pdf');
```

2.6 Rotation and Deformation of a Fluid Element

Recall from Section 1.1 that in a fluid, the *rates* of rotation and deformation are of interest, as opposed to the *amount* of rotation or deformation. Understanding how a fluid deforms is important for deriving the stress–strain relationship that will be presented in Chapter 3.

Key to the discussion in this section will be the concept of a fluid element, which will be drawn as a small square in two dimensions, as shown in Figure 2.1. At least this will be the *initial* shape of the fluid element. When placed in a non-uniform velocity field, the square fluid element will in general translate, rotate, and deform, so that at later times it will no longer be a square. We will mostly be concerned with the short-time behavior of the shape, which we will be able to describe as small perturbations from an initial square shape. However, before doing this, we first look at an example of tracking the fluid element shape exactly.

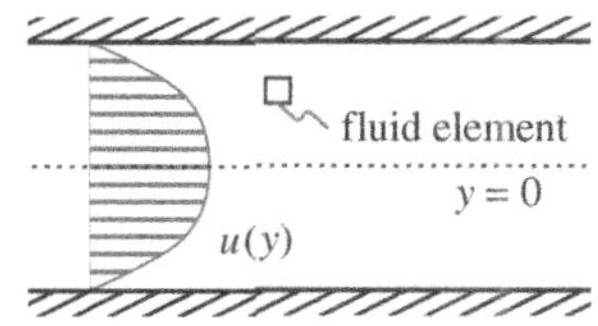

Figure 2.27 Fluid element in channel flow.

Example 2.12 Fluid element deformation in channel flow
The velocity field in pressure-driven flow of an incompressible fluid in a two-dimensional channel of half-width h is

$$\vec{v} = u_0 \left(1 - \frac{y^2}{h^2}\right) \hat{\imath}, \tag{2.93}$$

where u_0 is the centerline velocity. This formula can be derived using the equations presented in the next chapter. Note, this flow is rotational. Consider a small square fluid element of side length $0.2h$ at $y = 0.5h$, as shown in Figure 2.27. To see what this element looks like later, we can use the given velocity field to update the position of every point on the square boundary. The code in Listing 2.6 does this, using a simple forward Euler update, and it plots the fluid element at later times.

Listing 2.6 Draws fluid element shapes in pressure-driven channel flow.

```
function felemchan
% plots the shape of a square fluid element as it moves
% through a rotational velocity field

% create points on the boundary of a square fluid element
s = linspace(0,1,101); s = s(1:end-1); o = ones(size(s));
dx = .2; % square size
X = dx*[s, o, 1-s, 0*o]; X(end+1) = X(1);
Y = 0.5+dx*[0*o, s, o, 1-s]; Y(end+1) = Y(1);
```

Code notes:

- The square is defined by 100 points on each of its four sides.
- The time step for forward Euler is hard-coded at $\tau/100$.
- The fluid element is drawn at four equal intervals during the simulation, after the initial condition. The variable m stores the number of time steps in between successive drawings.
- The velocity is coded in a function, `getv`.
- As the fluid element deforms, it also moves downstream (to the right), so that the fluid elements will not be drawn on top of each other.

```
figure(1); clf;

% advance the points forward in time
tau = 1; dt = .01; N = ceil(2*tau/dt); m = ceil(N/4);
ctype = {'#000', '#444', '#666', '#888', '#AAA'};
j = 0;
for n = 0:N, % loop over times
  for i = 1:length(X), % loop over points
    [u, v] = getv(X(i), Y(i));
    X(i) = X(i) + dt*u;
    Y(i) = Y(i) + dt*v;
  end
  if (mod(n,m)==0)
    j = j+1;
    plot(X,Y, '-', 'linewidth', 1, 'color', ctype{j}); hold on;
    text(X(10), 0.45, sprintf('$t = %.1f \\tau$', n*dt), ...
         'interpreter', 'latex', 'fontsize', 10);
  end
end
axis equal; axis off;
exportgraphics(gcf, '../figs/felemchan.pdf');

%--------------------------------

function [u, v] = getv(x, y);
h = 1; u0 = 1;
u = u0*(1-y^2/h^2); % horizontal velocity
v = 0;              % vertical velocity
```

Note, we use $\tau \equiv h/u_0$ as the relevant time scale based on the maximum flow speed and channel half-width. Figure 2.28 shows the shape of the fluid element at various times. We see that the square first deforms into what looks like a parallelogram before the curvature along two of its sides becomes more pronounced. That is, an initially square fluid element does not in general remain a square, and its sides may not even remain linear as the fluid element is transported by the flow. However, in the remainder of this section, we will be concerned with short-time behavior of small fluid elements, and in these conditions, a linear approximation of the sides is justified.

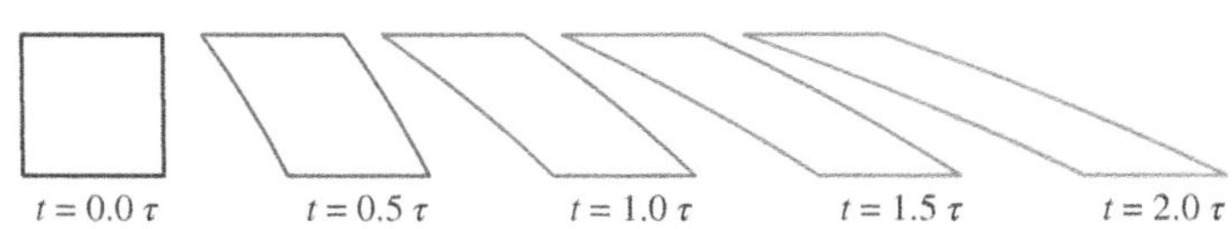

Figure 2.28 Fluid element shapes at various times in pressure-driven channel flow.

Consider a very small fluid element at two (very close) instances in time. Although the fluid can change position by translating to another location due to its velocity, the material derivative already takes this motion into account. The remaining motion of the fluid can be described as a superposition of rotation and deformation, as illustrated in Figure 2.29. Note that deformation is subdivided into two categories: normal compression/dilation and shearing. In pure rotation, the fluid element rotates as a rigid body, which means that the distance between neighboring points does not change, while the angle does according to the rate of rotation. On the other hand, deformation is synonymous with strain: it refers to any changes in length between two adjacent points within the fluid element. Specifically, a *normal strain* is a compression or dilation along a single axis, whereas a *shear strain* is related to the angle change between a pair of axes.

All of the information on the relative motion of nearby points, including the effects of rotation and deformation, is contained in spatial derivatives of the velocity vector.* Since there are, in three dimensions, three components of velocity and

* If the velocity is constant, then fluid elements do not change shape. Rotation and deformation can only occur in the presence of a velocity gradient.

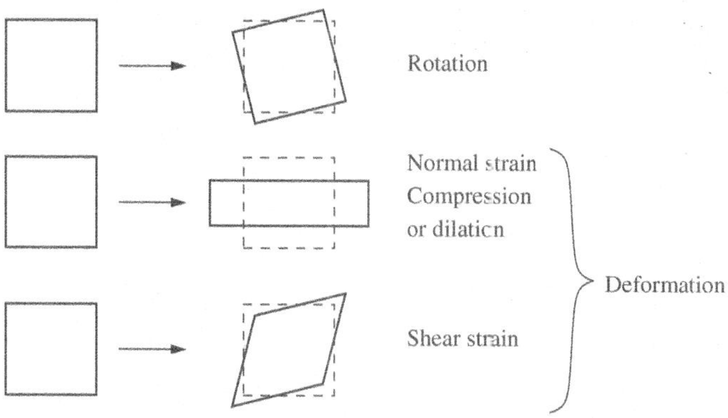

Figure 2.29 Rotation and deformation of a square fluid element.

three spatial coordinates, a total of nine derivatives are possible. The *velocity gradient* provides a way to organize this information in a matrix form,

$$\nabla\vec{v} = \partial_i v_j = \begin{bmatrix} \frac{\partial u}{\partial x} & \frac{\partial v}{\partial x} & \frac{\partial w}{\partial x} \\ \frac{\partial u}{\partial y} & \frac{\partial v}{\partial y} & \frac{\partial w}{\partial y} \\ \frac{\partial u}{\partial z} & \frac{\partial v}{\partial z} & \frac{\partial w}{\partial z} \end{bmatrix}. \tag{2.94}$$

Rotation arises from the antisymmetric part of $\nabla\vec{v}$, while deformation arises from the symmetric part of $\nabla\vec{v}$. That is, we can write,

$$\nabla\vec{v} = \underbrace{\frac{1}{2}\left[\nabla\vec{v} - (\nabla\vec{v})^T\right]}_{\Omega} + \underbrace{\frac{1}{2}\left[\nabla\vec{v} + (\nabla\vec{v})^T\right]}_{\epsilon}, \tag{2.95}$$

where the antisymmetric† *vorticity tensor* is

$$\boldsymbol{\Omega} = \Omega_{ij} = \frac{1}{2}\left(\partial_i v_j - \partial_j v_i\right), \tag{2.96}$$

and the symmetric *rate-of-strain tensor* is‡

$$\epsilon = \epsilon_{ij} = \frac{1}{2}\left(\partial_i v_j + \partial_j v_i\right). \tag{2.97}$$

In these expressions, the superscript ^{T} indicates the transpose.

† An antisymmetric tensor is one whose transpose is the negative of the tensor: it necessarily has zeros on the main diagonal. A symmetric tensor is one whose transpose equals the tensor: its main diagonal can be arbitrary.

‡ Not all authors use the factor of 1/2 in defining the off-diagonal components of ϵ, e.g. see Anderson (2016), but we include the 1/2 for symmetry in the definition of the vorticity and rate-of-strain tensors, and to allow the simple expression for the velocity gradient as the sum of these two tensors.

2.6.1 Rotation

To see how $\boldsymbol{\Omega}$ relates to rotation, we write out its components,

$$\begin{aligned} \boldsymbol{\Omega} &= \frac{1}{2}\begin{bmatrix} 0 & \partial_1 v_2 - \partial_2 v_1 & \partial_1 v_3 - \partial_3 v_1 \\ \partial_2 v_1 - \partial_1 v_2 & 0 & \partial_2 v_3 - \partial_3 v_2 \\ \partial_3 v_1 - \partial_1 v_3 & \partial_3 v_2 - \partial_2 v_3 & 0 \end{bmatrix} \\ &= \begin{bmatrix} 0 & \omega_3 & -\omega_2 \\ -\omega_3 & 0 & \omega_1 \\ \omega_2 & -\omega_1 & 0 \end{bmatrix}, \end{aligned} \tag{2.98}$$

where $\vec{\omega}$ is the *angular velocity* vector,

$$\vec{\omega} = \omega_1\hat{\imath} + \omega_2\hat{\jmath} + \omega_3\hat{k} = \frac{1}{2}\nabla\times\vec{v}. \tag{2.99}$$

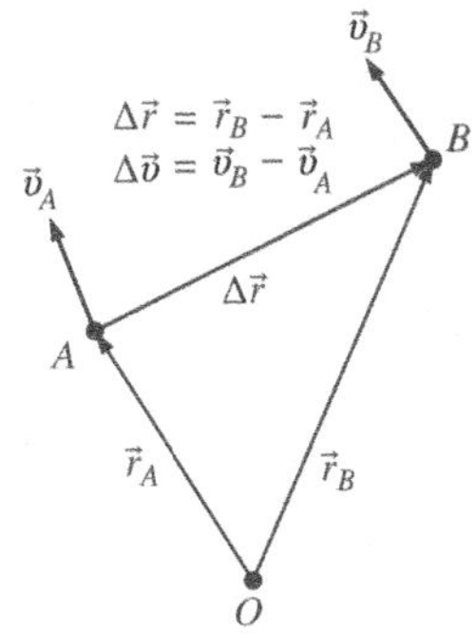

Figure 2.30 Two points A and B used for interpretation of rotation and deformation.

In rigid body motion, the relative velocity $\Delta\vec{v} = \vec{v}_B - \vec{v}_A$ of two points separated by $\Delta\vec{r} = \vec{r}_B - \vec{r}_A$ (see Figure 2.30) can be expressed purely in terms of $\boldsymbol{\Omega}$:

$$\Delta\vec{v} = \vec{\omega} \times \Delta\vec{r} = \Delta\vec{r} \cdot \boldsymbol{\Omega} \quad \left(\Delta v_j = \Delta r_i \Omega_{ij}\right).$$

2.6.2 Deformation

To see that $\boldsymbol{\epsilon}$ relates to deformation, consider two points A and B separated by $\Delta\vec{r}$, as in Figure 2.30, in the limit $|\Delta\vec{r}| \to 0$. The rate of change in $|\Delta\vec{r}|$ can be derived as follows:

$$|\Delta\vec{r}|\frac{d|\Delta\vec{r}|}{dt} = \frac{1}{2}\frac{d}{dt}\left(|\Delta\vec{r}|^2\right) = \frac{d\Delta\vec{r}}{dt} \cdot \Delta\vec{r} = \Delta\vec{v} \cdot \Delta\vec{r}. \tag{2.100}$$

Using a Taylor-series expansion of $\vec{v}$ around A in the direction of B, we have

$$\Delta\vec{v} = \Delta\vec{r} \cdot \nabla\vec{v} + \mathcal{O}(|\Delta\vec{r}|^2) = \Delta\vec{r} \cdot \boldsymbol{\Omega} + \Delta\vec{r} \cdot \boldsymbol{\epsilon} + \mathcal{O}(|\Delta\vec{r}|^2), \tag{2.101}$$

where the second equality follows from Equation 2.95. Since $\Delta\vec{r} \cdot \boldsymbol{\Omega} \cdot \Delta\vec{r} = 0$ for any antisymmetric $\boldsymbol{\Omega}$,* Equation 2.100 becomes, after dropping the $\mathcal{O}(|\Delta\vec{r}|^2)$ terms,

$$|\Delta\vec{r}|\frac{d|\Delta\vec{r}|}{dt} = \Delta\vec{r} \cdot \boldsymbol{\epsilon} \cdot \Delta\vec{r}.$$

*This can be shown using index notation,

$$\Delta\vec{r} \cdot \boldsymbol{\Omega} \cdot \Delta\vec{r} = \Delta r_i \Omega_{ij} \Delta r_j.$$

In the implied double sum, the (j, i) contribution cancels the (i, j) contribution because $\Omega_{ij} = -\Omega_{ji}$.

The percentage rate of change in $|\Delta\vec{r}|$ is the strain rate, and it is obtained by dividing by $|\Delta\vec{r}|^2$,

$$\frac{1}{|\Delta\vec{r}|}\frac{d|\Delta\vec{r}|}{dt} = \frac{\Delta\vec{r} \cdot \boldsymbol{\epsilon} \cdot \Delta\vec{r}}{|\Delta\vec{r}|^2} \quad \left(\frac{\Delta r_i\, \epsilon_{ij}\, \Delta r_j}{\Delta r_k\, \Delta r_k} \text{ in index notation}\right). \tag{2.102}$$

Therefore, deformation, the change in length of $|\Delta\vec{r}|$, is fully described by $\boldsymbol{\epsilon}$. Writing out the components in a chosen set of coordinates,

$$\boldsymbol{\epsilon} = \begin{bmatrix} \epsilon_{11} & \epsilon_{12} & \epsilon_{13} \\ \epsilon_{21} & \epsilon_{22} & \epsilon_{23} \\ \epsilon_{31} & \epsilon_{32} & \epsilon_{33} \end{bmatrix}. \tag{2.103}$$

The diagonal entries, $\epsilon_{11}, \epsilon_{22}$, and ϵ_{33}, are the normal strain rates.† The off-diagonal components, $\epsilon_{ij}, i \neq j$, represent the shear rates of strain. One can show that these shear rates correspond to 1/2 the rate of decrease of the angle between coordinate axes (Anderson 2016).

† For two points separated by a small distance Δx_1 along the $\hat{i}$ axis, $\Delta\vec{r} = \Delta x_1 \hat{i}$, and Equation 2.102 gives $(1/\Delta x_1)(d\Delta x_1/dt) = \epsilon_{11} = \partial_1 v_1$. So ϵ_{11} is related to compression/dilation: it tells us how quickly the x distance between two points is changing.

Since $\boldsymbol{\epsilon}$ is symmetric, a result from linear algebra is that there exist orthogonal directions, called the *principal directions of strain*, for which all of the shear strain rates are zero. In this coordinate system, $\boldsymbol{\epsilon}$ becomes

$$\boldsymbol{\epsilon} = \begin{bmatrix} \epsilon_I & 0 & 0 \\ 0 & \epsilon_{II} & 0 \\ 0 & 0 & \epsilon_{III} \end{bmatrix}, \tag{2.104}$$

where $\epsilon_I, \epsilon_{II}$, and ϵ_{III} are the principal normal rates of strain. Mathematically, these principal strains are the eigenvalues of the strain rate tensor, Equation 2.103, and the corresponding eigenvectors give the principal directions of strain. Computing these amounts to diagonalizing $\boldsymbol{\epsilon}$.

Example 2.13 Vorticity and rate-of-strain tensors
In this example, we calculate the vorticity tensor $\mathbf{\Omega}$ and the rate-of-strain tensor, ϵ, for flow that has a velocity vector given by

$$u(x,y) = y, \qquad v(x,y) = x. \tag{2.105}$$

The vorticity tensor is

$$\mathbf{\Omega} = \frac{1}{2}\begin{bmatrix} 0 & \dfrac{\partial v}{\partial x} - \dfrac{\partial u}{\partial y} \\ \dfrac{\partial u}{\partial y} - \dfrac{\partial v}{\partial x} & 0 \end{bmatrix} = \begin{bmatrix} 0 & 0 \\ 0 & 0 \end{bmatrix}. \tag{2.106}$$

The fact that the vorticity tensor is zero tells us that the flow is irrotational. We would have arrived at the same conclusion if we had calculated the curl, $\nabla \times \vec{v}$.

The rate-of-strain tensor is given by

$$\epsilon = \begin{bmatrix} \dfrac{\partial u}{\partial x} & \dfrac{1}{2}\left(\dfrac{\partial u}{\partial y} + \dfrac{\partial v}{\partial x}\right) \\ \dfrac{1}{2}\left(\dfrac{\partial u}{\partial y} + \dfrac{\partial v}{\partial x}\right) & \dfrac{\partial v}{\partial y} \end{bmatrix} = \begin{bmatrix} 0 & 1 \\ 1 & 0 \end{bmatrix}. \tag{2.107}$$

This rate of strain tensor describes pure shear. So an initially square fluid element undergoes deformation shown in Figure 2.31 a short time later. Note that positive off-diagonal entries in ϵ correspond to a *decrease* in the lower-left corner angle in the fluid element. In this pure shear case, u increases with y, and v increases with x.

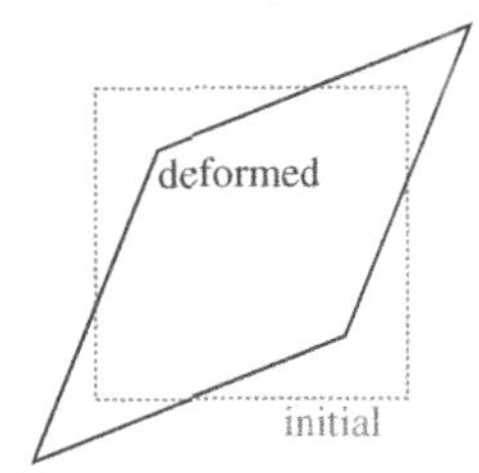

Figure 2.31 Illustration of pure shear deformation.

Example 2.14 Fluid element kinematics in a vortex
Consider a vortex flow with the following velocity field,

$$v_r = 0, \qquad v_\theta = \frac{b}{r}\left(\frac{r^2}{r^2 + a^2}\right), \tag{2.108}$$

where $a > 0$, $b > 0$ are constants. A plot of $v_\theta(r)$ is shown in Figure 2.32. The maximum is $v_\theta(a) = b/(2a)$.

We can obtain the angular velocity of a fluid element by calculating the vorticity and halving it,

$$\begin{aligned} \vec{\omega} &= \frac{1}{2}\nabla \times \vec{v} \\ &= \frac{1}{2}\left(\frac{1}{r}\frac{\partial(rv_\theta)}{\partial r} - \frac{1}{r}\frac{\partial v_r}{\partial \theta}\right)\hat{k} \\ &= \frac{1}{2}\frac{1}{r}\frac{\partial}{\partial r}\left(\frac{br^2}{r^2 + a^2}\right)\hat{k} \\ &= \frac{1}{2}\frac{1}{r}\left(\frac{2bra^2}{(r^2 + a^2)^2}\right)\hat{k} = \frac{ba^2}{(r^2 + a^2)^2}\hat{k}. \end{aligned} \tag{2.109}$$

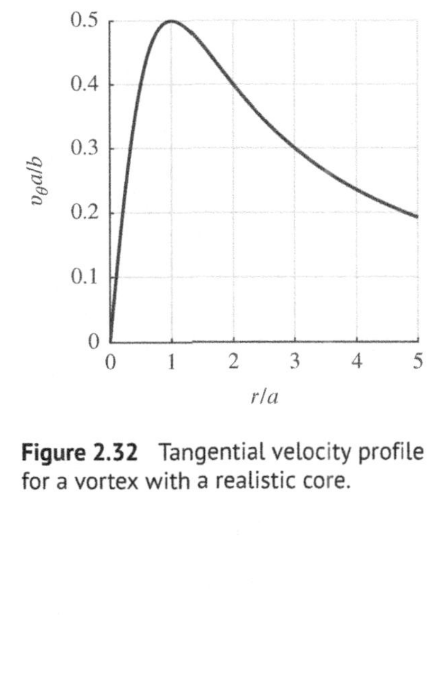

Figure 2.32 Tangential velocity profile for a vortex with a realistic core.

For $r \ll a$, this expression becomes b/a^2. For $r \gg a$, the expression is $ba^2/r^4 \rightarrow 0$. We take a closer look at these two regimes.

First, we consider the rate-of-strain tensor. To calculate $\epsilon = \frac{1}{2}(\nabla\vec{v} + \nabla\vec{v}^T)$, we can convert the velocity field to Cartesian coordinates. However, this can get messy, and a quicker approach in this case is to use the following expressions for the velocity gradient tensor components in cylindrical coordinates:

$$(\nabla\vec{v})_{rr} = \frac{\partial v_r}{\partial r}, \qquad (\nabla\vec{v})_{r\theta} = \frac{\partial v_\theta}{\partial r},$$
$$(\nabla\vec{v})_{\theta r} = \frac{1}{r}\frac{\partial v_r}{\partial \theta} - \frac{v_\theta}{r}, \qquad (\nabla\vec{v})_{\theta\theta} = \frac{1}{r}\frac{\partial v_\theta}{\partial \theta} + \frac{v_r}{r}. \tag{2.110}$$

The only nonzero component for this vortex flow is

$$\begin{aligned} \epsilon_{r\theta} = \epsilon_{\theta r} &= \frac{1}{2}\left(\frac{\partial v_\theta}{\partial r} - \frac{v_\theta}{r}\right) \\ &= \frac{1}{2}\left(\frac{b(r^2+a^2) - br(2r)}{(r^2+a^2)^2} - \frac{b}{r^2+a^2}\right) \\ &= \frac{-br^2}{(r^2+a^2)^2}. \end{aligned} \tag{2.111}$$

For $r \ll a$, this expression evaluates to $-br^2/a^4 \to 0$. For $r \gg a$, the expression evaluates to $-b/r^2$. We note that this is a shear strain in $\hat{r}, \hat{\theta}$; the normal strains are zero.

Based on the above results, an initially square fluid element at $r \ll a$ experiences rotation but no strain. Assuming that a and b are positive, the fluid element will rotate as shown in Figure 2.33. Note, the angle of rotation is $|\vec{\omega}|\Delta t$.

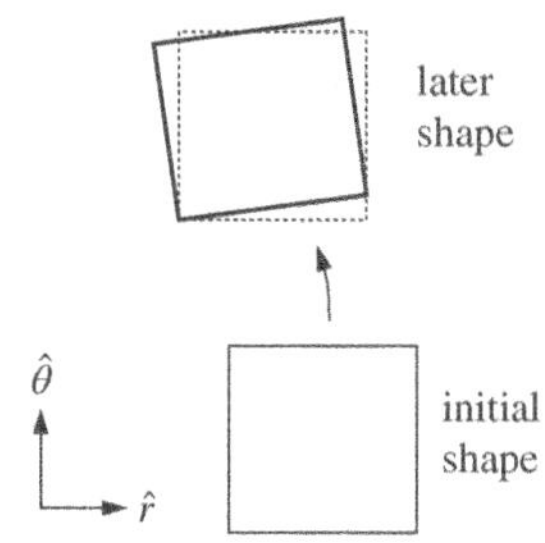

Figure 2.33 Fluid element undergoing only rotation, with no strain.

An initially square (in $\hat{r}$, $\hat{\theta}$ axes) fluid element at $r \gg a$ experiences no rotation and only shear strain. Since $b > 0$, the shear strain is negative, $\epsilon_{r\theta} < 0$, which means that the lower left angle is increasing. The fluid element therefore deforms as shown in Figure 2.34. Note, the change in the lower left corner angle is $-\epsilon_{r\theta}/2$.

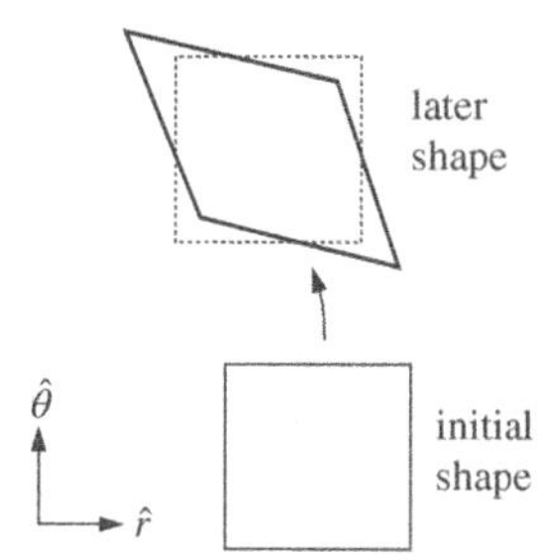

Figure 2.34 Fluid element undergoing only strain, with no rotation.

Finally, for this flow, a stream function exists because the flow is incompressible,

$$\nabla \cdot \vec{v} = \frac{1}{r}\frac{\partial (rv_r)}{\partial r} + \frac{1}{r}\frac{\partial v_\theta}{\partial \theta} = 0.$$

To obtain the stream function, we use

$$v_\theta = -\frac{\partial \psi}{\partial r}.$$

Integrating with respect to r,

$$\psi(r) = -\int v_\theta(r)dr = -\int \frac{br}{r^2+a^2}dr = -\frac{b}{2}\ln(r^2+a^2) + C.$$

Note, the constant C has to be independent of θ since $v_r = \frac{1}{r}\frac{\partial \psi}{\partial \theta} = 0$. The stream function can be used to visualize the flow by plotting the streamlines, which will be concentric circles due to the fact that $v_r = 0$. Such a plot can still be useful to show where the flow speed is high, based on the spacing between adjacent streamlines. Listing 2.7 shows a code that draws the streamlines, and Figure 2.35 shows the result.

Code notes:

- The choices of a and b are arbitrary, as the code scales the drawing window based on a and the stream function contours based on b.
- There is no singularity at the origin because the flow speed is zero there.

Listing 2.7 Streamline plotting for a vortex flow.

```
function vortexstream
% plots streamlines for a vortex flow

% coefficients defining the stream function
a = 1; b = 1;
```

```
% Cartesian coordinates (for plotting contours)
[x,y] = ndgrid(linspace(-5*a,5*a,200));

% corresponding r
r = sqrt(x.^2 + y.^2);

figure(1); clf; hold on;

% streamlines = contours of the stream function
psi = -(b/2)*log(r.^2 + a^2);
contour(x,y,psi, linspace(-1.6*b,0,21), 'k-', 'linewidth', 1);
set(gca, 'xtick', linspace(-4,4,5));
set(gca, 'ytick', linspace(-4,4,5));
axis equal; set(gcf, 'position', 400*[1,1,1,1]);
writeplot('$x/a$', '$y/a$', '../figs/vortexstream.pdf', 4);
```

The streamlines are closest together near $r = a$, which is the location of peak v_θ from Figure 2.32. By conservation of mass, when the streamlines are drawn at equal intervals of the stream function, the spacing between the streamlines is inversely proportional to the flow speed. So streamlines that are close together indicate fast flow.

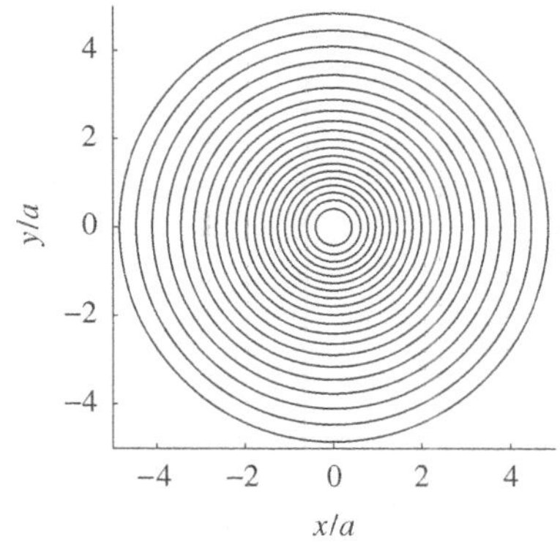

Figure 2.35 Streamlines for a vortex flow with zero velocity at the origin.

2.7 Summary and Concepts

This chapter presented various ways of describing the motion of a fluid. Whereas this task is simple for a rigid solid, which can translate and rotate with a small number of degrees of freedom, it is more challenging for a fluid, which consists of molecules continuously sliding past each other. Considering fluid elements, which are small control volumes, gives rise to the Lagrangian frame and the material derivative for quantifying rates of change. Flow lines help in visualizing the motion of the fluid elements in general Eulerian frames. In the special case, when fluid elements do not rotate, the velocity field can be obtained from a scalar velocity potential field. Similarly, if fluid elements cannot be compressed, a scalar stream function field fully describes the flow. In the most general case, the gradient of the velocity vector contains all of the information about fluid elements' relative motion, and it can be decomposed into vorticity and rate-of-strain tensors, which describe rotation and deformation, respectively.

The key concepts covered in this chapter are as follows:

- A fluid element is a small control volume that helps to describe the motion of a fluid.
- Different frames of reference lead to different perceived motions: a Lagrangian frame is one fixed to a fluid element, whereas an Eulerian frame is that of an observer watching the fluid.
- The material derivative is the rate of change of a quantity measured by a sensor moving with the fluid. It includes effects of both temporal and spatial variation of the quantity.
- Pathlines, streamlines, and streaklines are flow lines that can be generated by releasing one or more markers into the flow and tracking their positions.
- Vorticity measures the rotation of fluid elements. Its integral counterpart, the circulation, measures the net rotational velocity around the perimeter of a closed contour. The two are related through Stokes's theorem.
- Mass conservation applied to a control volume states that an imbalance of mass entering/leaving a control volume results in its accumulation inside.

- In differential form, mass conservation relates the time rate of change of density to the divergence of the mass flux. If the fluid is incompressible, this equation simplifies to the divergence of the velocity being zero.
- In an irrotational flow, a scalar velocity potential fully describes the velocity field as the gradient of the potential.
- In an incompressible, two-dimensional flow, a scalar stream function also fully describes the velocity, through a curl operation.
- Streamlines are lines of constant stream function, everywhere tangent to the velocity. The difference in the values of the stream function between streamlines measures the area flow rate in between them.
- Different relative velocities of adjacent fluid elements can lead to their rotation and deformation. The velocity gradient contains information about these differences.
- The vorticity and rate-of-strain tensors together contain the same information as the velocity gradient. The former describes rotation and the latter deformation, including normal and shear strain rates.

2.8 Problems

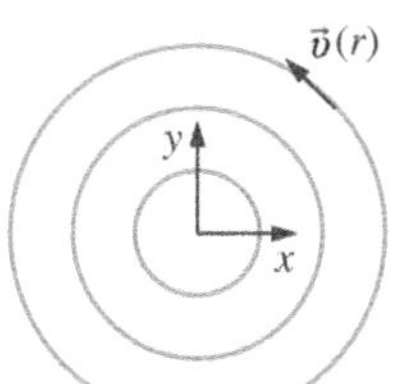

Figure 2.36 Circular streamlines.

Figure 2.37 A fluid element at two times.

1. Justify each answer to the below questions with an explanation.
  a What are the dimensions of the stream function?
  b A steady two-dimensional flow has circular streamlines about the origin and a velocity independent of θ as shown in Figure 2.36. In which direction does $\frac{D\vec{v}}{Dt}$ point?
  c True or false: if the circulation is zero for a closed curve, then the vorticity is zero everywhere inside the closed curve.
  d True or false: if pathlines and streamlines coincide, then the flow must be steady in an Eulerian frame.
  e True or false: a streamline can intersect itself for a flow with $\vec{v}$ everywhere nonzero.
  f If the velocity is $\vec{v} = 2y\hat{\imath}$, what is the circulation for a unit circle contour around the origin?
  g True or false: the fluid element shown in Figure 2.37 is undergoing rotation.

2. Calculate the material derivative of the velocity field $\vec{v} = \sin(r)\hat{\theta}$. Express your answer in cylindrical coordinates and basis.

3. Given the following steady velocity field,

$$\vec{v} = -y\hat{\imath} + x\hat{\jmath}, \tag{2.112}$$

use the forward Euler and predictor–corrector methods to draw a pathline, which is also a streamline, through the point $(x, y) = (2, 0)$. Show streamlines for time step sizes $\Delta t = 0.2$ and 0.1. How quickly do your flow lines converge as Δt is lowered? Derive an analytical form for the stream function and compare your numerical results to the analytical streamline through the same point.

4. Given the following velocity field,

$$\vec{v} = (2x - y - t)\hat{\imath} + (2y - x + t)\hat{\jmath}, \tag{2.113}$$

determine equations for a pathline released at $t = 0$ from the point $(x, y) = (1,\ 2)$, a streamline through the same point at $t = 0$, and a streakline at $t = 0.5$ formed by a continuous release of markers from $(x, y) = (1,\ 2)$ between $t = 0$ and $t = 0.5$. Plot all of these flow lines in one figure.

5. A vortical flow has the velocity field

$$\vec{v} = \frac{a}{r}\left[e^{-(r/R)^2} - 1\right]\hat{\theta}, \tag{2.114}$$

where a and R are constants. Determine the vorticity $\vec{\xi}$ as a function of r, and calculate the circulation for a contour that is a circle around the origin with radius $2R$. Integrate the vorticity vector dotted with $\hat{z}$ over the interior of this circle and show that your answer agrees with the circulation.

6. Which of the following velocity fields satisfies conservation of mass for an incompressible flow? Justify your answers and determine the stream function when possible.

a. $u = x^3 \sin y,\ v = 3x^2 \cos y$
b. $u = x^3 \sin y,\ v = -3x^2 \cos y$
c. $v_r = 2r \sin\theta \cos\theta,\ v_\theta = -r\sin^2\theta$

7. An incompressible fluid flowing over a horizontal wall at $y = 0$ has the following horizontal velocity:

$$u(x, y) = U_\infty\left(1 - e^{-y/\sqrt{\ell x}}\right), \tag{2.115}$$

where U_∞ and ℓ are constants. Determine the vertical velocity, $v(x, y)$, and calculate the stream function, $\psi(x, y)$, assuming that $\psi = 0$ at the wall. Also calculate the vorticity vector, $\vec{\xi}$, at the wall.

8. A two-dimensional steady flow has a rate-of-strain tensor given by

$$\epsilon = \begin{bmatrix} -\sin x & \frac{1}{2}y\cos x + y + \cos x \\ \frac{1}{2}y\cos x + y + \cos x & \sin x \end{bmatrix}. \tag{2.116}$$

Determine the velocity field given that $\vec{v} = 0$ at $\vec{x} = 0$. Show that the flow is rotational but incompressible, and determine the stream function. For an initially square fluid element at $(x, y) = (0, 2\pi)$, sketch the shape of the fluid element a short time later and describe the type of motion/deformation that the element is undergoing.

3

Dynamics

Dynamics is the study of the forces on a fluid and how the fluid responds. In contrast, kinematics deals only with the description of the motion of the fluid, without accounting for the forces that create the motion. Whereas the primary concepts covered in the previous chapter were the conservation of mass and the constraint of irrotationality, the primary dynamics concept is the conservation of momentum. This concept leads directly to the Navier–Stokes equations, which are used to solve a wide variety of flows in aerodynamics.

3.1 Conservation of Momentum

The momentum of an object with mass m and velocity $\vec{v}$ is the product $m\vec{v}$. Considering fluid elements as objects, each fluid element of a small volume dV has a mass of ρdV, and if it is moving with velocity $\vec{v}$, a momentum of $\rho\vec{v}dV$. Hence, the quantity $\rho\vec{v}$ is the momentum per unit volume, and this vector can vary in space and time in a fluid.

Force is the time rate of change of momentum, so a force applied to an object changes its momentum. If no net force acts on a system of objects, the total momentum of the objects does not change. If forces act on all or some of the objects, then the rate of change of the total momentum of the system equals the sum of the forces. Objects leaving or entering the system also naturally lead to a change in the momentum that depends on the mass, velocity, and rate at which the objects are entering or leaving. Here, we apply this reasoning to a control volume of fluid, where individual fluid elements are the objects.

3.1.1 Integral Form

Consider a control volume V of fluid enclosed by a surface S that the fluid may cross, as illustrated in Figure 3.1. S need not represent the wall of a containing vessel, although it could. The principle of momentum conservation extends the discussion of the previous paragraph to the control volume of fluid elements. It states that

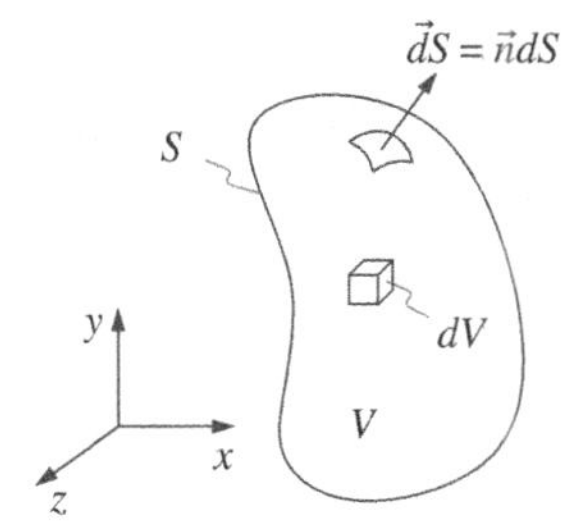

Figure 3.1 Control volume used for conservation of momentum analysis.

Law 3.1.1 (Conservation of momentum.) *The time rate of change of momentum in the control volume equals the sum of the net momentum flow rate into the control volume and the forces acting on the fluid inside the control volume.*

Aerodynamics: A Computational Introduction, First Edition. Krzysztof J. Fidkowski.

© 2026 John Wiley & Sons, Inc. Published 2026 by John Wiley & Sons, Inc.

The mathematical expressions corresponding to the individual terms in this statement are

$$\text{time rate of momentum increase in } V = \frac{d}{dt}\int_V \rho\vec{v}\,dV$$
$$\text{net momentum flow rate into } V = -\int_S \rho\vec{v}(\vec{v}\cdot\vec{n}\,)\,dS$$
$$\text{forces acting on } V = \vec{F}.$$

The minus sign on the second term is due to $\vec{n}$ being an outward-pointing normal. The integrand in this term is the *momentum flux*,

$$\rho\vec{v}(\vec{v}\cdot\vec{n}\,) = \text{momentum flux} \quad \left[\frac{\text{momentum}}{\text{time}\cdot\text{area}}\right]. \tag{3.1}$$

Putting these expressions together yields the general *integral form of the conservation of momentum equation*,

$$\frac{d}{dt}\int_V \rho\vec{v}\,dV + \int_S \rho\vec{v}(\vec{v}\cdot\vec{n}\,)\,dS = \vec{F}. \tag{3.2}$$

Note, this is a *vector equation*, and each term has dimensions of force. At a high level, this equation states that net force acting on the fluid in the control volume, $\vec{F}$ on the right-hand side, changes the net momentum of the fluid in the control volume, as measured by the left-hand side.*

*The change in momentum in V is a combination of the fluid in V experiencing a change in momentum and fluid entering/leaving V and bringing/taking its momentum with it.

We can simplify the left-hand side through a few manipulations, by assuming that V does not change in time,

$$\begin{aligned} LHS &= \frac{d}{dt}\int_V \rho\vec{v}\,dV + \int_S \rho\vec{v}(\vec{v}\cdot\vec{n}\,)\,dS \\ &= \int_V \left[\frac{\partial(\rho\vec{v}\,)}{\partial t} + \nabla\cdot(\rho\vec{v}\,\vec{v}\,)\right] dV \\ &= \int_V \left[\left(\frac{\partial\rho}{\partial t} + \nabla\cdot(\rho\vec{v}\,)\right)\vec{v} + \rho\frac{\partial\vec{v}}{\partial t} + \rho\vec{v}\cdot\nabla\vec{v}\right] dV \\ &= \int_V \rho\frac{D\vec{v}}{Dt}\,dV, \end{aligned} \tag{3.3}$$

where we used continuity† and, in the first step, the divergence theorem to convert the surface integral to a volume integral, and where $\nabla\cdot(\rho\vec{v}\,\vec{v}\,) = \partial_i(\rho v_i v_j)$ in index notation. Therefore, we can re-write Equation 3.2 using the material derivative,

† The last step in this sequence of equations uses the differential form of the mass conservation equation, i.e. continuity, which states

$$\frac{\partial\rho}{\partial t} + \nabla\cdot(\rho\vec{v}\,) = 0.$$

$$\int_V \rho\frac{D\vec{v}}{Dt}\,dV = \vec{F} = \vec{F}_{\text{body}} + \vec{F}_{\text{surf}}. \tag{3.4}$$

This equation is the fluid-dynamics equivalent of Newton's second law, which states that mass times acceleration is force. On the left-hand side, the material derivative of the velocity is like the acceleration, and the density combined with the integral is like the mass. Equivalently, the integral can be interpreted as a sum over fluid elements, and $(\rho dV)\frac{D\vec{v}}{Dt}$ is the mass times acceleration of each fluid element.

3.1.2 Forces

We will often separate the net force $\vec{F}$ into surface and body forces,

$$\vec{F} = \vec{F}_{\text{body}} + \vec{F}_{\text{surf}}. \tag{3.5}$$

Body forces act at a distance on the fluid inside the control volume. The net body force is therefore given by a volume integral,

$$\vec{F}_{\text{body}} = \int_V \rho \vec{f}_{\text{body}}(\vec{x})\, dV, \tag{3.6}$$

where $\vec{f}_{\text{body}}(\vec{x})$ is a force per unit mass, i.e. an acceleration. This force could in general depend on position, and this is why we put $\vec{x}$ in the argument.*

*An example of a body force is gravity, in which $\vec{f}_{\text{body}}(\vec{x}) = \vec{g}$. Other body forces, such as those arising from electric fields, can be included as well, although for our purposes, gravity will be the most important. In many flows, we will also even neglect gravity.

Surface forces act on the control volume boundary, S, and they include both pressure and viscous forces. The net surface force is

$$\vec{F}_{\text{surf}} = \int_S \vec{f}_{\text{surf}}(\vec{x}, \vec{n})\, dS, \tag{3.7}$$

where $\vec{f}_{\text{surf}}(\vec{x}, \vec{n})$ is the *stress vector* on the control volume surface. As illustrated in Figure 3.2, $\vec{f}_{\text{surf}}(\vec{x}, \vec{n})$ does not necessarily act along $\vec{n}$. Shear forces can be present, and these introduce a tangential component to the surface force. In addition, the surface force may be caused by the wall of an object, such as a containing vessel, or by fluid outside of the control volume.

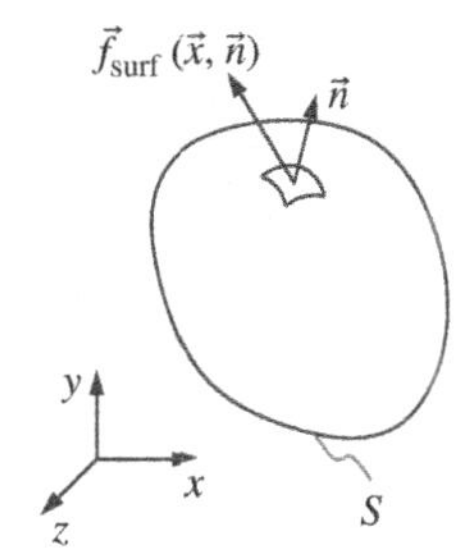

Figure 3.2 Stress vector acting on the control volume surface.

Substituting the expressions for the body and surface forces, the integral form of momentum conservation, Equation 3.4, becomes

$$\int_V \rho \frac{D\vec{v}}{Dt}\, dV = \int_V \rho \vec{f}_{\text{body}}(\vec{x})\, dV + \int_S \vec{f}_{\text{surf}}(\vec{x}, \vec{n})\, dS. \tag{3.8}$$

This equation can be used in the analysis of control volumes to compute the force acting on a fluid or an adjacent object.

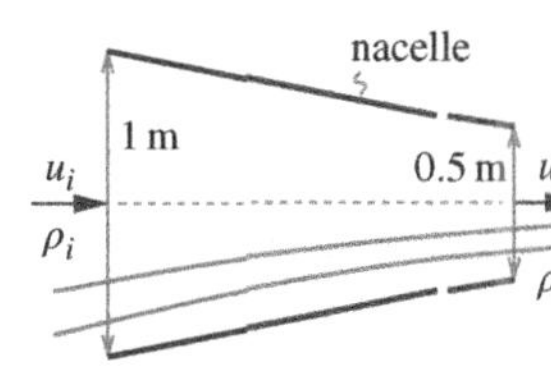

Figure 3.3 An axisymmetric engine in its baseline configuration.

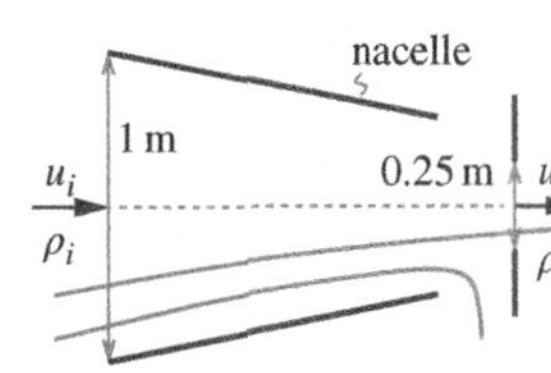

Figure 3.4 An axisymmetric engine with thrust reversers deployed.

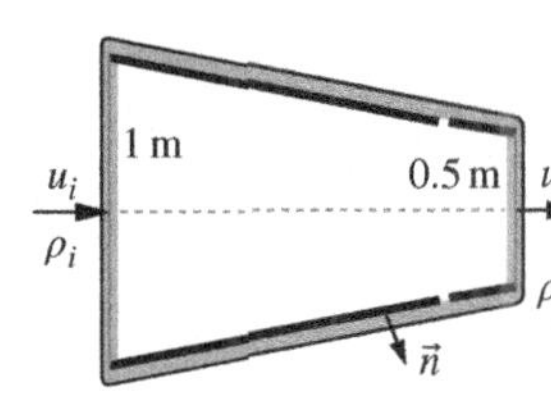

Figure 3.5 Control volume used for the analysis of an axisymmetric engine.

Example 3.1 An axisymmetric jet engine with thrust reversers

Consider an axisymmetric jet engine in two configurations: one baseline, and one with thrust reversers deployed, as illustrated in Figure 3.3 and Figure 3.4. Note that the thrust reversers deflect some of the flow perpendicular to the center axis. We assume that the air at the inlet and exhaust is at atmospheric pressure in both cases, that the flows are steady, and that the exhaust density, ρ_e, is the same for both configurations. The inlet conditions are $\rho_i = 1.23\,\text{kg/m}^3$, $u_i = 60\,\text{m/s}$, and the horizontal exit speed is $u_e = 200\,\text{m/s}$ in both cases. We would like to calculate the thrust produced by the engine in each configuration.

Consider the baseline configuration first, using the control volume shown in Figure 3.5. This control volume is drawn just outside of the nacelle of the engine, so that it includes the nacelle. We apply the integral form of conservation of momentum,

$$\frac{d}{dt}\int_V \rho \vec{v}\, dV + \int_S \rho \vec{v}(\vec{v} \cdot \vec{n})\, dS = \vec{F}, \tag{3.9}$$

where $\vec{F}$ is the force on the fluid in the control volume. This is opposite of the thrust force that we are after. Note, atmospheric pressure has no net effect on this force because the pressure everywhere outside the engine is atmospheric. Since the flow is steady, the only contribution to $\vec{F}$ is from the net flux of momentum through the control volume boundary. The mass flow rate is given from the inlet information,

$$\dot{m} = \rho_i u_i A_i = \left(1.23\frac{\text{kg}}{\text{m}^3}\right)\left(60\frac{\text{m}}{\text{s}}\right)\pi(0.5\,\text{m})^2 = 57.962\frac{\text{kg}}{\text{s}}. \tag{3.10}$$

The horizontal momentum flowing into the control volume is $\dot{m}u_i$, whereas the momentum flowing out is $\dot{m}u_e$. The thrust force, T, is the difference,†

† Note that this method of obtaining the thrust is equivalent to calculating the surface integral term in Equation 3.9, since $\vec{v} \cdot \vec{n}$ is u_e at the exit and $-u_i$ at the inflow. In addition, $\dot{m} = \rho u A$ is the horizontal component of the integral of $\rho\vec{v}$ over either the exit or the inflow.

$$T_{\text{baseline}} = \dot{m}(u_e - u_i) = 8114.8\,\text{N}. \tag{3.11}$$

Next, with the thrust reversers deployed, some of the exhaust flow is deflected radially outward and does not contribute to the axial thrust force. We can follow the same procedure for finding the thrust as in the baseline case, except that we need to use a lower value for the mass flow leaving the control volume in the axial direction. Since the exhaust density and velocity are the same for both configurations, the mass flow is just proportional to the exhaust area. From Figure 3.4, we see that when the thrust reversers are deployed, the area of the axial exhaust is reduced by a factor of 4, since the diameter is cut in half. So the thrust is‡

‡ The negative value indicates that the thrust has indeed reversed with the reversers deployed.

$$T_{\text{reversers}} = \frac{1}{4}\dot{m}u_e - \dot{m}u_i = -579.62\,\text{N}. \tag{3.12}$$

Example 3.2 Stream impinging on a cylinder
An inviscid fluid stream of density $\rho = 1\,\text{kg/m}^3$ impinges on a cylinder and is turned due to the contact, as shown in Figure 3.6. Assume that the flow is steady, that there are no body forces, and that the ambient pressure is constant. Using the integral form of conservation of momentum, we can calculate the force (per unit depth) on the cylinder.

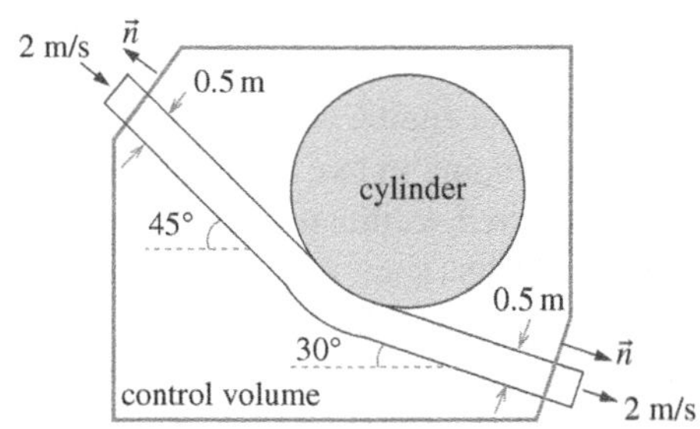

Figure 3.6 A fluid stream impinging on a cylinder in two dimensions.

Consider a control volume around the fluid stream illustrated in Figure 3.6. Since the flow is steady, inviscid, and without body forces, the integral momentum equation in Equation 3.2 reduces to a balance between the net momentum leaving the control volume and the net force on the fluid in the control volume,

$$\underbrace{\oint_S \rho\vec{v}\,\vec{v}\cdot\vec{n}\,dS}_{\text{momentum leaving CV}} = \underbrace{\vec{F'}}_{\text{net force on fluid in CV}}. \tag{3.13}$$

The prime on $\vec{F}'$ indicates that we are looking at a sectional force, per unit depth out of a page, in this two-dimensional problem. This also means that S is the contour representing the perimeter instead of an actual surface. The only contributions to the left-hand side are due to the inflow and outflow of the stream. At the inflow,§

$$\rho\vec{v} = \left(1\frac{\text{kg}}{\text{m}^3}\right)\left(\sqrt{2}\frac{\text{m}}{\text{s}}\hat{\imath} - \sqrt{2}\frac{\text{m}}{\text{s}}\hat{\jmath}\right), \quad \vec{v}\cdot\vec{n} = -2\frac{\text{m}}{\text{s}}. \tag{3.14}$$

Integrating over the inflow,

$$\begin{aligned}\int_{\text{inflow}} \rho\vec{v}\,\vec{v}\cdot\vec{n}dS &= \left(1\frac{\text{kg}}{\text{m}^3}\right)\left(\sqrt{2}\frac{\text{m}}{\text{s}}\hat{\imath} - \sqrt{2}\frac{\text{m}}{\text{s}}\hat{\jmath}\right)\left(-2\frac{\text{m}}{\text{s}}\right)0.5\,\text{m}\\ &= \left(-\sqrt{2}\hat{\imath} + \sqrt{2}\hat{\jmath}\right)\frac{\text{N}}{\text{m}}.\end{aligned} \tag{3.15}$$

At the outflow,¶

$$\rho\vec{v} = \left(1\frac{\text{kg}}{\text{m}^3}\right)\left(1\frac{\text{kg}}{\text{m}^3}\right)\left(\sqrt{3}\frac{\text{m}}{\text{s}}\hat{\imath} - 1\frac{\text{m}}{\text{s}}\hat{\jmath}\right), \quad \vec{v}\cdot\vec{n} = 2\frac{\text{m}}{\text{s}}. \tag{3.16}$$

Integrating over the outflow,

$$\begin{aligned}\int_{\text{outflow}} \rho\vec{v}\,\vec{v}\cdot\vec{n}dS &= \left(\sqrt{3}\frac{\text{m}}{\text{s}}\hat{\imath} - 1\frac{\text{m}}{\text{s}}\hat{\jmath}\right)\left(2\frac{\text{m}}{\text{s}}\right)0.5\,\text{m}\\ &= \left(\sqrt{3}\hat{\imath} - 1\hat{\jmath}\right)\frac{\text{N}}{\text{m}}.\end{aligned} \tag{3.17}$$

Summing the two contributions gives

$$\begin{aligned}\oint_S \rho\vec{v}\,\vec{v}\cdot\vec{n}dS &= \int_{\text{inflow}} \rho\vec{v}\,\vec{v}\cdot\vec{n}dS + \int_{\text{outflow}} \rho\vec{v}\,\vec{v}\cdot\vec{n}dS\\ &= \left((\sqrt{3}-\sqrt{2})\hat{\imath} + (\sqrt{2}-1)\hat{\jmath}\right)\frac{\text{N}}{\text{m}}\\ &\approx \left(.318\hat{\imath} + .414\hat{\jmath}\right)\frac{\text{N}}{\text{m}}.\end{aligned} \tag{3.18}$$

By Equation 3.13, this is the net (sectional) force on the fluid in the control volume. Since, in the control volume, force is exerted on the fluid only by the cylinder, by Newton's third law, this force must be equal in magnitude and opposite in direction to the force on the cylinder. Therefore, the force (per unit depth) on the cylinder is||

$$\vec{F}'_{\text{cylinder}} = -\vec{F}' = -\oint_S \rho\vec{v}\,\vec{v}\cdot\vec{n}dS \approx \left(-.318\hat{\imath} - .414\hat{\jmath}\right)\frac{\text{N}}{\text{m}}. \tag{3.19}$$

§ The inflow velocity is 2 m/s down and to the right, which, resolved into x and y components given the 45° angle, becomes $\sqrt{2}\hat{\imath} - \sqrt{2}\hat{\jmath}$ m/s.

¶ The outflow velocity magnitude is 2 m/s, the same as the inflow, but the angle is 30° below the horizontal. Resolving into x and y components gives $\sqrt{3}\hat{\imath} - \hat{\jmath}$ m/s.

|| This force is downward and to the left, which should be expected given the shape of the water stream. The angle of the impinging flow becomes more shallow after contact with the cylinder, which means that the cylinder exerted a force up and to the right on the fluid. The fluid therefore exerted a force down and to the left on the cylinder.

3.2 Surface Forces

Recall that the surface force $\vec{F}_{\text{surf}}$ is the force exerted on the fluid inside the control volume by the surrounding fluid or objects. By Equation 3.7, this force is a surface integral of the stress vector, $\vec{f}_{\text{surf}}$. This section dives deeper into properties of the stress vector, introduces the stress tensor, and presents a constitutive relationship that relates the stress to the rate of strain.

3.2.1 The Stress Tensor

An important principle involving the stress vector, $\vec{f}_{\text{surf}}$, is *stress equilibrium*, which states that the net surface stress on an infinitesimal volume, V, is zero. If S denotes the surface of this volume, stress equilibrium states that

$$\lim_{S\to 0} \frac{1}{S} \int_S \vec{f}_{\text{surf}}(\vec{x}, \vec{n}\,)\, dS = 0. \tag{3.20}$$

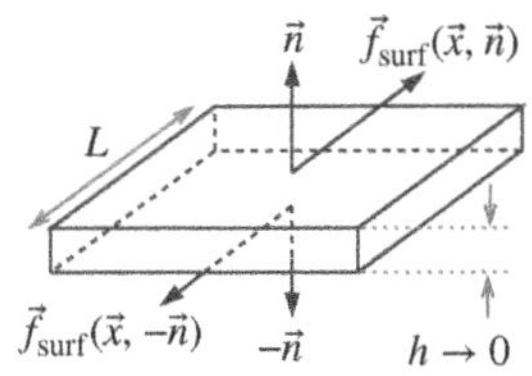

Figure 3.7 Control volume for showing $\vec{f}_{\text{surf}}(\vec{x}, -\vec{n}\,) = -\vec{f}_{\text{surf}}(\vec{x}, \vec{n}\,)$.

This equation follows from the integral form of momentum conservation, Equation 3.8, by letting V, and hence S, go to zero. The volume integrals decrease more quickly than the surface integral because the volume scales as L^3 whereas the surface scales as L^2, where L is some measure of the control volume diameter. So in the limit of a very small control volume, only the surface integral of $\vec{f}_{\text{surf}}$ remains in Equation 3.8, and the result is Equation 3.20.

We can use stress equilibrium to obtain properties of the stress vector. First, consider applying stress equilibrium to a control volume that is thin in one direction, as shown in Figure 3.7. If the height h of the control volume goes to zero faster than the length or width, L, i.e. $h/L \to 0$, then the only contributions to Equation 3.20 will come from the horizontal upper and lower surfaces. Since these are effectively at the same $\vec{x}$ location, the result of stress equilibrium is that

$$\vec{f}_{\text{surf}}(\vec{x}, -\vec{n}\,) = -\vec{f}_{\text{surf}}(\vec{x}, \vec{n}\,). \tag{3.21}$$

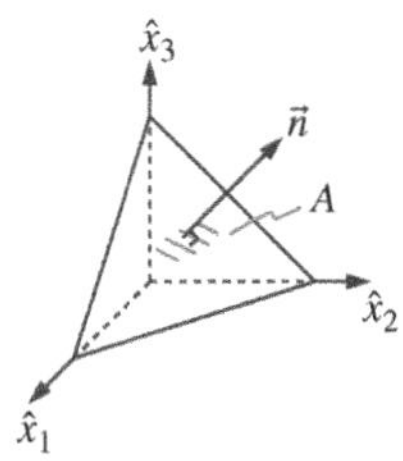

Figure 3.8 Control volume for decomposing the stress vector on a skewed face into principal stress vectors on coordinate-aligned faces.

That is, flipping the normal negates the stress vector.

Next, consider the small tetrahedron control volume shown in Figure 3.8. Applying stress equilibrium allows us to write $\vec{f}_{\text{surf}}(\vec{x}, \vec{n}\,)$ for arbitrary $\vec{n}$ as,

$$\vec{f}_{\text{surf}}(\vec{x}, \vec{n}\,) = \sum_i n_i \vec{f}_{\text{surf}}(\vec{x}, \hat{x}_i) = \vec{n} \cdot \boldsymbol{\sigma}(\vec{x}\,) = n_i \sigma_{ij}(\vec{x}\,), \tag{3.22}$$

where $\hat{x}_i$ is the unit vector in the ith direction. This equation is based on the fact that the projection of the triangular area A onto planes formed by two coordinate axes gives right-triangle areas of $n_i A$. Setting the sum over all faces to zero yields the above expression.

In Equation 3.22, $\boldsymbol{\sigma}(\vec{x}\,)$ is the *stress tensor*. σ_{ij} is the jth component of the $\vec{f}_{\text{surf}}(\vec{x}, \hat{x}_i)$ vector, as illustrated in Figure 3.9. Equation 3.22 is an important result: it says that at a given point, the stress vector on a surface with arbitrary normal, $\vec{n}$, can be computed from a single stress tensor, which has at most nine independent entries.

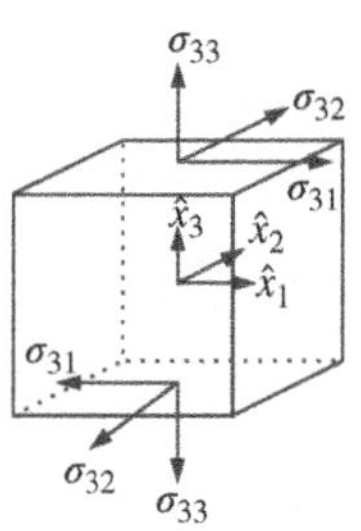

Figure 3.9 Stress vectors on an infinitesimal cube of fluid. σ_{ij} is the stress in the $\hat{x}_j$ direction on a face with normal in the $\hat{x}_i$ direction. Note, for a face with normal in the $-\hat{x}_i$ direction, the stress vectors are negated, since $\vec{f}_{\text{surf}}(\vec{x}, -\vec{n}\,) = -\vec{f}_{\text{surf}}(\vec{x}, \vec{n}\,)$.

For most fluids, including the broad category of Newtonian fluids (Batchelor 2000), the stress tensor $\boldsymbol{\sigma}$ *is symmetric*, so that $\boldsymbol{\sigma}$ has only six independent entries. This symmetry can be proven by conservation of angular momentum. Fluids for which $\boldsymbol{\sigma}$ is not symmetric include some suspensions such as ferrofluids, in which additional torques can exist.

Example 3.3 Stress tensor calculations

Consider the following stress tensor for a two-dimensional flow:

$$\boldsymbol{\sigma} = \begin{bmatrix} -3 & -1 \\ -1 & -1 \end{bmatrix}. \tag{3.23}$$

Note that $\boldsymbol{\sigma}$ is independent of position, $\vec{x}$. We can calculate the stress vector on an edge with an arbitrary normal, e.g. $\vec{n} = 0.6\hat{x}_1 + 0.8\hat{x}_2$, by applying Equation 3.22,

$$\vec{f}_{\text{surf}} = \vec{n} \cdot \boldsymbol{\sigma} = \begin{bmatrix} 0.6 & 0.8 \end{bmatrix} \begin{bmatrix} -3 & -1 \\ -1 & -1 \end{bmatrix}$$
$$= (-3(0.6) - 1(0.8))\,\hat{x}_1 + (-1(0.6) - 1(0.8))\,\hat{x}_2 \qquad (3.24)$$
$$= -2.6\hat{x}_1 - 1.4\hat{x}_2.$$

We can also use the stress tensor to calculate the net (integrated) surface force on a unit square fluid element. Since the stress tensor is independent of position, the net surface force on any fluid element will be zero: the contribution from the right edge will cancel that from the left, and the top will cancel the bottom.*

*Another approach is to use the divergence theorem, Equation 3.26, which tells us that the surface integral of the stress vector is the volume integral of the divergence of the stress tensor. For constant $\boldsymbol{\sigma}$, the divergence $\nabla \cdot \boldsymbol{\sigma} = 0$, so the force is zero.

Substituting the definition of the stress tensor from Equation 3.22 into Equation 3.8, we have

$$\int_V \rho \frac{D\vec{v}}{Dt}\, dV = \int_V \rho \vec{f}_{\text{body}}(\vec{x})\, dV + \int_S \vec{n} \cdot \boldsymbol{\sigma}(\vec{x})\, dS. \qquad (3.25)$$

We can use the divergence theorem to convert the surface integral term into a volume integral,

$$\int_S \vec{n} \cdot \boldsymbol{\sigma}\, dS = \int_V \nabla \cdot \boldsymbol{\sigma}\, dV, \qquad (3.26)$$

where we dropped the dependence on $\vec{x}$ for clarity, and where $\nabla \cdot \boldsymbol{\sigma} = \partial_i \sigma_{ij}$ in index notation. This conversion leaves just volume integrals, and using the fact that the equation must hold for arbitrary control volumes gives the *differential form of momentum conservation*,

$$\rho \frac{D\vec{v}}{Dt} = \rho \vec{f}_{\text{body}} + \nabla \cdot \boldsymbol{\sigma}. \qquad (3.27)$$

Finally, it is common to separate the stress tensor into pressure and viscous components by writing

$$\boldsymbol{\sigma} = \underbrace{-p\boldsymbol{\delta}}_{\text{pressure stress}} + \underbrace{\boldsymbol{\tau}}_{\text{viscous stress}}, \qquad (3.28)$$

where $\boldsymbol{\delta}$ is the identity tensor.† In writing the above equation, the pressure, p, is the thermodynamic pressure of the fluid. The second term, $\boldsymbol{\tau}$, is the *viscous stress tensor*; it is symmetric by virtue of the fact that $\boldsymbol{\sigma}$ is symmetric. Substituting Equation 3.28 into Equation 3.27 yields the *Cauchy momentum equation* (Acheson 1990):

† Dotting the pressure stress with the normal gives $\vec{n} \cdot (-p\boldsymbol{\delta}) = -p\vec{n}$, since $\vec{n} \cdot \boldsymbol{\delta} = \vec{n}$. This result is consistent with the expectation that pressure acts inward on the control volume, along $-\vec{n}$.

$$\rho \frac{D\vec{v}}{Dt} = \rho \vec{f}_{\text{body}} - \nabla p + \nabla \cdot \boldsymbol{\tau}. \qquad (3.29)$$

The physical significance of each of the terms is as follows:

$\frac{D\vec{v}}{Dt}$ = acceleration of an infinitesimal fluid element

$\rho \vec{f}_{\text{body}}$ = body force per unit volume on the fluid element

$-\nabla p$ = pressure force per unit volume on the fluid element

$\nabla \cdot \boldsymbol{\tau}$ = viscous force per unit volume on the fluid element

So Equation 3.29 parallels Newton's second law: mass times acceleration equals force, with the force split into three physical contributions: body, pressure, and viscous. Also, for clarity, in three dimensions, the components of the vector $\nabla \cdot \boldsymbol{\tau}$ are

$$\nabla \cdot \boldsymbol{\tau} = \partial_i \tau_{ij} = \begin{bmatrix} \dfrac{\partial \tau_{xx}}{\partial x} + \dfrac{\partial \tau_{yx}}{\partial y} + \dfrac{\partial \tau_{zx}}{\partial z} \\ \dfrac{\partial \tau_{xy}}{\partial x} + \dfrac{\partial \tau_{yy}}{\partial y} + \dfrac{\partial \tau_{zy}}{\partial z} \\ \dfrac{\partial \tau_{xz}}{\partial x} + \dfrac{\partial \tau_{yz}}{\partial y} + \dfrac{\partial \tau_{zz}}{\partial z} \end{bmatrix}. \tag{3.30}$$

3.2.2 Stress–Strain Relationship for Newtonian Fluids

The Cauchy momentum equation, Equation 3.29, is a general expression of momentum conservation. However, without a model for the viscous stress tensor, $\boldsymbol{\tau}$, the equation cannot be used in a solution process because of the lack of closure: different values of the stress tensor will yield different flows. A model for $\boldsymbol{\tau}$ comes in the form of a constitutive equation between the viscous stress tensor, $\boldsymbol{\tau}$, and the rate-of-strain tensor, $\boldsymbol{\epsilon}$. Since $\boldsymbol{\epsilon}$ is a function of only the velocity gradient, and $\vec{v}$ is already a variable, a closed set of equations can be formed.*

* We focus on the viscous stress tensor because the pressure term in $\boldsymbol{\sigma}$ can usually be computed from an equation of state, whereas thus far we have no model for $\boldsymbol{\tau}$.

A relatively simple relation between $\boldsymbol{\tau}$ and $\boldsymbol{\epsilon}$ exists for *Newtonian fluids*. A Newtonian fluid is one which satisfies the following assumptions (Batchelor 2000):

1. The viscous stress tensor, $\boldsymbol{\tau}$, is at most a linear function of the rate-of-strain tensor, $\boldsymbol{\epsilon}$.
2. The fluid is isotropic, which means that it has no preferred direction.
3. For zero deformation, $\epsilon_{ij} = 0$, the viscous stresses are zero, $\tau_{ij} = 0$.
4. The fluid is homogenous: $\boldsymbol{\tau}$ does not depend explicitly on position.

In our study of aerodynamics, we will assume Newtonian fluids. Another way to state the isotropy assumption is to say that a normal stress does not produce a shearing deformation. This statement implies that the principal directions of $\boldsymbol{\tau}$ must coincide with the principal directions of $\boldsymbol{\epsilon}$, where principal directions are the coordinate axes for which the tensors are diagonal.

Based on assumptions 1 and 4, we can have no more than 36 degrees of freedom in the relation between τ_{ij} and ϵ_{ij}. This is because the six independent components of ϵ_{ij} must map linearly to the six components of τ_{ij}, without a constant term. Note that there are only six independent components in a symmetric 3×3 matrix. However, the isotropy assumption reduces this count to only two degrees of freedom. To see this, we work in the principal direction coordinates, $\hat{x}_1, \hat{x}_2,$ and $\hat{x}_3$, in which both $\boldsymbol{\tau}$ and $\boldsymbol{\epsilon}$ must be diagonal,

$$\boldsymbol{\tau} = \begin{bmatrix} \overline{\tau}_1 & 0 & 0 \\ 0 & \overline{\tau}_2 & 0 \\ 0 & 0 & \overline{\tau}_3 \end{bmatrix}, \quad \boldsymbol{\epsilon} = \begin{bmatrix} \overline{\epsilon}_1 & 0 & 0 \\ 0 & \overline{\epsilon}_2 & 0 \\ 0 & 0 & \overline{\epsilon}_3 \end{bmatrix}.$$

Already, we can have at most nine degrees of freedom, since the mapping between the three principal values must take the form:

$$\overline{\tau}_i = C_{ij} \overline{\epsilon}_j . \tag{3.31}$$

However, by isotropy, permuting the order of the $\bar{\epsilon}_j$ must produce the same permutation in the $\bar{\tau}_i$. This means that

$$C_{11} = C_{22} = C_{33} \equiv \lambda + 2\mu, \tag{3.32}$$

$$C_{12} = C_{21} = C_{13} = C_{31} = C_{23} = C_{32} \equiv \lambda\,. \tag{3.33}$$

The physical significance of μ and λ will be stated shortly – for now, they are just two constants. Therefore, Equation 3.31 becomes

$$\bar{\tau}_i = 2\mu\bar{\epsilon}_i + \lambda\left(\bar{\epsilon}_1 + \bar{\epsilon}_2 + \bar{\epsilon}_3\right)\,. \tag{3.34}$$

Transforming to a general coordinate system and using the fact that τ and ϵ have the same principal axes yield

$$\tau = 2\mu\epsilon + \lambda\delta\nabla\cdot\vec{v}, \tag{3.35}$$

where δ is the identity tensor, with ones on the diagonal and zeros elsewhere. μ is the familiar dynamic viscosity, while λ is the second, or dilational, viscosity. λ is known to be small, but for many fluids, its precise value is still disputed, as it is hard to measure (Gad-el-Hak 1995). Stokes's hypothesis is that $\lambda = -2\mu/3$, which is convenient because for this value, the mean normal viscous stress, τ_{ii}, is zero.† For lack of a better theory, this value is commonly used in practice. In most cases, the exact value is not important. For example, in incompressible or nearly-incompressible flows, $\nabla\cdot\vec{v}\approx 0$, and the contribution of λ to Equation 3.35 drops out.

† Specifically, with this choice, the *bulk viscosity*, $\lambda + \frac{2}{3}\mu$ is zero. The bulk viscosity relates to the dissipation associated with volumetric changes of a fluid element.

Example 3.4 Stress tensor in Newtonian fluids
Could the stress tensor given in Example 3.3 correspond to a Newtonian fluid with the following velocity field?

$$u = -x + 2y, \quad v = y - 4x\,. \tag{3.36}$$

We have to check if it is possible to decompose the stress tensor as

$$\sigma = -p\delta + \tau, \tag{3.37}$$

with $\tau = 2\mu\epsilon + \lambda\delta\nabla\cdot\vec{v}$. The rate-of-strain tensor is‡

$$\epsilon = \begin{bmatrix} \dfrac{\partial u}{\partial x} & \dfrac{1}{2}\left(\dfrac{\partial u}{\partial y} + \dfrac{\partial v}{\partial x}\right) \\ \dfrac{1}{2}\left(\dfrac{\partial u}{\partial y} + \dfrac{\partial v}{\partial x}\right) & \dfrac{\partial v}{\partial y} \end{bmatrix} = \begin{bmatrix} -1 & -1 \\ -1 & 1 \end{bmatrix}. \tag{3.38}$$

‡ The flow is incompressible, $\nabla\cdot\vec{v} = 0$, as seen by direct substitution of the given velocity or by summing the diagonal of the rate-of-strain tensor.

Therefore, we need the following to hold:

$$\begin{bmatrix} -3 & -1 \\ -1 & -1 \end{bmatrix} = \begin{bmatrix} -p & 0 \\ 0 & -p \end{bmatrix} + 2\mu\begin{bmatrix} -1 & -1 \\ -1 & 1 \end{bmatrix}. \tag{3.39}$$

This is possible by setting $2\mu = 1$ and $p = 2$, so the answer is yes, the stress tensor of Example 3.3 corresponds to a Newtonian fluid with the given velocity.

If we were instead given the pressure and the viscosities μ and λ, we could use the stress tensor decomposition to solve for the rate-of-strain tensor. In general, the flow could be compressible, and writing expressions for the three independent entries of the stress tensor gives

$$\sigma_{xx} = -p + 2\mu u_x + \lambda(u_x + v_y), \tag{3.40}$$

$$\sigma_{yy} = -p + 2\mu u_y + \lambda(u_x + v_y), \tag{3.41}$$

$$\sigma_{xy} = 2\mu\epsilon_{xy}. \tag{3.42}$$

This is a set of three equations for three unknowns, u_x, v_y, and ϵ_{xy}, from which the rate-of-strain tensor can be constructed.

3.2.3 Force Integration

When working with the integral form of conservation of momentum, the integral of the surface stress appears as the net surface force on the fluid or an object. Combining Equation 3.7, Equation 3.22, and Equation 3.28, the force on the fluid acting through a surface S is

$$\vec{F}_{\text{surf}} = \int_S \vec{n} \cdot \boldsymbol{\sigma}\, dS = \int_S (-p\vec{n} + \vec{n} \cdot \boldsymbol{\tau})\, dS, \tag{3.43}$$

where $\vec{n}$ is the normal pointing out of the fluid. The pressure stress acts into the fluid along the direction -$\vec{n}$, and it is isotropic: changing the direction of the normal does not affect the magnitude of the stress. On the other hand, the viscous stress $\vec{n} \cdot \boldsymbol{\tau}$ is generally anisotropic and a little harder to calculate. In three dimensions, its components are

$$\vec{n} \cdot \boldsymbol{\tau} = n_i\tau_{ij} = \begin{bmatrix} n_x\tau_{xx} + n_y\tau_{yx} + n_z\tau_{zx} \\ n_x\tau_{xy} + n_y\tau_{yy} + n_z\tau_{zy} \\ n_x\tau_{xz} + n_y\tau_{yz} + n_z\tau_{zz} \end{bmatrix}. \tag{3.44}$$

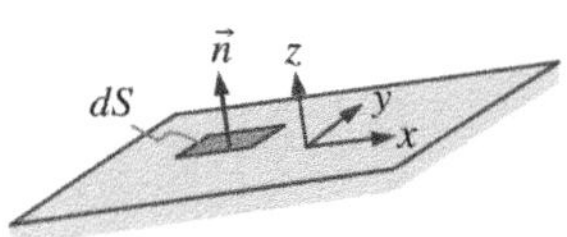

Figure 3.10 Wall-aligned coordinate system for shear stress calculations. $\vec{n}$ points out of the fluid.

This general expression simplifies in certain specific cases. One of these is a solid wall, through which no flow can pass. In a wall-aligned coordinate system, shown in Figure 3.10, the viscous stress on the fluid is

$$\vec{n} \cdot \boldsymbol{\tau} = \tau_{zx}\hat{i} + \tau_{zy}\hat{j} + \tau_{zz}\hat{k}. \tag{3.45}$$

If the fluid is incompressible, Equation 3.35 gives the relevant components of $\boldsymbol{\tau}$,

$$\tau_{zx} = \mu\left(\frac{\partial u}{\partial z} + \frac{\partial w}{\partial x}\right), \tag{3.46}$$

$$\tau_{zy} = \mu\left(\frac{\partial v}{\partial z} + \frac{\partial w}{\partial y}\right), \tag{3.47}$$

$$\tau_{zz} = 2\mu\frac{\partial w}{\partial z}. \tag{3.48}$$

Since $w = 0$ everywhere on the wall,* $\frac{\partial w}{\partial x} = 0$ and $\frac{\partial w}{\partial y} = 0$ on the wall. Also, continuity for incompressible flow requires

$$\nabla \cdot \vec{v} = \frac{\partial u}{\partial x} + \frac{\partial v}{\partial y} + \frac{\partial w}{\partial z} = 0. \tag{3.49}$$

Since at the wall u and v are constant,† $\frac{\partial u}{\partial x} = \frac{\partial v}{\partial y} = 0$, so that $\frac{\partial w}{\partial z} = 0$. Therefore, the viscous stress on the fluid at the wall simplifies to

$$\vec{n} \cdot \boldsymbol{\tau} = \mu\frac{\partial u}{\partial z}\hat{i} + \mu\frac{\partial v}{\partial z}\hat{j} = \mu\frac{\partial \vec{v}}{\partial z}. \tag{3.50}$$

*This is the flow-tangency condition: no flow can cross a solid wall.

† This is the no-slip condition. The velocity is zero if the wall is stationary, or it matches the wall velocity, which we assume has no spatial variation along the wall.

Note that this is a shear stress: it acts tangentially along the surface, with no $\hat{k}$ component. Also, the stress on the wall is the opposite of that on the fluid. We can compute it either by negating the quantity in Equation 3.50 or by flipping the z axis in the calculation.

Example 3.5 Force on an airfoil
In this example, we compute the force coefficients on an airfoil from pressure and shear stress data on the airfoil's surface. We assume that we have a text file that contains information about the location of each sensor and the corresponding pressure and shear stress measurements. Figure 3.11 shows the setup for this problem, and Listing 3.1 shows the first few lines of this text file.

Listing 3.1 Airfoil pressure and shear stress data.

```
% Sensor      x[m]        y[m]      p[Pa]   tau[Pa]
      1     0.9830      0.0008   101966.6    -10.38
      2     0.9638      0.0012   102046.2     -8.24
      3     0.9287      0.0008   102091.5     -7.04
      4     0.8721     -0.0026   102065.6     -7.28
      5     0.8188     -0.0077   101989.8     -7.98
      6     0.7671     -0.0139   101891.0     -8.22
      7     0.7413     -0.0173   101846.4     -6.80
    ...        ...         ...        ...       ...
```

The sensors are ordered clockwise around the airfoil, starting near the trailing edge, which is at $(x, y) = (1\text{m}, 0)$. The convention is that positive shear stress, $\tau_w > 0$, corresponds to shear that is directed clockwise on the airfoil's surface.

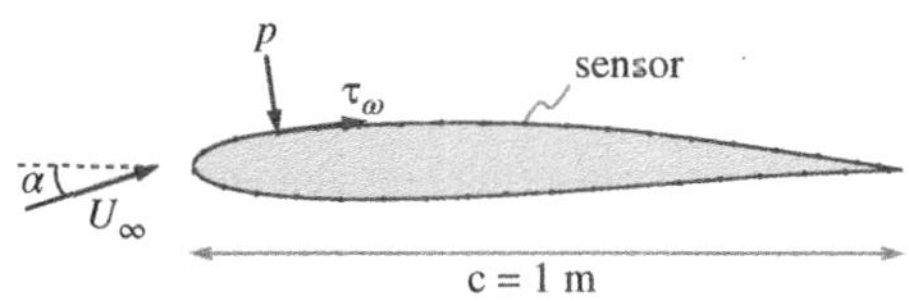

Figure 3.11 Force calculation on an airfoil from pressure and shear stress data.

We write a program that reads in the text file and computes the sectional‡ lift and drag coefficients,

$$c_\ell = \frac{L'}{\frac{1}{2}\rho U_\infty^2 c}, \quad c_d = \frac{D'}{\frac{1}{2}\rho U_\infty^2 c},$$

where the sectional drag, D', and sectional lift, L', are components of the sectional force aligned with the freestream velocity and perpendicular to it, respectively.

‡ Sectional forces, also called two-dimensional forces, are per unit depth out of the page, or, on an aircraft, per unit span. They have dimensions of force/length, and they are typically denoted with a prime symbol.

The total force per unit span on the airfoil is the integral of the pressure and shear stress,

$$\vec{F}' = \int_{\text{airfoil}} \left(p\vec{n} + \tau_w \vec{t}\right) dl, \tag{3.51}$$

where $\vec{n}$ is the normal vector pointing into the airfoil, and $\vec{t}$ is the clockwise tangent vector on the airfoil surface.

To carry out the surface integral numerically, we break the airfoil surface up into N intervals, where N is the number of sensors. The endpoints of the intervals are the midpoints between the sensor coordinates, as shown in Figure 3.12. For the first and last sensors, the trailing edge serves as the other endpoint.

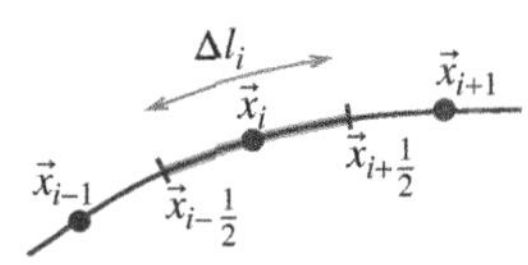

Figure 3.12 An interval around a sensor for force integration.

§ This sum is an approximation of the surface integral. The small length Δl_i replaces the infinitesimal length dl, and the integrand is evaluated in the middle of each finite interval.

Then the force calculation becomes§

$$\vec{F}' \approx \sum_{i=1}^{N} \left(p_i \vec{n}_i + \tau_{w,i} \vec{t}_i\right) \Delta l_i, \tag{3.52}$$

where $\vec{n}_i$ is the normal at sensor i, $\vec{t}_i$ is the tangent at sensor i, and Δl_i is the interval length for sensor i. These are given by

$$\Delta l_i = \sqrt{(x_{i+\frac{1}{2}} - x_{i-\frac{1}{2}})^2 + (y_{i+\frac{1}{2}} - y_{i-\frac{1}{2}})^2}, \tag{3.53}$$

$$\vec{t}_i = \frac{x_{i+\frac{1}{2}} - x_{i-\frac{1}{2}}}{\Delta l_i} \hat{\imath} + \frac{y_{i+\frac{1}{2}} - y_{i-\frac{1}{2}}}{\Delta l_i} \hat{\jmath}, \tag{3.54}$$

$$\vec{n}_i = \frac{y_{i+\frac{1}{2}} - y_{i-\frac{1}{2}}}{\Delta l_i} \hat{\imath} - \frac{x_{i+\frac{1}{2}} - x_{i-\frac{1}{2}}}{\Delta l_i} \hat{\jmath}, \tag{3.55}$$

where the half-indices denote the midpoints between the sensors, as shown in Figure 3.12. Also, p_i and $\tau_{w,i}$ are the pressure and shear stress at sensor i.

Once $\vec{F}'$ is known, the sectional drag and lift are computed as components of $\vec{F}'$,

$$D' = \vec{F}' \cdot (\cos\alpha\hat{\imath} + \sin\alpha\hat{\jmath}), \tag{3.56}$$

$$L' = \vec{F}' \cdot (-\sin\alpha\hat{\imath} + \cos\alpha\hat{\jmath}). \tag{3.57}$$

The coefficients are obtained by non-dimensionalizing the drag and lift forces, as given in Equation 3.51. Listing 3.2 presents a code that reads in the text file and calculates the requested coefficients.

Code notes:

- The dynamic pressure is defined as $q_\infty = \frac{1}{2}\rho U_\infty^2$, and the code uses $\rho = 1.2\,\text{kg/m}^3$, $U_\infty = 60\,\text{m/s}$.
- Variables related to the sensors are stored as vectors, so that the dot (".") is used before operations.
- The code first evaluates forces in the x and y directions, and it then rotates them to obtain the sectional drag and lift.

Listing 3.2 Force calculation from airfoil pressure and shear stress data.

```
function [cl, cd] = fairfoil
% computes lift and drag coefficients from sensor data

% load data
A = load('airfoildata.txt');

c = 1; % chord [m]
qinf = 0.5*1.2*60^2; % dynamic pressure [Pa]
alpha = 5; % angle of attack [deg]

% process sensors
x = A(:,2); y = A(:,3);  % sensor x and y coordinates
xm = 0.5*(x(1:end-1) + x(2:end)); % midpoints: x
ym = 0.5*(y(1:end-1) + y(2:end)); % midpoints: y
xm = [c;xm;c]; ym = [0;ym;0]; % include trailing edge as endpoint
dx = xm(2:end)-xm(1:end-1); % interval delta x
dy = ym(2:end)-ym(1:end-1); % interval delta y
ds = sqrt(dx.^2 + dy.^2);  % interval lengths
t = [dx./ds, dy./ds]; % tangent vectors
n = [dy./ds, -dx./ds]; % normal vectors

% process data
p = A(:,4); % pressure at sensors
tau = A(:,5); % shear stress at sensors
Fx = sum((p.*n(:,1) + tau.*t(:,1)).*ds); % x force
Fy = sum((p.*n(:,2) + tau.*t(:,2)).*ds); % y force
D = Fx*cosd(alpha) + Fy*sind(alpha); % sectional drag force
L = -Fx*sind(alpha) + Fy*cosd(alpha); % sectional lift force
cl = L/(qinf*c);  % lift coefficient
cd = D/(qinf*c);  % drag coefficient
```

For the particular data file used, the results are

$$c_\ell = 1.0129, \quad c_d = 0.013363. \tag{3.58}$$

These are reasonable coefficients for an airfoil: the drag coefficient is generally much lower than the lift coefficient for a nominal operating condition.

3.3 The Navier–Stokes Equations

Substituting the stress–strain model, Equation 3.35, into the Cauchy momentum equation, Equation 3.29 yields the *compressible Navier–Stokes equations* (Stokes 1845; Acheson 1990)

$$\rho \frac{D\vec{v}}{Dt} = \rho \vec{f}_{\text{body}} - \nabla p + \nabla \cdot \left(2\mu\epsilon + \lambda\delta\nabla \cdot \vec{v}\right). \tag{3.59}$$

For incompressible flows, $\nabla \cdot \vec{v} = \partial_i v_i = 0$. If also μ is constant, then

$$\begin{aligned} \nabla \cdot \tau = \mu\nabla \cdot (2\epsilon) &= \mu\partial_i \left(\partial_i v_j + \partial_j v_i\right) \\ &= \mu\partial_i\partial_i v_j + \partial_j \underbrace{\partial_i v_i}_{=0} = \mu\nabla^2\vec{v}, \end{aligned} \tag{3.60}$$

where $\nabla^2\vec{v} = \partial_i\partial_i v_j$ is the *Laplacian* of the velocity.* Therefore, for incompressible and constant viscosity flows, Equation 3.59 reduces to the *incompressible Navier–Stokes equations*

$$\rho \frac{D\vec{v}}{Dt} = \rho \vec{f}_{\text{body}} - \nabla p + \mu\nabla^2\vec{v}. \tag{3.61}$$

If we further neglect viscosity, we obtain the *Euler equations*,

$$\rho \frac{D\vec{v}}{Dt} = \rho \vec{f}_{\text{body}} - \nabla p, \tag{3.62}$$

which are valid for both compressible and incompressible flows. As a final remark, note that all of the above equations are variations of the statement of conservation of momentum. To obtain flow solutions, we also need to use the conservation of mass, and, sometimes the conservation of energy, for example if ρ is not constant or if the fluid properties depend on temperature.

* The Laplacian operator $\nabla^2 = \nabla \cdot \nabla$ is a scalar operator that acts separately on the components of a vector. In three dimensions, the Laplacian of a vector $\vec{v}$ has components

$$\nabla^2\vec{v} = \begin{bmatrix} \frac{\partial^2 u}{\partial x^2} + \frac{\partial^2 u}{\partial y^2} + \frac{\partial^2 u}{\partial z^2} \\ \frac{\partial^2 v}{\partial x^2} + \frac{\partial^2 v}{\partial y^2} + \frac{\partial^2 v}{\partial z^2} \\ \frac{\partial^2 w}{\partial x^2} + \frac{\partial^2 w}{\partial y^2} + \frac{\partial^2 w}{\partial z^2} \end{bmatrix}.$$

Example 3.6 Couette flow
Consider the flow of a fluid between two infinite horizontal walls, where the bottom wall is stationary while the top wall is moving in the $+x$ direction with speed U_{wall},

We make the following assumptions about the flow:

- steady, $\frac{\partial(\cdot)}{\partial t} = 0$
- fully developed, $\frac{\partial \vec{v}}{\partial x} = 0^\dagger$
- incompressible, $\nabla \cdot \vec{v} = 0$
- two dimensional, $\frac{\partial(\cdot)}{\partial z} = 0$
- zero pressure gradient, $\nabla p = 0$
- no body forces, $\vec{f}_{\text{body}} = 0$
- constant viscosity, μ =const.

The boundary conditions are no slip and no flow through the walls,

- at $y = h$: $u = U_{\text{wall}}$, $v = 0$
- at $y = -h$: $u = 0$, $v = 0$.

† Fully developed means that the velocity profile looks the same at all x locations. The term arises from pipe flows, where sufficiently far downstream from the entrance of the pipe, the velocity establishes or develops itself into a constant profile that does not change with distance along the pipe. In this case, there is no entrance to the channel, and the fully developed assumption is justified by the fact that the flat plates are infinite.

We first turn to the continuity equation,

$$\frac{\partial u}{\partial x} + \frac{\partial v}{\partial y} = 0 \quad \Rightarrow \quad \frac{\partial v}{\partial y} = 0 \quad \Rightarrow \quad v = 0.$$

The first step used the fully developed assumption, $\frac{\partial u}{\partial x} = 0$, and the second step used the no flow through condition, $v = 0$ at the wall. This equation tells us that the vertical velocity is everywhere zero in this problem.‡ The acceleration term in the Navier–Stokes equations is then zero,

‡ For low viscosities, the flow could become unstable to small disturbances and eventually turbulent, which would create unsteady fluctuations in all velocity components. Our assumption of steady flow precludes this possibility.

$$\frac{D\vec{v}}{=} \underbrace{\frac{\partial \vec{v}}{\partial t}}_{=0} + u \underbrace{\frac{\partial \vec{v}}{\partial x}x}_{=0} + \underbrace{v}_{=0} \frac{\partial \vec{v}}{\partial y} = 0.$$

Without acceleration, pressure force, or body force, the x-component of the Navier–Stokes equations in Equation 3.61 becomes

$$0 = \mu \frac{\partial^2 u}{\partial y^2}. \tag{3.63}$$

Integrating twice in y and using the boundary conditions on u give

$$u(y) = \frac{U_{\text{wall}}}{2}\left(1 + \frac{y}{h}\right). \tag{3.64}$$

This is a linear velocity profile, as illustrated in Figure 3.13. Suppose that we are interested in the viscous stresses on the bottom and top walls. Recall from Equation 3.22 that the surface stress on the fluid is given by the dot product of the outward pointing normal vector and the stress tensor. Looking at only the viscous stresses, we have

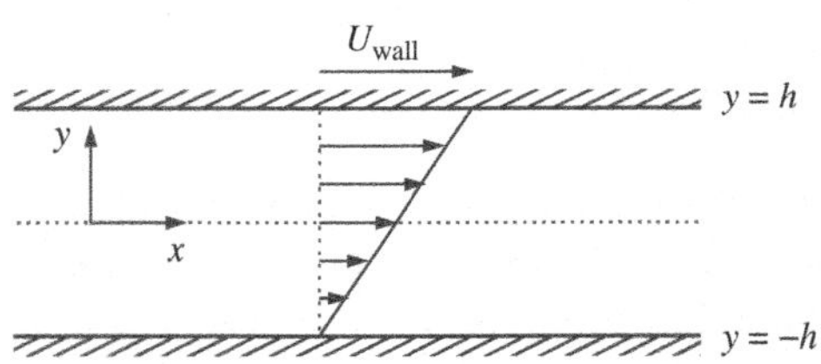

Figure 3.13 Setup for planar Couette flow.

$$\text{surface stress on fluid} = \vec{n} \cdot \tau, \tag{3.65}$$

where the normal points down on the bottom wall and up at the top wall: both are outward from the fluid, as shown in Figure 3.14.

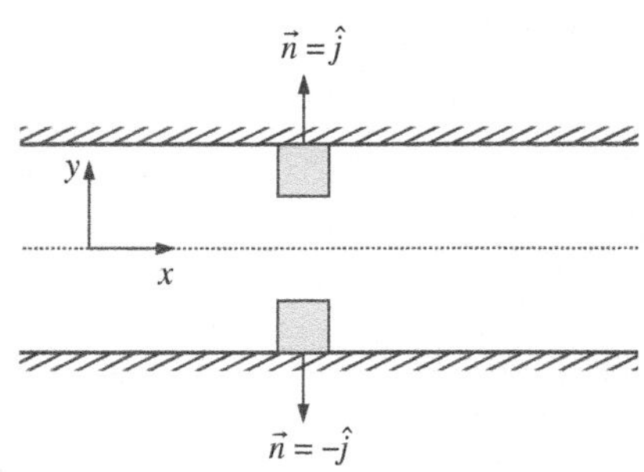

Figure 3.14 Stress calculation in Couette flow.

The viscous stress tensor for this incompressible flow is

$$\tau = 2\mu\epsilon = 2\mu \begin{bmatrix} \frac{\partial u}{\partial x} & \frac{1}{2}\left(\frac{\partial u}{\partial y} + \frac{\partial v}{\partial x}\right) \\ \frac{1}{2}\left(\frac{\partial u}{\partial y} + \frac{\partial v}{\partial x}\right) & \frac{\partial v}{\partial y} \end{bmatrix} = \begin{bmatrix} 0 & \frac{\mu U_{\text{wall}}}{2h} \\ \frac{\mu U_{\text{wall}}}{2h} & 0 \end{bmatrix}. \tag{3.66}$$

On the top wall, $\vec{n} = \hat{j} = [0,1]^T$, so that the surface stress on the fluid is

$$\vec{n} \cdot \tau\Big|_{\text{top}} = \begin{bmatrix} n_x\tau_{xx} + n_y\tau_{yx} \\ n_x\tau_{xy} + n_y\tau_{yy} \end{bmatrix} = \begin{bmatrix} \frac{\mu U_{\text{wall}}}{2h} \\ 0 \end{bmatrix}.$$

This is the shear stress on a fluid element adjacent to the top wall. The stress on the top wall is equal and opposite: in this case, the stress on the top wall is in the $-x$ direction, for positive U_{wall}. On the bottom wall, the normal is in the opposite direction, so the result is that the stress on the bottom wall is in the $+x$ direction.

Example 3.7 Poiseuille flow
Consider steady viscous flow through a circular pipe of radius R, known as Poiseuille flow, as shown in Figure 3.15

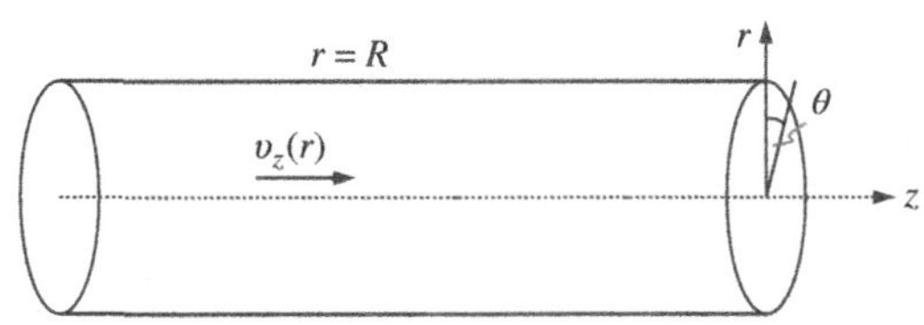

Figure 3.15 Pipe flow setup.

We assume that the flow is incompressible, fully developed, axisymmetric, and free from body forces. We also assume a constant pressure gradient in the axial direction, $\frac{dp}{dz}$, and a constant dynamic viscosity, μ. To solve for the flow in this pipe, we need the axial component of the Navier–Stokes equations in cylindrical coordinates, which is

$$\begin{aligned} &\rho\left(\frac{\partial v_z}{\partial t} + v_r\frac{\partial v_z}{\partial r} + \frac{v_\theta}{r}\frac{\partial v_z}{\partial \theta} + v_z\frac{\partial v_z}{\partial z}\right) = \\ &-\frac{dp}{dz} + \mu\left(\frac{1}{r}\frac{\partial}{\partial r}\left(r\frac{\partial v_z}{\partial r}\right) + \frac{1}{r^2}\frac{\partial^2 v_z}{\partial \theta^2} + \frac{\partial^2 v_z}{\partial z^2}\right). \end{aligned} \tag{3.67}$$

This equation simplifies given our assumptions. First, the flow is steady and fully developed, so derivatives with respect to t and z are zero. Also, because the flow is axisymmetric, derivatives with respect to θ are zero. Finally, by conservation of mass,

$$\nabla \cdot \vec{v} = \frac{1}{r}\frac{\partial(rv_r)}{\partial r} + \frac{1}{r}\frac{\partial v_\theta}{\partial \theta} + \frac{\partial v_z}{\partial z} = 0 \quad \Rightarrow \quad \frac{1}{r}\frac{\partial(rv_r)}{\partial r} = 0. \tag{3.68}$$

The solution to this equation is $v_r = c/r$, but since $v_r(r = R) = 0$, $c = 0$, and hence $v_r = 0$ everywhere.

Therefore, the axial component of the Navier–Stokes equation simplifies to

$$0 = -\frac{dp}{dz} + \mu\frac{1}{r}\frac{\partial}{\partial r}\left(r\frac{\partial v_z}{\partial r}\right). \tag{3.69}$$

Next, we look at the boundary conditions for the axial velocity, $v_z(r)$. Since the fluid is viscous, we need no slip at the pipe walls, which means that $v_z(R) = 0$. At $r = 0$, symmetry of the problem dictates that there must be zero stress, so that $\left.\frac{\partial v_z}{\partial r}\right|_{r=0} = 0$.§

§ If the zero-stress condition were not satisfied, an infinitesimal fluid element at the centerline would experience an infinite acceleration.

We can now integrate Equation 3.69 with respect to r to obtain the axial velocity profile, $v_z(r)$. Integrating once gives

$$\begin{aligned}\int \frac{\partial}{\partial r}\left(r\frac{\partial v_z}{\partial r}\right) dr &= \int \frac{r}{\mu}\frac{dp}{dz}\, dr \\ r\frac{\partial(v_z)}{\partial r} &= \frac{r^2}{2\mu}\frac{dp}{dz} + C_1.\end{aligned} \tag{3.70}$$

Evaluating at $r = 0$ gives $C_1 = 0$. Integrating again,

$$v_z = \frac{r^2}{4\mu}\frac{dp}{dz} + C_2. \tag{3.71}$$

Since $v_z(r = R) = 0$, $C_2 = -\frac{R^2}{4\mu}\frac{dp}{dz}$, so that

$$v_z(r) = -\frac{R^2 - r^2}{4\mu}\frac{dp}{dz}. \tag{3.72}$$

The presence of the minus sign can be reasoned by considering the pressure derivative: if the pressure decreases along the pipe, $\frac{dp}{dz} < 0$, the pressure gradient pushes the fluid to the right and $v_z > 0$.

Knowing the velocity profile, we can determine the volumetric flow rate, Q, through the pipe. Q is obtained by integrating the velocity profile through a cross-section of the pipe,¶

¶ The steps in this integration are as follows:

$$\begin{aligned}Q &= \int_0^{2\pi}\int_0^R v_z(r) r dr d\theta \\ &= -2\pi \int_0^R \frac{R^2 r - r^3}{4\mu}\frac{dp}{dz} dr \\ &= -\frac{2\pi}{4\mu}\frac{dp}{dz}\int_0^R (R^2 r - r^3) dr \\ &= -\frac{2\pi}{4\mu}\frac{dp}{dz}\left[R^4/2 - R^4/4\right] \\ &= -\frac{\pi R^4}{8\mu}\frac{dp}{dz}.\end{aligned}$$

$$Q = \int_0^{2\pi}\int_0^R v_z(r) r dr d\theta = -\frac{\pi R^4}{8\mu}\frac{dp}{dz}.$$

We see that the flow rate is directly proportional to the pressure gradient and the fourth power of the radius, and inversely proportional to the viscosity.

3.4 A Finite-Difference Solver

Solving the Navier–Stokes, or even Euler, equations analytically is not easy except in certain special cases. Numerical techniques such as the finite-difference, finite-volume, or finite-element method are therefore often used. Most of such techniques involve extensive numerical methods background in solvers and mesh generation, and they are beyond the scope of this text. However, for the incompressible Navier–Stokes equations in two dimensions, there is a relatively simple solution approach that involves two variables, the vorticity and the stream function. This is the approach that we present in this section.

3.4.1 Equations

For incompressible flow, the density is assumed to be given and constant, and the unknowns are the velocity and the pressure. Solving for these unknowns requires the same number of equations, which are conservation of mass and momentum.

The former is a scalar equation, whereas the latter is a vector one, so we have enough equations to solve for the vector velocity and scalar pressure.

In two dimensions, the combination of continuity and the incompressible Navier–Stokes equations is*

*We place the pressure gradient term on the left-hand side but leave the viscous force on the right in the momentum equations.

$$\frac{\partial u}{\partial x} + \frac{\partial v}{\partial y} = 0, \tag{3.73}$$

$$\frac{\partial u}{\partial t} + u\frac{\partial u}{\partial x} + v\frac{\partial u}{\partial y} + \frac{1}{\rho}\frac{\partial p}{\partial x} = \nu\nabla^2 u, \tag{3.74}$$

$$\frac{\partial v}{\partial t} + u\frac{\partial v}{\partial x} + v\frac{\partial v}{\partial y} + \frac{1}{\rho}\frac{\partial p}{\partial y} = \nu\nabla^2 v. \tag{3.75}$$

These three equations can be reduced to two by using the following two independent unknowns:†

† Vorticity here refers to the z component of the vorticity vector, i.e. $\vec{\xi} = \xi\hat{k}$, since a two-dimensional flow can only have a z vorticity component.

$$\text{vorticity}: \quad \xi = \frac{\partial v}{\partial x} - \frac{\partial u}{\partial y}, \tag{3.76}$$

$$\text{stream function}: \quad \frac{\partial \psi}{\partial y} = u, \quad \frac{\partial \psi}{\partial x} = -v. \tag{3.77}$$

By construction, defining a stream function enforces continuity, since we automatically obtain $\frac{\partial u}{\partial x} + \frac{\partial v}{\partial y} = 0$. The x and y momentum equations can be combined to eliminate the pressure, resulting in the *vorticity transport equation*,

$$\frac{\partial}{\partial x}(\text{y} - \text{mom}) - \frac{\partial}{\partial y}(\text{x} - \text{mom}) \Rightarrow \frac{\partial \xi}{\partial t} + u\frac{\partial \xi}{\partial x} + v\frac{\partial \xi}{\partial y} = \nu\nabla^2\xi, \tag{3.78}$$

$$\text{where}\nabla^2\xi = \frac{\partial^2 \xi}{\partial x^2} + \frac{\partial^2 \xi}{\partial y^2}. \tag{3.79}$$

Substituting ψ for u and v in the vorticity transport equation and the vorticity definition gives‡

‡ The vorticity definition yields a Poisson equation involving the Laplacian of the stream function.

$$\frac{\partial \xi}{\partial t} + \frac{\partial \psi}{\partial y}\frac{\partial \xi}{\partial x} - \frac{\partial \psi}{\partial x}\frac{\partial \xi}{\partial y} = \nu\nabla^2\xi, \tag{3.80}$$

$$\nabla^2\psi = -\xi. \tag{3.81}$$

Thus, the three equations for u, v, and p have been converted into two equations for ξ and ψ. Note that the first equation has a time derivative, while the second one does not. This difference affects the solution process, as we will soon see.

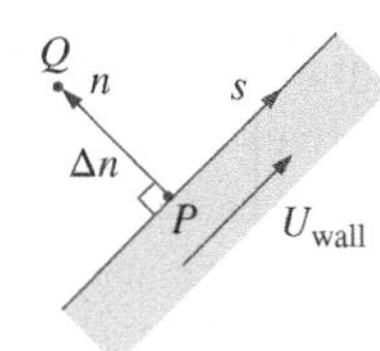

Figure 3.16 Quantities used in enforcing a moving wall boundary condition.

3.4.2 Boundary Conditions

In addition to the governing differential equations, boundary conditions are required to obtain a solution to a particular problem. The boundary conditions of interest here will be stationary and moving walls, the former being a special case of the latter.

On a moving wall, using a wall-aligned coordinate system, s and n, as shown in Figure 3.16, the stream function definition gives, at point P,*

*A subscript on the field variable is shorthand notation for a derivative, e.g. $\psi_n = \frac{\partial \psi}{\partial n}$.

$$\psi_n(P) = \frac{\partial \psi}{\partial n}(P) = U_{\text{wall}} \quad \text{no slip}, \tag{3.82}$$

$$\psi_s(P) = \frac{\partial \psi}{\partial s}(P) = 0 \quad \text{no flow} - \text{through}. \tag{3.83}$$

For a closed domain with no flow-through, the second equation means that ψ is constant on the boundary, and it is simplest to set $\psi(P) = 0$, assuming a simply connected single wall in the domain. Any other constant value of the stream function on the boundary would also work, since a constant offset in Ψ does not affect the velocity. The vorticity on the boundary can be related to the stream function through Equation 3.81

$$\xi(P) = -\nabla^2\psi(P) = -\psi_{nn}(P), \tag{3.84}$$

where we used the fact that ψ is constant along the wall, so that $\psi_{ss}(P) = 0$. A Taylor-series expansion for ψ to a point Q on the interior gives

$$\psi(Q) = \psi(P) + \psi_n(P)\Delta n + \frac{1}{2}\psi_{nn}(P)\Delta n^2 + \mathcal{O}(\Delta n^3),$$
$$\Rightarrow -\psi_{nn}(P) = \frac{2}{\Delta n^2}(\psi(P) - \psi(Q)) + \frac{2}{\Delta n}\psi_n(P) + \mathcal{O}(\Delta n).$$

Since $\psi_n(P) = U_{\text{wall}}$, a first-order accurate boundary condition† for the vorticity is

† Higher-order accurate conditions can be derived, using more data for ψ from the interior, but instabilities can set in at high Reynolds numbers, $Re = U_{\text{wall}}L/\nu$, where L is a characteristic length.

$$\xi(P) = \frac{2}{\Delta n^2}(\psi(P) - \psi(Q)) + \frac{2}{\Delta n}U_{\text{wall}} + \mathcal{O}(\Delta n). \tag{3.85}$$

The sign in front of the U_{wall} term is the sign of $(\vec{U}_{\text{wall}} \times \vec{n}) \cdot \hat{k}$, which is positive as shown in Figure 3.16, since $\hat{k}$ points out of the page.

3.4.3 Finite-Difference Discretization

Given a square geometry of length L, consider a grid of $N \times N$ intervals with uniform spacing, $\Delta x = \Delta y = h \equiv L/N$, and $\xi_{i,j}$, $\psi_{i,j}$ storing the solution at point $(i, j) \in [1 \ldots N]$, with coordinates $(x = ih, y = jh)$. We use a forward-time centered-space (FTCS) discretization* of the parabolic vorticity transport equation, Equation 3.80,

* This means that the time derivative is discretized with a forward difference, while spatial derivatives are discretized with a central difference, according to Equation 1.57.

$$\begin{aligned}\xi_{i,j}^{n+1} = \xi_{i,j}^n - \Delta t\Bigg[&\frac{(\psi_{i,j+1}^n - \psi_{i,j-1}^n)}{2h}\frac{(\xi_{i+1,j}^n - \xi_{i-1,j}^n)}{2h} \\ &-\frac{(\psi_{i+1,j}^n - \psi_{i-1,j}^n)}{2h}\frac{(\xi_{i,j+1}^n - \xi_{i,j-1}^n)}{2h} \\ &-\nu\frac{(\xi_{i+1,j}^n + \xi_{i-1,j}^n + \xi_{i,j+1}^n + \xi_{i,j-1}^n - 4\xi_{i,j}^n)}{h^2}\Bigg],\end{aligned} \tag{3.86}$$

and a centered-space discretization,† applied to the stream-function Poisson equation, Equation 3.81,

† This is Equation 1.60 for both the x and y derivatives summed together.

$$\frac{\psi_{i+1,j}^n + \psi_{i-1,j}^n + \psi_{i,j+1}^n + \psi_{i,j-1}^n - 4\psi_{i,j}^n}{h^2} = -\xi_{i,j}^n. \tag{3.87}$$

3.4.4 Solution Procedure

The vorticity transport and stream-function Poisson equations constitute a coupled nonlinear system. A steady-state or implicit solution would require some form of nonlinear solver, such as Newton–Raphson. Alternatively, with the time derivative included, the system of equations could be marched forward in time explicitly, even to steady state if desired, with a large number of iterations. We take the latter approach, resulting in the following unsteady solution procedure:

1. Initialize the vorticity field to $\xi^0 = 0$.
2. Solve the stream function Poisson equation using ξ^n.
3. Calculate the vorticity on the boundary using ψ on the boundary and the interior.
4. Update the vorticity on interior nodes using FTCS.
5. Increment n and return to step 2.

The time step restriction for forward Euler is

$$\frac{\nu \Delta t}{\Delta x^2} \leq \frac{1}{4}, \qquad \frac{(|u| + |v|)^2 \Delta t}{\nu} \leq 4, \tag{3.88}$$

where u and v can be calculated from the stream function.* This solution procedure will be used to march a problem to steady state. A convergence check on the changes in ξ and ψ can be used as a termination criterion.

*These restrictions come from a stability analysis of a convection-diffusion equation.

Example 3.8 The lid-driven cavity problem
Consider a square domain with static walls on three of its boundaries, and a moving wall on the fourth, top boundary, as shown in Figure 3.17. The square cavity is filled with a fluid of kinematic viscosity ν. No-slip boundary conditions are imposed on all of the walls so that the velocity is zero on all boundaries except the top wall, where $u = U_{\text{wall}}$. Since no flow enters or leaves the domain, $\psi = \text{constant} = 0^{\dagger}$ on the boundary. Figure 3.17 indicates these boundary conditions in terms of ψ and ξ.

† This choice of ψ is arbitrary: any constant will work as long as the value is the same for all boundaries, since the perimeter of the square is a streamline of the flow.

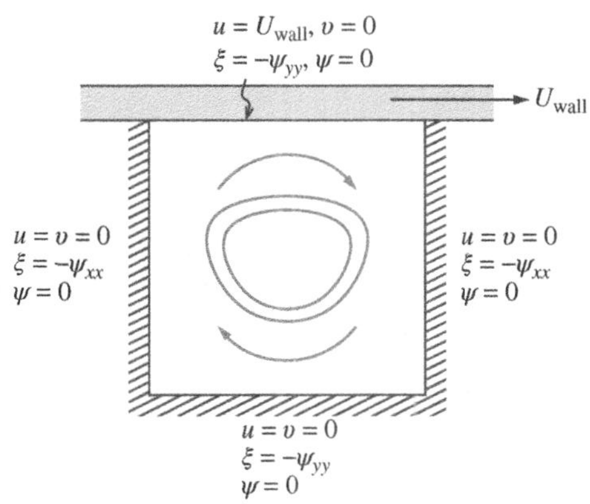

Figure 3.17 Setup of the lid-driven cavity problem.

We expect that due to viscous shear, the moving top wall will impart a force to the right on the fluid, causing it to rotate clockwise in the square cavity domain. Listing 3.3 shows a code that implements the vorticity-stream function formulation of the incompressible Navier–Stokes equations for this problem. The code takes as input the number of intervals in the grid, the Reynolds number based on the square side length, and the target simulation time. The square length, L, and the wall velocity are hard-coded as unity. The goal of the simulation is to achieve steady state, and the longer the simulation time, the closer the solution will be to steady state.

After solving for the stream function, the code plots streamlines as contours of the stream function. Figure 3.18 shows the streamlines, which are contours of the stream function, for the case $Re = 100$ and $N = 32$. These indicate that the flow rotates (clockwise) in response to the moving upper wall. The center of rotation is not at the center of the cavity, and the flow speed is low near the bottom corners, as evidenced by the large spacing of the streamlines there. Note that the streamlines are drawn equally spaced over the minimum-to-maximum range, so the distance between adjacent streamlines indicates the flow speed (large distance means a low speed).

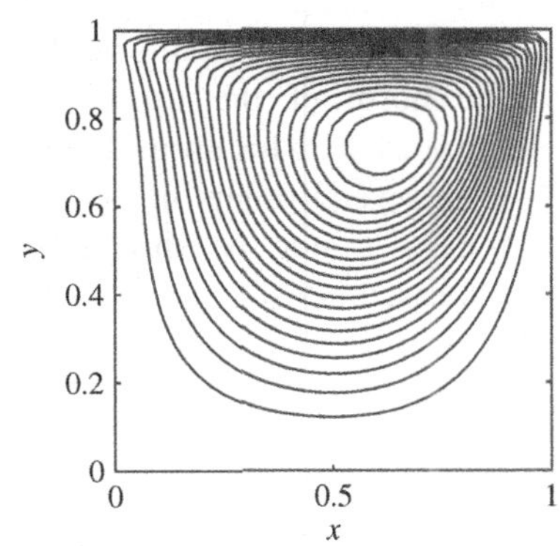

Figure 3.18 Streamlines for $Re = 100$, $N = 32$.

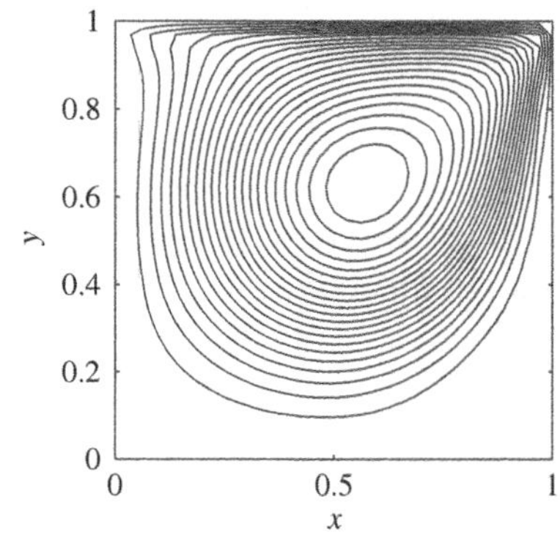

Figure 3.19 Streamlines for $Re = 400$, $N = 32$.

Code notes:

- N intervals means $N + 1$ points in each direction.
- The given Reynolds number sets the viscosity.
- The Poisson equation for the stream function is solved with a direct sparse solver.
- The `A` matrix for the Poisson solver is sparse and built once before the time steps.
- At each Poisson solution, the `Psi` matrix is reshaped into a vector, then back to a matrix after the solution.
- The time step is computed at every solver iteration, since the velocity changes during the iterations.
- Convergence is monitored by the norm of the update to the vorticity. This norm should drop several orders of magnitude – if not, N_t should be increased.

At a higher Reynolds number of 400, Figure 3.19 shows that the location of the center of rotation shifts, and that the region of slowest flow toward the bottom of the cavity decreases in size. We could also plot the vorticity field, although this is a little less interesting, as large values of the vorticity concentrate in the top corners.

Listing 3.3 Finite-difference code for solving the lid-driven cavity problem in a vorticity-stream function formulation.

```
function cavitystream(Re, N, Nt)
% Solves the incompressible Navier-Stokes equations in a
% vorticity-stream function formulation, for Reynolds number
% Re, with NxN intervals and Nt steps.

L = 1;            % domain size
Uwall = 1;        % top wall velocity
nu = Uwall*L/Re;  % kinematic viscosity
[X,Y] = ndgrid(linspace(0,L,N+1));  % mesh
h = L/N; h2 = h*h;   % mesh size
Psi  = zeros(N+1,N+1); % stream function
Xi   = zeros(N+1,N+1); % vorticity
W    = zeros(N+1,N+1); % temporary matrix

% matrix for Poisson solver (5 point stencil)
A = sparse(N-1,N-1);
for ix=1:N-1,
  for iy = 1:N-1,
    k = (iy-1)*(N-1)+ix;
    A(k,k)   = -4/h2;
    if (ix> 1), A(k,k-1) = 1/h2; end;
    if (ix<N-1), A(k,k+1) = 1/h2; end;
    if (iy> 1), A(k,k-(N-1)) = 1/h2; end;
    if (iy<N-1), A(k,k+(N-1)) = 1/h2; end;
  end
end

% start loop over time steps
for it = 1:Nt,

 % Poisson solve for stream function, Psi
 F = -reshape(Xi(2:N, 2:N),(N-1)*(N-1), 1);
 P = A\F;
 Psi(2:N, 2:N) = reshape(P,N-1,N-1);

 % vorticity update on boundary nodes
 Xi(2:N,    1) = -2*Psi(2:N,    2)/(h*h);                  % Bottom
 Xi(2:N,  N+1) = -2*Psi(2:N,    N)/(h*h)-2.0/h*Uwall;     % Top
 Xi(1   , 2:N) = -2*Psi(2   , 2:N)/(h*h);                  % Left
 Xi(N+1 , 2:N) = -2*Psi(N   , 2:N)/(h*h);                  % Right

 % vorticity update on interior nodes, no dt yet
 for i=2:N
   for j=2:N
     W(i,j) = -0.25*((Psi(i,j+1)-Psi(i,j-1))*(Xi(i+1,j)-Xi(i-1,j))-(
     Psi(i+1,j)-Psi(i-1,j))*(Xi(i,j+1)-Xi(i,j-1)))/h2 + nu*(Xi(i+1,j)
     +Xi(i-1,j)+Xi(i,j+1)+Xi(i,j-1)-4.0*Xi(i,j))/h2;
     end;
  end;

 if (mod(it,100)==0),
    fprintf(1, 'it = %4d, W norm = %.5e\n', it, norm(W, 'fro'));
 end

 % max velocity for time step
 U = abs(Psi(:,2:N+1)-Psi(:,1:N))/h;
```

```
V = abs(Psi(2:N+1,:)-Psi(1:N,:))/h;
vmax = max(max(max(U)) + max(max(V)), Uwall);
dt = min(.25*h2/nu, 4.0*nu/vmax^2);

% forward Euler time step
Xi(2:N,2:N) = Xi(2:N,2:N)+dt*W(2:N,2:N);
end

figure(1); clf;
contour(X,Y,Psi,21,'k'), axis('equal');
writeplot('$x$', '$y$', sprintf('../figs/cavity_Re%d_N%d.pdf', Re,N),
    4);
```

3.5 Bernoulli's Equation

For inviscid, incompressible flow, we can obtain a useful relationship between the fluid velocity, pressure, and vertical position (in the presence of gravity). Starting with the Euler equations, Equation 3.62, we put all of the terms on the left-hand side,

$$\frac{D\vec{v}}{Dt} + \frac{1}{\rho}\nabla p - \vec{f}_{\text{body}} = 0. \tag{3.89}$$

Expanding the material derivative, $\frac{D\vec{v}}{Dt} = \frac{\partial \vec{v}}{\partial t} + \vec{v}\cdot\nabla\vec{v}$, using the definition of vorticity, $\vec{\xi} = \nabla\times\vec{v}$, and a vector identity,* we obtain

$$\frac{\partial \vec{v}}{\partial t} + \frac{1}{2}\nabla|\vec{v}|^2 + \frac{1}{\rho}\nabla p - \vec{f}_{\text{body}} = \vec{v}\times\vec{\xi}. \tag{3.90}$$

We now assume:

- Steady flow, $\frac{\partial \vec{v}}{\partial t} = 0$
- Incompressible flow, $\rho = \text{const.} \Rightarrow \frac{1}{\rho}\nabla p = \nabla\left(\frac{p}{\rho}\right)$
- Gravity as the only body force, $\vec{f}_{\text{body}} = -g\hat{k} \Rightarrow \vec{f}_{\text{body}} = -\nabla(gz)$

Therefore, Equation 3.90 becomes

$$\begin{aligned}\frac{1}{2}\nabla|\vec{v}|^2 + \nabla\left(\frac{p}{\rho}\right) + \nabla(gz) &= \vec{v}\times\vec{\xi},\\ \nabla\underbrace{\left[\frac{1}{2}|\vec{v}|^2 + \frac{p}{\rho} + gz\right]}_{H\,=\,\text{total head}} &= \vec{v}\times\vec{\xi}.\end{aligned} \tag{3.91}$$

Consider the change in H† as we move along a path with tangent $\vec{t}$, through a directional derivative,

$$\frac{dH}{ds} = \nabla H\cdot\vec{t}, \tag{3.92}$$

where s is the distance along the path. Using the definition of H in Equation 3.91,

$$\frac{dH}{ds} = \nabla\left[\frac{1}{2}|\vec{v}|^2 + \frac{p}{\rho} + gz\right]\cdot\vec{t} = (\vec{v}\times\vec{\xi})\cdot\vec{t}. \tag{3.93}$$

In particular, if $(\vec{v}\times\vec{\xi})\cdot\vec{t} = 0$, $\frac{dH}{ds} = 0$. This is *Bernoulli's equation,*‡

$$\frac{1}{2}|\vec{v}|^2 + \frac{p}{\rho} + gz = \text{constant when } (\vec{v}\times\vec{\xi})\cdot\vec{t} = 0. \tag{3.94}$$

* The relevant vector identity is $\vec{v}\cdot\nabla\vec{v} = \frac{1}{2}\nabla(\vec{v}\cdot\vec{v}) - \vec{v}\times(\nabla\times\vec{v})$, which can be proven using index notation.

† The total head H is a measure of the energy of the fluid, taking into account kinetic energy through $\frac{1}{2}|\vec{v}|^2$, internal energy through p/ρ, and gravitational potential energy through gz.

‡ Note that $(\vec{v}\times\vec{\xi})\cdot\vec{t} = 0$ when the vector $\vec{t}$ points along any of the following lines:

1. a streamline, $\vec{t}\,\|\,\vec{v}$
2. a vortex line, $\vec{t}\,\|\,\vec{\xi}$
3. a line where $\vec{v} = 0$
4. a line where $\vec{\xi} = 0$
5. a line where $\vec{v}\,\|\,\vec{\xi}$

Out of these cases, the first and fourth will be the most relevant in our study of aerodynamics.

In this equation, $\vec{t}$ is the tangent direction along a path, which means that we can apply Bernoulli's equation either between points in a flow if we connect them with certain line such a streamline, or globally in the flow field if $\vec{v}$ or $\vec{\xi}$ are identically zero. Finally, recall the main assumptions that led to Bernoulli's equation were that the flow is steady, inviscid, and incompressible.

Example 3.9 Hydrostatics, $\vec{v} = 0$
What is the pressure 5 km below the surface of the ocean? If we assume that the water is stationary, $\vec{v} = 0$, then Bernoulli's equation reduces to

$$\frac{p}{\rho} + gz = \text{const.}$$

Using two points, one at the ocean surface, $z = 0$, and one 5 km below the surface, Equation 3.94 reduces to

$$\frac{p_{z=0}}{\rho} + g(0) = \frac{p_{z=-5\,\text{km}}}{\rho} + g(-5000\,\text{m})$$

$$\Rightarrow \quad p_{z=-5\,\text{km}} = p_{z=0} + 5000\rho g.$$

Assuming $\rho = 1000\,\text{kg/m}^3$, and $g = 9.8\,\text{m/s}^2$, we obtain§

$$p_{z=-5\,\text{km}} - p_{z=0} = 4.9 \times 10^7\,\text{Pa} = 484\,\text{atm}.$$

§ This calculation assumes a constant density, but, in reality, ocean density is not constant due to temperature variations.

Example 3.10 Irrotational flow, $\vec{\xi} = 0$
In irrotational flow, H is constant everywhere, because the right-hand side of Equation 3.94 is identically zero. If, for two points 1 and 2, $z_1 \approx z_2$, Bernoulli's equation reduces to

$$\frac{p_1}{\rho} + \frac{1}{2}|\vec{v}_1|^2 = \frac{p_2}{\rho} + \frac{1}{2}|\vec{v}_2|^2. \tag{3.95}$$

If one of the points is in the *farfield*, i.e. away from the body in an external flow, then

$$\frac{p}{\rho} + \frac{1}{2}|\vec{v}|^2 = \frac{p_\infty}{\rho} + \frac{1}{2}|\vec{v}_\infty|^2, \tag{3.96}$$

where p_∞ is the farfield pressure and $\vec{v}_\infty$ is the farfield velocity. Note that the above equation holds at every point in the flow.¶ In Chapter 4, we will calculate the velocity for many irrotational flows. Equation 3.96 will let us calculate the pressure, which we will integrate to obtain forces.

¶ Ignoring z changes, which we do in most aerodynamics calculations when considering an aircraft and the air near it.

Example 3.11 Jet impinging on a flat plate
A two-dimensional jet impinges on a flat plate, which is at an angle θ as shown below.

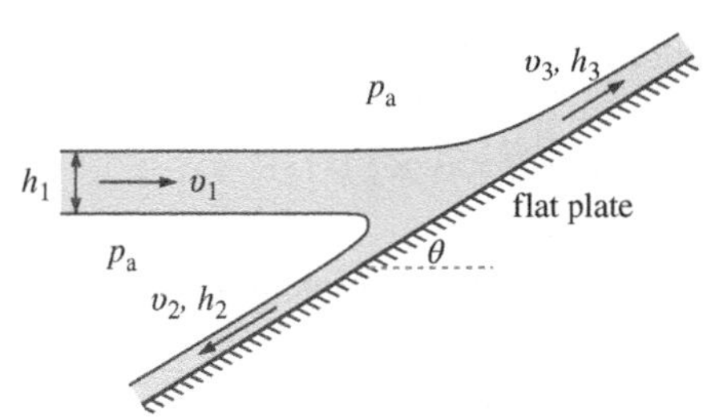

Figure 3.20 Jet impinging on a flat plate.

We assume the flow is inviscid, incompressible, irrotational, steady, and free from body forces. The atmospheric pressure, p_a, is constant, and the fluid density is ρ. Our goal is to determine the velocities and widths of the two resulting streams.

Once the two resulting streams are moving in a straight line, there can be no pressure gradient in the fluid,‖ and so the pressure in the fluid is just the atmospheric pressure. This means that, by Bernoulli's equation, the velocity magnitudes are the same for the two outgoing streams as for the incoming stream,

$$v_2 = v_3 = v_1. \tag{3.97}$$

‖ There can be no pressure changes perpendicular to the flow; otherwise, the streamlines would curve, as will be discussed in Section 3.6. There is also no viscosity and hence no friction along the wall, so no streamwise pressure gradient is required to balance friction losses.

Conservation of mass then tells us that $h_2 + h_3 = h_1$. Conservation of momentum in the direction parallel to the flat plate tells us that, since there are no forces on the fluid parallel to the flat plate, due to lack of friction between the fluid and plate, the sum of the momenta of the resulting streams equals the momentum of the incoming stream,

$$\begin{aligned} (h_1 \rho v_1) v_1 \cos(\theta) &= (h_2 \rho v_2)(-v_2) + (h_3 \rho v_3) v_3, \\ h_1 \cos(\theta) &= -h_2 + h_3. \end{aligned} \tag{3.98}$$

Solving for h_3 and h_2, we have

$$h_3 = h_1 \frac{1 + \cos(\theta)}{2}, \quad h_2 = h_1 \frac{1 - \cos(\theta)}{2}.$$

The flat plate imparts a force on the fluid in a direction perpendicular to the plate. We can calculate this force by considering the conservation of momentum in the perpendicular direction. The incoming stream has a momentum flow (per unit depth out of the page) of $\rho v_1^2 h_1 \sin(\theta)$ in the plate-perpendicular direction (down and right in the diagram), while the resulting two streams have zero momentum in this direction. Therefore, the flat plate must impart a force (per unit depth out of the page) of $\rho v_1^2 h_1 \sin(\theta)$ to the fluid, and this is in the "up and left" direction in Figure 3.20. Specifically, the sectional force on the fluid is

$$\vec{F}'_{\text{on-fluid}} = \rho v_1^2 \sin(\theta) h_1 \left(-\sin(\theta)\hat{\imath} + \cos(\theta)\hat{\jmath}\right). \tag{3.99}$$

Example 3.12 A centrifugal pump

An incompressible, inviscid fluid is pumped to a height h above a large reservoir using a thin spinning L-shaped pipe, as illustrated in Figure 3.21. Our task is to determine the pipe angular speed required to operate the pump.

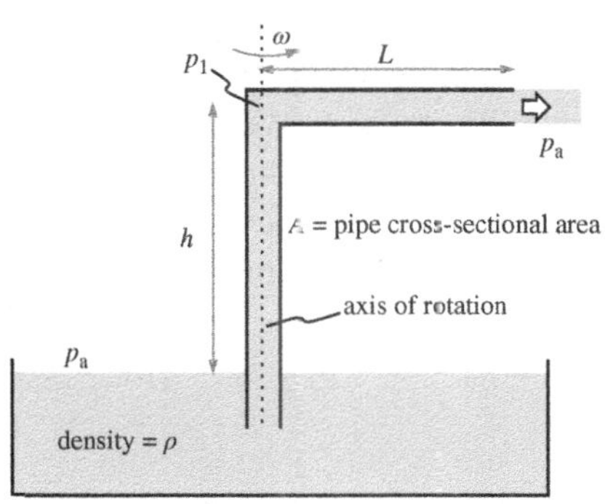

Figure 3.21 A centrifugal pump.

First, we assume that the angular speed is large enough, $\omega = \omega_{crit}$, such that fluid in the pipe remains in place.** Using hydrostatics in the vertical portion of the pipe, we can determine the pressure at the elbow, p_1.

** One can imagine placing a cap on the free end to prevent fluid trickling out, with the cap loose so that it does not support a significant pressure difference.

From hydrostatics, fluid that is deeper has a higher pressure by an amount ρgh, where h is the depth, or height. p_1 is a distance h above the surface of the reservoir, where the pressure is p_a. Therefore,

$$p_1 = p_a - \rho gh. \tag{3.100}$$

Next, still assuming $\omega = \omega_{crit}$, we relate the pressure difference in the horizontal part of the pipe, $p_1 - p_a$, to the angular speed. In the horizontal section, the fluid must be subjected to a centripetal acceleration that keeps the fluid moving in a circular path. This acceleration comes from the pressure gradient.†† To solve for the pressure, we use the Euler equations (since the fluid is inviscid). The horizontal (x) component is

†† Another way of thinking about this from the (non-inertial) frame of the pipe is that the pressure gradient counters a fictitious body force, the centrifugal acceleration.

$$\rho \frac{Du}{Dt} = \rho f_{body} - \frac{\partial p}{\partial x}. \tag{3.101}$$

Since we are still considering the static condition, $\frac{Du}{Dt} = 0$. Also, the centrifugal acceleration comes in through the body force, $f_{body} = \omega^2 x$. Therefore, the momentum equation is

$$0 = \rho\omega^2 x - \frac{\partial p}{\partial x} \implies \frac{\partial p}{\partial x} = \rho\omega^2 x. \tag{3.102}$$

Integrating with respect to x from $x = 0$ to $x = L$ gives

$$p_a - p_1 = \frac{1}{2}\rho\omega^2 L^2. \tag{3.103}$$

Using these results, we can solve for ω_{crit}, which is the critical angular speed required to operate the pump. Equating the expressions for p_1 from Equation 3.100 and Equation 3.103,

$$p_1 = p_a - \rho gh = p_a - \frac{1}{2}\rho\omega^2 L^2. \tag{3.104}$$

Solving for ω gives the desired ω_{crit},‡‡

‡‡ Note that this answer does not depend on how deep the vertical portion of the pipe extends into the fluid reservoir, as long as the height above the reservoir remains h. This is because at any depth in the reservoir, the fluid pressure inside the pipe (in static conditions) is equal to the fluid pressure outside the pump. So adding length to the pipe extending into the reservoir does not change any pressures in the system. Only the length h affects the pressure p_1 in the hydrostatic case.

$$\omega_{crit} = \sqrt{\frac{2gh}{L^2}}. \tag{3.105}$$

If the pump is spinning at a faster rate, $\omega > \omega_{crit}$, fluid will flow up through the pipe and out of the reservoir. We can determine the volumetric flow rate as follows.

First, this is no longer a hydrostatic problem, so our first pressure calculation above does not hold. Instead, we modify the calculation in the vertical portion of the pipe using Bernoulli's equation. We denote by $\bar{v}$ the velocity in the pipe, and we assume that it is constant throughout the pipe cross-section.§§ Connecting via a streamline the elbow of the pipe to the surface of the reservoir, where the velocity is zero, yields the following pressure–velocity relationship:

§§ $\bar{v}$ is constant throughout the length of the pipe since the fluid is incompressible.

$$p_1 + \frac{1}{2}\rho\bar{v}^2 + \rho gh = p_a \quad \Rightarrow \quad p_1 = p_a - \frac{1}{2}\rho\bar{v}^2 - \rho gh. \tag{3.106}$$

In the horizontal section of the pipe, the calculation does not change, because in steady state $\frac{\partial u}{\partial t} = 0$ and $\frac{\partial u}{\partial x} = 0$ ($\bar{v}$ is constant), so that $\frac{Du}{Dt} = 0$. Equating that expression, Equation 3.103, with the expression for p_1 in Equation 3.106 gives

$$p_1 = p_a - \frac{1}{2}\rho\bar{v}^2 - \rho gh = p_a - \frac{1}{2}\rho\omega^2 L^2. \tag{3.107}$$

Solving for $\bar{v}$,

$$\bar{v} = \sqrt{\omega^2 L^2 - 2gh/\rho}. \tag{3.108}$$

The volumetric flow rate is $A\bar{v}$, where A is the cross-sectional area of the pipe,¶¶

$$\text{volumetric flow rate} = A\sqrt{\omega^2 L^2 - 2gh}. \tag{3.109}$$

¶¶ Note that this formula only makes sense as long as $\omega > \omega_{crit}$, since otherwise the quantity inside the square root is negative.

3.5.1 A Pitot-Static Probe

One application of Bernoulli's equation is in the analysis of a pitot-static probe, which is a tube-shaped device used to measure the speed of an aircraft (Anderson 2012). This probe is illustrated in Figure 3.22.

The hole facing the flow measures *stagnation pressure*, which is the sum of the *static pressure*, p, and *dynamic pressure*, $\frac{1}{2}\rho U_\infty^2$. This follows from applying Bernoulli's equation along the *stagnation streamline.** Consider two points on the streamline: one far away from the tube, where $|\vec{v}| = U_\infty$, and one at the *stagnation point*, where $|\vec{v}| = 0$, which is the definition of a stagnation point. Assuming that the stagnation streamline is nearly horizontal, we can safely ignore changes in z. Bernoulli's equation† then states that

$$\frac{1}{2}U_\infty^2 + \frac{p_\infty}{\rho} = \frac{1}{2}0^2 + \frac{p_{stag}}{\rho} \quad \Rightarrow \quad p_{stag} = p_\infty - \frac{1}{2}\rho U_\infty^2. \tag{3.110}$$

* The stagnation streamline divides the flow going around the tube in different directions.

† We assume incompressible flow, so that ρ is constant.

On the other hand, the hole whose face is parallel to the flow, called the static port, measures the static pressure p_∞.‡ This is because if the pressure were higher or lower than p_∞, air would move out of or into the hole.

Knowing both p_{stag} and p_∞ lets us calculate the airspeed. From Equation 3.110,

$$U_\infty = \sqrt{\frac{2(p_{stag} - p_\infty)}{\rho}}. \tag{3.111}$$

‡ An assumption here is that the pitot tube is small in cross-section, so that the air going around it is moving at approximately U_∞ at the location of the static port.

The airspeed computed this way is known as the *true airspeed (TAS)*. To use this equation, we need to know the density, ρ, which is not always available. If we instead use sea-level density, ρ_0, assumed a known reference quantity, we obtain the *equivalent airspeed (EAS)*,

$$U_\infty^{EAS} = \sqrt{\frac{2(p_{stag} - p_\infty)}{\rho_0}}. \tag{3.112}$$

At altitude, the equivalent airspeed will generally be less than the true airspeed, since atmospheric density decreases as altitude increases.

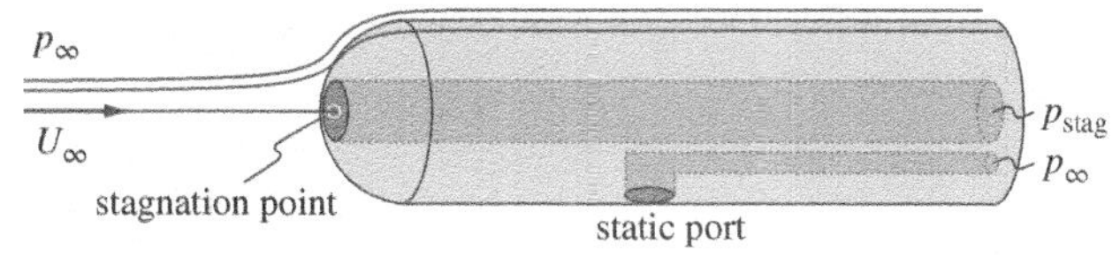

Figure 3.22 A pitot-static probe for measuring airspeed.

Example 3.13 Pitot-static probe calculations

A pitot-static tube on an aircraft measures a stagnation pressure of 82,300 Pa and a static pressure of $p_\infty = 79{,}507$ Pa. A temperature probe measures the outside air temperature at 280 K. What are the true and equivalent airspeeds?

Since we know the ambient temperature and static pressure, we can obtain the ambient density,

$$\rho = \frac{p}{RT} = \frac{79{,}507\ \text{Pa}}{[287\text{J}/(\text{kg K})]280\text{K}} = 0.9894\frac{\text{kg}}{\text{m}^3}. \tag{3.113}$$

Using the measured stagnation pressure, the true airspeed is then calculated as

$$U_\infty = \sqrt{\frac{2(82{,}300\ \text{Pa} - 79{,}507\ \text{Pa})}{0.9894\ \text{kg/m}^3}} = 75.1\frac{\text{m}}{\text{s}}. \tag{3.114}$$

The equivalent airspeed is obtained by using the sea-level density of $\rho_0 = 1.225\ \text{kg/m}^3$,

$$U_\infty^{EAS} = \sqrt{\frac{2(82{,}300\ \text{Pa} - 79{,}507\ \text{Pa})}{1.225\ \text{kg/m}^3}} = 67.5\frac{\text{m}}{\text{s}}. \tag{3.115}$$

3.6 Streamline Curvature

Bernoulli's equation is a powerful tool for computing speed given pressure or vice versa. Its assumptions are not overly restrictive, and corrections can be included for compressibility. Missing from the equation, however, is an accounting for the direction of the velocity. A fluid element whose speed is constant but whose velocity changes direction experiences an acceleration perpendicular to the velocity vector. Bernoulli's equation, which uses the velocity magnitude, does not provide information about this acceleration. Indeed, the pressure at the fluid element remains constant if the speed does not change.*

*This is by Equation 3.95, assuming negligible changes in z and irrotational flow.

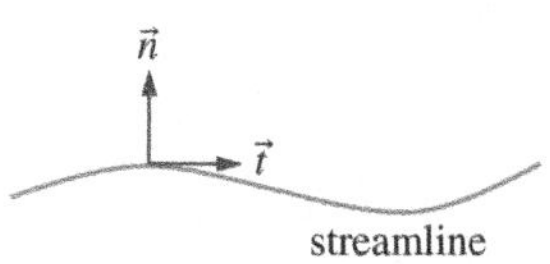

Figure 3.23 A streamline-aligned coordinate system.

In certain cases, such as flow over a lifting airfoil, it is instructive to relate the force on a fluid element to the deflection, or curvature, of its streamline. Consider a flow that is steady, inviscid, and free from body forces. The Euler equation, Equation 3.62, reduces to

$$\vec{v} \cdot \nabla\vec{v} = -\frac{\nabla p}{\rho}. \tag{3.116}$$

This equation states that changes in the flow velocity are related to changes in the pressure. To analyze this relationship, we adopt a *streamline-aligned coordinate system*, as illustrated in Figure 3.23. Note that the direction vectors $\vec{t}$ and $\vec{n}$ are not necessarily constant: if the streamline curves, $\vec{t}$ and $\vec{n}$ will change. In this coordinate system, the velocity is

$$\vec{v} = V\vec{t},$$

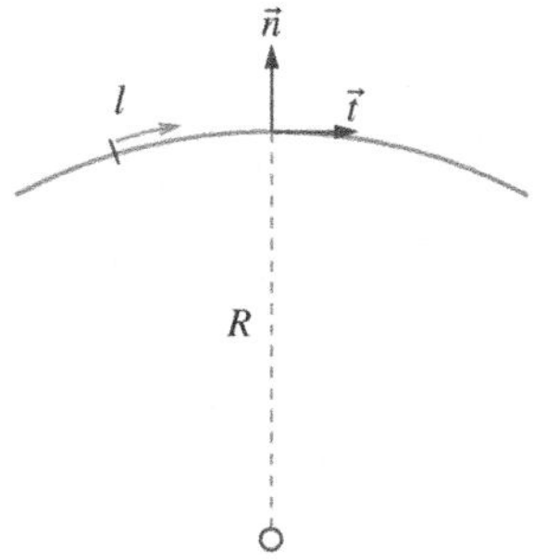

Figure 3.24 A curving streamline: $\vec{t}$ and $\vec{n}$ are the tangent and normal vectors, R is the radius of curvature, and l is the distance along the streamline.

where V is the velocity magnitude, and both V and $\vec{t}$ can change along a streamline. Consider now a curving streamline, as shown in Figure 3.24. We evaluate the left-hand side of Equation 3.116 by considering the rate of change in the l direction, along the streamline,

$$\vec{v}\cdot\nabla\vec{v} = (V\vec{t}\,)\cdot\nabla(V\vec{t}\,)$$
$$= V(\vec{t}\cdot\nabla)(V\vec{t}\,)$$
$$= V\frac{\partial}{\partial l}(V\vec{t}\,) \tag{3.117}$$
$$= V\frac{\partial V}{\partial l}\vec{t} + V^2\frac{\partial \vec{t}}{\partial l},$$

where we used the definition of the directional derivative to write $\vec{t}\cdot\nabla = \frac{\partial}{\partial l}$, and the product rule on $V\vec{t}$ in the last step. Since $\vec{t}$ is a unit vector, its magnitude cannot change. Thus, $\frac{\partial \vec{t}}{\partial l}$ must be purely in the direction perpendicular to $\vec{t}$, which is $\vec{n}$. From Figure 3.25, assuming a small angle change $\Delta\alpha$ measured in degrees, we have

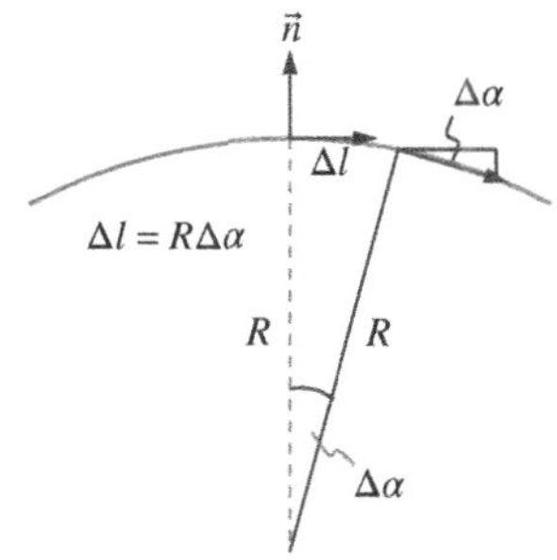

Figure 3.25 Calculation of $\frac{\partial \vec{t}}{\partial l}$.

$$\Delta\vec{t} = \Delta\alpha(-\vec{n}\,) = -\frac{\Delta l}{R}\vec{n} \quad\Rightarrow\quad \frac{\partial \vec{t}}{\partial l} = -\frac{\vec{n}}{R}. \tag{3.118}$$

Using this result in Equation 3.117, and combining with Equation 3.116, we have

$$\vec{v}\cdot\nabla\vec{v} = V\frac{\partial V}{\partial l}\vec{t} - \frac{V^2}{R}\vec{n} = -\frac{\nabla p}{\rho} = -\frac{1}{\rho}\left(\frac{\partial p}{\partial l}\vec{t} + \frac{\partial p}{\partial n}\vec{n}\right). \tag{3.119}$$

This is a vector equation, written in the streamline-aligned coordinate system. Setting the components equal,

$$\vec{t}\text{ direction:}\quad V\frac{\partial V}{\partial l} = -\frac{1}{\rho}\frac{\partial p}{\partial l},$$
$$\vec{n}\text{ direction:}\quad \frac{V^2}{R} = \frac{1}{\rho}\frac{\partial p}{\partial n}. \tag{3.120}$$

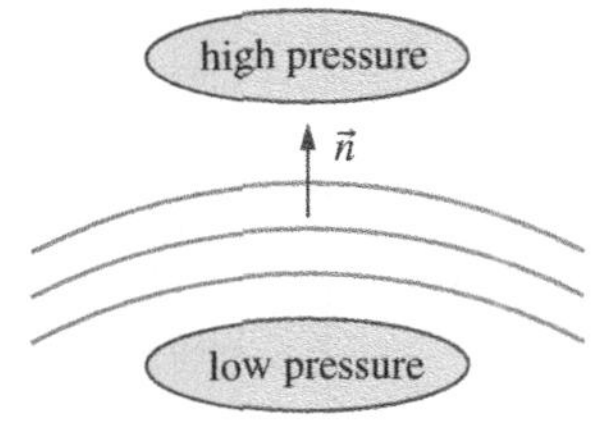

Figure 3.26 Pressure increases in the normal direction for a set of curving streamlines, where the normal vector points away from the center of curvature.

For constant ρ, the $\vec{t}$ equation can be rewritten as

$$\frac{\partial}{\partial l}\left[\frac{1}{2}V^2 + \frac{p}{\rho}\right] = 0, \tag{3.121}$$

which is a re-statement of Bernoulli's equation along a streamline. On the other hand, the $\vec{n}$ equation is new. It states

$$\frac{\partial p}{\partial n} = \frac{\rho V^2}{R}. \tag{3.122}$$

This equation relates pressure changes *normal* to the streamlines when the streamlines are curved. As illustrated in Figure 3.25, the pressure increases in the $\vec{n}$ direction, where $\vec{n}$ points away from the center of curvature. Another way to remember the sign is to think about the force required to make an element on the streamline follow a curved path. The only forces present under our assumption are pressure forces. So, if the a streamline is turning down, as shown in Figure 3.26, there must be higher pressure above the streamline than below it.

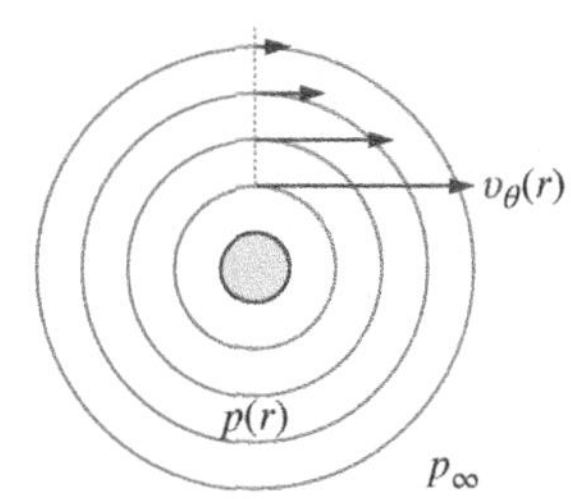

Figure 3.27 Vortex streamlines and pressures.

Example 3.14 Pressure in a vortex

What is the pressure inside a vortex, illustrated in Figure 3.27? We will see in the next chapter that the velocity field in an irrotational vortex is given by

$$v_\theta = -\frac{\Gamma}{2\pi r}, \qquad v_r = 0. \tag{3.123}$$

where Γ is a measure of the vortex strength. This velocity field is representative of physical vortices such as those that occur in aircraft wakes.† Assuming that

† The exception is at $r \to 0$, for which Equation 3.123 gives an infinite speed: a real vortex has a core region over which the flow velocity goes to zero at the center.

the ambient pressure far away from the vortex is p_∞ and ignoring effects of gravity, we can calculate the pressure inside the vortex in two ways.

First, since this vortex flow is irrotational, we can relate the total pressure head at any two arbitrary points using Bernoulli's equation. We pick the first point at a location r away from the vortex, and the second point in the farfield ($r \to \infty$). Bernoulli's equation states that

$$p(r) + \frac{1}{2}\rho(v_\theta(r))^2 = p_\infty + \frac{1}{2}\rho\, 0^2$$
$$p(r) = p_\infty - \frac{1}{2}\rho(v_\theta(r))^2 \tag{3.124}$$
$$= p_\infty - \frac{\rho\Gamma^2}{8\pi^2 r^2}.$$

Second, we use the streamline curvature equation to determine the pressure gradient, $\frac{\partial p}{\partial r}$, and integrate from ∞ to r. The streamline curvature equation states that,

$$\frac{dp}{dr} = \rho\frac{|\vec{v}|^2}{r}.$$

Substituting for $\vec{v}$ and integrating yield

$$\frac{dp}{dr} = \rho\frac{\Gamma^2}{4\pi^2 r^3},$$
$$dp = \rho\frac{\Gamma^2}{4\pi^2 r^3}dr,$$
$$\int_{p_\infty}^{p(r)} dp = \int_{\infty}^{r} \rho\frac{\Gamma^2}{4\pi^2 r'^3}dr',$$
$$p(r) - p_\infty = \rho\frac{\Gamma^2}{4\pi^2}\left[-\frac{1}{2r^2}\right]_\infty^r$$
$$p(r) = p_\infty - \rho\frac{\Gamma^2}{8\pi^2 r^2}.$$

Note that the results from the two approaches agree.

Example 3.15 Streamline curvature applied to an airfoil

This example applies streamline curvature to the qualitative analysis of flow over an airfoil. Given the streamlines in Figure 3.28, and assuming steady, inviscid, irrotational, incompressible flow, is the velocity magnitude higher over the upper or the lower surface of the airfoil?

Figure 3.28 Streamlines over an airfoil, drawn for flow moving from left to right over the airfoil.

Deducing the velocity magnitude from streamlines is not always possible.‡ Turning to pressure instead, we can apply streamline curvature to determine how the pressure above and below the airfoil compares to the freestream pressure, p_∞.

The streamlines are concave down above and below the airfoil, which indicates that the pressure gradient is in the upward direction both above and below the airfoil, as shown in Figure 3.29. This is due to streamline curvature, which states that pressure increases away from the center of curvature.

‡ However, if the streamlines are drawn equally spaced in the freestream, so at equal intervals of the stream function, then we can deduce velocity magnitude by looking at the distance between the streamlines: streamlines close together mean faster flow by continuity.

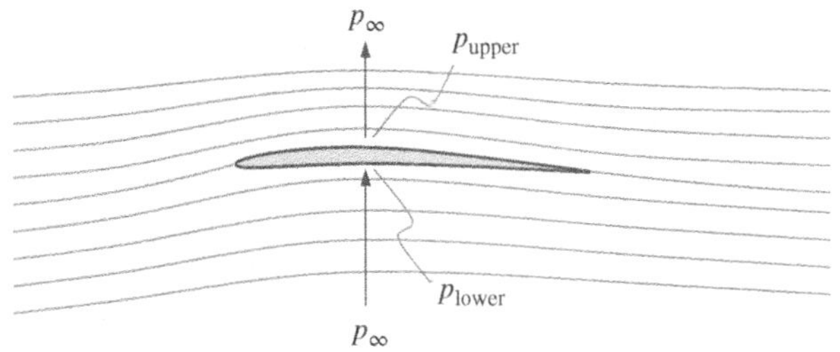

Figure 3.29 Pressure gradients across streamlines. The arrows indicate the pressure gradient, which is the direction in which the pressure is increasing.

The pressure approaches atmospheric far above and far below the airfoil. Looking below the airfoil first, we start from atmospheric pressure far below and see that the pressure must increase to p_{lower} just below the airfoil. This means that $p_{\text{lower}} > p_\infty$. Next, above the airfoil, by the direction of the pressure gradient, the pressure increases from p_{upper} just above the airfoil to p_∞ far above it. This means that $p_{\text{upper}} < p_\infty$. In summary,

$$p_{\text{upper}} < p_\infty < p_{\text{lower}}. \tag{3.125}$$

This pressure difference creates an upward, lift force on the airfoil. To compare the velocity magnitudes, we can apply Bernoulli's equation to this pressure result and conclude that the velocity magnitude is higher on the upper surface than on the lower surface.§

§ These pressure and velocity statements are qualitative observations for a generic point in the middle of the thin, cambered airfoil shown in the figures. In reality, the pressure and velocity vary across both upper and lower surfaces.

3.7 Summary and Concepts

This chapter derived the equations governing the motion of a fluid by relating forces acting on a fluid to its acceleration. The physical principle behind these equations was conservation of momentum, applied first in an integral form but then converted to the differential form using continuity and the divergence theorem. The force acting on the fluid was separated into body and surface contributions, and the latter gave rise to the stress tensor, which was related to both pressure and rate of strain in the fluid. Various forms of the governing equation were presented, using different assumptions about compressibility, vorticity, and the role of viscosity. These equations included the well-known Navier–Stokes, Euler, and Bernoulli equations, and they all derived from the same principle of momentum conservation.

The key concepts covered in this chapter are as follows:

- The integral form of momentum conservation states that forces acting on the fluid in a control volume lead to a net acceleration of the fluid inside the volume or an outflow of momentum through the surface.
- In differential form, momentum conservation states that the forces acting on a fluid element lead to its acceleration.
- The volumetric forces on a fluid can be caused by pressure gradients, body forces, and variations in the stress tensor.
- The stress tensor includes pressure and viscous contributions.
- In a Newtonian fluid, the viscous stress tensor depends linearly on the rate-of-strain tensor, through the dynamic and bulk viscosities.
- The compressible Navier–Stokes equations are the most general form of differential conservation of momentum for a Newtonian fluid.
- For an incompressible flow, the viscous stress tensor in the Navier–Stokes equations simplifies to the product of the dynamic viscosity and the Laplacian of the velocity.
- The Navier–Stokes equations can be solved numerically using a finite-difference method applied to a grid of the domain.
- In the absence of viscosity, the Navier–Stokes equations become the Euler equations.
- Under additional assumptions, such as irrotationality, the Euler equations yield Bernoulli's equation, which can be used to relate pressure and speed in a flow.
- The component of the Euler equations perpendicular to a streamline yields the streamline curvature equation, which dictates how transverse pressure gradients deflect streamlines.
- An object in a flow, such as airfoil, can produce a lift force by deflecting the incoming flow: a force equal in magnitude but opposite in direction to that required to deflect the flow acts on the object.
- Streamline curvature helps explain the pressure differences on an object that deflects the flow.

3.8 Problems

1. Justify each answer to the below questions with an explanation.

a In Couette flow, what is the direction of the viscous force, per unit volume, on a fluid element?

b True or false: the viscous force on an element is zero if and only if $\tau = 0$.

c When the radius of a pipe doubles, how does the maximum speed in a pressure-driven pipe flow change? Assume the pressure gradient remains constant.

d True or false: in a steady flow, the momentum leaving a control volume must balance the momentum entering it.

e The two inviscid flows in Figure 3.30 are identical upstream of the ramps. Which pressure is greater, p_1 or p_2?

f True or false: in a flow with constant velocity, the viscous stress tensor is zero.

g True or false: Bernoulli's equation can be used to relate pressure to velocity in Poiseuille flow.

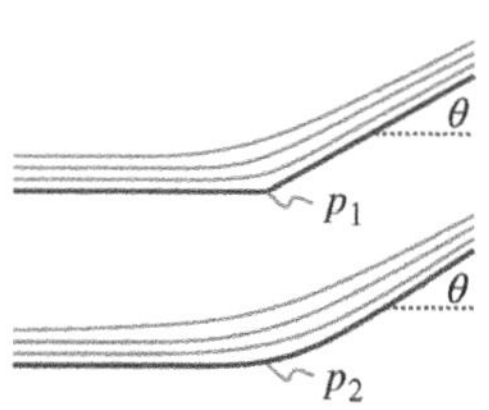

Figure 3.30 Two ramps.

2. Consider a unit right triangle control volume in a two-dimensional flow field, shown in Figure 3.31, with the given stress tensor:

$$\sigma = \begin{bmatrix} -x^2 & \frac{xy}{2} \\ \frac{xy}{2} & -xy \end{bmatrix}.$$

Determine the stress vector, $\vec{f}_{\text{surf}}$, on each edge of the triangle, and calculate the force on the triangle by (1) integrating the stress vector over its perimeter and (2) by integrating $\nabla \cdot \boldsymbol{\sigma}$ over the interior of the triangle. Verify that you obtain the same answer using both approaches.

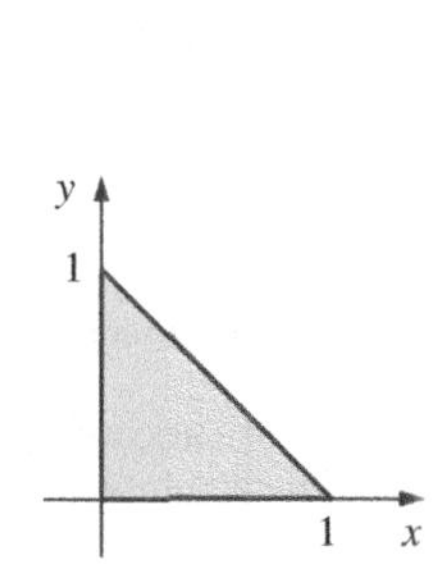

Figure 3.31 Unit-triangle control volume.

3. An incompressible fluid of density ρ flows steadily through a two-dimensional nozzle, shown in Figure 3.32, at a flow rate of Q (area/time). Assume irrotational, quasi-1D flow with no body forces. Use Bernoulli's equation to compute the pressure at the start of the nozzle, and calculate the force on the nozzle by applying the integral form of conservation of momentum. Note that the pressure everywhere outside the nozzle is p_∞.

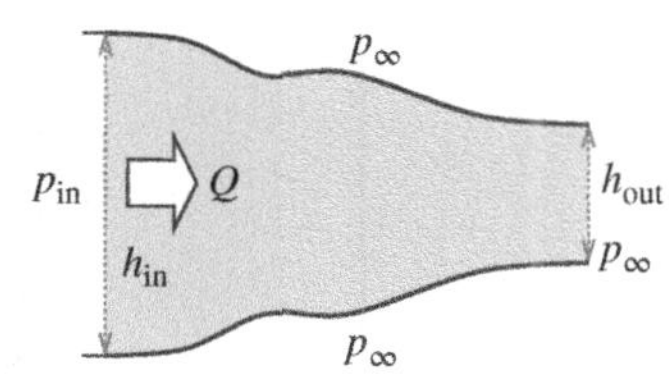

Figure 3.32 Flow through a 2D nozzle.

4. An incompressible fluid of density ρ and viscosity μ flows down a plane inclined at an angle α, as shown in Figure 3.33. Assume constant gravitational acceleration downward, fully developed flow, constant pressure in the air outside the fluid, and zero shear stress exerted by the air on the fluid. Solve for the velocity profile, $u(y)$, and compute the mass flow rate, per unit depth out of the page, of the water down the incline.

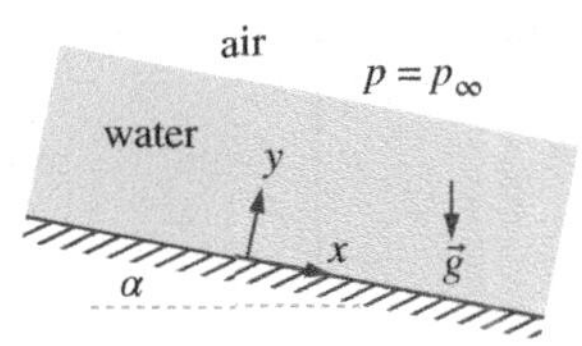

Figure 3.33 Gravity-driven flow down an incline.

5. A two-dimensional fluid stream of density ρ is split in half by a curved plate, and the two resulting streams come off at different angles, as shown in Figure 3.34. Assume the fluid is steady, incompressible, and that the velocity is uniform across the cross-section of each stream. Calculate the force, per unit depth, on the curved plate due to the interaction with the fluid streams.

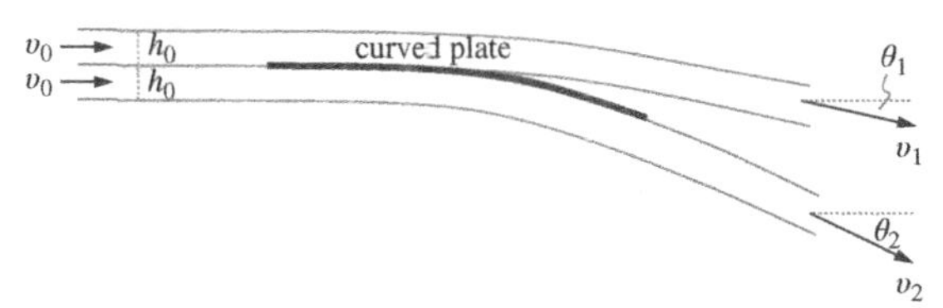

Figure 3.34 A two-dimensional fluid stream split by a curved flat plate.

6. A thin, steady stream of width h of inviscid, incompressible fluid of density ρ impinges on a block and makes a 90° turn of radius $R \gg h$, as shown in Figure 3.35. Assume a constant ambient pressure and no body forces. Use the integral form of conservation of momentum to determine the force on the block. Then apply the streamline curvature equation to determine the pressure on the surface of the block, and integrate this pressure to determine the force. Verify that your force calculations agree.

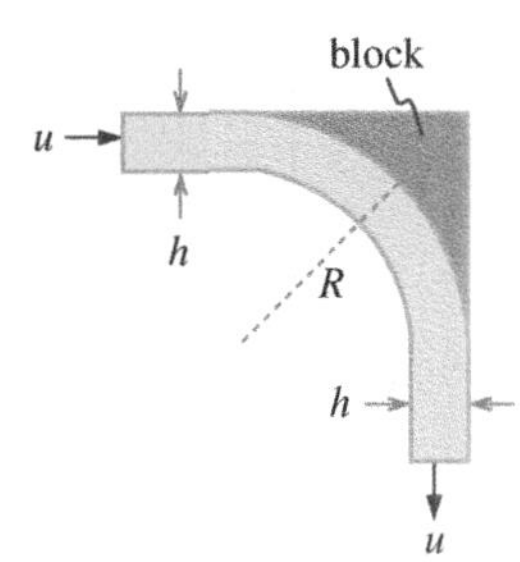

Figure 3.35 Flow turned by a block.

7. Calculate the surface force on the cylindrical control volume shown in Figure 3.36, given the stress tensor below. Perform the calculation twice: once by integrating the stress vector over the surface, and once by integrating the

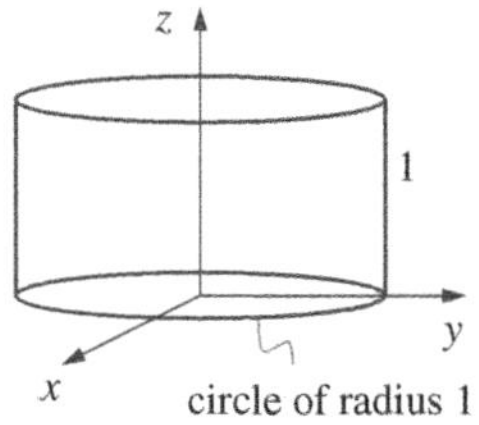

Figure 3.36 Cylindrical control volume for a surface force calculation.

divergence of the stress tensor over the volume. Verify that you obtain the same answer.

$$\sigma = \begin{bmatrix} 3 & x & z \\ x & 2 & y \\ z & y & 1 \end{bmatrix}. \tag{3.126}$$

8. Consider steady, two-dimensional, viscous, incompressible flow between two horizontal walls, with a stationary divider plate, starting half-way down the channel, as shown in Figure 3.37. Assume fully developed flow in the single channel before the divider plate and the two channels after. Ignore entrance effects at the start of the divider plate. The flow has viscosity μ and is driven by a difference of pressures, $p_0 > p_1$, with no body forces. Determine the total flow rate, the pressure at the start of the divider, and the force on the divider.

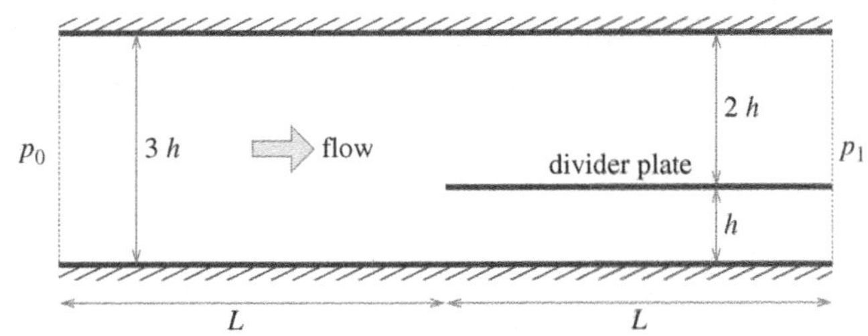

Figure 3.37 Planar flow with a divider plate.

4

Potential Flow

Air flowing at low speeds and with negligible viscous effects can often be approximated as incompressible and irrotational. We use the term "potential flow" to describe these flows, although strictly speaking only the irrotational property is required to define a velocity potential. For simplicity in exposition, we also assume incompressibility.

These two approximations hold for a large range of aerodynamic applications, including general aviation and commercial transport aircraft. The irrotational assumption relies on viscous effects not being significant, and it is valid at high Reynolds numbers when viscosity only affects a thin layer of fluid near no-slip boundaries, which we will analyze in Chapter 6. It breaks down when either the Reynolds number is low, which is typically not the case for aircraft, or when a shape is not very streamlined, e.g. a bluff body like a landing gear or an airfoil stalling at a high angle of attack. In the cases of bluff bodies and stall, large regions of separated flow appear, as illustrated in Figure 4.1, and these cannot be modeled by potential flow.

The incompressible assumption is valid as long as the Mach number, the ratio of the flow speed to the speed of sound, is well below 1. This is the case for low-speed aircraft flying at Mach numbers less than about 0.3. Higher Mach numbers can still be handled with corrections to potential flow, as we will see in Chapter 8. The corrections fail once shocks appear, so potential flow cannot be applied to high-speed transonic or supersonic aircraft.

This chapter presents the mathematical formulation of potential flow in two dimensions, as well as analysis tools for visualizing the flows, computing pressures and velocity fields, and calculating forces. Later chapters build on potential flow, applying it to airfoils, including viscous effects, extending it to three-dimensional wings, and accounting for compressibility.

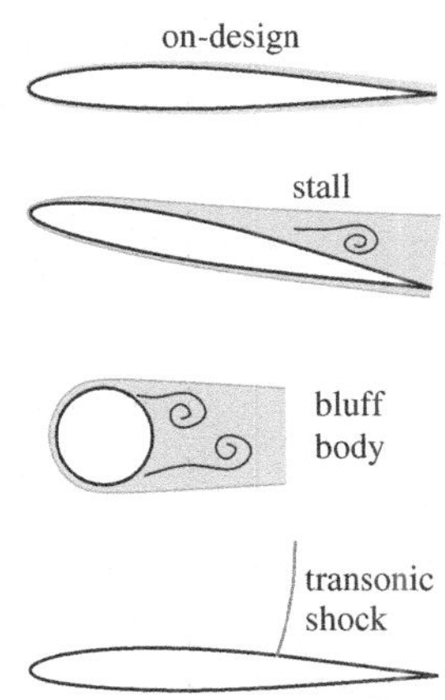

Figure 4.1 Potential flow assumptions break down at stall, for bluff bodies, and in transonic conditions.

4.1 Formulation

This section presents key mathematical modeling aspects of potential flows, starting from the assumptions of irrotational and incompressible flow in two dimensions. The complex potential is introduced as a useful mathematical construct that combines both the velocity potential and the stream function into one complex-number field. Working with complex numbers, while perhaps intimidating at first, often simplifies the description of potential flows and the calculation of the velocity field.

Aerodynamics: A Computational Introduction, First Edition. Krzysztof J. Fidkowski.

© 2026 John Wiley & Sons, Inc. Published 2026 by John Wiley & Sons, Inc.

4.1.1 Incompressible, Irrotational, Flow

The velocity field $\vec{v}(\vec{x})$ in potential flow satisfies two equations,

$$\begin{aligned} \nabla \times \vec{v} &= 0 \qquad \text{(irrotational)}, \\ \nabla \cdot \vec{v} &= 0 \qquad \text{(incompressible)}. \end{aligned} \tag{4.1}$$

As described in Section 2.5, we can automatically satisfy irrotationality by setting the velocity to be the gradient of a scalar potential field,*

$$\vec{v} = \nabla\phi, \quad \phi = \text{velocity potential}. \tag{4.2}$$

Incompressibility then requires

$$\nabla \cdot \vec{v} = 0 \Rightarrow \nabla^2\phi = 0. \tag{4.3}$$

Alternatively, we can automatically satisfy the two-dimensional incompressibility constraint by setting the velocity to be the curl of a specific vector field,†

$$\vec{v} = \nabla \times (\psi\hat{k}), \quad \psi(x, y) = \text{stream function}.$$

Irrotationality then requires

$$\nabla \times \vec{v} = 0 \Rightarrow \nabla^2\psi = 0.$$

Since ϕ and ψ describe the same velocity field, we must have

$$\begin{aligned} u &= \frac{\partial\phi}{\partial x} = \frac{\partial\psi}{\partial y}, \\ v &= \frac{\partial\phi}{\partial y} = -\frac{\partial\psi}{\partial x}. \end{aligned} \tag{4.4}$$

These two relationships between ϕ and ψ are known as the *Cauchy–Riemann* equations (d'Alembert 2012; McBain 2012).

*Due to the gradient, a constant potential yields a zero velocity, so the field $\phi(x, y)$ is only defined up to an additive constant.

† Likewise, due to the curl operator, a constant stream function yields a zero velocity, so the field $\psi(x, y)$ is only defined up to an additive constant.

4.1.2 The Complex Potential

In complex analysis, a consequence of the Cauchy–Riemann equations is that the complex function

$$F(z) = \phi(x, y) + i\psi(x, y), \quad z = x + iy, \tag{4.5}$$

is *analytic*, which means that its derivative,

$$F'(z) \equiv \lim_{|\Delta z| \to 0} \frac{F(z + \Delta z) - F(z)}{\Delta z}, \tag{4.6}$$

exists and is independent of the path along which $|\Delta z| \to 0$. Since Δz is a complex number, the path along which it can approach zero is not unique, as illustrated in Figure 4.2. There are many non-analytic complex functions, for which the derivative in Equation 4.6 does not exist. For example, the function $\Re(z)$, the real part of z, is not analytic since purely real and purely imaginary Δz yield different results in Equation 4.6. In particular, if Δz is real, the derivative of $\Re(z)$ is 1, but if Δz is imaginary, the derivative is 0. An analytic function would have the same value of the derivative, independent of the path.

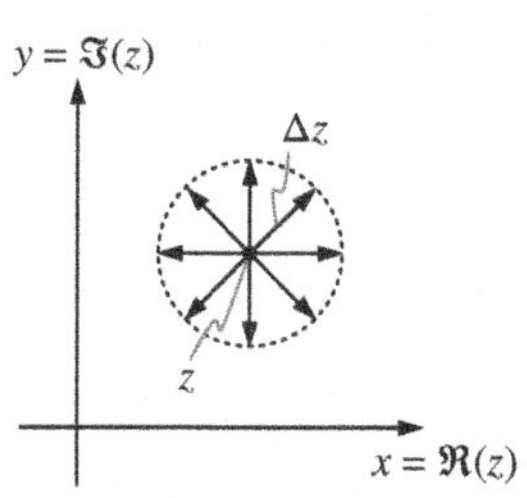

Figure 4.2 Δz can be offset from z in an arbitrary direction in the complex number plane.

We now show how the Cauchy–Riemann equations imply that $F(z)$ in Equation 4.5 is analytic. Consider $\Delta z = \Delta x + i\Delta y$ going to zero along some path, so that $\Delta x \to 0$ and $\Delta y \to 0$. Substituting Equation 4.5 into Equation 4.6, we have

$$F'(z) = \lim_{\substack{\Delta x \to 0 \\ \Delta y \to 0}} \frac{[\phi(x+\Delta x, y+\Delta y) + i\psi(x+\Delta x, y+\Delta y)] - [\phi(x,y) + i\psi(x,y)]}{\Delta x + i\Delta y}. \tag{4.7}$$

Now,

$$\phi(x+\Delta x, y+\Delta y) = \phi(x,y) + \Delta x\frac{\partial \phi}{\partial x} + \Delta y\frac{\partial \phi}{\partial y} + \text{H.O.T.}$$

$$\psi(x+\Delta x, y+\Delta y) = \psi(x,y) + \Delta x\frac{\partial \psi}{\partial x} + \Delta y\frac{\partial \psi}{\partial y} + \text{H.O.T.}$$

where H.O.T. indicates higher-order terms in Δx and Δy, which we can ignore since we are taking the limit $\Delta x \to 0$ and $\Delta y \to 0$.* Substituting these expressions into Equation 4.7, we obtain

* As $\Delta x \to 0$, $\Delta x^2 \ll \Delta x$, and similarly for Δy.

$$F'(z) = \lim_{\substack{\Delta x \to 0 \\ \Delta y \to 0}} \frac{\Delta x\frac{\partial \phi}{\partial x} + i\Delta x\frac{\partial \psi}{\partial x} + \Delta y\frac{\partial \phi}{\partial y} + i\Delta y\frac{\partial \psi}{\partial y}}{\Delta x + i\Delta y}. \tag{4.8}$$

For arbitrary ϕ and ψ, this expression will generally not be unique with respect to how $\Delta z = \Delta x + i\Delta y$ approaches 0.† However, consider the case when the Cauchy–Riemann equations in Equation 4.4 are satisfied. Using $\frac{\partial \psi}{\partial y} = \frac{\partial \phi}{\partial x}$ and $\frac{\partial \phi}{\partial y} = -\frac{\partial \psi}{\partial x}$ to eliminate the y derivatives in Equation 4.8 yields

† For example, if $\phi = x$ and $\psi = 0$, the quantity in the limit reduces to $\Delta x/(\Delta x + i\Delta y)$, which is 0 along a vertical path for which $\Delta x = 0$ and 1 along a horizontal path for which $\Delta y = 0$.

$$F'(z) = \lim_{\substack{\Delta x \to 0 \\ \Delta y \to 0}} \frac{(\Delta x + i\Delta y)\frac{\partial \phi}{\partial x} + i(\Delta x - i\Delta y)\frac{\partial \psi}{\partial x}}{\Delta x + i\Delta y} = \frac{\partial \phi}{\partial x} + i\frac{\partial \psi}{\partial x}. \tag{4.9}$$

Alternatively, using the Cauchy–Riemann equations to eliminate the x derivatives in Equation 4.8 gives

$$F'(z) = \frac{\partial \psi}{\partial y} - i\frac{\partial \phi}{\partial y}. \tag{4.10}$$

Note that Δx and Δy dropped out in the above expressions, which shows that the derivative $F'(z)$ is independent of the path along which $|\Delta z| \to 0$, and hence, that $F(z)$ is analytic. Also, the two expressions are indeed the same by the Cauchy–Riemann equations. This is clear when we write $F'(z)$ in terms of the velocity components,

$$F'(z) = u - iv. \tag{4.11}$$

This derivative is called the *complex velocity*,

$$w(z) = F'(z) = \text{complex velocity} = u - iv. \tag{4.12}$$

Our analysis shows that the velocity potential and stream function can be combined into one complex potential function, $F(z)$, that is analytic. Conversely, *any analytic function $F(z)$ can be decomposed into ϕ (real part) and ψ (imaginary part) that satisfy the Cauchy–Riemann equations and hence describe some two-dimensional, incompressible, irrotational flow.* If we find a complex potential for which the velocity satisfies certain given boundary conditions, it will constitute a solution to the potential flow problem.

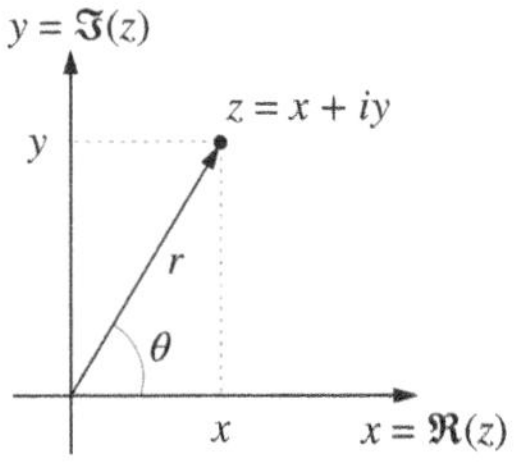

Figure 4.3 Definition of the complex coordinate z.

Example 4.1 Review of complex numbers

A complex number has both a real part and an imaginary part, so it can be thought of as two numbers combined into one. For the complex coordinate $z = x + iy$, the real part gives the horizontal, x, location, and the imaginary part gives the vertical, y, location. This coordinate can be visualized on the complex number plane, as shown in Figure 4.3.

Just like a vector in real space can be represented using different bases, so a complex number can be represented in different forms. One is the Cartesian representation, using x and y. Another is a polar representation, using the distance r and the angle θ. Some useful relationships between these representations are

$$z \equiv x + iy = r\cos\theta + ir\sin\theta = re^{i\theta}, \tag{4.13}$$

$$z^n = \left(re^{i\theta}\right)^n = r^n e^{in\theta} = r^n\left(\cos n\theta + i\sin n\theta\right), \tag{4.14}$$

$$\ln(z) = \ln\left(re^{i\theta}\right) = \ln r + \ln\left(e^{i\theta}\right) = \ln r + i\theta. \tag{4.15}$$

Example 4.2 Working with the complex potential

This example shows how to compute fields in a potential flow starting with the complex potential. Given the following complex potential,

$$F(z) = e^z, \tag{4.16}$$

we can first calculate the velocity potential and stream function by separating $F(z)$ into real and imaginary parts,[‡]

‡ The steps in this separation are

$$\begin{aligned} F(z) &= e^z \\ &= e^{x+iy} \\ &= e^x e^{iy} \\ &= e^x(\cos(y) + i\sin(y)). \end{aligned}$$

$$F(z) = \underbrace{e^x\cos(y)}_{\phi} + \underbrace{e^x\sin(y)}_{\psi}i. \tag{4.17}$$

This separation gives the velocity potential as $\phi = e^x\cos(y)$, and the stream function as $\psi = e^x\sin(y)$, both in terms of the real coordinates, x and y. We can calculate the velocity vector from either ϕ or ψ, or we can obtain it from the complex potential, by differentiating $F(z)$. Taking the latter approach, by Equation 4.12,

$$w = F'(z) = e^z = \underbrace{e^x\cos(y)}_{u} + \underbrace{e^x\sin(y)}_{-v}i. \tag{4.18}$$

So the Cartesian components of the velocity vector are[§]

§ Note that the vertical velocity is the *negative* of the imaginary component of the complex velocity,

$$v = -\Im(w).$$

$$u = e^x\cos(y), \quad v = -e^x\sin(y). \tag{4.19}$$

Example 4.3 From ϕ to F and more

Given a velocity potential, we can compute the complex potential if the flow is also incompressible. Suppose that we are given

$$\phi(x, y) = 2x^2 + 6xy - 2y^2. \tag{4.20}$$

Before calculating $F(z)$, we first verify that the flow is incompressible. This is indeed the case, since

$$\nabla^2\phi = \frac{\partial^2\phi}{\partial x^2} + \frac{\partial^2\phi}{\partial y^2} = 4 - 4 = 0. \quad (4.21)$$

The velocity components are

$$u = \frac{\partial\phi}{\partial x} = 4x + 6y, \quad v = \frac{\partial\phi}{\partial y} = 6x - 4y. \quad (4.22)$$

Combining these into a complex velocity,¶

$$\begin{aligned} w(z) = u - iv &= 4x + 6y - i(6x - 4y) \\ &= 4(x + iy) - 6i(x + iy) \\ &= (4 - 6i)z. \end{aligned} \quad (4.23)$$

¶ If the flow were not both incompressible and irrotational, then we would not be able to write w solely in terms of z.

Integrating with respect to z then gives the complex potential,‖

‖ C is an arbitrary *complex* constant, e.g. $C = C_r + iC_i$.

$$F(z) = \int w(z)dz = (2 - 3i)z^2 + C. \quad (4.24)$$

Separating into real and imaginary parts

$$F(z) = 2x^2 - 2y^2 + 6xy + C_r + i(4xy - 3x^2 + 3y^2 + C_i),$$

we obtain the stream function $\psi = \Im(F)$,

$$\psi(x, y) = 4xy - 3x^2 + 3y^2 + C_i. \quad (4.25)$$

4.1.3 Visualization

Every potential flow can be described by its velocity, stream function, velocity potential, complex potential, or complex velocity. Given one of these fields, it is possible to obtain the others, up to an additive constant in the potential and stream function.* These quantities also provide a way to visualize the flow field. In particular, the velocity field can be visualized by vector arrows, and streamlines are contours of the stream function.

* Recall that such a constant does not affect the velocity field, and hence, it does not affect the streamlines.

Example 4.4 Visualizing velocity vectors
Consider the velocity field given by the components

$$u = -x, \quad v = y. \quad (4.26)$$

This velocity field, known as *stagnation-point flow*, models horizontal flow moving right along the negative x axis, hitting a vertical wall at $x = 0$, and spreading out. It satisfies both the irrotational and incompressible requirements given in Equation 4.1. We can visualize it in a certain region by plotting velocity vectors as arrows at many points in that region. Listing 4.1 shows a code that implements this visualization using the `quiver` command.

Listing 4.1 Code for plotting velocity vectors in stagnation-point flow.

```
function stagv
[X,Y] = ndgrid(linspace(-2,0,11), linspace(-1,1,11));
U = -X; V = Y;
figure(1); clf; set(gcf, 'position', [0,0,400,400]);
h = quiver(X,Y,U,V, 'color', '#888888');
axis equal; box off;
writeplot('x', 'y', '../figs/stagv.pdf', 4);
```

Code notes:

- `ndgrid` makes two repetitive matrices that store the (x, y) coordinates of a lattice of points, based on the input x and y vectors.
- The matrices `U` and `V` store the x and y velocity components.
- Using `axis equal` makes the figure correctly proportioned.

Figure 4.4 shows the velocity vectors produced by this code. The domain of interest is $x < 0$, as that is where the horizontal velocity is positive. Along the y axis, i.e. on the line $x = 0$, the horizontal velocity is zero, and so this line models a vertical wall that obstructs the flow. The flow slows down as it approaches the wall and spreads out: fluid elements with $y > 0$ go upward, whereas those with $y < 0$ go downward. Near the origin, the velocity approaches zero, and the vectors become dots.

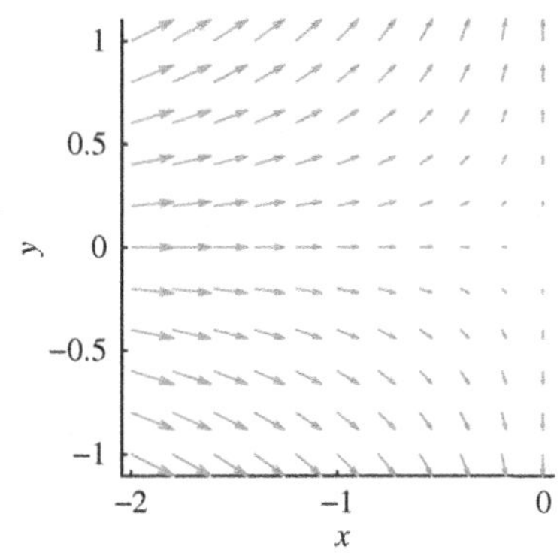

Figure 4.4 Velocity vectors for stagnation-point flow.

Example 4.5 Visualizing streamlines

We can visualize the same flow field as in the previous example using streamlines, which are contours of the stream function, ψ. One way to obtain the stream function is to integrate the velocity components,

$$\psi = \int u\,dy = -xy + f(x), \tag{4.27}$$

$$\psi = \int -v\,dx = -xy + g(y). \tag{4.28}$$

Comparing these expressions, $f(x) = g(y) = C$, and we set $C = 0$ since an additive constant does not affect the velocity, so that $\psi = -xy$. To plot the streamlines, we use the `contour` command, which takes as input the stream function evaluated on a grid of points. Listing 4.2 shows a code that follows these steps to make the streamlines.

Listing 4.2 Code for plotting streamlines in stagnation-point flow.

```
function stagpsi
[X,Y] = ndgrid(linspace(-2,0,41), linspace(-1,1,41));
Psi = -X.*Y;
figure(1); clf; set(gcf, 'position', [0,0,400,400]);
contour(X,Y,Psi, 24, 'color', '#888888');
axis equal; box off;
writeplot('x', 'y', '../figs/stagpsi.pdf', 2);
```

Code notes:

- The more points used in `ndgrid`, the smoother and more accurate the contours.
- The matrix `Psi` stores the stream function at each point.
- `contour` plots the contours using scalar data on a grid of points.

Figure 4.5 shows the resulting streamlines, which are consistent with expectations after seeing the velocity vectors. Flow moves from left to right and spreads out vertically as it approaches the vertical wall at $x = 0$. The streamlines actually do not tell us in which direction the flow is going: we would obtain the same picture if the velocity field were negated. They also do not directly show the velocity magnitude, but we can infer it from the spacing of the streamlines. The flow rate between two adjacent streamlines is constant, since no flow crosses the streamlines. So by conservation of mass, the speed is high where the streamlines are close together and low where they are far apart. The slowing down of the flow as it approaches the origin is then apparent in the large spacing of the streamlines there.

Instead of integrating the velocity components, we could also obtain the stream function by using the complex velocity and potential. The advantage here is that integration with respect to the complex coordinate z is sometimes easier than separate x and y integrations. From the given u and v, the complex velocity is

$$w = u - iv = -x - iy = -z. \tag{4.29}$$

The complex potential is the integral of w,

$$F = \int_z w\,dz = -\frac{1}{2}z^2 + C = \frac{1}{2}(y^2 - x^2) - xyi + C. \tag{4.30}$$

The stream function is the imaginary part of the complex potential,

$$\psi = \Im(F) = -xy + \Im(C). \tag{4.31}$$

This expression is consistent up to an additive constant with the previous result obtained by integrating the velocity components separately.

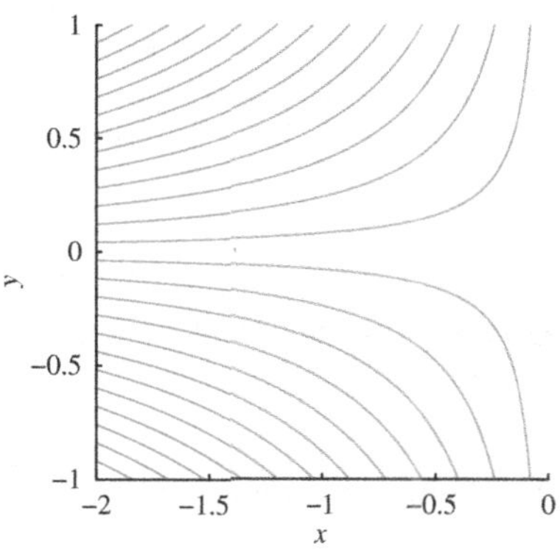

Figure 4.5 Streamlines for stagnation-point flow.

4.2 Elementary Flows

Any analytic function of z can be written using a power series,

$$F(z) = \sum_{n=1}^{\infty} A_n z^n + A_0 \ln z + \sum_{n=1}^{\infty} a_n z^{-n}, \tag{4.32}$$

where $\ln z$ is included to obtain a $1/z$ term in the expansion for the complex velocity,

$$w = F'(z) = \sum_{n=1}^{\infty} nA_n z^{n-1} + \frac{A_0}{z} - \sum_{n=1}^{\infty} na_n z^{-(n+1)}. \tag{4.33}$$

Not all of these terms are always necessary. For external flows,* we need to satisfy the farfield boundary condition,

$$\vec{v} \rightarrow \vec{V}_\infty \quad \text{as} \quad |\vec{x}| \rightarrow \infty.$$

In the power series for the complex velocity, Equation 4.33, the terms $z, z^2, z^3, \ldots$ all grow unbounded as $|z| = |\vec{x}| \rightarrow \infty$. Hence, we need

$$A_2 = A_3 = A_4 = \ldots = 0,$$

since otherwise the velocity field would not satisfy the farfield boundary condition. It turns out that just a few of the remaining terms in the series describe or approximate many aerodynamic flows of interest. These terms produce the *elementary flows*, and these are considered individually in the next sections.

*An external flow is flow over an object in an otherwise constant freestream, for example flow over an airfoil. If the object is at or near the origin, $|z| \rightarrow \infty$ indicates being far from the object, in the freestream flow.

4.2.1 $F(z) = A_1 z$: Uniform Flow

This flow has a constant complex velocity

$$w = F'(z) = A_1, \tag{4.34}$$

where A_1 is a complex number. Recall that $w(z) = u - iv$, so if we separate A_1 into real and imaginary components by writing $A_1 = U_\infty - iV_\infty$, then $u = U_\infty$ and $v = V_\infty$. This is uniform flow, as illustrated in Figure 4.6 We can calculate ϕ and ψ by separating $F(z)$ into real and imaginary parts,

$$\begin{aligned} F(z) = A_1 z &= (U_\infty - iV_\infty)(x + iy) \\ &= \underbrace{(U_\infty x + V_\infty y)}_{\phi(x,y)} + i\underbrace{(U_\infty y - V_\infty x)}_{\psi(x,y)}. \end{aligned} \tag{4.35}$$

The velocity potential and stream function are linear functions of the position, x and y.

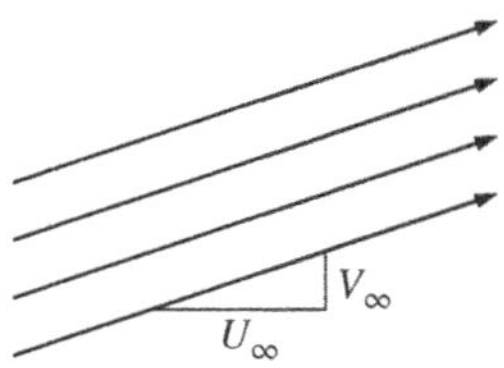

Figure 4.6 Streamlines for uniform flow, $F(z) = A_1 z$.

Uniform flow summary

$$F(z) = (U_\infty - iV_\infty)z$$
$$\phi = U_\infty x + V_\infty y$$
$$\psi = U_\infty y - V_\infty x$$
$$u = U_\infty$$
$$v = V_\infty$$

4.2.2 $F(z) = A_0 \ln z$: Source and Vortex Flow

The complex velocity for this flow is

$$w = F'(z) = \frac{A_0}{z} = \frac{A_0}{r} e^{-i\theta}, \tag{4.36}$$

where we used the polar form for z,

$$z = x + iy = re^{i\theta} \quad \Rightarrow \quad \frac{1}{z} = \frac{1}{r} e^{-i\theta}. \tag{4.37}$$

A_0 can be a complex number, and we consider the real and imaginary components separately.

A_0 is real

In anticipation of the result, we write

$$A_0 = \frac{Q}{2\pi}, \tag{4.38}$$

where Q is real and has a physical meaning. The complex velocity is

$$w = \frac{Q}{2\pi r} e^{-i\theta} = \frac{Q}{2\pi r}(\cos\theta - i\sin\theta) = u - iv. \tag{4.39}$$

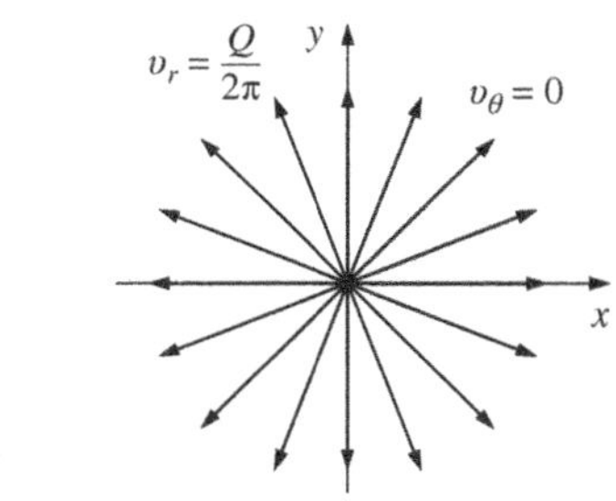

Figure 4.7 Streamlines for source flow, $F(z) = \frac{Q}{2\pi} \ln z$. v_r and v_θ are the radial and tangential components of the velocity.

So the velocity components are $u = \frac{Q}{2\pi r}\cos\theta$ and $v = \frac{Q}{2\pi r}\sin\theta$. As illustrated in Figure 4.7, this is a *purely radial flow*: the velocity vectors point into or out from the origin, depending on the sign of Q.

We can calculate ϕ and ψ for this flow,

$$F(z) = \frac{Q}{2\pi} \ln z = \frac{Q}{2\pi}(\ln r + i\theta) = \phi + i\psi, \tag{4.40}$$

where we used $\ln(re^{i\theta}) = \ln r + \ln(e^{i\theta}) = \ln r + i\theta$. Note that $\psi = \frac{Q}{2\pi}\theta$ is consistent with radial streamlines, in that $\theta = \text{constant} \Rightarrow \psi = \text{constant}$. We can also obtain the velocities in polar form from ϕ or ψ. For example, using ϕ,*

$$v_r = \frac{\partial \phi}{\partial r} = \frac{Q}{2\pi r}, \quad v_\theta = \frac{1}{r}\frac{\partial \phi}{\partial \theta} = 0. \tag{4.41}$$

* Recall from Equation 1.31 that the gradient operator in polar coordinates is

$$\nabla = \hat{r}\frac{\partial}{\partial r} + \frac{\hat{\theta}}{r}\frac{\partial}{\partial \theta}.$$

This flow is a point *source* ($Q > 0$) or *sink* ($Q < 0$) centered at the origin. In this flow, Q has physical significance. Considering a circular control volume of radius R around the origin, the two-dimensional volumetric (i.e. area) outflow is[****]

$$\oint \vec{v} \cdot \vec{n} = \int_0^{2\pi} v_r R d\theta = \int_0^{2\pi} \frac{Q}{2\pi R} R d\theta = Q. \tag{4.42}$$

For $Q > 0$, we have a source, and for $Q < 0$, we have a sink. Q is the source/sink strength, and it has units of area per time.

[****] **Point source/sink summary**

$$F(z) = \frac{Q}{2\pi} \ln z$$
$$\phi = \frac{Q}{2\pi} \ln r$$
$$\psi = \frac{Q}{2\pi} \theta$$
$$v_r = \frac{Q}{2\pi r}$$
$$v_\theta = 0$$

A_0 is imaginary

Second, we consider the case when A_0 is purely imaginary. Again, in anticipation of the result, we write

$$A_0 = \frac{i\Gamma}{2\pi}, \tag{4.43}$$

where Γ is real and has a physical meaning. The complex velocity is

$$w = \frac{i\Gamma}{2\pi r} e^{-i\theta} = \frac{\Gamma}{2\pi r}(\sin\theta + i\cos\theta) = u - iv. \tag{4.44}$$

So the velocity components are $u = \frac{\Gamma}{2\pi r}\sin\theta$ and $v = -\frac{\Gamma}{2\pi r}\cos\theta$. As illustrated in Figure 4.8, this is a *purely tangential flow*: the velocity vectors are tangent to circles centered at the origin.

We can calculate ϕ and ψ for this flow,

$$F(z) = \frac{i\Gamma}{2\pi} \ln z = \frac{\Gamma}{2\pi}(-\theta + i\ln r) = \phi + i\psi, \tag{4.45}$$

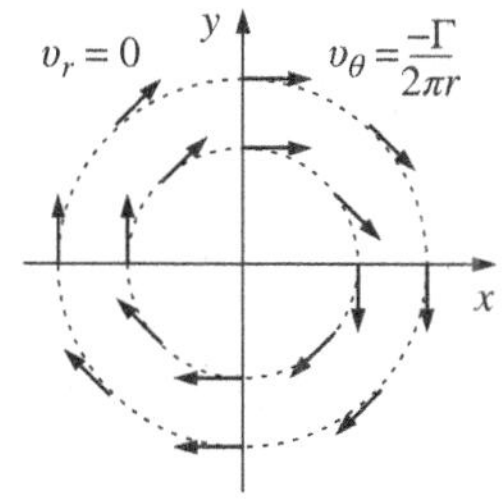

Figure 4.8 Streamlines for vortex flow, $F(z) = \frac{i\Gamma}{2\pi} \ln z$. Velocity vectors shown for $\Gamma > 0$.

Note that $\psi = \frac{\Gamma}{2\pi} \ln r$ is consistent with tangential streamlines, in that $r =$ constant $\Rightarrow \psi =$ constant. The velocities in polar form are

$$v_r = \frac{\partial\phi}{\partial r} = 0, \quad v_\theta = \frac{1}{r}\frac{\partial\phi}{\partial\theta} = -\frac{\Gamma}{2\pi r}. \tag{4.46}$$

This flow is a point *vortex* centered at the origin. The physical significance of Γ is that it is the *circulation* associated with the vortex. We can see this by calculating the circulation for a circular contour of radius R centered at the origin,

$$-\oint \vec{v} \cdot d\vec{l} = -\int_0^{2\pi} v_\theta R d\theta = \int_0^{2\pi} \frac{\Gamma}{2\pi R} R d\theta = \Gamma. \tag{4.47}$$

For $\Gamma > 0$, the vortex rotates clockwise. For $\Gamma < 0$, it rotates counter-clockwise. Γ is also known as the *vortex strength* and has units of area per time.

Point vortex summary

$$F(z) = \frac{i\Gamma}{2\pi} \ln z$$
$$\phi = -\frac{\Gamma}{2\pi} \theta$$
$$\psi = \frac{\Gamma}{2\pi} \ln r$$
$$v_r = 0$$
$$v_\theta = -\frac{\Gamma}{2\pi r}$$

4.2.3 $F(z) = a_1/z$: Doublet Flow

For this flow, we again treat separately the cases of real and imaginary a_1. For general complex a_1, the flow will be a combination of the real and imaginary results.

a_1 is real

First, for a_1 real, we write

$$a_1 = \frac{\kappa}{2\pi} \quad \Rightarrow \quad F(z) = \frac{\kappa}{2\pi r} e^{-i\theta} = \frac{\kappa}{2\pi r}(\cos\theta - i\sin\theta). \tag{4.48}$$

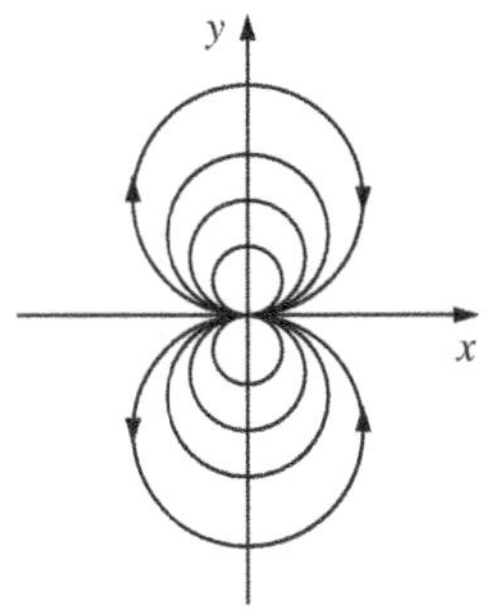

Figure 4.9 Streamlines for doublet flow in the x direction, $F(z) = \frac{\kappa}{2\pi z}$.

From this equation, the velocity potential and stream function are

$$\phi = \Re[F(z)] = \frac{\kappa}{2\pi r}\cos\theta, \quad \psi = \Im[F(z)] = -\frac{\kappa}{2\pi r}\sin\theta. \tag{4.49}$$

The polar components of the velocity are

$$v_r = \frac{\partial\phi}{\partial r} = -\frac{\kappa}{2\pi r^2}\cos\theta, \quad v_\theta = \frac{1}{r}\frac{\partial\phi}{\partial\theta} = -\frac{\kappa}{2\pi r^2}\sin\theta. \tag{4.50}$$

This flow is a *doublet aligned with the x axis*, and its streamlines are illustrated in Figure 4.9. Along the x axis, $v_\theta = 0$, so the flow is purely horizontal, and it points toward the negative real axis. Above and below the x axis, the streamlines are circular and pass through the origin.

a_1 is imaginary

For a_1 imaginary, we write

$$a_1 = \frac{i\kappa}{2\pi} \quad \Rightarrow \quad F(z) = \frac{i\kappa}{2\pi r}e^{-i\theta} = \frac{\kappa}{2\pi r}(\sin\theta + i\cos\theta). \tag{4.51}$$

x-**Doublet** summary

$$F(z) = \frac{\kappa}{2\pi z}$$
$$\phi = \frac{\kappa}{2\pi r}\cos\theta$$
$$\psi = -\frac{\kappa}{2\pi r}\sin\theta$$
$$v_r = -\frac{\kappa}{2\pi r^2}\cos\theta$$
$$v_\theta = -\frac{\kappa}{2\pi r^2}\sin\theta$$

The velocity potential and stream function are then

$$\phi = \Re[F(z)] = \frac{\kappa}{2\pi r}\sin\theta, \quad \psi = \Im[F(z)] = \frac{\kappa}{2\pi r}\cos\theta. \tag{4.52}$$

The polar components of the velocity are

$$v_r = \frac{\partial\phi}{\partial r} = -\frac{\kappa}{2\pi r^2}\sin\theta, \quad v_\theta = \frac{1}{r}\frac{\partial\phi}{\partial\theta} = \frac{\kappa}{2\pi r^2}\cos\theta. \tag{4.53}$$

This flow is a *doublet aligned with the y axis*, and its streamlines are illustrated in Figure 4.10. Along the y axis, $v_\theta = 0$, so the flow is purely vertical and it points down, along the negative y axis. To the left and right of the y axis, the streamlines are circular and pass through the origin.

When κ is complex, $\kappa = \kappa_r + i\kappa_i$, the doublet is aligned with a skew direction that is neither horizontal nor vertical. The direction of the doublet will instead point toward $\kappa_r + i\kappa_i$ in the complex number plane, and along this ray, the flow of the doublet will be in the opposite direction, i.e. toward $-\kappa_r - i\kappa_i$.

In all cases, κ, or its magnitude if it is complex, is the *doublet strength*, and it has a physical meaning. A doublet arises from the limit of a source/sink pair whose separation, δ, goes to zero but whose strength, Q, grows to infinity such that the product $\kappa \equiv \delta Q$ is constant.

y-**Doublet** summary

$$F(z) = \frac{i\kappa}{2\pi z}$$
$$\phi = \frac{\kappa}{2\pi r}\sin\theta$$
$$\psi = \frac{\kappa}{2\pi r}\cos\theta$$
$$v_r = -\frac{\kappa}{2\pi r^2}\sin\theta$$
$$v_\theta = -\frac{\kappa}{2\pi r^2}\cos\theta$$

For an x-aligned doublet, the source/sink pair is on the x axis. Centering the coordinate system on the sink, as shown in Figure 4.11, we can *superimpose* the effects of the source and sink to obtain the total complex potential for this system,

$$F^{\text{tot}}(z) = \frac{Q}{2\pi}\ln(z+\delta) - \frac{Q}{2\pi}\ln z = \frac{Q}{2\pi}\ln\left(1 + \frac{\delta}{z}\right). \tag{4.54}$$

Taking the limit $\delta \to 0$, $Q\delta = \text{constant} = \kappa$, we obtain

$$F^{\text{tot}}(z) = \frac{Q\delta}{2\pi z} = \frac{\kappa}{2\pi z}, \tag{4.55}$$

where we have used $\lim_{\epsilon\to 0}\ln(1+\epsilon) = \epsilon$. Note that $Q = \kappa/\delta$ grows unbounded as $\delta \to 0$. Also, if we had used δi, a vertical source/sink pair, we would have obtained $F^{\text{tot}}(z) = \frac{i\kappa}{2\pi z}$, a y-aligned doublet.

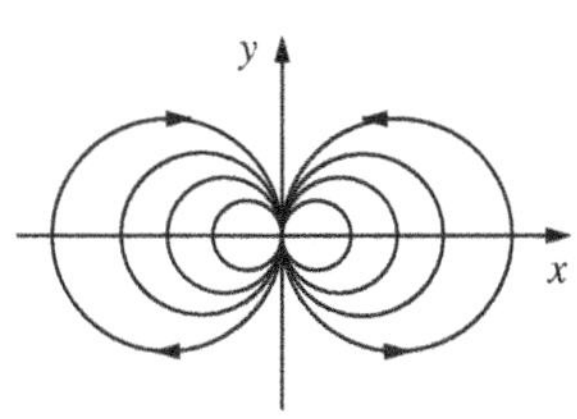

Figure 4.10 Streamlines for doublet flow in the y direction, $F(z) = \frac{i\kappa}{2\pi z}$.

A doublet came from the a_1/z term. All further terms are higher-order singularities that we will not use to construct the flows in this chapter.*

*This does not mean that they are not useful. For example, the term a_2/z^2 is a quadrupole that is relevant in acoustics.

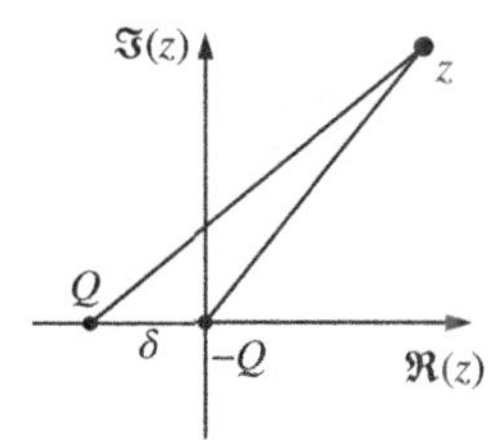

Figure 4.11 An x-aligned doublet as the limit of a source/sink pair. Diagram illustrating the calculation of the potential at $z = x + iy$.

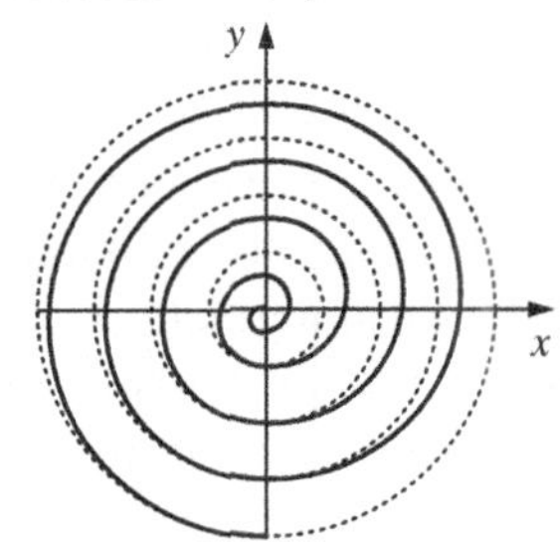

Figure 4.12 Streamlines in a potential-flow tornado model.

Example 4.6 Flow in a tornado

The flow in a two-dimensional slice of a tornado, shown in Figure 4.12, can be modeled by a complex potential that combines a sink and a vortex centered at the origin,

$$F(z) = \frac{Q}{2\pi}\ln(z) + \frac{\Gamma i}{2\pi}\ln(z), \tag{4.56}$$

where $Q < 0$ and $\Gamma > 0$. The velocity, $\vec{v}$, is a superposition of the purely radial sink velocity and the purely tangential vortex velocity,

$$\vec{v} = \underbrace{\frac{Q}{2\pi r}\hat{r}}_{\text{sink}} + \underbrace{\frac{\Gamma}{2\pi r}(-\hat{\theta})}_{\text{vortex}}. \tag{4.57}$$

The velocity potential and stream function are easy to obtain as well, as superpositions of the sink and vortex functions,

$$\phi = \frac{Q}{2\pi}\ln r - \frac{\Gamma}{2\pi}\theta, \quad \psi = \frac{Q}{2\pi}\theta + \frac{\Gamma}{2\pi}\ln r. \tag{4.58}$$

An additional quantity of interest is the pressure. Assuming $p \to p_\infty$ as $r \to \infty$, we can calculate the pressure as a function of r, where $r = \sqrt{x^2 + y^2}$, by using Bernoulli's equation,†

$$p(r) + \frac{1}{2}\rho|\vec{v}|^2 = p_\infty. \tag{4.59}$$

The right-hand side has no velocity term because the flow velocity due to both the sink and the vortex goes to zero as $r \to \infty$. Solving for the pressure at r yields‡

$$p(r) = p_\infty - \frac{\rho(Q^2 + \Gamma^2)}{8\pi^2 r^2}. \tag{4.60}$$

† Bernoulli's equation applies in potential flow because the flow is irrotational.

‡ The steps in this calculation are

$$\begin{aligned} p(r) &= p_\infty - \frac{1}{2}\rho|\vec{v}|^2 \\ &= p_\infty - \frac{1}{2}\rho\left(v_r^2 + v_\theta^2\right) \\ &= p_\infty - \frac{1}{2}\rho\left(\frac{Q^2 + \Gamma^2}{4\pi^2 r^2}\right) \\ &= p_\infty - \frac{\rho(Q^2 + \Gamma^2)}{8\pi^2 r^2}. \end{aligned}$$

The pressure is highest in the freestream and decreases toward the core of the tornado. It does not depend on θ because the velocity magnitude does not depend on θ in this flow. Note that the origin is a singularity, and in reality, the potential flow model breaks down here as viscous effects become relevant. A realistic model near the origin would have a finite-size core, or eye, that is rotational and in which $\nabla \cdot \vec{v} \neq 0$, as the flow is sucked upward.

4.3 Superposition

Elementary flows can be superimposed to yield more complex flows, since the sum of two analytic $F(z)$ is also analytic. Alternatively, we can think about summing ϕ or ψ, since for two velocity potentials, ϕ_1 and ϕ_2,

$$\nabla^2\phi_1 = 0, \quad \nabla^2\phi_2 = 0 \Rightarrow \nabla^2(\phi_1 + \phi_2) = 0, \tag{4.61}$$

and for two stream functions, ψ_1 and ψ_2,

$$\nabla^2\psi_1 = 0, \quad \nabla^2\psi_2 = 0 \Rightarrow \nabla^2(\psi_1 + \psi_2) = 0. \tag{4.62}$$

So the combination of two or more potential flows is also a potential flow. The velocity, velocity potential, stream function, complex potential, and complex velocity of the combination are the sums of the individual pieces.

4.3.1 Rankine Half Body

Figure 4.13 Superposition of a source and freestream models a Rankine half body.

In aerodynamics, superposition often involves the freestream, i.e. uniform flow, which allows for the modeling of flow over a stationary object. One simple combination is freestream and a source, as shown in Figure 4.13.

The source creates outward flow from the origin, the speed of which diminishes with distance form the origin. Eventually, for large distances from the origin, the freestream dominates, so that even flow that started moving to the left near the origin is turned to the right.

Before turning to equations, it is useful to visualize the flow by plotting the streamlines for an arbitrary choice of Q and U_∞. This is a quick calculation when working with complex numbers, since the complex potential is simple to write down for elementary flows and calculating the stream function consists of taking the imaginary part of the complex potential. Listing 4.3 shows a code that does this over a small range of x and y values near the origin.

Code notes:

- Different relative magnitudes of and Q will change the shape of the streamlines. U_∞
- The first set of contours shows the streamlines over the entire domain.
- The second contour plot shows the dividing and stagnation streamlines at $\psi = \pm Q/2$.

Listing 4.3 Code for plotting streamlines in a source and freestream combination.

```
function rhbpsi
Uinf = 1; % freestream speed
Q = 3;    % source strength
[X,Y] = ndgrid(linspace(-2,2,101));
Z = X + 1i*Y;
F = Uinf*Z + Q/(2*pi)*log(Z);
Psi = imag(F);
figure(1); clf; set(gcf, 'position', [0,0,400,400]);
contour(X,Y,Psi, 20, 'color', '#888888'); hold on;
contour(X,Y,Psi, [-Q/2, Q/2], 'k-', 'linewidth', 2);
axis equal; box off; set(gca, 'xtick', linspace(-2,2,5));
writeplot('$x$', '$y$', '../figs/rhbpsi.pdf', 4);
```

Figure 4.14 shows the resulting streamlines. Note that the flow from the source does not mix with the flow from the freestream. Streamlines emanating from the origin do not merge with ones coming from the freestream. Instead, the freestream flow goes around an "obstacle," known as the *Rankine half body*, created by the source. There is then a streamline that divides the two flows, and it is highlighted as a thick line in Figure 4.14.

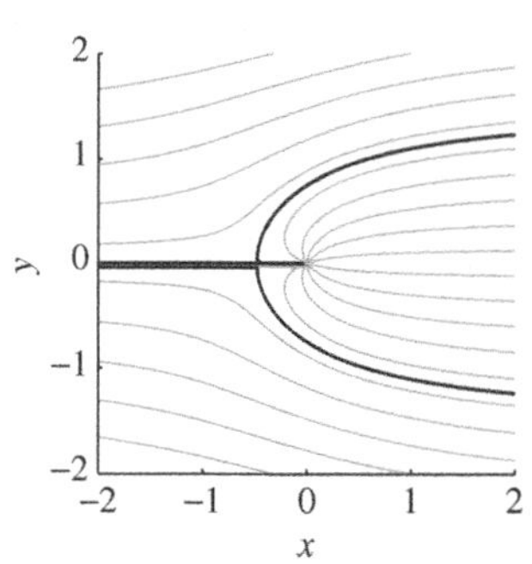

Figure 4.14 Streamlines for a source and freestream: flow over a Rankine half body.

What is the value of the stream function on this dividing streamline? This line intersects the stagnation streamline, which divides the flows going above and below the object. So the two lines share the same value of the stream function, and this is why they are both highlighted as thick in the streamline plot. The expression for the stream function follows from superimposing the freestream and source stream functions,

$$\psi = U_\infty y + \frac{Q}{2\pi}\theta. \tag{4.63}$$

The presence of θ makes the stream function discontinuous on an arbitrary radial line from the origin: θ jumps as we walk more than one full circle around the origin. We should expect this discontinuity due to the singularity of the source at the origin. Its location is arbitrary because we can choose the θ interval to be $[0,2\pi]$, $[-\pi,\pi]$, or anything in between. Physically, we can think of this line as the supply hose for the source at the origin: it runs along the positive x axis for the choice $\theta \in [0,2\pi]$ and along the negative x axis for $\theta \in [-\pi,\pi]$. With the latter choice, which is implicit in MATLAB's angle calculation from complex numbers, $\psi = \pm Q/2$ on the stagnation streamline, which is at $\theta = \pm\pi$.* So the code in Listing 4.3 uses these values to highlight the stagnation and dividing streamlines.

The Rankine half body is so named because the object is not closed. We should expect this when putting a source in a freestream: the extra fluid injected at the origin will propagate forever in the x direction. To obtain a closed body, we would need to remove that injected fluid, e.g. through a sink of the same strength, which would produce the Rankine oval, analyzed in Problem 4. Alternatively, we could use a doublet, as shown in the next section.

* The stagnation streamline is along the negative x axis and has $y = 0$ and $\theta = \pm\pi$, so by Equation 4.63,

$$\psi = U_\infty(0) + \frac{Q}{2\pi}(\pm\pi) = \pm\frac{Q}{2}.$$

4.3.2 Non-Lifting Flow over a Cylinder

A doublet does not produce any net flow, so if we put it in a freestream, we should expect to obtain a closed body. Consider a horizontal freestream and an x-aligned doublet, as shown in Figure 4.15. We choose an x-aligned doublet so that its flow emanating along the negative x axis counters the freestream and creates a dividing streamline.

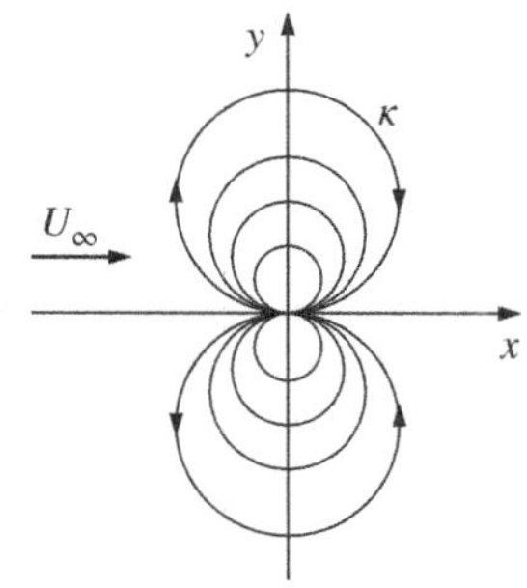

Figure 4.15 Superposition of uniform flow and a doublet.

How does this flow look? Again, before proceeding to the equations, it helps to visualize the flow numerically by plotting the streamlines for a particular choice of the freestream speed, U_∞, and the doublet strength, κ. Listing 4.4 shows a code that does this, using the same procedure as for the Rankine half body in the previous section.

Listing 4.4 Code for plotting streamlines in a doublet and freestream combination.

```
function cylnlpsi
Uinf = 1;   % freestream speed
kappa = 3; % doublet strength
[X,Y] = ndgrid(linspace(-2,2,101));
Z = X + 1i*Y;
F = Uinf*Z + kappa./(2*pi*Z);
Psi = imag(F);
figure(1); clf; set(gcf, 'position', [0,0,400,400]);
contour(X,Y,Psi, linspace(-2*Uinf,2*Uinf,17), 'color', '#888888');
hold on; contour(X,Y,Psi, [0, 0], 'k-', 'linewidth', 2);
axis equal; axis off; box off;
writeplot('x', 'y', '../figs/cylnlpsi.pdf', 2);
```

Code notes:

- The singularity at the origin is strong due to a doublet, creating large magnitude ψ there.
- The contour levels for the streamlines are thus explicitly given to avoid crowding at the origin.

Figure 4.16 shows the resulting streamlines. The streamline that divides the freestream and doublet flows is highlighted as a thick line in Figure 4.16. Its value is $\psi = 0$, as will be clear once we write the expression for ψ, which is also the value of ψ on the two stagnation streamlines, which are horizontal.

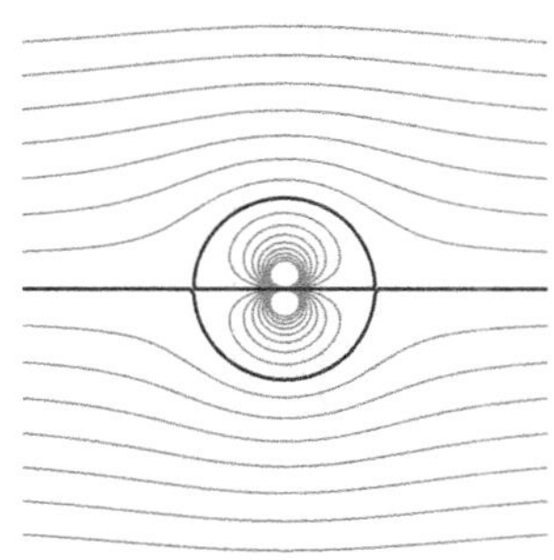

Figure 4.16 Streamlines for a doublet and freestream: flow over a cylinder.

We see from Figure 4.16 that this combination does indeed yield a closed body, and that its shape is a circle. The flow field is also symmetrical about the x axis, so we should not expect it to produce any lifting force. Hence, we have modeled non-lifting flow over a "cylinder," which is just a circle in two dimensions.

Turning now to equations, the total complex potential for this superposition is

$$F^{\text{tot}} = U_\infty z + \frac{\kappa}{2\pi z}.$$

The imaginary component yields the stream function, which in polar coordinates is

$$\psi(r, \theta) = \left(U_\infty r - \frac{\kappa}{2\pi r}\right)\sin\theta. \tag{4.64}$$

Recall that on streamlines, including solid boundaries, ψ is constant. Using the above expression, we see that ψ is zero, independent of θ when

$$U_\infty r - \frac{\kappa}{2\pi r} = 0 \quad \Rightarrow \quad r = \sqrt{\frac{\kappa}{2\pi U_\infty}}. \tag{4.65}$$

Therefore, $\psi = 0$ on a cylinder that has radius $r = a$, where

$$a = \sqrt{\frac{\kappa}{2\pi U_\infty}}. \tag{4.66}$$

Also, $\psi = 0$ independent of r when $\sin\theta = 0$, which occurs when $\theta = 0$ and when $\theta = \pi$. These two lines emanating from the cylinder are the *stagnation streamlines*, and they intersect the cylinder at *stagnation points*. Figure 4.16 shows these as highlighted thick lines, since they intersect the cylinder and have the same value. Note that the above equations let us compute the flow inside the cylinder, which is also shown in the figure. However, when describing flow over a cylinder, we treat the flow inside as fictitious. Potential flow theory gives us flow at all points in space, but only the flow outside the cylinder is physically relevant.*

*The interior flow is also not well-behaved, since it contains a singularity that makes the velocity magnitude infinite.

In reality, with viscous effects, flow over the cylinder does not look like that depicted in Figure 4.16. It is not symmetric fore and aft because the flow separates on the aft side of the cylinder, as illustrated schematically in Figure 4.17. Separation is a viscous effect, to be discussed in Chapter 6, and is an example of the limits of potential flow theory. In general, it is hard to achieve attached (not separated) flow for bluff bodies like the cylinder, so potential flow does not yield reasonable results. However, for streamlined bodies, such as airfoils, the flow predicted by potential flow theory is much closer to reality.

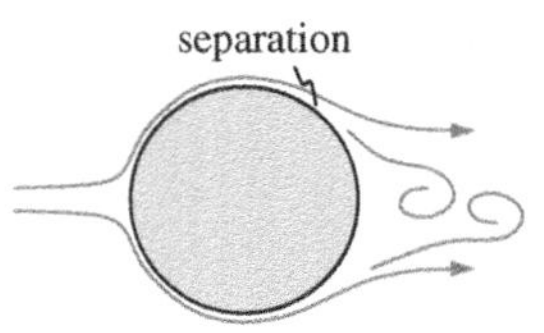

Figure 4.17 Viscous flow over a cylinder exhibits separation and is often unsteady.

4.3.3 Lifting Flow over a Cylinder

Lift is a force perpendicular to the freestream. To achieve it for the case of a cylinder, we need to break the upper/lower symmetry of the flow in the previous section, so that flow is deflected upward or downward. One way to do this is with a vortex, which, if placed at the origin, will not disturb the dividing streamline. This is because a vortex produces tangential flow, which has no component in the normal direction, and so it leaves a circular streamline intact. Figure 4.18 illustrates this combination.

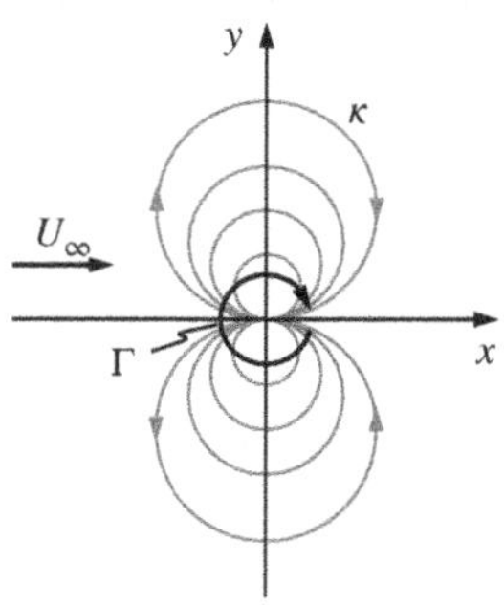

Figure 4.18 Lifting flow over a cylinder results from a superposition of uniform flow and a collocated doublet and vortex.

Listing 4.5 shows a code that plots the streamlines with the vortex included. Otherwise, it is similar to the code for the non-lifting cylinder (Listing 4.4). It applies a constant offset to the stream function, rolled into the vortex contribution, which as we will soon see keeps $\psi = 0$ on the cylinder surface.

Listing 4.5 Code for plotting streamlines for lifting flow over a cylinder.

```
function cylpsi
Uinf = 1;  % freestream speed
kappa = 3; % doublet strength
Gamma = 5; % vortex strength
a = sqrt(kappa/(2*pi*Uinf)); % cylinder radius
[X,Y] = ndgrid(linspace(-2,2,101));
Z = X + 1i*Y;
F = Uinf*Z + kappa./(2*pi*Z) + i*Gamma/(2*pi)*log(Z/a);
Psi = imag(F);
figure(1); clf; set(gcf, 'position', [0,0,400,400]);
contour(X,Y,Psi, linspace(-2*Uinf, 2*Uinf,17), 'color', '#888888');
hold on; contour(X,Y,Psi, [0, 0], 'k-', 'linewidth', 2);
axis equal; axis off; box off;
writeplot('x', 'y', '../figs/cylpsi.pdf', 2);
```

The shape of the streamlines depends on the value of Γ. Figure 4.19 shows the streamlines for positive Γ, which corresponds to a vortex spinning clockwise. As expected from this direction of rotation, the flow moving from left to right turns downward as it passes over the cylinder, and this turning creates positive lift. Figure 4.20 shows the streamlines for negative Γ, which corresponds to a vortex spinning counter-clockwise. This time, the flow turns upward as it passes over the cylinder, and this turning creates negative lift.

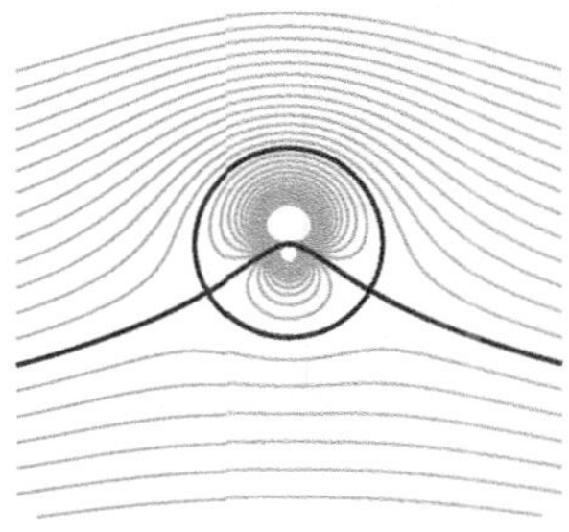

Figure 4.19 Streamlines for lifting flow over a cylinder, $\Gamma > 0$.

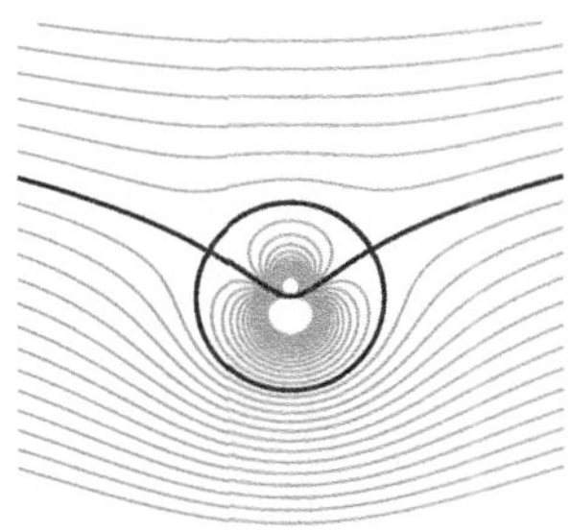

Figure 4.20 Streamlines for lifting flow over a cylinder, $\Gamma < 0$.

Considering now the equations, the stream function is obtained by summing together the stream functions for each of the elementary flows,

$$\begin{aligned}\psi &= \underbrace{U_\infty y}_{\text{uniform flow}} \underbrace{- \frac{\kappa}{2\pi r}\sin\theta}_{\text{doublet}} + \underbrace{\frac{\Gamma}{2\pi}\ln r + C}_{\text{vortex}} \\ &= U_\infty r \sin\theta\left(1 - \frac{\kappa}{2\pi U_\infty r^2}\right) + \frac{\Gamma}{2\pi}\ln r + C.\end{aligned} \tag{4.67}$$

We are allowed to add an arbitrary constant to the stream function because constants drop out when differentiating to obtain the velocity field. The purpose of the constant C is one of convenience: to make $\psi = 0$ on the cylinder surface.

ψ corresponds to flow over a cylinder because it is independent of θ when

$$1 - \frac{\kappa}{2\pi U_\infty r^2} = 0 \quad \Rightarrow \quad r = a = \sqrt{\frac{\kappa}{2\pi U_\infty}}. \tag{4.68}$$

So the cylinder radius is the same as in the non-lifting case. Setting $\psi = 0$ on the cylinder surface yields

$$0 = \psi = U_\infty a \sin\theta(0) + \frac{\Gamma}{2\pi}\ln a + C \quad \Rightarrow \quad C = -\frac{\Gamma}{2\pi}\ln a. \tag{4.69}$$

Hence, the stream function is*

$$\psi(r,\theta) = U_\infty r \sin\theta\left[1 - \frac{a^2}{r^2}\right] + \frac{\Gamma}{2\pi}\ln\left(\frac{r}{a}\right). \tag{4.70}$$

* Note that $\psi = 0$ when $r = a$, but no longer when $\theta = \pm\pi$, so the stagnation streamlines will have a different shape.

Locations of interest for flow over a lifting cylinder include stagnation points, defined as points where $\vec{v} = 0$. The velocity components in polar coordinates are

$$v_r = \frac{1}{r}\frac{\partial\psi}{\partial\theta} = U_\infty \cos\theta\left[1 - \frac{a^2}{r^2}\right], \tag{4.71}$$

$$v_\theta = -\frac{\partial\psi}{\partial r} = -U_\infty \sin\theta\left[1 + \frac{a^2}{r^2}\right] - \frac{\Gamma}{2\pi r}. \tag{4.72}$$

Considering first the cylinder surface,

$$r = a \Rightarrow \quad v_r = 0, \tag{4.73}$$

$$v_\theta = -U_\infty \sin\theta(2) - \frac{\Gamma}{2\pi a} = 0. \tag{4.74}$$

The first equation is reassuring, since it states that no flow crosses the cylinder surface. Solving for θ from the second equation yields

$$\sin\theta = -\frac{\Gamma}{4\pi U_\infty a}. \tag{4.75}$$

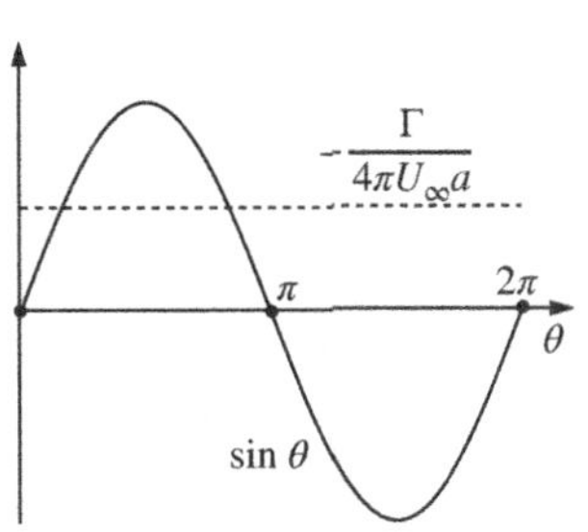

Figure 4.21 Graphical solution for stagnation points on the cylinder surface in lifting flow.

We can gain some insight into the solution to this equation by plotting the left-hand side and right-hand side versus θ, as shown in Figure 4.21.

From the graph, we observe the following:

- For $\Gamma = 0$, there are two stagnation points, at $\theta_1 = 0$ and $\theta_2 = \pi$ (see Figure 4.16).

- For $0 < \Gamma < 4\pi U_\infty a$, there are two stagnation points, at $\pi < \theta_1 < 3\pi/2 < \theta_2 < 2\pi$ (see Figure 4.19).
- For $-4\pi U_\infty a < \Gamma < 0$, there are two stagnation points, at $0 < \theta_1 < \pi/2 < \theta_2 < \pi$ (see Figure 4.20).
- For $\Gamma = 4\pi U_\infty a$, there is one stagnation point, at $\theta_1 = 3\pi/2$ (see Figure 4.22).
- For $\Gamma = -4\pi U_\infty a$, there is one stagnation point, at $\theta_1 = \pi/2$.
- For $\Gamma > 4\pi U_\infty a$ or $\Gamma < -4\pi U_\infty a$, there are no stagnation points on the cylinder (see Figure 4.23).

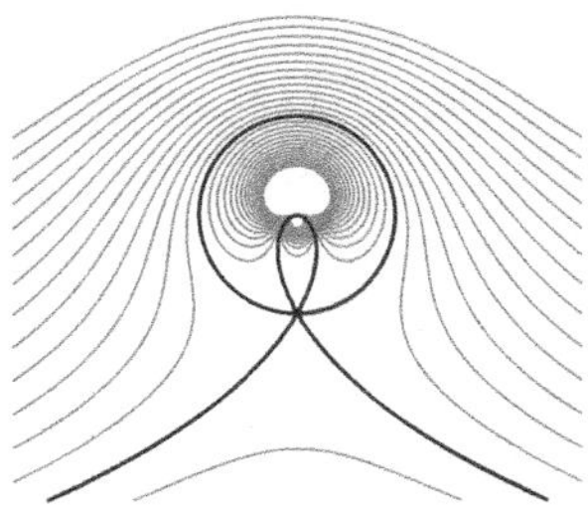

Figure 4.22 Streamlines for the case of one stagnation point on the cylinder surface.

Note that while $|\Gamma| > 4\pi U_\infty a$ results in no stagnation points on the cylinder surface, there is a stagnation point *off the cylinder*. We can obtain its location by looking for the solution to $v_r = 0$ and $v_\theta = 0$ when $r \neq a$. Using Equation 4.71 first,

$$v_r = 0 \quad \Rightarrow \quad \cos\theta = 0 \quad \Rightarrow \quad \theta = \frac{\pi}{2} \text{ or } \frac{3\pi}{2}. \tag{4.76}$$

This equation shows that the stagnation points off the cylinder must lie on the y axis, directly below or above the cylinder. To determine the r value of these stagnation points, we turn to the $v_\theta = 0$ equation, using Equation 4.72

$$v_\theta = 0 \quad \Rightarrow \quad -U_\infty(\pm 1)\left(1 + \frac{a^2}{r^2}\right) - \frac{\Gamma}{2\pi r} = 0. \tag{4.77}$$

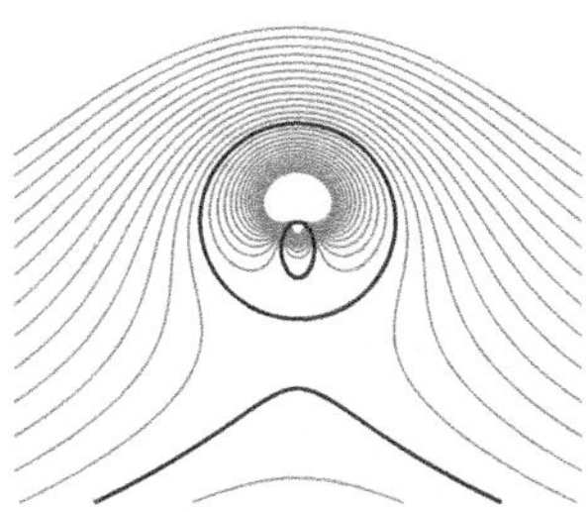

Figure 4.23 Streamlines for the case of no stagnation points on the cylinder surface.

This is a quadratic equation for r, with the ± 1 term arising from $\sin\theta$. For $\Gamma > 0$, only $\theta = 3\pi/2$ yields a solution, since otherwise both terms would be negative. Likewise, for $\Gamma < 0$, only $\theta = \pi/2$ yields a solution. In addition, two real solutions only occur when the vortex is sufficiently strong, $|\Gamma| > 4\pi U_\infty a$, consistent with the above discussion. In these conditions, one stagnation point will be inside the cylinder, in the fictitious flow, and one will be outside. Figure 4.23 illustrates such a case.

4.3.4 Pressure Distribution

Viscous effects are ignored in potential flow,* which means that pressure determines the force of the fluid on an object and vice versa. To calculate the pressure, we first obtain the velocity from the potential-flow model and then use Bernoulli's equation, which is applicable due to the irrotational flow assumption. Integrating the pressure over a surface then yields the force.

* We assumed irrotational flow, whereas viscous shear would cause fluid elements to rotate.

Obtaining pressure from Bernoulli's equation requires the velocity magnitude, which is available in several ways: from the velocity potential, the stream function, or the complex velocity. Numerically, the complex velocity offers a simple approach, since the formulas for the complex potential and its derivative are relatively straightforward for elementary flows. The velocity magnitude is†

† For complex numbers, $|\cdot|$ denotes the magnitude, which is the distance from the origin.

$$|\vec{v}| = |w| = |F'(z)|. \tag{4.78}$$

Suppose that far away from an object, the static pressure is p_∞, and the flow speed is U_∞. Let p and $\vec{v}$ be the pressure and velocity at any other point in the flow. Applying Bernoulli's equation, Equation 3.96, between this point and a point far away gives

$$\begin{aligned} p + \frac{1}{2}\rho|\vec{v}|^2 &= p_\infty + \frac{1}{2}\rho U_\infty^2, \\ p &= p_\infty + \frac{1}{2}\rho U_\infty^2 - \frac{1}{2}\rho|\vec{v}|^2. \end{aligned} \tag{4.79}$$

This is a dimensional pressure, and its value depends on the choice of units. A non-dimensional measure of the pressure is the *pressure coefficient*, which is defined as‡

$$c_p = \frac{p - p_\infty}{\frac{1}{2}\rho U_\infty^2}. \tag{4.80}$$

‡ The quantity $\frac{1}{2}\rho U_\infty^2$ is called the *dynamic pressure*, and its definition is only valid when a freestream is present.

Using Equation 4.79, the pressure coefficient in potential flow is§

$$c_p = 1 - \frac{|\vec{v}|^2}{U_\infty^2}. \tag{4.81}$$

§ In potential flow, c_p depends only on the ratio of the local speed to the freestream speed, not on the freestream pressure or the fluid density.

Recall flow over a lifting cylinder, for which the complex potential is

$$F(z) = U_\infty z + \frac{\kappa}{2\pi z} + \frac{i\Gamma}{2\pi} \ln \frac{z}{a}. \tag{4.82}$$

Differentiating with respect to z gives the complex velocity

$$w = F' = U_\infty - \frac{\kappa}{2\pi z^2} + \frac{i\Gamma}{2\pi z}. \tag{4.83}$$

Using Equation 4.78 and Equation 4.81, we can calculate the pressure coefficient at any point in the flow and make a contour plot. Listing 4.6 shows a code that does this.

Listing 4.6 Code for plotting pressure coefficient contours for flow over a cylinder.

```
function cylcp(Lambda, cprange, fname)
Uinf = 1;   % freestream speed
kappa = 3;  % doublet strength
a = sqrt(kappa/(2*pi*Uinf)); % cylinder radius
Gamma = 4*pi*a*Uinf*Lambda;
[X,Y] = ndgrid(linspace(-2,2,100));
Z = X + 1i*Y;
W = Uinf - kappa./(2*pi*Z.^2) + i*Gamma./(2*pi*Z);
cp = 1 - (abs(W)).^2/Uinf^2;
figure(1); clf; set(gcf, 'position', [0,0,600,600]);
contourf(X,Y,cp, cprange, 'linestyle', 'none'); hold on;
t = linspace(0,2*pi,100);
patch(a*cos(t), a*sin(t), 'w'); % plot a white circle
colormap(gray(20)); colorbar('southoutside'); % color map and bar
axis equal; axis off; box off;
writeplot('x', 'y', fname, 2);
```

Code notes:

- The input `Lambda` (Λ) sets the vortex strength.
- The function `abs` returns the magnitude of a complex number.
- `W` contains the complex velocity at every point.
- Filled contours are made by using `contourf` (note the f).
- A white filled circle is plotted on top of the contours to hide the fictitious flow inside the cylinder.

Figure 4.24 shows the resulting pressure coefficient field plot. By Equation 4.81, the largest value c_p can attain is 1, and this occurs at the stagnation points, where the velocity is zero. A negative c_p indicates that the flow speed is faster than U_∞, and inasmuch as the flow speed can be arbitrarily large, c_p can be arbitrarily low. A limit is that the assumption of incompressible flow will eventually break down when the flow speed is not small relative to the speed of sound.

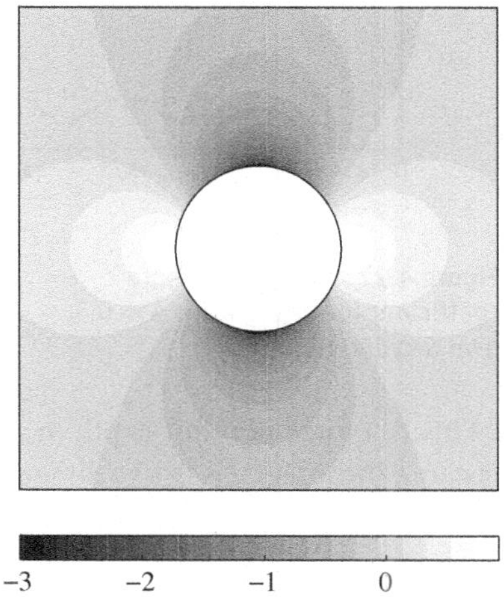

Figure 4.24 Pressure coefficient contours for non-lifting flow over a cylinder.

The quantity Λ that is an input into the function in Listing 4.6 is a *dimensionless circulation*, defined by

$$\Lambda \equiv \frac{\Gamma}{4\pi a U_\infty}. \tag{4.84}$$

When $\Lambda = 0$, the flow is non-lifting, while positive/negative values of Λ correspond to positive/negative lift. For $\Lambda = \pm 1$, only one stagnation point exists on the cylinder surface (see Figure 4.22), and for $|\Lambda| > 1$, there are no stagnation points on the cylinder surface (see Figure 4.23).

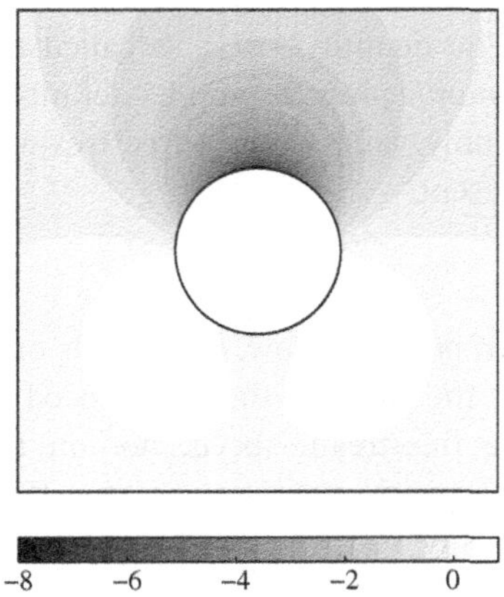

Figure 4.25 Pressure coefficient contours for $\Lambda = 0.5$.

Figure 4.25 shows the c_p field for $\Lambda = 0.5$. Note the more negative c_p on the top of the cylinder, which corresponds to faster flow there. The stagnation points, where $c_p = 1$, have also shifted down to the lower surface of the cylinder, consistent with the streamlines shown in Figure 4.19.

Of interest in a force calculation is the pressure, or equivalently c_p, on the surface of the object. We can determine c_p by first finding the speed and then using Bernoulli's equation. For the case of a cylinder, the simple geometry lends itself to an analytical calculation.

On the cylinder surface, $r = a$, so the velocity components are, using Equation 4.71 and Equation 4.72,

$$v_r = 0, \tag{4.85}$$

$$v_\theta = -2U_\infty \sin\theta - \frac{\Gamma}{2\pi a} = -2U_\infty(\sin\theta + \Lambda). \tag{4.86}$$

By Equation 4.81, the pressure coefficient is

$$c_p = 1 - \frac{|\vec{v}|^2}{U_\infty^2} = 1 - \frac{v_\theta^2}{U_\infty^2} = 1 - 4(\sin\theta + \Lambda)^2. \tag{4.87}$$

When $\Lambda = 0$, we have the case of a non-lifting cylinder, and

$$\Lambda = 0 \quad \Rightarrow \quad c_p(\theta) = 1 - 4\sin^2\theta. \tag{4.88}$$

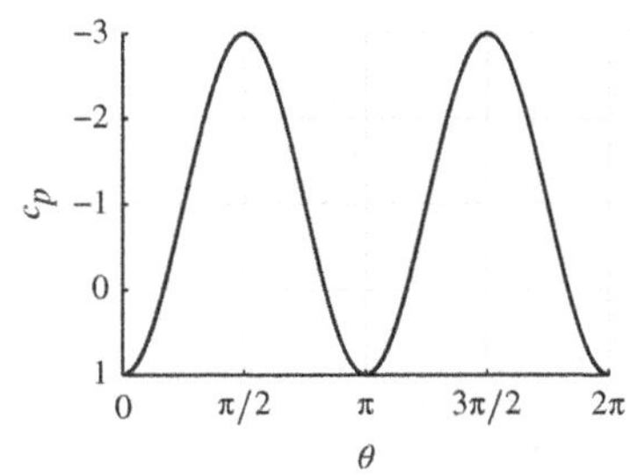

Figure 4.26 Surface pressure coefficient for non-lifting flow over a cylinder.

This pressure coefficient distribution is plotted in Figure 4.26. Note that the convention is to plot $-c_p$, or to use a reversed y axis, so that more negative c_p values, which correspond to lower pressures, appear higher on the c_p plot. At stagnation points, the velocity is zero, the pressure is the stagnation pressure, and $c_p = 1$. We can see this in Figure 4.26 at $\theta = 0$ and $\theta = \pi$, which are the stagnation points in the non-lifting case. For positive circulation, $\Gamma > 0$, the pressure decreases further on the top of the cylinder, which corresponds to $0 < \theta < \pi$. This is illustrated in Figure 4.27. Note that now the stagnation points lie between π and 2π, as that is where $c_p = 1$. This observation is consistent with our discussion of stagnation point locations in the previous section.

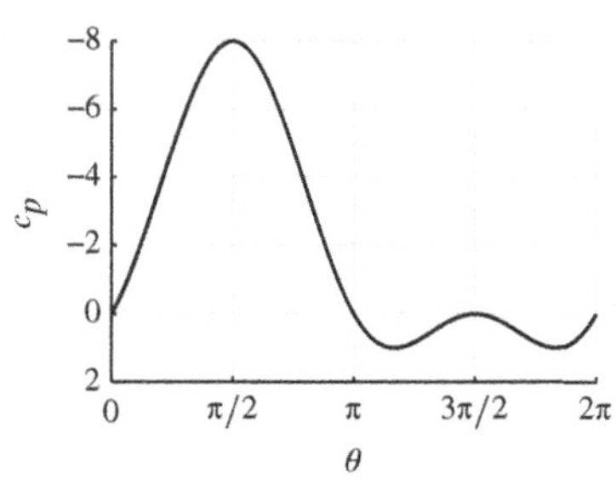

Figure 4.27 Surface pressure coefficient for lifting flow, $\Lambda = 0.5$, over a cylinder.

4.3.5 Force Calculation

In an inviscid flow, the force on a body is due to the pressure acting on the surface of the body. In this section, we calculate the force on the cylinder by integrating the pressure over the cylinder's surface. As illustrated in Figure 4.28, the force on the cylinder is*

$$\vec{F}' = \oint_{\text{cylinder}} p\vec{n}\, dl = \int_0^{2\pi} p(\theta)\vec{n}(\theta) a d\theta, \tag{4.89}$$

where the prime denotes that this is a force per unit depth, and $\vec{n}$ is the normal vector that points outward from the fluid toward the origin. This normal vector is

$$\vec{n} = -\cos\theta\hat{\imath} - \sin\theta\hat{\jmath}, \tag{4.90}$$

so that the components of $\vec{F}' = F'_x\hat{\imath} + F'_y\hat{\jmath}$ are

$$F'_x = -\int_0^{2\pi} p(\theta)\cos\theta\, a d\theta, \quad F'_y = -\int_0^{2\pi} p(\theta)\sin\theta\, a d\theta. \tag{4.91}$$

*This is a force per unit depth in z, because the integral is in two dimensions.

Considering first F'_x,

$$F'_x = -\int_0^{2\pi} p(\theta)\cos\theta\, a d\theta$$
$$= -a\int_0^{2\pi} p_\infty \cos\theta\, d\theta$$
$$-\frac{1}{2}\rho U_\infty^2 a \int_0^{2\pi}\left(1 - 4\sin^2\theta - 8\Lambda\sin\theta - 4\Lambda^2\right)\cos\theta\, d\theta,$$

where we used the definition of c_p to write $p = p_\infty + \frac{1}{2}\rho U_\infty^2 c_p$, and the expression for the surface c_p from Equation 4.87. Evaluating the integrals† yields the result that all terms vanish in the F'_x calculation,

$$F'_x = 0. \tag{4.92}$$

Since x is aligned with the freestream direction, F'_x is the drag (per unit depth) on the cylinder, and the above equation then states that there is *no drag on the cylinder*, regardless of how strong the vortex is. Looking at the y component of the force, which is the lift, we have‡

$$F'_y = -\int_0^{2\pi} p(\theta)\sin\theta\, a d\theta$$
$$= -a\int_0^{2\pi} p_\infty \sin\theta d\theta$$
$$-\frac{1}{2}\rho U_\infty^2 a \int_0^{2\pi}\left(1 - 4\sin^2\theta - 8\Lambda\sin\theta - 4\Lambda^2\right)\sin\theta\, d\theta$$
$$= \frac{1}{2}\rho U_\infty^2 a 8\Lambda\pi.$$

Using $\Lambda = \Gamma/(4\pi a U_\infty)$ from Equation 4.84, we have

$$F'_y = \rho U_\infty \Gamma. \tag{4.93}$$

That is, lift is proportional to the vortex strength. Putting these results together yields the following result for a cylinder:§

$$\begin{aligned} &\text{Drag per unit depth, } D' = 0,\\ &\text{Lift per unit depth, } L' = \rho U_\infty \Gamma. \end{aligned} \tag{4.94}$$

Example 4.7 Hangar on a windy day
Consider a long, semi-cylindrical hangar on flat ground as shown in Figure 4.29. Ignoring end effects, we model the flow as two-dimensional. Assume potential flow theory and a uniform freestream velocity of U_∞ in the $+x$ direction. Also assume that the air pressure inside the hangar is equal to the freestream static pressure, p_∞.

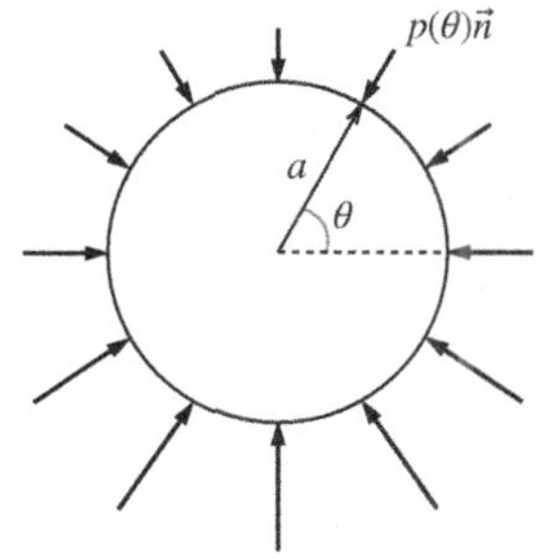

Figure 4.28 The force on a cylinder is obtained by integrating the pressure on the cylinder surface.

† The relevant trigonometric identities are

$$\int_0^{2\pi} \cos\theta\, d\theta = 0,$$
$$\int_0^{2\pi} \sin^2\theta\cos\theta\, d\theta = 0,$$
$$\int_0^{2\pi} \sin\theta\cos\theta\, d\theta = 0.$$

‡ The relevant trigonometric identities here are

$$\int_0^{2\pi} \sin\theta\, d\theta = 0,$$
$$\int_0^{2\pi} \sin^3\theta\, d\theta = 0,$$
$$\int_0^{2\pi} \sin^2\theta\, d\theta = \pi.$$

§ Although we have derived these expressions for a cylinder, we will see in Section 4.5 that they hold for general shapes as well.

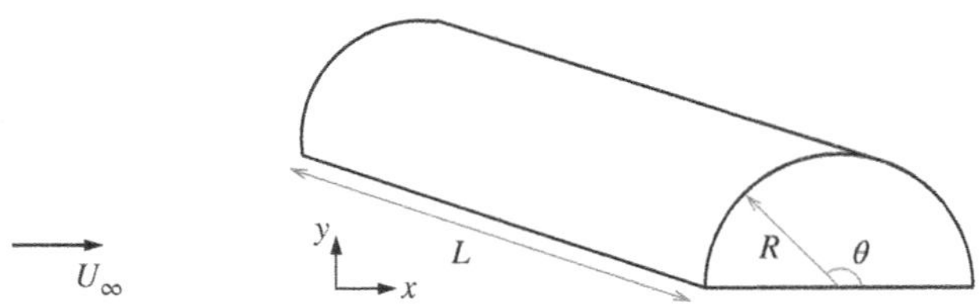

Figure 4.29 Semi-cylindrical hangar geometry.

We can calculate the force on the hangar in terms of ρ, U_∞, R, and L. This force will be positive in the vertical direction, tending to lift the hangar off the ground.

Since the ground must be a streamline, we are effectively looking at the top half of flow over a cylinder with zero circulation. The pressure coefficient over the top of the cylinder is, from Equation 4.88,

$$c_p(\theta) = \frac{p(\theta) - p_\infty}{\frac{1}{2}\rho U_\infty^2} = 1 - 4\sin^2\theta. \tag{4.95}$$

The stagnation points are at $\theta = 0, \theta = \pi$. The pressure inside the hangar is p_∞, and so integrating the pressure difference over the hangar surface gives the force¶

¶ The force vector on the hangar involves the integral of $p_\infty - p(\theta)$ times $\vec{n}$, where $\vec{n} = \hat{r} = \cos\theta\hat{\imath} + \sin\theta\hat{\jmath}$ is the normal vector pointing outward from the cylinder. So $\sin\theta$ is the weight that gives the vertical force. In addition, L accounts for the third dimension, turning the sectional force into a total force on the hangar.

$$\begin{aligned} F_y &= L\int_0^\pi (p_\infty - p(\theta))\sin(\theta)\,Rd\theta \\ &= LR\int_0^\pi \left(p_\infty - p_\infty - (1 - 4\sin^2\theta)\frac{1}{2}\rho U_\infty^2\right)\sin\theta d\theta \\ &= \frac{1}{2}\rho U_\infty^2 LR\int_0^\pi \left(4\sin^3\theta - \sin\theta\right)d\theta \\ &= \frac{1}{2}\rho U_\infty^2 LR\int_0^\pi (3 - 4\cos^2\theta)\sin\theta d\theta \\ &= \frac{5}{3}\rho U_\infty^2 LR. \end{aligned} \tag{4.96}$$

Using reasonable numbers, $\rho = 1.23$ kg/m^3, $U_\infty = 20$ m/s, $R = 15$ m, $L = 100$ m, this force evaluates to

$$F_y = \frac{5}{3}1.23\text{ kg/m}^3(20\text{ m/s})^2(100\text{ m})(15\text{ m}) = 1{,}230{,}000\text{ N}. \tag{4.97}$$

The horizontal force is zero because the pressure distribution is fore–aft symmetric. Again, this is due to a potential-flow idealization, since the hangar is a bluff body and in reality the flow will separate, as shown in Figure 4.17, which will lead to a horizontal drag force.

4.3.6 A Point-Source Numerical Method

Superposition can be used to model flows over general objects such as airfoils. One technique uses point sources distributed horizontally on the x axis, as shown in Figure 4.30.

The strengths of the N sources, Q_j, $1 \leq j \leq N$, each located at $\vec{x}_j$, are the unknowns in this method. We can solve for these unknowns by writing N equations that express the conditions we want the flow to satisfy. In this case, given a geometry of an airfoil, the physical condition is that no flow crosses the geometry surface. The flow must be tangent to the surface, making it a streamline, and this condition is known as flow tangency. It will be the basis of airfoil theory in the next chapter.

Since with N unknowns we can only enforce N equations, we cannot make the flow tangent to the airfoil at all points. Instead, we pick N points on the surface at which to satisfy flow tangency, and these are called *control points*. As shown in Figure 4.30, the control points are only on the upper surface of the airfoil. This is because the flow induced by a line of sources is symmetrical about the line, so only symmetrical airfoils can be modeled by this method.*

* Non-symmetrical airfoils generally produce lift, and as we will see in Section 4.5, vortices are needed to model lifting flow.

The numerical approach is to write down the flow tangency equation at each control point i, located at $\vec{x}_i^{cp}$, in terms of the unknown strengths Q_j. Each source affects the velocity at each control point. By superposition, the velocity at control point i is

$$\vec{v}_i = \sum_{j=1}^{N} \Delta\vec{v}_{ij} + U_\infty \hat{\imath}, \tag{4.98}$$

where $\Delta\vec{v}_{ij}$ is the velocity induced by source j on control point i, and where we have assumed a horizontal freestream. Flow tangency requires that $\vec{v}_i \cdot \vec{n}_i = 0$, where $\vec{n}_i$ is the normal vector, perpendicular to the geometry, at control point i.†

† In terms of the geometry function, $y(x)$, the normal vector depends on the slope,

$$\vec{n} \propto -\frac{dy}{dx}\hat{\imath} + 1\hat{\jmath}.$$

We can normalize the vector or leave it as is, since a constant in this vector only scales the equations.

To determine $\Delta\vec{v}_{ij}$, we use the elementary flow result for source flow. Source j induces a radial velocity outward from its center, of magnitude $Q/(2\pi r)$, where r is the distance from the source. This distance is between the control point and the source center,

$$r = \sqrt{(x_i^{cp} - x_j)^2 + (y_i^{cp} - y_j)^2}. \tag{4.99}$$

Since the direction is radially outward, the effect of source j on control point i is

$$\Delta\vec{v}_{ij} = \frac{Q_j}{2\pi r}\left(\frac{x_i^{cp} - x_j}{r}\hat{\imath} + \frac{y_i^{cp} - y_i}{r}\hat{\jmath}\right). \tag{4.100}$$

Dotting $\Delta\vec{v}_{ij}$ with the normal, $\vec{n}_i$, gives

$$\Delta\vec{v}_{ij} \cdot \vec{n}_i = \frac{1}{2\pi r}\left(\frac{x_i^{cp} - x_j}{r}n_{ix} + \frac{y_i^{cp} - y_i}{r}n_{iy}\right)Q_j = A_{ij}Q_j, \tag{4.101}$$

where n_{ix} and n_{iy} are the components of the normal vector $\vec{n}_i$, and A_{ij} is the influence coefficient of source j on control point i.‡

‡ Summation is not implied in this equation: this is the effect of one source, Q_j, on one control point, i.

Considering now all of the control points together, the flow tangency condition applied at the N control points gives a linear system of equations for the unknown source strengths. This equation is of the form

$$\mathbf{AQ} = \mathbf{b}, \tag{4.102}$$

where $\mathbf{A}$ is an N by N matrix with entries A_{ij}, $\mathbf{Q}$ is a vector of N unknown source strengths, and $\mathbf{b}$ is a right-hand-side vector containing N entries.

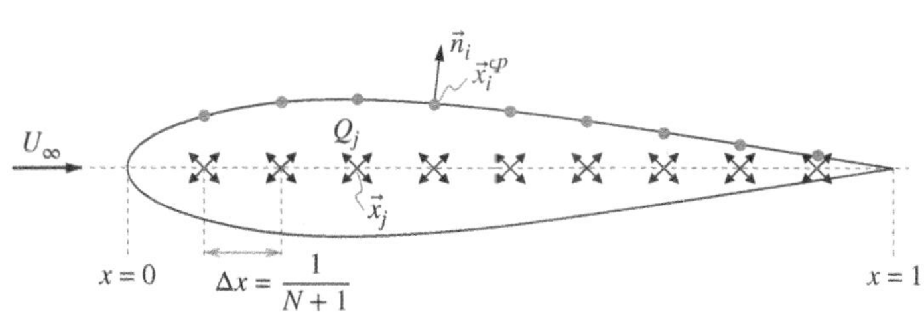

Figure 4.30 Model of non-lifting flow over a symmetrical airfoil using a line of sources.

To determine the entries of **b** and form the system, we substitute the $\vec{v}_i$ expression from Equation 4.98 into the flow tangency requirement, $\vec{v}_i \cdot \vec{n}_i = 0$,

$$\begin{aligned} \vec{v}_i \cdot \vec{n}_i = \left[\sum_{j=1}^{N} \Delta\vec{v}_{ij} + U_\infty \hat{i}\right] \cdot \vec{n}_i &= 0 \\ \sum_{j=1}^{N} \Delta\vec{v}_{ij} \cdot \vec{n}_i + U_\infty n_{ix} &= 0 \\ \sum_{j=1}^{N} A_{ij} Q_j &= -U_\infty n_{ix}. \end{aligned} \tag{4.103}$$

This set of equations can be written as a matrix system, with i indexing the rows and j the columns,

$$\begin{matrix} \text{point } 1 \rightarrow \\ \text{point } 2 \rightarrow \\ \vdots \\ \text{point } N \rightarrow \end{matrix} \begin{bmatrix} A_{1,1} & A_{1,2} & \cdots & A_{1,N} \\ A_{2,1} & A_{2,2} & \cdots & A_{2,N} \\ \vdots & \vdots & \ddots & \vdots \\ A_{N,1} & A_{N,2} & \cdots & A_{N,N} \end{bmatrix} \begin{bmatrix} Q_1 \\ Q_2 \\ \vdots \\ Q_N \end{bmatrix} = \begin{bmatrix} -U_\infty n_{1x} \\ -U_\infty n_{2x} \\ \vdots \\ -U_\infty n_{Nx} \end{bmatrix}. \tag{4.104}$$

Solving this system of equations gives the source strengths, which, once known, can be used to plot streamlines and compute pressures or forces. Note that the freestream velocity only enters into the right-hand side, **b**, and it does not affect the matrix **A**, which only depends on the airfoil geometry and the distribution of sources.

Example 4.8 Point-source method applied to an airfoil
This example shows how to apply the point-source method to analyze the flow over a specific airfoil, the NACA 0020. This is a four-digit series airfoil that has no camber and whose thickness is 20% of the chord. Chapter 5 discusses these airfoils further, and for this problem, we only need to know its geometry. Using a unit chord length, the upper and lower surfaces of the airfoil are given by

$$y = \pm\left[0.2969\sqrt{x} - 0.1260x - 0.3516x^2 + 0.2843x^3 - 0.1015x^4\right].$$

Following the steps outlined in this section, we write a code that builds and solves the linear system in Equation 4.104. Listing 4.7 shows such a code.

Code notes:

- The freestream speed $U_\infty = 1$ is chosen arbitrarily. It does not affect the shape of the streamlines or any pressure/-force coefficients, which are non-dimensional.
- The geometry function computes both $y(x)$ for placing the control points, and dy/dx, for computing the normal vector.
- Construction of the matrix **A** occurs in a double loop over indices i and j; at each i, an entry in the right-hand-side **b** vector is also filled in.

Listing 4.7 Code for solving for flow over a symmetrical airfoil using a line of point sources.

```
function linesource
% Models flow over a symmetric airfoil using a line of sources

Uinf = 1;              % freestream speed
N = 50;                % number of unknowns
x = linspace(0,1,N+2);
xj = x(2:end-1);       % source point x locations
xi = xj;               % control point x locations
[yi, y_x] = geom(xi); % geometry
ni = ones(N,2); ni(:,1) = -y_x; % normal vector
for i = 1:N, ni(i,:) = ni(i,:)/norm(ni(i,:)); end

% set up the matrix system
A = zeros(N,N); b = zeros(N,1);
for i = 1:N
  for j = 1:N
    dx = xi(i)-xj(j); dy = yi(i); r2 = dx^2 + dy^2;
```

```
    A(i,j) = 1/(2*pi*r2) * (dx*ni(i,1) + dy*ni(i,2));
  end
  b(i) = -Uinf*ni(i,1);
end
Q = A\b; % solve the system

%----------------------------
function [y,y_x] = geom(x)
% computes the geometry and its derivative for a NACA 0020 airfoil
y = (0.2969*sqrt(x) - 0.1260*x - 0.3516*x.^2 + 0.2843*x.^3 - 0.1015*x
    .^4);
y_x = 0.5*0.2969./sqrt(x) - 0.1260 - 2*0.3516*x + 3*0.2843*x.^2
    -4*0.1015*x.^3;
```

After solving for the source strengths, we can use superposition again to process the results. For example, we can visualize the streamlines by plotting contours of the stream function, which is the imaginary part of the complex potential. The code in Listing 4.8 does this, and Figure 4.31 shows the resulting streamlines. The streamlines appear tangent to the overlaid geometry, which verifies that the flow tangency equations are being satisfied.§

§ Tangency is only enforced at discrete control points, but with many control points, the flow will be close to tangent to the entire geometry.

Listing 4.8 Code for plotting streamlines from a line of sources.

```
% plot streamlines and the airfoil
figure(1); clf;
[X,Y] = ndgrid(linspace(-.2,1.2,101),linspace(-.5,.5,51));
Z = X + Y*1i; % complex coordinate
F = Uinf*Z;    % complex potential: freestream first
for j = 1:N, F = F + Q(j)/(2*pi) * log(Z - xj(j)); end
Psi = imag(F); % stream function
contour(X,Y,Psi,21, 'color', '#888888'); hold on;
xx = [0,xi,1]; yy = [0,yi,0];
plot(xx, yy,'k-', 'linewidth', 2);
plot(xx,-yy,'k-', 'linewidth', 2);
axis equal; axis off; box off;
writeplot('x', 'y', '../figs/linesourcestream.pdf', 2);
```

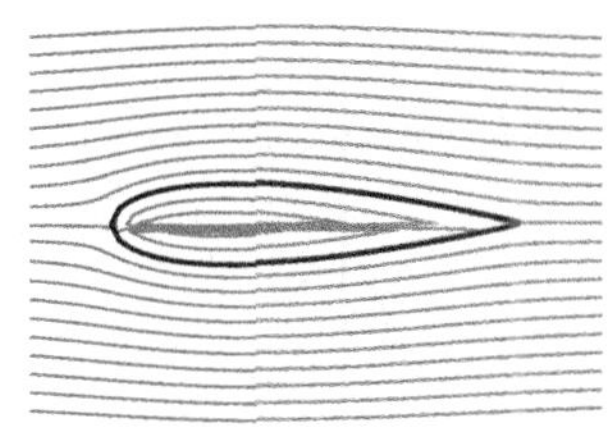

Figure 4.31 Streamlines for flow over a symmetric airfoil, generated using a line of point sources.

4.4 Images and Motion

So far we have considered elementary flows and their superposition to model steady flows over simple two-dimensional objects. This section considers the addition of an infinite solid surface, such as a ground, which is relevant to aerodynamics when studying low-level flight and ground effect. The steady assumption is also relaxed in the presentation of moving vortices, which are relevant to the understanding of vortices left behind in the wake of an aircraft.

4.4.1 The Method of Images

One use of superposition is the *method of images* which lets us model elementary flows in the vicinity of impenetrable walls. Suppose that a source of strength Q is a distance h from a wall, as shown in Figure 4.32. Since the wall is solid, there is no flow across it. This means that the vertical velocity at the wall must be zero.

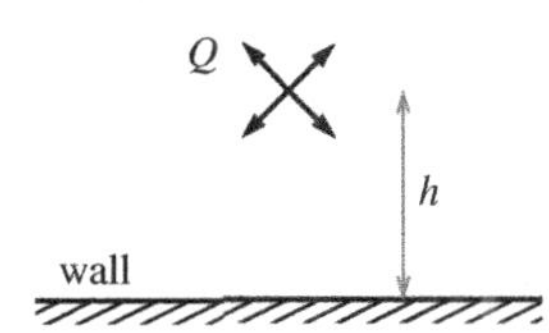

Figure 4.32 Source above a wall.

Elementary flows do not provide a solid wall: each of these flows can be evaluated at any position x and y. However, we can "simulate" the effect of a wall by using *image* elementary flows. For the current case, consider the flow that results

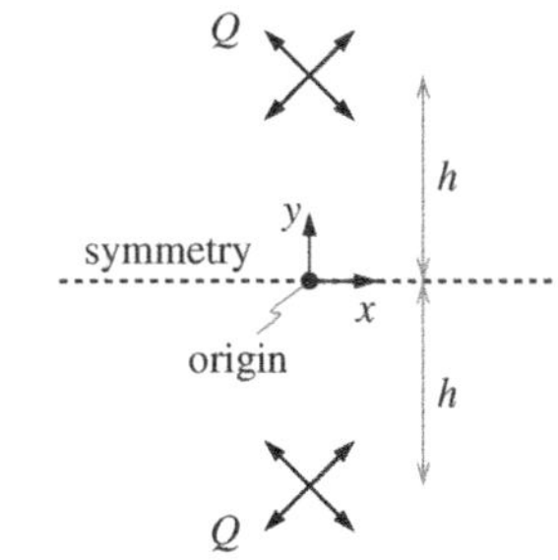

Figure 4.33 Source and wall modeled using the method of images.

* Note the $\pm hi$ in the argument to ln, which accounts for the vertical offset of the source and image relative to the origin.

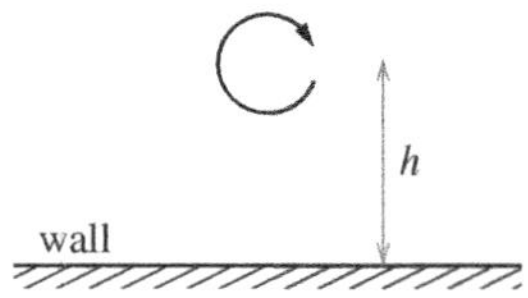

Figure 4.34 Vortex above a wall.

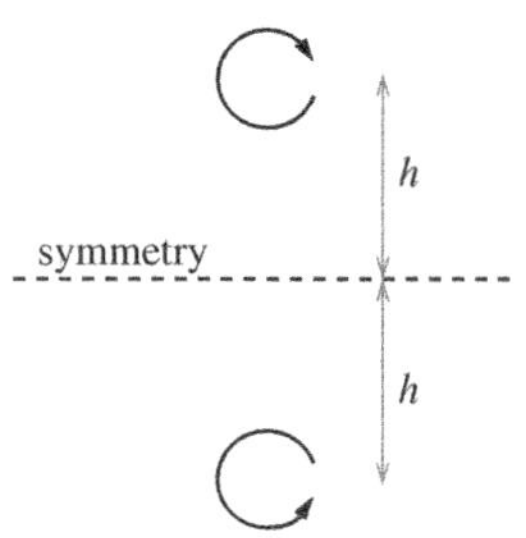

Figure 4.35 Vortex and wall modeled using the method of images.

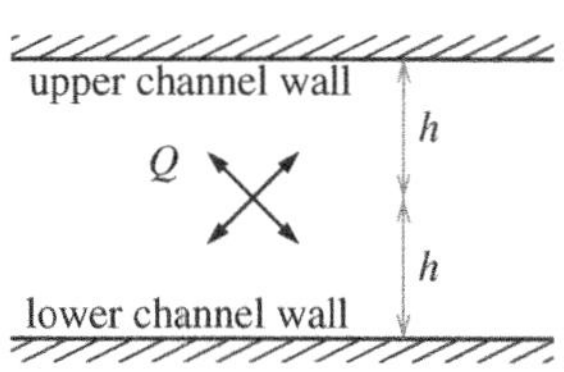

Figure 4.36 Using the method of images to model a source in a channel.

from two sources, placed as shown in Figure 4.33. Both sources have the same strength, so the resulting flow is symmetric about the dashed line shown. In particular, the vertical velocity on the symmetry line must be zero: the upper source induces a velocity with a downward component, which is balanced by the upward velocity component from the lower source. Also, far away from the sources, the velocity decays to zero for both flows.

Considering the domain above the wall in Figure 4.32 and above the symmetry line in Figure 4.33, we see that both flows have potentials that satisfy Laplace's equation, and both flows satisfy the same boundary conditions. Hence, mathematically by uniqueness of solutions to the Laplace equation, *they are the same flow*. Therefore, we can use the source pair system to model a source near a wall. The complex potential for the source pair, written for an origin placed on the wall between the sources, is*

$$F^{\text{tot}}(z) = \underbrace{\frac{Q}{2\pi}\ln(z - hi)}_{\text{upper source}} + \underbrace{\frac{Q}{2\pi}\ln(z + hi)}_{\text{lower source}}. \tag{4.105}$$

From $F^{\text{tot}}(z)$, we can compute $\phi(x, y)$, $\psi(x, y)$, and $\vec{v}$. Although we will be able to compute the velocity at all points (x, y), in the physical case of a source near a wall, we are only interested in the flow above the wall. The flow below it can be thought of as *fictitious*.

We can also model a vortex near a solid wall by using an image vortex, as shown in Figures 4.34 and 4.35. To enforce symmetry, the image vortex must be of equal strength, but it must be rotating in the *opposite direction* relative to the original vortex. As shown in Example 4.9, the opposite rotation is required in order to make the vertical velocity components from the two vortices cancel each other at the wall. This cancellation then enforces flow tangency at the wall. Just like in the source case, considering the flow above the wall, both the original flow and the flow with the image satisfy the same equation and the same boundary conditions. Thus, the image system models the effect of the wall.

The method of images becomes more complicated in the presence of multiple walls. For example, consider a source in a channel, shown in Figure 4.36. Treating the walls independently by placing sources above and below the channel is not sufficient: the presence of the top image source destroys the symmetry condition on the bottom wall and vice versa. We can keep adding sources above and below the channel to fix this, and we will end up with an infinite number of sources. The complex potential will be, for a coordinate system centered at the physical source,

$$F^{\text{tot}}(z) = \sum_{n=-\infty}^{\infty} \frac{Q}{2\pi}\ln(z - 2nhi). \tag{4.106}$$

Note that the distance between the sources is $2h$, hence the factor of $2hi$ multiplying n in the offset for z. The imaginary unit i is present because the sources are displaced in the vertical, y, direction. From this velocity potential, velocities and pressures can be computed. This procedure is a standard approach for computing corrections to wind tunnel experiments, to account for wall effects. Using only a few terms from the infinite series is often sufficient, as the effect of images decreases with increasing distance.

Example 4.9 A vortex near the ground
An airfoil flying horizontally at speed U_∞ near the ground can be modeled with potential flow using a single vortex. Let the strength of the vortex be $\Gamma = \pi h_0 U_\infty$, and let $\beta = h/h_0$ be a dimensionless height of the vortex above the

ground. We assume that the two-dimensional ground extends infinitely in the $\pm x$ directions. Figure 4.37 shows the setup for this problem, in the frame of the airfoil, so that there is a freestream flow of speed U_∞.

Of interest in this problem is the pressure coefficient, $c_p(x)$, on the ground ($y = 0$) as a function of x/h_0, and this answer will be in terms of β only. By Equation 4.81, the pressure coefficient depends on the ratio of the local speed to the freestream speed. Hence, we need the velocity at the ground as a function of x. To model the ground, we use an image vortex, as shown in Figure 4.38.

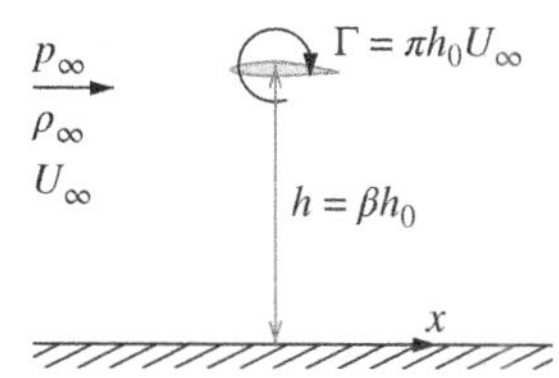

Figure 4.37 An airfoil in freestream near the ground, modeled by a single vortex.

Note that the orientation of the image vortex is flipped to maintain the condition of no flow through the ground. At a location x on the ground, each vortex induces both a vertical and a horizontal velocity component. The vertical components cancel,† but the horizontal components $-\Gamma/(2\pi r)\cos(\theta)$ from each vortex, add together. So the velocity at the ground is horizontal and has magnitude:

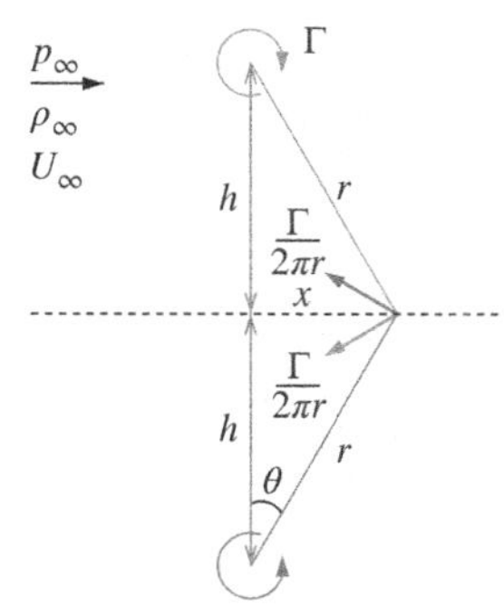

Figure 4.38 Velocity calculation with an image vortex.

$$\begin{aligned} u(x, y = 0) &= U_\infty - 2\left(\frac{\Gamma}{2\pi r}\cos(\theta)\right) \\ &= U_\infty - \frac{\Gamma}{\pi h}\cos^2(\theta) \\ &= U_\infty\left(1 - \frac{\cos^2(\theta)}{\beta}\right), \end{aligned} \tag{4.107}$$

where we used $\cos(\theta) = h/r, \beta = h/h_0$, and the given $\Gamma = \pi h_0 U_\infty$. Now, the pressure coefficient at $y = 0$ is

$$\begin{aligned} c_p(x) &= 1 - \left(\frac{u(x, y = 0)}{U_\infty}\right)^2 \\ &= 1 - \left(1 - \frac{\cos^2(\theta)}{\beta}\right)^2 \\ &= 2\frac{\cos^2(\theta)}{\beta} - \frac{\cos^4(\theta)}{\beta^2}, \end{aligned} \tag{4.108}$$

where $\theta = \tan^{-1}(x/h)$ so that

$$\cos^2\theta = \frac{h^2}{h^2 + x^2} = \frac{\beta^2}{\beta^2 + (x/h_0)^2}. \tag{4.109}$$

These expressions allow us to plot c_p as a function of x/h_0, with β as the only parameter. Figure 4.39 shows the plots for $\beta = 1, 2, 10$. Note that the pressure distribution is more "spiky" when the vortex is closer to the ground (lower β). Also, the vortex is close enough to the ground for $\beta = 1$ to make the flow stagnate at $x = 0$.

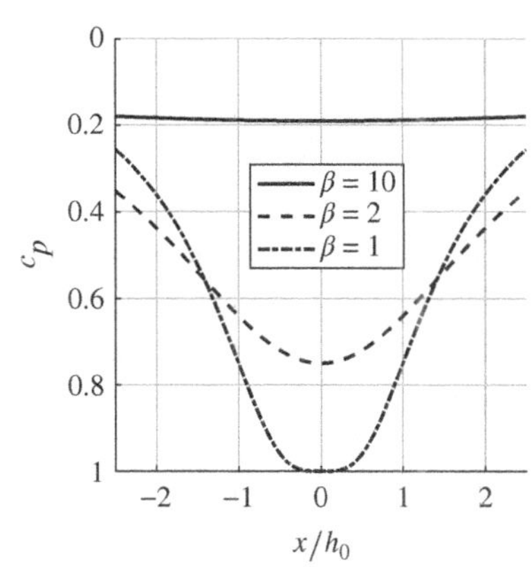

Figure 4.39 Pressure coefficient distributions on the ground.

The vertical force on the ground due to the vortex can be calculated by looking at the pressure difference that results from the presence of the vortex. Specifically, the extra *downward* vertical force on the ground due to the vortex has magnitude‡

$$F'_y = \int_{-\infty}^{\infty} (p(x) - p_\infty)dx. \tag{4.110}$$

† They would not cancel if the orientations of both vortices were the same. The image vortex is flipped to maintain flow tangency.

‡ If the vortex were not present, $p(x)$ would be p_∞ and F'_y would evaluate to zero as expected.

By the definition of c_p, the integrand is

$$p(x) - p_\infty = \frac{1}{2}\rho U_\infty^2 c_p(x), \tag{4.111}$$

so that

$$F_y' = \int_{-\infty}^{\infty} (p(x) - p_\infty)dx = \frac{1}{2}\rho U_\infty^2 \int_{-\infty}^{\infty} c_p(x)dx. \tag{4.112}$$

This integral is easier to calculate with respect to θ instead of x, by using the change of variables, $x/h = \tan\theta$.§ Substituting for c_p using Equation 4.108, the integral is

§ Differentiating the transformation gives

$$dx = h\sec^2\theta d\theta = \frac{h}{\cos^2\theta}d\theta.$$

$$\begin{aligned} F_y' &= \frac{1}{2}\rho U_\infty^2 \int_{-\pi/2}^{\pi/2} \left(2\frac{\cos^2(\theta)}{\beta} - \frac{\cos^4(\theta)}{\beta^2}\right)\frac{h}{\cos^2\theta}d\theta \\ &= \rho U_\infty^2 h_0 \int_{-\pi/2}^{\pi/2} \left(1 - \frac{\cos^2(\theta)}{2\beta}\right)d\theta \\ &= \rho U_\infty^2 h_0 \left(\pi - \frac{\pi}{4\beta}\right). \end{aligned} \tag{4.113}$$

The limiting values of this expression are interesting. First, as $\beta \to 0$, the force switches sign and then grows unbounded. In reality, the inviscid vortex model is not valid near the vortex core, so the result for very small β is not physical. On the other hand, if the vortex is very far away from the ground, i.e. $\beta \to \infty$, the force on the ground does not go to zero. Rather it approaches $\pi\rho U_\infty^2 h_0 = \rho U_\infty \Gamma$. This is actually an application of the Kutta–Joukowsky theorem, which we prove in Section 4.5.

4.4.2 Moving Vortices

So far, the potential flows that we have created have been steady, in that in any Eulerian frame there was no time dependence in the flow field. We can, however, also model unsteady flow fields, or at least quasi-steady ones, by letting elementary flows move. A realistic application of such motion is in the movement of vortices. A vortex that is not bound to any object and hence is free to move will do so with the local velocity of the fluid. The assumption is that the velocity field around the vortex re-establishes itself immediately after the core of the vortex moves. This would require an infinite speed of sound, which is actually consistent with the incompressibility assumption.*

* Recall that no fluid is actually incompressible, but the approximation is good when the speed of sound is large relative to the flow speed.

The velocity of a vortex is the velocity of the flow at its center. Mathematically, the center is a singularity with an undefined velocity, but a real vortex has a finite-size core, at the center of which the velocity is zero.† The center then moves according to the velocity field in which it is placed. This velocity can be due to a freestream or other nearby singularities.

† That the vortex center's velocity is zero can also be argued by symmetry.

For example, consider the case of two side-by-side vortices of equal magnitude but opposite direction strengths $\pm\Gamma$, as illustrated in Figure 4.40. For the directions shown, each vortex induces a downward velocity on the other. The magnitude of this velocity is $\Gamma/(2\pi d)$, and so both vortices descend at a speed that is directly proportional to their vortex strength and inversely proportional to their separation distance.

The effect of one vortex on the other is through an induced velocity, which is not a force. That is, one vortex does not "push" down on the other. Forces in the fluid were important in the creation of the vortices, e.g. by a wing with pressure differences. However, once established, the resulting field can be described approximately as potential flow if it satisfies the assumptions of incompressible and irrotational velocity. This potential flow solution has associated velocities that we can superimpose and evaluate: this is what we do when we look at the effect of one vortex on the other.

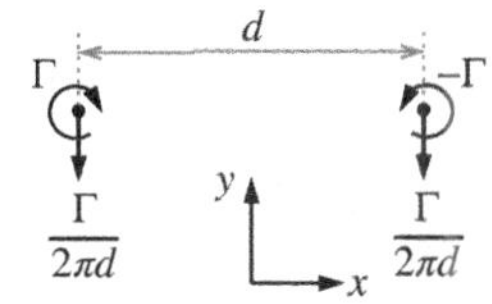

Figure 4.40 Two side-by-side vortices will move downward.

Example 4.10 Four moving vortices

When more than just a couple of singularities are present, the task of finding their velocities can be cumbersome. In this example, we consider four vortices, two pairs of two different strengths, as shown in Figure 4.41. Such a configuration can arise behind an aircraft, with the main wing producing the stronger vortices, and the tail producing the weaker ones.

The four vortices will move due to their mutual induced velocities. A numerical approach, e.g. forward Euler or predictor corrector, can be used to track their positions. To do this, we cast the problem as a dynamical system,

$$\dot{\mathbf{u}} = \mathbf{f}, \tag{4.114}$$

where $\mathbf{u}$ is a state vector storing the vortex positions, and $\mathbf{f}$ is a vector of velocities evaluated at the vortex centers. For this example, $\mathbf{u}$ contains eight numbers: four (x, y) pairs. Similarly, $\mathbf{f}$ contains the four corresponding $(\dot{x}, \dot{y})$ pairs.

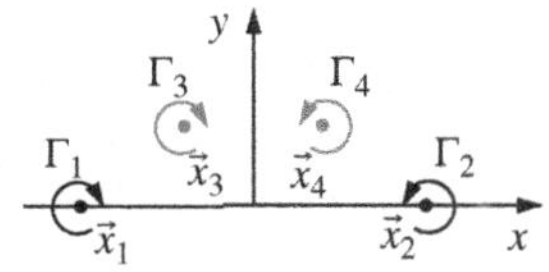

Figure 4.41 Initial condition for studying the motion of four vortices.

Each vortex induces a velocity on all of the other vortices. Consider the center of one vortex, at position $\vec{x}_i$. The velocity at $\vec{x}_i$ induced by one vortex of strength Γ_j located at position $\vec{x}_j$ is

$$\vec{v}_{ij} = \frac{\Gamma_j}{2\pi r_{ij}}\left(\frac{(y_i - y_j)}{r_{ij}}\hat{i} - \frac{(x_i - x_j)}{r_{ij}}\hat{j}\right), \tag{4.115}$$

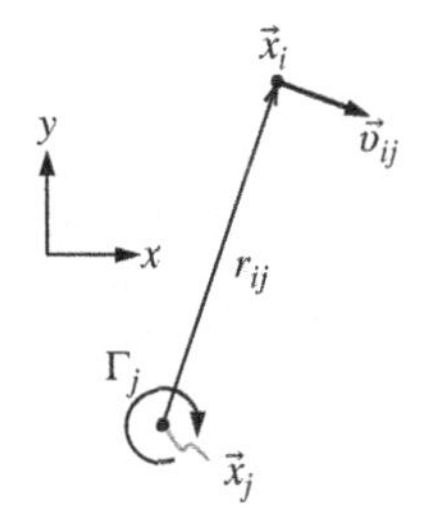

Figure 4.42 Velocity at point $\vec{x}_i$ induced by a vortex at $\vec{x}_j$.

where r_{ij} is the distance between $\vec{x}_i$ and $\vec{x}_j$, as shown in Figure 4.42. This formula comes from two properties of the velocity induced by a vortex: first, that its magnitude at a distance r_{ij} is $\Gamma_j/(2\pi r_{ij})$; and second, that its direction is clockwise tangential. The term in the large parentheses in Equation 4.115 is a unit vector in this clockwise direction.

The total velocity at $\vec{x}_i$ is the sum of the contributions from all vortices,

$$\vec{v}_i = \sum_{j \neq i} \vec{v}_{ij}. \tag{4.116}$$

Listing 4.9 shows a code that evaluates this velocity for each vortex and wraps a predictor–corrector time integration around this evaluation to obtain the solution to Equation 4.114 over a given time horizon.

Listing 4.9 Code for simulating the motion of four vortices.

```
function U = fourvort(T,N)
% tracks position of four vortices
% T = time horizon [s], N = # time steps

G = [50,-50, 5,-5]; % vortex strengths [m^2/s]
u0 = [-8,0, 8,0, -2,1, 2,1]; % initial positions [m]
```

Code notes:

- The initial state consists of the positions of the four vortices as (x, y) pairs.
- The stronger vortices are further apart and on the x axis.
- A separate function evaluates the velocities, so that it can be called with different states.
- The state at each time step is saved into a matrix for plotting of the vortex paths.

```
dt = T/N; % time step
U = [u0; zeros(N,8)]; % matrix that will store all states

% predictor-corrector time integration
for n = 1:N,
  f0 = getf(U(n,:), G);
  f1 = getf(U(n,:) + dt*f0, G);
  U(n+1,:) = U(n,:) + dt*0.5*(f0+f1);
end

%-----------------------------------------
function f = getf(u, G)
% computes the velocities of all vortices
f = zeros(size(u));
M = length(G); % number of vortices
for i = 1:M, % loop over vortices, i
  vi = zeros(1,2);
  xi = u([2*i-1,2*i]); % location of vortex i
  for j = 1:M, % loop over vortices, j
    if (i==j), continue; end
    xj = u([2*j-1,2*j]); % location of vortex j
    r = norm(xi-xj);
    vi = vi + G(j)/(2*pi*r^2)*[xi(2)-xj(2),-(xi(1)-xj(1))];
  end
  f([2*i-1,2*i]) = vi;
end
```

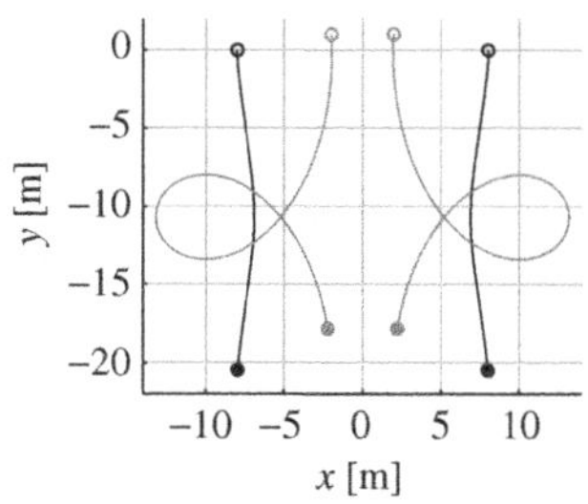

Figure 4.43 Traces of the paths of four vortices. Empty/filled circles are the initial/final positions.

Figure 4.43 shows the paths of the vortices for a simulation time of $T = 35$ units and $N = 350$ time steps. The large vortices move primarily down, whereas the smaller vortices loop around the large vortices while also traveling down on average. The vortices stay together, and the pattern of loops continues indefinitely with our potential-flow assumptions.

This same process of analyzing moving vortices applies in the presence of a ground. When unrestrained, real and image vortices affect each other and both move. For example, in Figure 4.35, both vortices will move to the left at speed $\Gamma/(4\pi h)$.

4.5 The Kutta–Joukowsky Theorem

This section presents a theorem that relates vortex strength, or equivalently circulation, to lift production in potential flow. We will use this theorem to calculate lift for both airfoils and wings, after modeling the flow with elementary singularities. We have already seen the result for a cylinder, in Equation 4.94, and here we look at the case of a body of a general shape.

Consider potential flow around an arbitrary body, constructed using various sources, sinks, vortices, doublets, and higher-order singularities. If we assume that the body is *closed*, there can be no net source or sink. That is, the sum of all of the source strengths must be zero. If we had a net source, then we could not draw a closed streamline around it, since the fluid is incompressible and no flow crosses streamlines.

Now, take a bigger-picture view of the model. From far away, the collection of elementary flows reduces to a single vortex,* and we will denote by Γ the strength of this vortex. Γ is the sum of all the vortex strengths used to model flow around the body. In this section, we show that given Γ, we can calculate the net force on the body.

Consider a donut-shaped two-dimensional control volume with one boundary a circle of radius R centered on the body, where R is much larger than the dimension

*Although other singularities can be present, their effect on the velocity field decays faster than the effect of the vortex. For example, the velocity induced by a doublet decays as $1/r^2$, where r is the distance from the doublet. The net strength of sources must match that of sinks, so their effect is similar to a doublet from far away. Higher-order singularities induce velocities that decay even faster with r. A vortex, however, has a velocity field that decays as $1/r$, and so the effect of the vortex will dominate for large enough r.

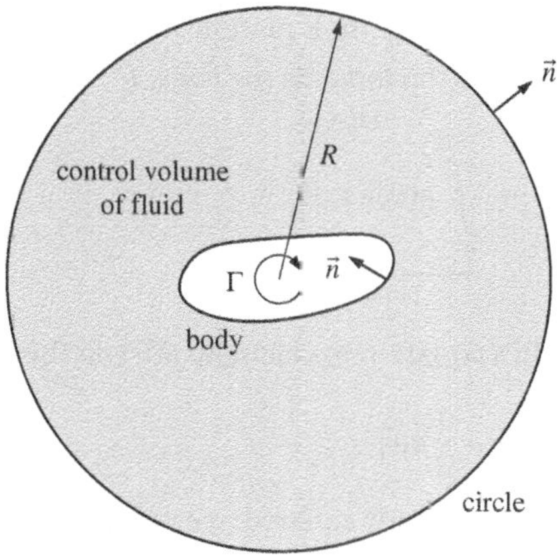

Figure 4.44 Control volume used for proving the Kutta–Joukowsky theorem.

of the body. The other boundary is formed by the body, which can be thought of as cut out from the larger circle, like a donut hole. Figure 4.44 shows such a control volume.

Conservation of momentum, with steady flow and no viscosity or external forces, applied to this control volume is

$$\oint_C p\vec{n}\,dl + \oint_C \rho\vec{v}\,\vec{v}\cdot\vec{n}\,dl = \vec{0}, \tag{4.117}$$

where C is the boundary of the control volume and includes both the circle of radius R and the body boundary. $\vec{n}$ is a normal vector that points outward from the control volume. It points outward on the circular boundary, and it points into the body, which is still out of the fluid, on the body boundary. Note that without viscosity, only a pressure force is present, and in Equation 4.117, it appears on the left-hand side.

We consider the two terms in Equation 4.117 separately, starting with $\oint_C \rho\vec{v}\,\vec{v}\cdot\vec{n}\,dl$. Since $\vec{v}\cdot\vec{n} = 0$ on the body boundary, due to flow tangency, the only contribution to this integral will be from the outer circular boundary, where the velocity is†

$$\vec{v} = \underbrace{U_\infty\hat{\imath}}_{\text{freestream}} + \underbrace{\frac{\Gamma}{2\pi R}(\sin\theta\hat{\imath} - \cos\theta\hat{\jmath})}_{\text{vortex}}. \tag{4.118}$$

† For positive Γ, the vortex spins clockwise, and $\sin\theta\hat{\imath} - \cos\theta\hat{\jmath}$ is the unit tangential clockwise direction in cylindrical coordinates.

Note that we have assumed that the freestream is in the $\hat{\imath}$ direction. On the circular boundary, the outward normal is $\vec{n} = \cos\theta\hat{\imath} + \sin\theta\hat{\jmath}$, so

$$\vec{v}\cdot\vec{n} = U_\infty\cos\theta. \tag{4.119}$$

Therefore, the momentum flux integral becomes‡

$$\begin{aligned}\oint C\rho\vec{v}\,\vec{v}\cdot\vec{n}\,dl &= \int_0^{2\pi} \rho\vec{v}\,U\infty\cos\theta\,Rd\theta \\ &= \int 0^{2\pi}\rho\left[\left(U\infty + \frac{\Gamma}{2\pi R}\sin\theta\right)\hat{\imath}\right. \\ &\quad \left. -\frac{\Gamma}{2\pi R}\cos\theta\hat{\jmath}\right] U_\infty\cos\theta\,Rd\theta \\ &= -\rho U_\infty\frac{\Gamma}{2\pi R}\pi R\hat{\jmath} \\ &= -\frac{1}{2}\rho U_\infty\Gamma\hat{\jmath}.\end{aligned} \tag{4.120}$$

‡ The trigonometric identities needed for this integration are

$$\int_0^{2\pi}\cos\theta\,d\theta = 0,$$

$$\int_0^{2\pi}\cos\theta\sin\theta\,d\theta = 0,$$

$$\int_0^{2\pi}\cos^2\theta d\theta = \pi.$$

Next consider the other term in Equation 4.117, $\oint_C p\vec{n}dl$. The part of this integral over the body yields the force *on the body*, $\vec{F}'$, so that

$$\oint_C p\vec{n}dl = \underbrace{\vec{F}'}_{\text{body}} + \underbrace{\int_0^{2\pi} p(\theta)\vec{n}\,Rd\theta.}_{\text{circle}} \tag{4.121}$$

We now use Bernoulli's equation to calculate $p(\theta)$ on the farfield,

$$p(\theta) = p_\infty + \frac{1}{2}\rho U_\infty^2 - \frac{1}{2}\rho|\vec{v}|^2. \tag{4.122}$$

To obtain the velocity magnitude squared, we use Equation 4.118. Squaring and summing the $\hat{\imath}$ and $\hat{\jmath}$ components give

$$\begin{aligned}|\vec{v}|^2 &= U_\infty^2 + \left(\frac{\Gamma}{2\pi R}\right)^2 \sin^2\theta + 2U_\infty\frac{\Gamma}{2\pi R}\sin\theta + \left(\frac{\Gamma}{2\pi R}\right)^2\cos^2\theta \\ &= U_\infty^2 + \left(\frac{\Gamma}{2\pi R}\right)^2 + \frac{U_\infty\Gamma}{\pi R}\sin\theta.\end{aligned}$$

Substituting into the equation for $p(\theta)$, we obtain

$$p(\theta) = p_\infty - \frac{1}{2}\rho\left(\frac{\Gamma}{2\pi R}\right)^2 - \frac{1}{2}\rho\frac{U_\infty\Gamma}{\pi R}\sin\theta. \tag{4.123}$$

Finally, integrating over the outer circular boundary yields[§]

$$\begin{aligned}\int_0^{2\pi} p\vec{n}\,Rd\theta &= \int_0^{2\pi} [p(\theta)]\,(\cos\theta\hat{\imath} + \sin\theta\hat{\jmath})\,Rd\theta \\ &= \int_0^{2\pi}\left[-\frac{1}{2}\rho\frac{U_\infty\Gamma}{\pi R}\sin^2\theta\right] \\ &= -\frac{1}{2}\rho U_\infty\Gamma\hat{\jmath}.\end{aligned} \tag{4.124}$$

§ The identity used in the last step is

$$\int_0^{2\pi}\sin^2\theta d\theta = \pi.$$

Putting everything together,

$$\begin{aligned}\oint_C p\vec{n}\,dl + \oint_C \rho\vec{v}\,\vec{v}\cdot\vec{n}\,dl &= 0, \\ \vec{F}' - \frac{1}{2}\rho U_\infty\Gamma\hat{\jmath} - \frac{1}{2}\rho U_\infty\Gamma\hat{\jmath} &= 0, \\ \vec{F}' &= \rho U_\infty\Gamma\hat{\jmath}.\end{aligned} \tag{4.125}$$

Note, the direction $\hat{\jmath}$ is perpendicular to the freestream, so that the above force is purely a lift force! This result tells us that for potential flow around *an arbitrary body*,

$$\begin{aligned}\text{Drag per unit depth,}\ D' &= 0, \\ \text{Lift per unit depth,}\ L' &= \rho U_\infty\Gamma.\end{aligned} \tag{4.126}$$

The result that $D' = 0$ is referred to as *d'Alembert's paradox*,[¶] while the result that $L' = \rho U_\infty\Gamma$ is referred to as the *Kutta–Joukowsky theorem*.

¶ The result seems paradoxical because we cannot achieve zero drag in reality. It is the lack of viscosity, along with the two-dimensional nature and absence of shocks, that makes the drag zero in potential flow. Note also that we are not violating any physical laws by generating lift without drag, since the lift force is perpendicular to the freestream velocity and hence does no net work on the fluid.

Example 4.11 Additional analysis of the vortex near the ground
We can apply the Kutta–Joukowsky theorem to the case of the vortex near the ground presented in Example 4.9. First, we argue that the lift on the airfoil (the vortex) is equal to F'_y, the extra downward force on the ground caused by the vortex system. If we draw a control volume (CV) around the airfoil, with the ground for the lower boundary, and the other boundaries very far away from the vortex (in the $\pm x$, $+y$ directions), conservation of momentum reduces to

$$\begin{aligned}\text{net momentum outflow} &= \text{sum of forces on fluid in CV}\\ &= \text{pressure force on CV} - \text{lift force}\end{aligned}$$

Note, we put the lift as an extra force acting on the CV.|| The net momentum outflow is zero, since there is no flow through the ground and since the flow is uniform in the farfield. Similarly, the pressure integral yields a zero horizontal force, as p_∞ on the left and right boundaries cancels, and it yields F'_y as the upward vertical force on the CV due to the ground integral.** Therefore, the above momentum conservation equation states that the lift force is equal to F'_y in the upward direction.

An airfoil flying far away from the ground has its lift given by the Kutta–Joukowsky theorem,

$$L'_{\text{far}} = \rho U_\infty \Gamma. \tag{4.127}$$

Substituting $\Gamma = \pi h_0 U_\infty$, the value given in the example problem statement yields

$$L'_{\text{far}} = \rho U_\infty^2 h_0 \pi. \tag{4.128}$$

Using the answer from Example 4.9, the ratio of the airfoil lift near the ground to L'_{far} is:

$$\frac{L'}{L'_{\text{far}}} = \frac{\rho U_\infty^2 h_0 \left(\pi - \frac{\pi}{4\beta}\right)}{\rho U_\infty^2 h_0 \pi} = 1 - \frac{1}{4\beta}. \tag{4.129}$$

As expected, when $\beta \to \infty$, the ratio approaches one. Finally, we note that the expression for the lift on the vortex near the ground,

$$L' = \rho U_\infty^2 h_0 \left(\pi - \frac{\pi}{4\beta}\right), \tag{4.130}$$

can be written as

$$L' = \rho U_\infty^2 h_0 \left(\pi - \frac{\pi}{4\beta}\right) = \rho \underbrace{\pi U_\infty h_0}_{\Gamma} \underbrace{\left(U_\infty - \frac{U_\infty}{4\beta}\right)}_{u_{\text{net}}} = \rho \Gamma u_{\text{net}}, \tag{4.131}$$

where u_{net} is the net horizontal velocity induced on the top vortex by the combination of the freestream and the image. We can see this more clearly by writing

$$u_{\text{net}} = U_\infty - \frac{\pi U_\infty h_0}{4\pi h_0 (h/h_0)} = U_\infty - \frac{\pi U_\infty h_0}{2\pi(2h)} = U_\infty - \frac{\Gamma}{2\pi(2h)}.$$

The first term is just the effect of the freestream, while the second term is the velocity induced by the image vortex, which is $(2h)$ below the real vortex. Equation 4.131 is then a more general form of the Kutta–Joukowsky theorem

|| The minus sign is necessary because the conservation of momentum is applied to the fluid, and the right-hand side consists of all forces acting on the fluid.

** To be precise, the velocity and pressure perturbations in the farfield decay as $1/r^2$ because from far away, the combination of the vortex and its image cancel each other out. However, the surface integrals only scale with r, and so the integral result diminishes to zero with increasing control volume distance from the vortex.

that relates the force on a vortex to the velocity induced at the vortex center by the freestream and other vortices, instead of just by the freestream itself. We'll use this idea when we study horseshoe vortices in Chapter 7.

4.6 Summary and Concepts

This chapter introduced potential flow, which forms the foundation for most of the flows considered in this text. Such a flow ignores viscous effects, which introduce rotation to the fluid. However, for many aerodynamic flows, viscous effects are confined to thin regions near walls, so that the potential flow approximation is very good for the majority of the flow field. The complex-number-based mathematical framework for potential flow presented in this chapter is a powerful tool that lets us easily characterize several flows fundamental to the study of lift-producing objects.

The key concepts covered in this chapter are as follows:

- Potential flow is irrotational and, under our assumptions, incompressible, and it is described by both a velocity potential and a stream function.
- The velocity components in potential flow satisfy the Cauchy–Riemann equations, which lead to the definition of a complex potential that combines the velocity potential and stream function.
- Streamlines can be visualized as contours of the stream function, which in potential flow is the imaginary part of the complex potential.
- Elementary flows arise from the first few terms of a power-series expansion of a general complex potential.
- A source and vortex are almost everywhere incompressible and irrotational, except at the center, where the velocity field is singular.
- Multiple potential flows can be combined by superposition: velocities and potentials add together.
- Uniform flow and a doublet combine to create flow over a cylinder, and the addition of a vortex introduces lift.
- Bernoulli's equation can be used to compute the pressure coefficient in potential flow.
- The Kutta–Joukowsky theorem and d'Alembert's paradox tell us that an object in potential flow can experience lift, directly proportional to the net vortex strength, but not drag.

4.7 Problems

1. Justify each answer to the below questions with an explanation.

a True or false: the complex potential $F(z) = x - iy$ is analytic.

b True or false: the complex potential $F(z) = \sin(z) + i$ is analytic.

c True or false: the velocity field $\vec{v} = y\hat{i} + x\hat{j}$ models potential flow.

d Determine the final width, far downstream of the source, of the Rankine half body.

e True or false: in the point-source numerical method, a change in the airfoil geometry affects only the right-hand side of the linear system of equations.

f In which direction will an unrestrained vortex spinning clockwise near a horizontal ground move?

g True or false: the pressure coefficient in potential flow is always greater than zero.

2. A flow is governed by the complex potential $F(z) = \sqrt{z}$. Determine the velocity potential, stream function, complex velocity, and velocity vector.

3. Prove that the streamlines above and below the x axis of an x-aligned doublet are circles.

4. The Rankine oval results from superimposing a uniform freestream, a source of strength Q at $x = -L$, and a sink of the same magnitude at $x = L$, as shown in Figure 4.45. Determine the velocity potential, stream function, and the velocity field. Find an implicit equation for the dividing streamline and plot it for the case $U_\infty L/Q = 1$.

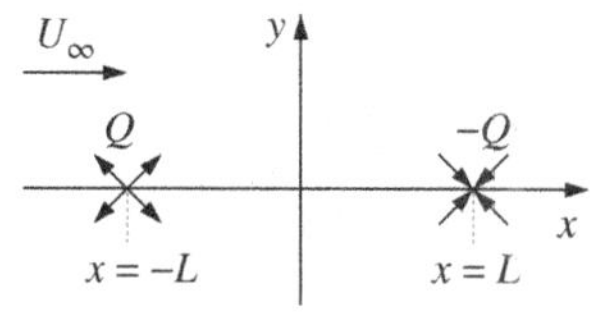

Figure 4.45 Source and sink setup for modeling a Rankine oval.

5. A source/sink pair of equal magnitude Q is positioned as shown in Figure 4.46. Write down an expression for the complex potential and complex velocity for this flow, and determine the velocity vector at $(x, y) = (-h, h)$ and at $(h, -h)$. Show that in the limit $h \to 0$ but $Qh = C =$ constant, this configuration becomes a doublet, and determine the doublet complex potential, its strength, and its alignment.

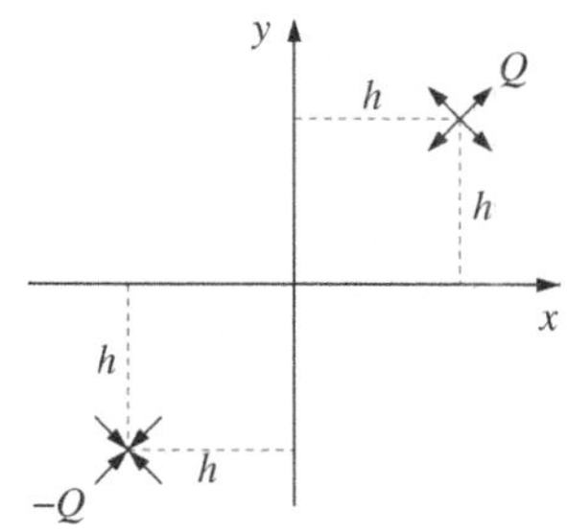

Figure 4.46 A source and sink in proximity along a diagonal.

6. Consider again the hangar analyzed in Example 4.7. Suppose that the hangar consists of two sections riveted together, as shown in Figure 4.47. These rivets can break if the force on them (compression or tension) is too large. Assuming the pressure inside the hangar is p_{in}, calculate the force on the rivets. At what interior pressure p_{in} is the force magnitude on the rivets minimized?

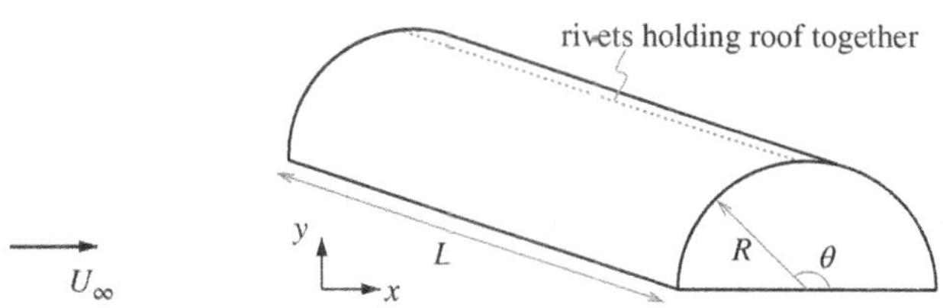

Figure 4.47 Hangar geometry with rivets holding the front and back sections together.

7. Extend the point-source method code in Example 4.8 to plot the negative pressure coefficient, -c_p, versus x for the upper surface. Overlay results from a panel solver such as mfoil, and discuss the agreement for various numbers of sources, N.

8. A vortex of strength $\Gamma = 10\,\text{m}^2/\text{s}$ is situated a distance of $h = 1$ m from the walls of a corner, as shown in Figure 4.48. Draw the image system that models the resulting flow, write down the complex potential, and plot streamlines to verify that the flow is tangent to the walls. If the vortex is not restrained, it will move. In this case, plot its location from $t = 0$ to $t = 1$ s.

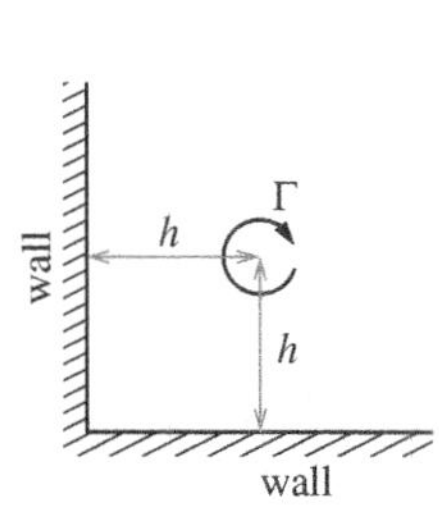

Figure 4.48 A vortex in a 90° corner between two walls.

5

Airfoil Theory

This chapter discusses potential flow over wing cross-sections, or airfoils. These shapes are prototypical aerodynamic bodies that serve as building blocks for wings and play an important role in aircraft performance. Two goals in aerospace engineering are (1) analyzing airfoil shapes to determine their performance and (2) designing airfoils to meet given requirements. This chapter addresses both of these topics.

We start with numerical solution techniques for simple geometries and the general case of arbitrary two-dimensional airfoil shapes. Following the numerical analysis, and with certain simplifying assumptions on the airfoil shape and operating conditions, we obtain analytical solutions for the forces and moments. The relationships between the airfoil geometry and forces and moments are used to develop airfoil design techniques.

5.1 Airfoil Nomenclature

An *airfoil* is a cross-section of a wing, as shown in Figure 5.1. By looking at two-dimensional slices only, we assume that the wing section does not vary in y, which effectively means that the wing has infinite span. This is a pretty good assumption away from the wing tips, for wings with long spans relative to the chord. Note that in the coordinate system shown, the wing cross-section is in the $x - z$ plane, whereas y is the spanwise coordinate, pointing out the right wing.

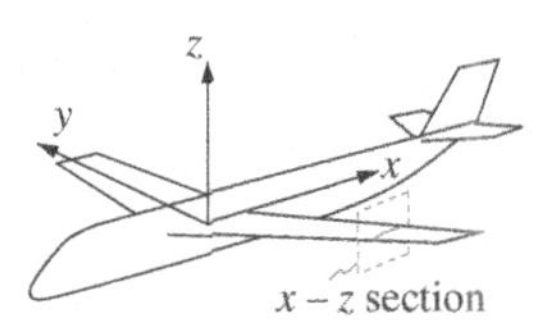

Figure 5.1 Aerodynamic coordinate system for an aircraft. An airfoil is a wing section in the $x - z$ plane.

Figure 5.2 identifies some of the common terms relevant to airfoils. The *camber line*, $z_c(x)$, is the average of the upper and lower surface lines,

$$z_c(x) = \frac{1}{2}(z_u(x) + z_l(x)). \quad (5.1)$$

The *thickness*, $t(x)$, is the distance between the upper and lower surfaces,

$$t(x) = z_u(x) - z_l(x). \quad (5.2)$$

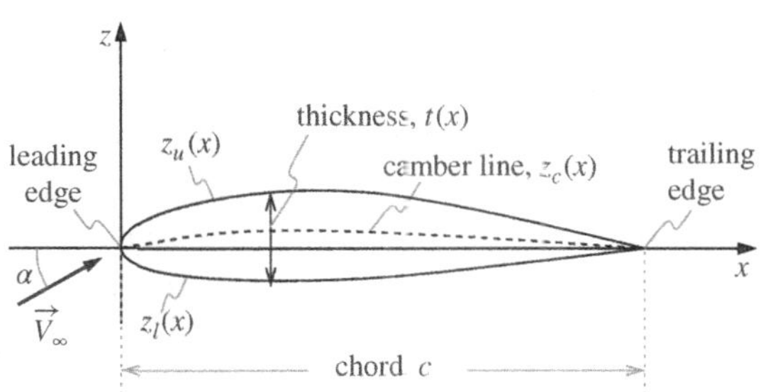

Figure 5.2 Nomenclature for an airfoil.

Aerodynamics: A Computational Introduction, First Edition. Krzysztof J. Fidkowski.
© 2026 John Wiley & Sons, Inc. Published 2026 by John Wiley & Sons, Inc.

The *chord line* is the straight line from the leading edge to the trailing edge, which for convenience we will often show as aligned with the x axis. However, *this is not a requirement*, as we could define the x axis arbitrarily. We will measure the angle of attack, α as the angle between the freestream direction and either the chord line or the x axis – again, there is arbitrariness in this definition.* The freestream velocity is $\vec{V}_\infty$, and to simplify notation, we will denote its magnitude by V_∞.

*Unless stated otherwise, in the remainder of this chapter, the x axis will be aligned with the chord line, and the angle of attack will be measured relative to this x axis/chord line, as shown in Figure 5.2.

There are a wide variety of airfoil shapes. The National Advisory Committee for Aeronautics (NACA) (predecessor of NASA) studied several *series* of airfoil shapes, named in a systematic manner as described below. Anderson (2016) provides additional background on NACA and airfoil nomenclature.

- NACA 4-digit series: The first digit in this series is the value of maximum camber in hundredths of chord. The second digit is the location of maximum camber, in tenths of chord. The final two digits give the maximum thickness of the airfoil, in hundredths of chord. For example, a NACA 2412 airfoil, shown in Figure 5.3, has a maximum camber of $0.02c$, located at $0.4c$, and a maximum thickness of $0.12c$. A symmetric airfoil, that is without camber, will have "00" for the first two digits.
- NACA 5-digit series: In this series, the first digit is 20/3 times the design lift coefficient. The second two digits give twice the location of maximum camber, in hundredths of chord. The last two digits give the thickness of the airfoil, in hundredths of chord. So a NACA 23012 airfoil has a design lift coefficient of 0.3, a location of maximum camber at $0.15c$, and a thickness of $0.12c$.
- NACA 6-digit series: The first digit in this series is a 6, and it identifies the series. The next digit is the location along the chord of minimum pressure at zero lift, in tenths of chord. The third digit is the design lift coefficient times 10, and the last two digits give the thickness in hundredths of chord. For example, a NACA 65-218 airfoil has the location of minimum pressure at $0.5c$, a design lift coefficient of 0.2, and a maximum thickness of $0.18c$.

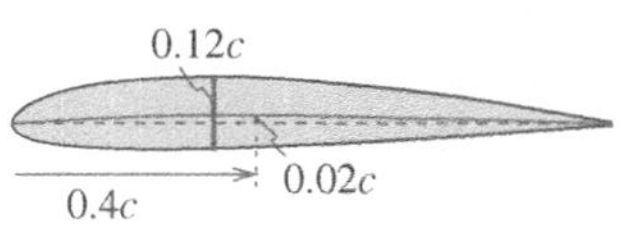

Figure 5.3 A NACA 2412 airfoil.

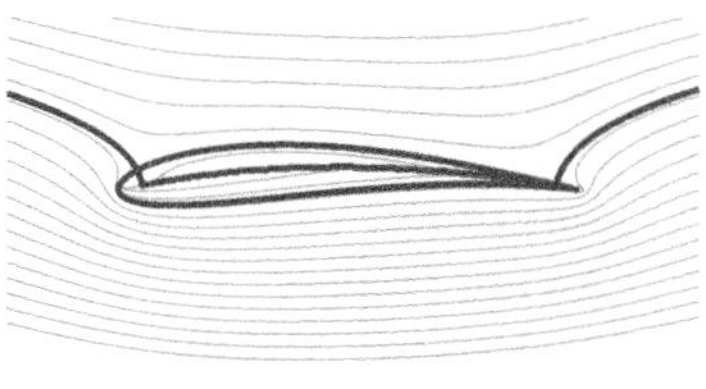
Figure 5.4 Streamlines for flow over an airfoil with a low value of circulation, Γ_1.

5.2 The Kutta Condition

In this chapter, we are concerned with potential flow over airfoils. In Chapter 4, we solved for potential flow over a cylinder and found that there exist an infinite number of solutions, since each value of the circulation, Γ, yields a different solution. The same holds true for other shapes, including airfoils. In fact, there are transformations that map the flow over a cylinder to flow over certain airfoils, one of them the Joukowsky transformation, covered in Section 5.8. So theoretically, there are an infinite number of solutions for flow over an airfoil.

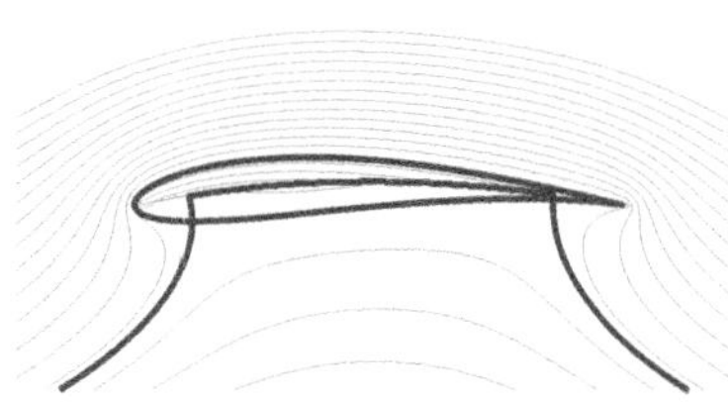
Figure 5.5 Streamlines for flow over an airfoil with a high value of circulation, Γ_2.

What do these possible solutions look like? Figures 5.4–5.6 show three possibilities, for three different Γ. The streamlines for these airfoils were generated using a Joukowsky mapping from a circle, and the flow inside the airfoil can be regarded as fictitious. How do we know which Γ is the correct physical one?

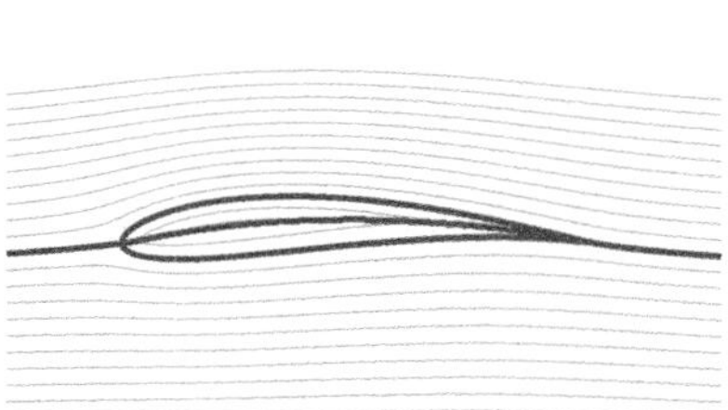
Figure 5.6 Streamlines for flow over an airfoil with a medium value of circulation, Γ_3.

To answer this question, we first need to define what we mean by physically correct. Potential flow is an idealization that assumes zero viscosity, and this lack of viscosity allows for an infinite number of solutions. A real flow with viscosity will not have this non-uniqueness, as discussed below, so by a "correct" solution, we mean one that approaches the viscous solution in the limit of viscosity going to zero.

A key difference among the possibilities shown in Figures 5.4–5.6 is the behavior of the flow at the trailing edge. For both Γ_1 and Γ_2, the flow "wraps around" the trailing edge. To negotiate a sharp corner, very large pressure gradients would be required, since the radius of curvature is very small at the trailing edge, and Equation 3.122 tells us that the pressure gradient is inversely proportional to the radius of curvature. We will see in the following chapter that in the presence of large pressure gradients, specifically adverse pressure gradients, a flow with viscosity separates from the surface. Therefore, given the sharp corner at an airfoil's trailing edge, the flow will separate from the surface at the trailing edge and not wrap around it. This observation is the *Kutta condition* (Anderson 2016):

Observation 5.2.1 (The Kutta condition.) *A physical flow, with viscosity, will smoothly leave a sharp trailing edge.*

Although the Kutta condition is based on considerations of a viscous flow, we can still use it for our inviscid analysis of airfoils. This is because the condition is precisely the extra information we need to determine the circulation. There is only one choice of Γ that makes the flow leave the trailing edge smoothly, and hence, the problem is well-posed. Given the airfoil shape and the freestream flow conditions, we can now solve for the flow past an airfoil.

We take a closer look at the consequences of the Kutta condition for the flow at the trailing edge of an airfoil. First, consider the case when the trailing edge has a *nonzero angle*, as shown in Figure 5.7. The velocities on the upper and lower surfaces must still be tangent to their respective surfaces. However, the nonzero angle between the upper and lower streams implies two different velocity directions at the trailing edge. The only resolution is that

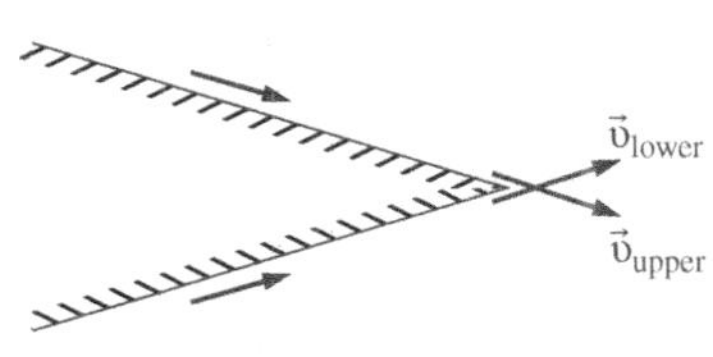

Figure 5.7 A nonzero trailing-edge angle.

$$\vec{v}_{TE} = \vec{v}_{\text{lower}} = \vec{v}_{\text{upper}} = 0. \tag{5.3}$$

In this case, *the trailing edge (TE) is a stagnation point*, and Bernoulli's equation tells us that the pressure there is the total pressure,

$$p_{TE} = p_\infty + \frac{1}{2}\rho U_\infty^2 - \frac{1}{2}\rho|\vec{v}_{TE}|^2 = p_\infty + \frac{1}{2}\rho U_\infty^2. \tag{5.4}$$

Substituting this pressure into the pressure coefficient definition, Equation 4.80, gives $c_p = 1$ at the trailing edge.

For a *cusped trailing edge*, illustrated in Figure 5.8, the tangent vectors on the upper and lower surfaces match, so we do not have the same requirement of zero velocity as in the case of the nonzero trailing-edge angle. However, we still need the pressures to match in order for the flow to come off smoothly. By Bernoulli's equation, this means that the velocity magnitudes must be equal, and so

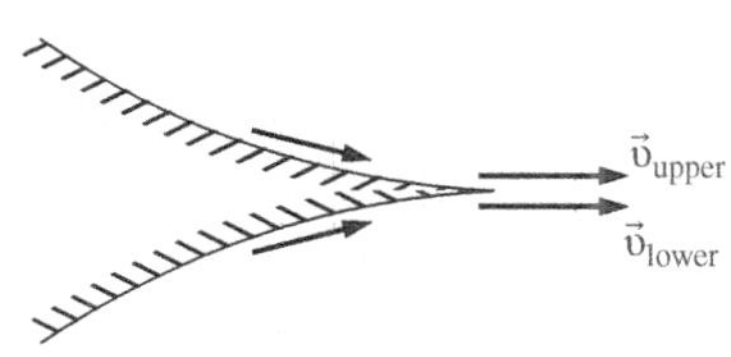

Figure 5.8 A cusped trailing edge.

$$\vec{v}_{TE} = \vec{v}_{\text{lower}} = \vec{v}_{\text{upper}}. \tag{5.5}$$

Example 5.1 The Kutta condition for an atypical airfoil
We wish to obtain an inviscid solution for the airfoil shown in Figure 5.9 – note the double trailing edge. Is there an equivalent of a Kutta condition for this geometry?

To answer this question, we consider how the flow would behave if it were a real fluid with a little bit of viscosity, since inviscid flow is a modeling idealization. With a little bit of viscosity, the flow would separate off the sharp corners. An inviscid solution cannot model this situation, since we can only have

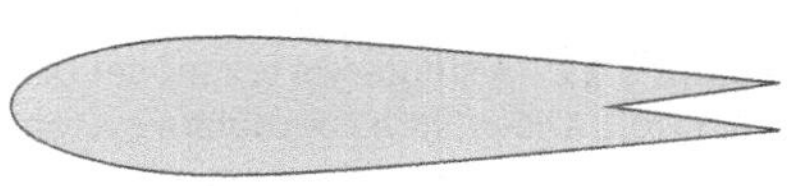

Figure 5.9 An airfoil with a double trailing edge.

one trailing-edge stagnation streamline,* and so the flow would have to wrap around at least one of the corners. Hence, seeking an inviscid solution for this geometry is not a good idea, since it would not be representative of reality, and so there is no reasonable Kutta condition for this geometry.

*One can show this using the fact that the stream function satisfies Laplace's equation, or using the fact that we can map a flow over a two-dimensional closed object to a flow over a cylinder, which is well-understood.

The problem to be solved in the analysis of an airfoil is the calculation of aerodynamic forces and moments on an airfoil at a given angle of attack, α. In the next sections, we approach this problem first numerically and then analytically, with simplifying assumptions.

5.3 Discrete Vortex Models

We saw in Chapter 4 that a potential flow model of flow over a body producing lift requires one or more vortices. Airfoils are made to generate lift, and so a natural starting point for modeling flow over them is with vortices. The question becomes how much vortex strength to use and where to put it for a particular airfoil at particular flight conditions. To answer these questions, we seek a *predictive* model that will be able to determine how much lift an airfoil produces just from the geometry, angle of attack, and freestream speed.

We already know from the Kutta–Joukowsky theorem that a vortex of strength Γ produces a lift force of magnitude $\rho V_\infty \Gamma$. What should the strength of the vortex be? The answer should depend on the geometry of the airfoil and its orientation relative to the oncoming flow. A key physical principle that will capture this information is the idea that no flow crosses the airfoil surface. This is called *flow tangency*, and it will give rise to the airfoil equations in this chapter.* The unknowns in these equations will be the strengths of the vortices in the model.

*Flow tangency is one boundary condition, on the airfoil surface. The other boundary condition is that $\vec{v} \to \vec{V}_\infty$ far away from the airfoil. Together with the potential-flow equations, these conditions provide a mathematically well-posed problem.

5.3.1 Flat-Plate Single-Vortex Model

As a starting point, consider a simple airfoil formed by a flat plate of chord length c at an angle of attack α, as shown in Figure 5.10. In our analysis, the plate has no thickness and no camber, and this approximation is reasonable for a thin, symmetric airfoil. We proceed to model the flow around the airfoil with vortices. For the simplest model, we consider just one vortex, of strength Γ. For reasons that will become clear later in the chapter, we place the vortex at the quarter-chord location of the airfoil, $x = c/4$, as shown in Figure 5.11.

Figure 5.10 A flat-plate airfoil approximation at angle of attack α.

Flow tangency requires that the flow does not cross the airfoil's surface. For our flat plate, this means that the flow is horizontal along the chord. However, it turns out that this is asking too much of the simple model. The potential flow that results from a superposition of the uniform freestream and a single vortex cannot make the flow horizontal along a continuous stretch in space. For instance, the vertical velocity will become unbounded close to one side of the vortex because of the inverse distance dependence of the velocity magnitude. So we have to relax our

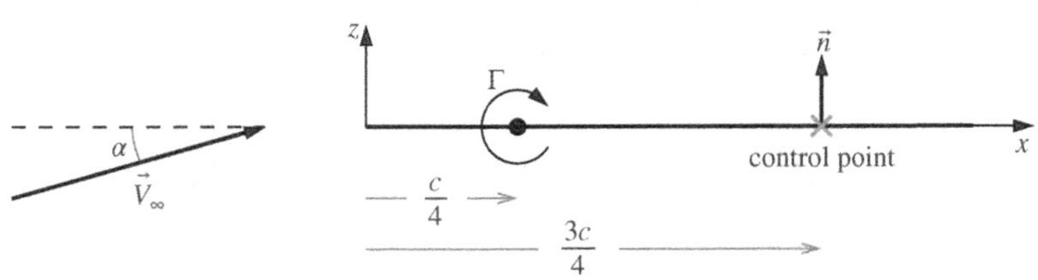

Figure 5.11 A single-vortex model of an airfoil, shown in a coordinate system where the horizontal x axis is aligned with the chord.

flow-tangency requirement and only enforce it at a single point. We call this the *control point*, and as shown in Figure 5.11, the location of this point is at the three-quarter-chord, $x = 3c/4$.*

Flow tangency at the control point means that the velocity is parallel to the surface, i.e. that it has no component in the direction *normal* to the surface. If $\vec{n}$ denotes the normal vector, the flow-tangency equation is

$$\vec{v} \cdot \vec{n} = 0 \text{ at the control point.} \tag{5.6}$$

For the flat plate geometry, $\vec{n} = \hat{k}$ (note, the airfoil section is in the $x - z$ plane). Also, the velocity at the control point is a superposition of the velocity due to the freestream, and the velocity due to the vortex,

$$\vec{v} = \underbrace{V_\infty \cos\alpha\, \hat{i} + V_\infty \sin\alpha\, \hat{k}}_{\text{freestream}} - \underbrace{\frac{\Gamma}{2\pi(3c/4 - c/4)}\hat{k}}_{\text{vortex}}. \tag{5.7}$$

The magnitude of the vortex velocity depends on the distance away from the vortex, in this case $3c/4 - c/4 = c/2$, and its direction is tangential to the radius, in this case in the -$\hat{k}$ direction. Enforcing flow tangency means

$$0 = \vec{v} \cdot \vec{n} = V_\infty \sin\alpha - \frac{\Gamma}{2\pi(c/2)}. \tag{5.8}$$

Solving for the vortex strength, we obtain

$$\Gamma = \pi c V_\infty \sin\alpha, \tag{5.9}$$

and by the Kutta–Joukowsky theorem, the sectional lift† is

$$L' = \rho \Gamma V_\infty = \pi \rho V_\infty^2 \sin\alpha. \tag{5.10}$$

In aerodynamics, it is common practice to non-dimensionalize forces using the dynamic pressure, $\frac{1}{2}\rho V_\infty^2$, and a reference length or area, depending on whether the force is sectional or not. Using the chord length c for the present case of sectional lift, we define the *lift coefficient*, c_ℓ, as

$$c_\ell = \frac{L'}{\frac{1}{2}\rho V_\infty^2 c} = 2\pi \sin\alpha. \tag{5.11}$$

Furthermore, if the angle of attack is small and measured in radians, $\sin\alpha \approx \alpha$, and the above formula becomes

$$c_\ell \approx 2\pi\alpha. \tag{5.12}$$

We will see the same result come out of thin-airfoil theory, presented later in this chapter.

5.3.2 Flat-Plate Multiple-Vortex Model

With one vortex, we can only enforce flow tangency at one point. If we judiciously choose that point and the location of the vortex, we can end up with a realistic result for a flat plate. But without advance knowledge of the expected result, we do not have additional information that dictates where to put the two points. We can do better with more vortices and more control points, since if we enforce flow tangency at more and more points, the flow will better and better approximate the true flow over the airfoil.

* This choice seems arbitrary, and it is made with foresight into how much lift a symmetric airfoil should produce. Without such foresight, this simple model is not predictive.

† The sectional lift is the lift per unit span, i.e. a force per unit length.

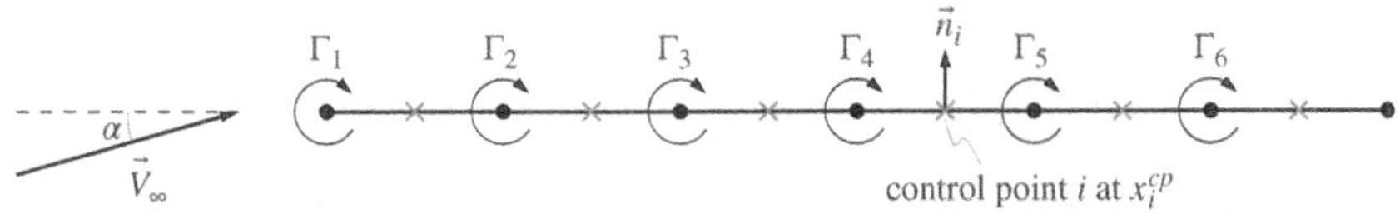

Figure 5.12 A multiple-vortex model of an airfoil, shown with $N = 6$ panels.

Returning to the flat plate of the previous section, we now split the plate into N equal intervals, or *panels*. On each panel, we place a control point, and without additional information, we choose the panel midpoint as the location of the control point. This choice will give us N equations. We now need N unknowns, which will be the strengths of N vortices. There are various choices for the locations of the vortices, and one is to put them at the *nodes* of the paneling, which means at the points that form the panel junctures. There are $N + 1$ such points, but to respect the Kutta condition, which requires no wrapping around of the flow at the trailing edge, we do not place a vortex at the last point. Figure 5.12 illustrates the resulting model.

Let x_i^{cp} denote the location of the ith control point. The flow-tangency equation at this control point reads

$$\vec{v}|_{x_i^{cp}} \cdot \vec{n}_i = 0, \tag{5.13}$$

where $\vec{n}_i = \hat{k}$ is the normal vector (same for all points), and where $\vec{v}|_{x_i^{cp}}$ is the velocity at point x_i^{cp} due to the freestream and all vortices. We can write this velocity as

$$\vec{v}|_{x_i^{cp}} = \vec{V}_\infty + \sum_{j=1}^{N} \frac{\Gamma_j}{2\pi(x_j - x_i^{cp})}\hat{k}, \tag{5.14}$$

where x_j denotes the location of the jth vortex. The contribution of each vortex in the above formula uses the $\Gamma/(2\pi r)$ result for the velocity magnitude at a distance r from a vortex. These contributions are all in the $\pm\hat{k}$ direction, as determined by the sign of $x_j - x_i^{cp}$.* Dotting Equation 5.14 with the normal vector, $\hat{k}$, and rearranging gives

$$\sum_{j=1}^{N} \frac{\Gamma_j}{2\pi(x_j - x_i^{cp})} = -\vec{V}_\infty \cdot \vec{n}_i. \tag{5.15}$$

This equation can be written as a matrix system for the unknowns Γ_j,

$$\sum_{j=1}^{N} A_{ij}\Gamma_j = b_i \quad \Rightarrow \quad \mathbf{A\Gamma} = \mathbf{b}, \tag{5.16}$$

where the matrix and right-hand side entries are

$$A_{ij} = \frac{1}{2\pi(x_j - x_i^{cp})}, \qquad b_i = -\vec{V}_\infty \cdot \vec{n}_i. \tag{5.17}$$

In matrix form, this system is

$$\underbrace{\begin{bmatrix} A_{11} & A_{1,2} & \cdots & A_{1,N} \\ A_{21} & A_{2,2} & \cdots & A_{2,N} \\ \vdots & \vdots & \ddots & \vdots \\ A_{N1} & A_{N,2} & \cdots & A_{N,N} \end{bmatrix}}_{\mathbf{A}} \underbrace{\begin{bmatrix} \Gamma_1 \\ \Gamma_2 \\ \vdots \\ \Gamma_N \end{bmatrix}}_{\mathbf{\Gamma}} = \underbrace{\begin{bmatrix} -\vec{V}_\infty \cdot \vec{n}_1 \\ -\vec{V}_\infty \cdot \vec{n}_2 \\ \vdots \\ -\vec{V}_\infty \cdot \vec{n}_N \end{bmatrix}}_{\mathbf{b}}. \tag{5.18}$$

* Whether the induced velocity is up or down depends on whether the vortex is to the right or left of the control point, and the sign of $x_j - x_i^{cp}$ automatically gives us this information. So no additional sign checks are necessary.

The matrix **A** is *dense* in that it does not contain any zeros, due to every vortex affecting the velocity at every control point. The right-hand side vector is constant in this case, but for more interesting geometries, its entries will change due to a varying normal vector.

When working with many panels, it makes sense to build and solve the matrix system numerically, using a computer program. Listing 5.1 presents a code that does this for a given N and α.

Listing 5.1 Discrete vortex code for a flat plate airfoil.

```
function cl = vortex_flatplate(N, alpha)
% Uses discrete vortices to model flow over a flat plate.
% N is the number of panels, and alpha is the angle of attack [deg].
% The output is the lift coefficient, cl.

c    =  1;   % chord length
Vinf =  1;   % free-stream velocity
rho  =  1;   % density

dx = c/N;                    % interval size
xj = linspace(0,c,N+1);  % locations of vortices (last one is TE)
xi = xj + 0.5*dx;          % locations of control points

% Build linear system and right-hand side
A = zeros(N,N); b = zeros(N,1);
for i=1:N,
  for j=1:N,
    A(i,j) = 1/(2*pi*(xj(j)-xi(i)));
  end
  b(i) = -Vinf*(alpha*pi/180);
end

% solve linear system
Gamma = A\b;

% calculate the sectional lift and lift coefficient
L  = rho*Vinf*sum(Gamma)*c;  % Kutta-Joukowsky
cl = L/(0.5*rho*Vinf^2*c);
```

Code notes:

- A double loop builds the matrix **A**: i indexes the rows, while j indexes the columns.
- The code makes use of a small-angle approximation, where $\vec{V}_\infty \cdot \vec{n}_i = V_\infty \sin\alpha \approx V_\infty \alpha$, although these are not necessary.
- The choices of density, free-stream speed, and airfoil chord length are arbitrary: their values do not affect the final lift coefficient.

Running the code for different angles of attack with $N = 10$ gives the lift curve shown in Figure 5.13. The lift coefficient increases linearly at a slope of 2π, which is consistent with thin-airfoil theory, to be introduced later. In fact, the slope does not depend on the number of panels for this simple case with no camber.

Figure 5.14 shows the streamlines for $N = 2$ and $N = 4$, at $\alpha = 10°$, plotted using contours of the stream function. The locations of some of the vortices are visible as the velocity field is singular at these locations. By the shape of the contours, we also see that flow tangency is satisfied at the control point locations on the panel midpoints. More panels means more places at which the flow is tangent to the flat plate, which translates into a more accurate solution.

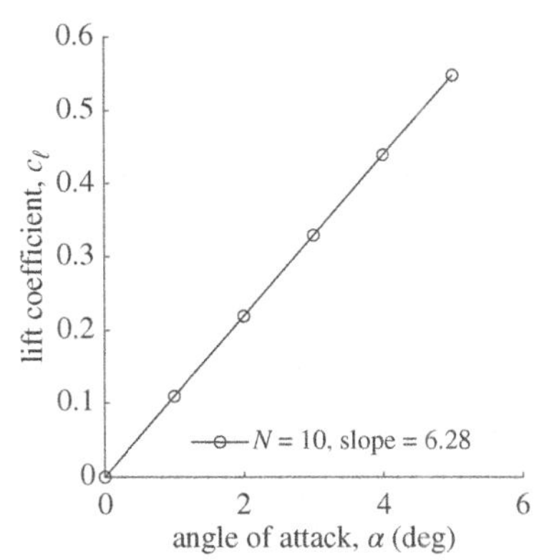

Figure 5.13 Lift coefficient results of the multi-vortex flat-plate case.

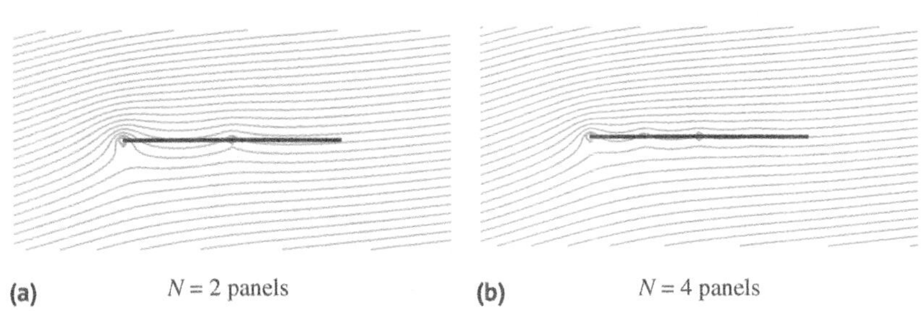

Figure 5.14 Streamlines for the multi-vortex flat-plate case.

5.3.3 The Discrete Vortex-Panel Method

For airfoils with nonzero thickness and camber, we can extend the discrete vortex model by placing vortices on the airfoil's surface, as shown in Figure 5.15. This placement assumes that the trailing edge has a gap, so that points 1 and $N+1$ are not coincident.*

Key to the method is the flow-tangency equation. We are simulating flow over a closed object, and we do not want flow going through its boundary, since we assume that the wall is impenetrable. So we must require that the normal component of the velocity, the component perpendicular to the wall, is zero at the wall,†

$$\vec{v} \cdot \vec{n} = 0, \tag{5.19}$$

where $\vec{n}$ is the normal vector at the wall.

We cannot satisfy this requirement everywhere on the airfoil's surface. For example, very close to one of the vortices, the velocity magnitude will grow unbounded, and Equation 5.19 cannot be satisfied. Nor do we want to satisfy it everywhere: with a finite number of unknown vortex strengths, we need the same number of equations in order to obtain a solvable system. So we enforce flow tangency at control points, which are the midpoints on the panels, as shown in Figure 5.15.

The jth vortex at (x_j, z_j) induces a velocity at the ith control point at (x_i^{cp}, z_i^{cp}) given by‡

$$\Delta\vec{v}_j(\vec{x}_i) = \frac{\Gamma_j}{2\pi r_{ij}} \left(\frac{z_i^{cp} - z_j}{r_{ij}} \hat{\imath} + \frac{x_j - x_i^{cp}}{r_{ij}} \hat{k} \right), \tag{5.20}$$

where $r_{ij}^2 = (x_i^{cp} - x_j)^2 + (z_i^{cp} - z_j)^2$. Each control point feels the effect of all of the vortices and the freestream. This means that the total velocity at control point i is

$$\vec{v}_i = \vec{V}_\infty + \sum_{j=1}^{N} \Delta\vec{v}_j(\vec{x}_i). \tag{5.21}$$

Equation 5.19 applied at control point i gives

$$\sum_{j=1}^{N} \underbrace{\Delta\vec{v}_j(\vec{x}_i) \cdot \vec{n}_i}_{A_{ij}\Gamma_j} = \underbrace{-\vec{V}_\infty \cdot \vec{n}_i}_{b_i}. \tag{5.22}$$

This is the same form as Equation 5.16, which means that the linear system looks like the one in Equation 5.18, except with different matrix entries. In addition, we have $N+1$ nodes and N control points, so we need one more equation.

* If there is no gap, then no vortex is placed on the trailing edge, to satisfy the Kutta condition, and one control point can be removed to keep the number of equations equal to the number of unknowns.

† For a fluid with viscosity, the requirement would be no slip, which for a stationary wall would mean $\vec{v} = \vec{0}$. However, here we assume inviscid flow, so a nonzero tangential velocity (slip) is allowed, and indeed all that we can mathematically impose.

‡ This equation comes from using $\Gamma_j/(2\pi r_{ij})$ for the magnitude of the velocity induced by a vortex. The term in parentheses is the clockwise tangential direction vector.

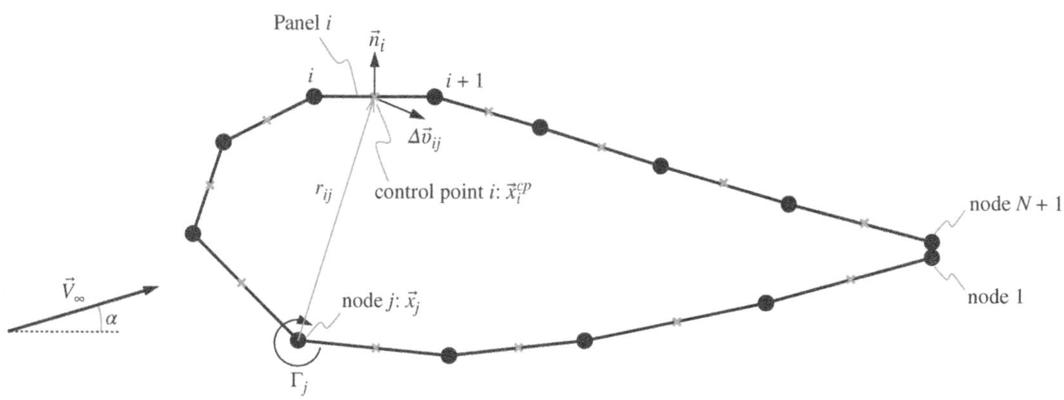

Figure 5.15 Placement of vortices and control points in a discrete vortex model of an airfoil.

This equation comes from the Kutta condition, which, to make the flow leave the trailing edge smoothly, requires that the net vortex strength at the trailing edge be zero,§

$$\Gamma_1 + \Gamma_{N+1} = 0. \tag{5.23}$$

§ Although there is a gap, for enforcing the Kutta condition, we treat the two nodes at the trailing edge as nearly coincident.

Adding this equation into the system and solving it yield the vortex strengths Γ_j, which can be summed to obtain the total lift by the Kutta–Joukowsky formula. The tangential velocities at the control points can also be converted to pressure and c_p by using Bernoulli's equation.

Example 5.2 Lift coefficient of a cambered airfoil

As an example of the discrete vortex method, consider an isosceles triangle airfoil with geometry as shown in Figure 5.16.

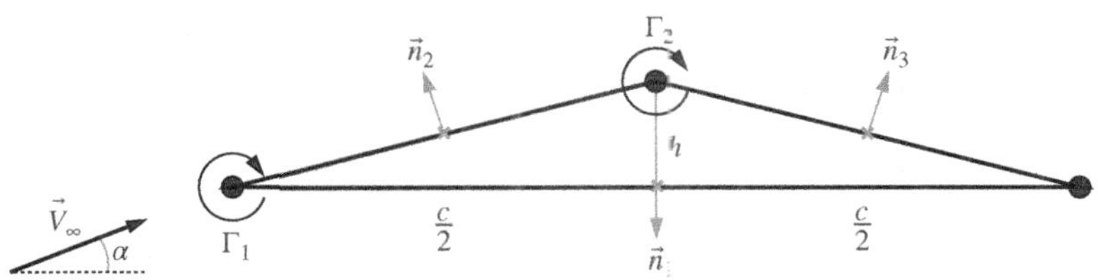

Figure 5.16 A three-panel model of an airfoil.

Applying the discrete vortex method to this geometry consists of placing vortices at the vertices of the triangle. This airfoil does not have a trailing-edge gap, which means that there is only one node at the trailing edge, and its vortex strength must be zero by the Kutta condition. So the only two unknowns are Γ_1 and Γ_2. Out of the three possible control points, we choose the first two to obtain the same number of equations.¶

The normal vectors at the two control points are

$$\vec{n}_1 = -1\hat{k}, \quad \vec{n}_2 = -\frac{h}{l}\hat{\imath} + \frac{c}{2l}\hat{k}, \tag{5.24}$$

where $l = \sqrt{c^2/4 + h^2}$. Using Equations 5.20 and 5.22, the matrix entries are‖

$$\begin{aligned}
A_{11} &= \frac{1}{2\pi r_{11}^2}\left[(z_1^{cp} - z_1)n_{1x} + (x_1 - x_1^{cp})n_{1z}\right] \\
&= \frac{1}{2\pi(c/2)^2}\left[(0)(0) + \left(\frac{-c}{2}\right)(-1)\right] = \frac{1}{\pi c}, \\
A_{12} &= \frac{1}{2\pi r_{12}^2}\left[(z_1^{cp} - z_2)n_{1x} + (x_2 - x_1^{cp})n_{1z}\right] \\
&= \frac{1}{2\pi(c/2)^2}\left[(-h)(0) + (0)(-1)\right] = 0, \\
A_{21} &= \frac{1}{2\pi r_{21}^2}\left[(z_2^{cp} - z_1)n_{2x} + (x_1 - x_2^{cp})n_{2z}\right] \\
&= \frac{1}{2\pi(l/2)^2}\left[\left(\frac{h}{2}\right)\left(\frac{-h}{l}\right) + \left(\frac{-c}{4}\right)\left(\frac{c}{2l}\right)\right] = \frac{-1}{\pi l}, \\
A_{22} &= \frac{1}{2\pi r_{22}^2}\left[(z_2^{cp} - z_2)n_{2x} + (x_2 - x_2^{cp})n_{2z}\right] \\
&= \frac{1}{2\pi(l/2)^2}\left[\left(\frac{-h}{2}\right)\left(\frac{-h}{l}\right) + \left(\frac{c}{4}\right)\left(\frac{c}{2l}\right)\right] = \frac{1}{\pi l}.
\end{aligned}$$

¶ The choice of which control point to drop is arbitrary, and different choices will yield slightly different results. The difference diminishes as the number of panels increases.

‖ Manually calculating these coefficients becomes tedious for more nodes or more complex geometries, and hence, the matrix system is most easily constructed using a computer program.

The right-hand-side vector entries are

$$\begin{aligned} b_1 &= -V_\infty(\cos\alpha\hat{i} + \sin\alpha\hat{k})\cdot(-\hat{k}) = V_\infty \sin\alpha, \\ b_2 &= -V_\infty(\cos\alpha\hat{i} + \sin\alpha\hat{k})\cdot\left[\frac{-h}{l}\hat{i} + \frac{c}{2l}\hat{k}\right] \\ &= V_\infty\left(\cos\alpha\frac{h}{l} - \sin\alpha\frac{c}{2l}\right). \end{aligned}$$

The 2×2 system of equations for the unknown vortex strengths is then**

$$\begin{bmatrix} 1/(\pi c) & 0 \\ -1/(\pi l) & 1/(\pi l) \end{bmatrix}\begin{bmatrix} \Gamma_1 \\ \Gamma_2 \end{bmatrix} = V_\infty \begin{bmatrix} \sin\alpha \\ h\cos\alpha/l - c\sin\alpha/(2l) \end{bmatrix}. \tag{5.25}$$

** If we included the third vortex at the trailing edge, and the third control point, this system of equations would be singular, indicating an infinite number of solutions due to not enforcing the Kutta condition.

The first equation tells us that $\Gamma_1 = \pi c V_\infty \sin\alpha$. Substituting this result into the second equation gives $\Gamma_2 = \pi V_\infty(h\cos\alpha + (c/2)\sin\alpha)$. Knowing the vortex strengths, we can calculate the lift coefficient, c_ℓ, as a function of the angle of attack, α. From the Kutta–Joukowsky theorem,

$$\begin{aligned} c_\ell &= \frac{L'}{\frac{1}{2}\rho V_\infty^2 c} = \frac{\rho V_\infty \sum_j \Gamma_j}{\frac{1}{2}\rho V_\infty^2 c} \\ &= 2\pi\sin\alpha + \pi\left(\frac{2h}{c}\cos\alpha + \sin\alpha\right). \end{aligned} \tag{5.26}$$

Using the small angle approximations $\cos\alpha \approx 1$ and $\sin\alpha \approx \alpha$, the lift coefficient becomes

$$c_\ell \approx 2\pi\alpha + \pi\left(\frac{2h}{c} + \alpha\right) = 3\pi\left(\alpha + \frac{2h}{3c}\right). \tag{5.27}$$

From this expression, the lift slope is 3π and the angle of zero lift is $\alpha_{L=0} = -\frac{2h}{3c}$. The lift slope is not very accurate, as we will soon see that thin airfoils have a slope of 2π. This inaccuracy is due to the small number of panels used to represent the geometry of the triangular airfoil.

The zero-lift angle of attack is, as its name suggests, the angle of attack of the airfoil that produces a zero lift coefficient. For example, if the airfoil is 10% thick ($h/c = 0.1$), the zero-lift angle of attack is

$$\alpha_{L=0} = -\frac{2h}{3c} = -.067\text{rad} = -3.82°. \tag{5.28}$$

Given the formulas for Γ_1 and Γ_2, we can plot the streamlines of the flow for a particular choice of chord length, thickness, freestream speed, and angle of attack. A simple way to do this is to evaluate the stream function, ψ, due to the superposition of the freestream and the two vortices at a grid of points, and then to plot contours of the stream function. Listing 5.2 shows a code that does this calculation and makes the streamline plot for the 10% thick geometry at a given angle of attack.

Code notes:

- The magnitude of the freestream speed is arbitrary because it does not change the shape of the streamlines.
- The code uses complex coordinates and calculates the complex potential before taking the imaginary part for plotting the streamlines.
- The triangular airfoil is overlaid on the streamlines, but streamlines inside the airfoil are not hidden. The flow inside the airfoil is physically fictitious but present in the mathematical model.

Listing 5.2 Code for plotting streamlines for the three-panel airfoil example.

```
function streamlines3v(alpha)
% alpha = angle of attack, radians

Vinf   =  1;                   % freestream speed
c      =  1;                   % chord
h      = .1;                   % thickness
x1     = 0;                    % Node 1 complex coordinate
```

```
x2    = c/2 + h*i;         % Node 2 complex coordinate

% Calculate vortex strengths at Node 1 and 2
G1 = pi*c*Vinf*alpha;
G2 = pi*Vinf*(h + (c/2)*alpha);

% Calculate the stream function
[X,Y] = ndgrid(linspace(-c/4,1+c/4,50),linspace(-3*c/4, 3*c/4,50));
Z = X + i*Y;
F = Vinf*Z + i*G1/(2*pi)*log(Z-x1) - i*G2/(2*pi)*log(Z-x2);
Psi = imag(F);

% Plot contours of the streamfunction
figure(1); clf;
contour(X,Y,Psi, linspace(-.5,.5,18), '-', 'color', '#888888');

% Plot the airfoil
xy = [c,0; 0,0; c/2,h; c,0];
hold on; plot(xy(:,1), xy(:,2), 'k-', 'linewidth', 2);
axis equal; axis([-c/4, 1+c/4, -3*c/4, 3*c/4]); axis off;
```

The streamlines for $\alpha = 0°$ and $\alpha = 2°$ are shown in Figures 5.17 and 5.18. Note, in both cases, the streamlines are tangent to control points 1 and 2, as required by the two equations that we kept. The flow is not exactly tangent to control point 3, but the deviation from tangency is not overly large.

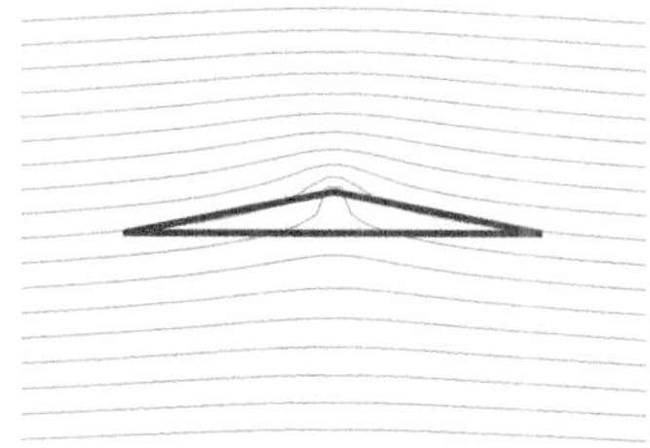

Figure 5.17 Streamlines for the three-panel airfoil at $\alpha = 0°$.

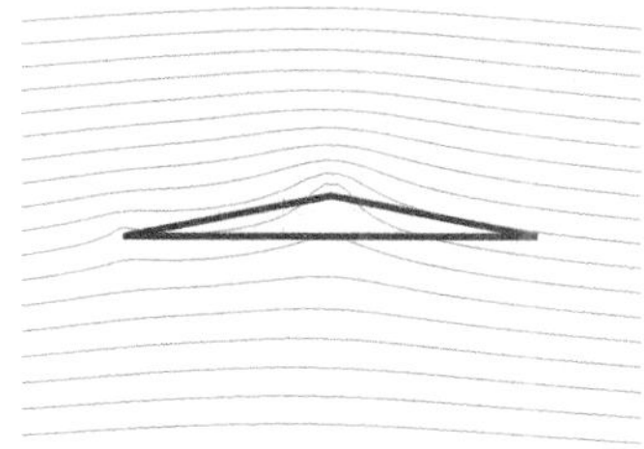

Figure 5.18 Streamlines for the three-panel airfoil at $\alpha = 2°$.

5.4 The Vortex-Panel Method

The discrete vortex models presented so far work reasonably well in predicting total forces, but they have a shortcoming. The concentration of vorticity at points does not allow for an accurate evaluation of the velocity near these points, which impedes pressure calculations and coupling with a boundary-layer solver. A solution to this problem is to use continuous vortex sheets instead of discrete vortices. The result is the continuous *vortex-panel method.*

Compared to the analytical thin-airfoil theory, to be presented next, the vortex-panel method is more general and makes fewer assumptions. In particular, it can model flow over thick and highly cambered airfoils, up to the assumptions of potential flow.* It is a numerical technique that, like the previous discrete vortex models, lends itself to solution and post-processing using a computer program.

*Viscous effects such as separation become more pronounced as thickness and camber increase.

5.4.1 The Vortex Sheet

In the continuous vortex-panel method, we place a *vortex sheet* on each panel of the geometry. Just like mass can be concentrated at a point or distributed along a line, so can vorticity. The resulting distribution of vorticity is called a sheet because in three dimensions, each infinitesimal vortex in the distribution corresponds to a vortex line perpendicular to the page, and the collection of these lines forms a surface, or sheet.

Consider one such sheet, on a panel of length c, aligned with the x axis for convenience, as shown in Figure 5.19. The sheet strength is $\gamma(x)$, and it need not be constant. The total vortex strength of the sheet is the integral of the vortex strength per unit length,

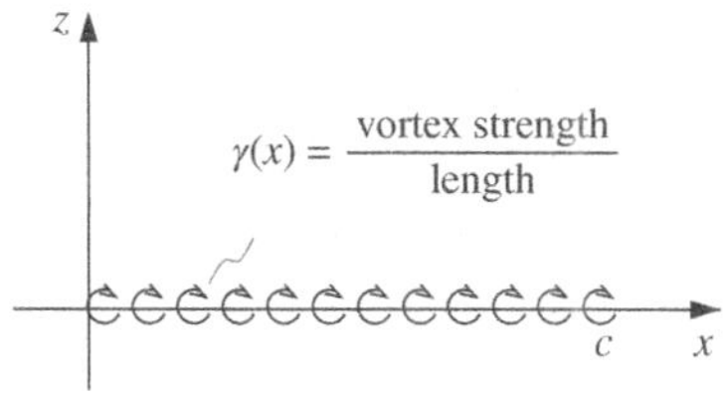

Figure 5.19 A vortex sheet of length c.

$$\Gamma = \int_0^c \gamma(x)\, dx. \tag{5.29}$$

The velocity at an arbitrary point, $\vec{x}_i = (x_i, z_i)$, induced by the vortex sheet is given by an integral,

$$\vec{v}(\vec{x}_i) = \int_0^c \frac{\gamma(x)dx}{2\pi r(x)} \underbrace{\left[\frac{z_i - 0}{r(x)}\hat{i} - \frac{x_i - x}{r(x)}\hat{k}\right]}_{\vec{t}(x)}, \tag{5.30}$$

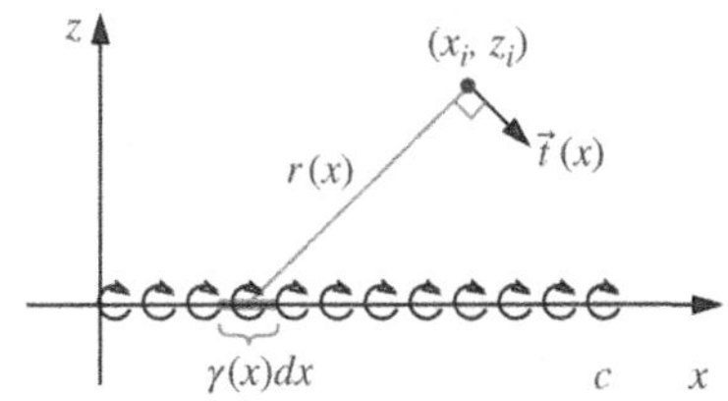

Figure 5.20 Calculation of the velocity induced by a vortex sheet.

where $r(x)$ is the distance between the point and a location on the vortex sheet, and $\vec{t}(x)$ is the direction of the velocity contribution due to the infinitesimal vortex at location x on the vortex sheet, as shown in Figure 5.20. This integral can be done analytically for certain forms of $\gamma(x)$, such as constant or linear, or numerically in general.

We can also compute the stream function at an arbitrary point by superimposing stream functions due to each infinitesimal portion of the vortex sheet. The integral is similar to that for the velocity, but now using the formula for the stream function at an arbitrary distance r from a vortex, Equation 4.45,

$$\psi(\vec{x}_i) = \int_0^c \frac{\gamma(x)dx}{2\pi} \ln r\,(x). \tag{5.31}$$

The stream function is useful for plotting streamlines of the flow induced by the vortex sheet.

At the vortex sheet, the tangential component of the velocity is discontinuous.* Consider a small section, Δx, of the vortex sheet, and a rectangular control volume around this section, as depicted in Figure 5.21. The circulation around the contour C bounding this control volume is equal to the enclosed vorticity,

* On the other hand, by conservation of mass, the normal component is continuous across the sheet.

$$\text{circulation around } C = \text{enclosed vorticity}$$

$$-\oint_C \vec{v} \cdot d\vec{l} = \gamma(x)\Delta x \tag{5.32}$$

$$\Delta x \left(u_{\text{upper}} - u_{\text{lower}}\right) + \Delta z \left(w_{\text{left}} - w_{\text{right}}\right) = \gamma(x)\Delta x$$

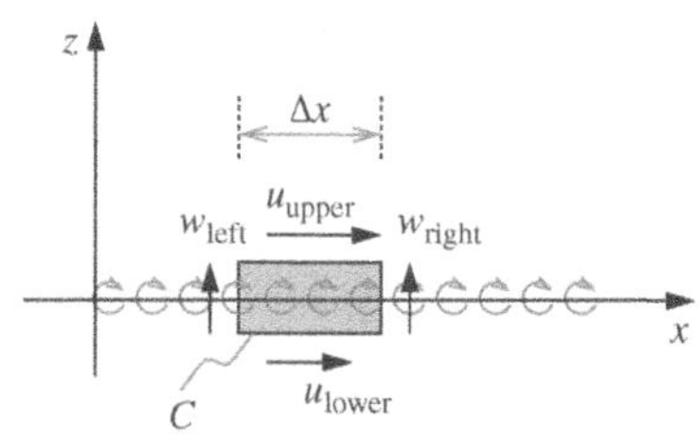

Figure 5.21 Rectangular control volume of size $\Delta x by \Delta z$ used for interpreting $\gamma(x)$ as the horizontal velocity jump across a vortex sheet.

where $u_{\text{upper}}/u_{\text{lower}}$ is the horizontal velocity on the upper/lower edge of the contour, respectively, and $w_{\text{left}}/w_{\text{right}}$ is the vertical velocity on the left/right edge. Shrinking the box vertically, $\Delta z \to 0$, we obtain

$$\gamma(x) = u_{\text{upper}} - u_{\text{lower}}. \tag{5.33}$$

That is, the vortex strength per unit length is the tangential velocity jump across the vortex sheet.

Example 5.3 Vortex sheet streamline

In this example, we use contours of the stream function to visualize the streamlines around a horizontal vortex sheet of unit length, $c = 1$. Equation 5.31 gives the stream function, and we approximate the integral by dividing it into N intervals of size $\Delta x = c/N$ and performing the sum

$$\psi(\vec{x}) \approx \sum_{j=1}^{N} \frac{\gamma(x_j)\Delta x}{2\pi} \ln |\vec{x} - \vec{x}_j|, \tag{5.34}$$

where $\vec{x}_j = x_j\hat{i}$ is the midpoint of the jth interval. Listing 5.3 presents a code that constructs the stream function on a grid of points and plots the streamlines. It does so for two types of vortex strength distributions: constant and linear.

Listing 5.3 Code for plotting streamlines around a vortex sheet.

```
function vsheetviz(gtype)
% Plots the streamlines around a vortex sheet
% gtype = 1 for constant, 2 for linear gamma(x)
c = 1;   % chord length
[X,Y] = ndgrid(linspace(-c/2,3*c/2,101), linspace(-c/2, c/2, 51));
N = 50; dx = c/N; xj = linspace(0,c,N+1)+0.5*dx;
Psi = zeros(size(X));
for j = 1:N,
  R = sqrt( (X-xj(j)).^2 + Y.^2 );
  Psi = Psi + gam(xj(j), gtype)*dx/(2*pi)*log(R);
end
figure(1); clf;
contour(X,Y, Psi, 21, 'linewidth', 1, 'color', '#888888'); hold on;
plot([0,c], [0,0], 'k-', 'linewidth', 2);
axis equal; axis off;
%--------------------------
function g = gam(x, gtype)
if (gtype == 2), g = x; else, g = 1; end;
```

Code notes:

- `ndgrid` returns matrices of coordinates for a lattice.
- The matrix `Psi` is computed incrementally by summing contributions of each vortex piece of size `dx`.
- The function `gam` computes $\gamma(x)$ for a given `gtype`, which dictates the type of distribution, constant or linear.

The associated streamlines in Figure 5.22 indicate flow circulating around the sheet, flattened toward the sheet and approaching circular further away. In the vortex-panel method, multiple vortex sheets together with a freestream superimpose to construct a flowfield that meets tangency conditions at specified points.

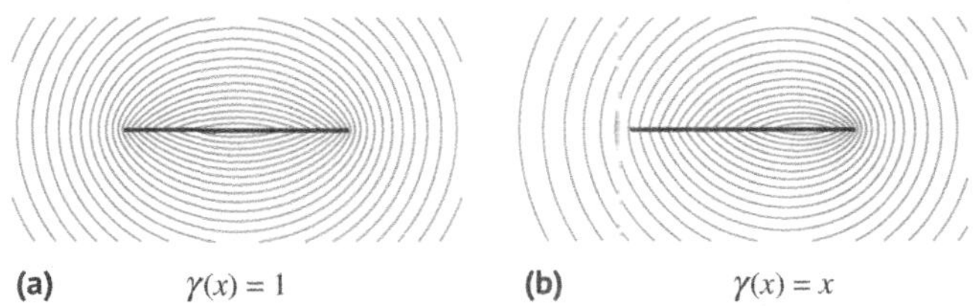

Figure 5.22 Streamlines for vortex sheets with constant and linear distribution of vortex strength per unit length, $\gamma(x)$.

5.4.2 Discretization

In the vortex-panel method, the geometry is approximated by linear (straight) panels. For N panels, there are $N + 1$ nodes, assuming that the first and last points are distinct. The node numbering starts with 1 on the lower side of the trailing edge, proceeds clockwise around the airfoil, and ends with $N + 1$ on the upper side of the trailing edge, as shown in Figure 5.23. If there is no trailing-edge gap, the first and last points are coincident, but they are still distinct for the purpose of defining the unknowns. The number of panels affects both cost and accuracy of the solution. More panels means a better approximation of the geometry and the flow, and hence improved accuracy. However, it also increases the cost of the numerical solution by making the linear system larger. For reasonable N, up to several hundred, this computational cost remains minimal compared to that of full computational fluid dynamics simulations.

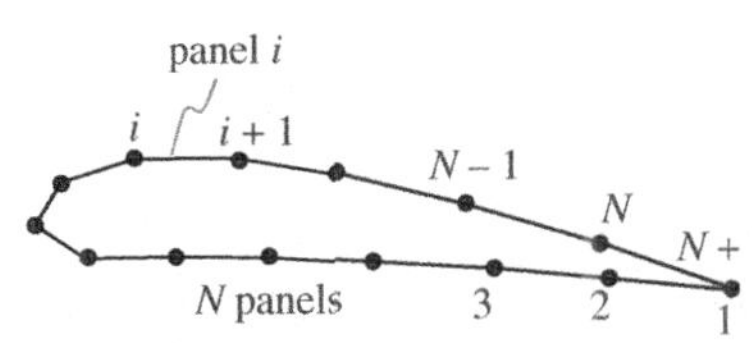

Figure 5.23 Discretization of an airfoil into N panels in the vortex-panel method.

The next step is to distribute vortex singularities on each panel, varying linearly from node to node, as illustrated in Figure 5.24. Denote by $\gamma_j(s)$ the vorticity per unit length on panel j, where $0 < s < \Delta s_j$ is the distance along the panel, which has length Δs_j. A linearly-varying vorticity takes the form

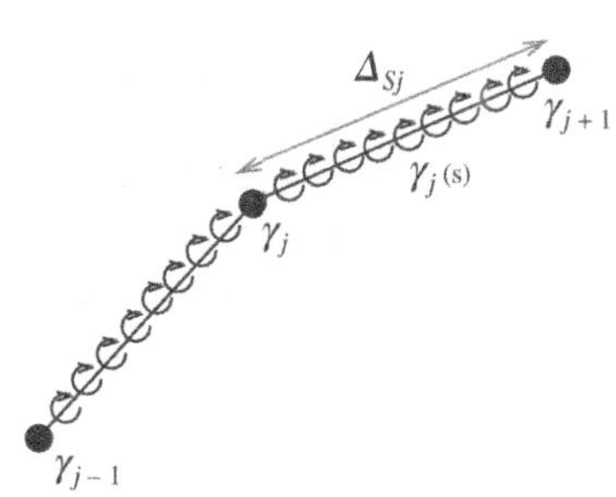

Figure 5.24 Distribution of linearly varying vorticity on each panel.

$$\gamma_j(s) = \gamma_j + (\gamma_{j+1} - \gamma_j)\frac{s}{\Delta s_j}, \tag{5.35}$$

where γ_j and γ_{j+1} are the values of the vorticity per unit length at the nodes adjacent to panel j. These nodal values are the unknowns in the method. Since there are $N+1$ nodes, there are $N+1$ unknowns: γ_j, $1 \leq j \leq N+1$.

5.4.3 Flow-Tangency Equations

The equations come from enforcing flow tangency at N control points, which are the panel midpoints, as in the discrete vortex-panel method presented in Section 5.3.3. Flow tangency is based on the physical requirement that the fluid does not cross the airfoil boundary. Although in reality, we would like the flow to be tangent everywhere to the airfoil surface, we cannot generally accomplish this with a finite number of unknowns. Hence, at locations that are not control points, the flow may not be tangent to the airfoil. However, as we increase the number of panels, we will converge closer to the physical solution.

The velocity at control point i is affected by the freestream and by the vorticity on all of the other panels. Consider the effect of one panel. The velocity induced by panel j at control point i is

$$\Delta\vec{v}_{ij} = \int_0^{\Delta s_j} \frac{\gamma_j(s)ds}{2\pi r_{ij}(s)} \left[\frac{z_i - z_j(s)}{r_{ij}(s)}\hat{i} - \frac{x_i - x_j(s)}{r_{ij}(s)}\hat{k} \right], \tag{5.36}$$

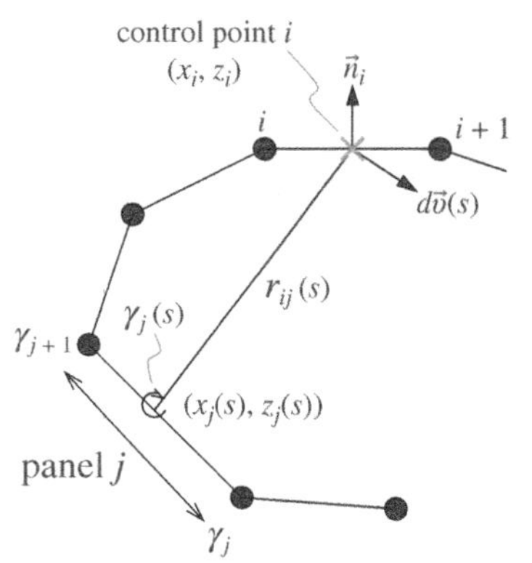

Figure 5.25 The vortex distribution on panel j influences the velocity at control point i.

where $r_{ij}(s)$ is the distance from location s along the panel to the control point, as shown in Figure 5.25. Summing over all of the panels and including the freestream, we obtain $\vec{v}_i$, the velocity at control point i,

$$\vec{v}_i = \vec{V}_\infty + \sum_{j=1}^{N} \Delta\vec{v}_{ij}. \tag{5.37}$$

Flow tangency at control point i, for which the normal vector is $\vec{n}_i$, requires

$$\vec{v}_i \cdot \vec{n}_i = \left(\vec{V}_\infty + \sum_{j=1}^{N} \Delta\vec{v}_{ij} \right) \cdot \vec{n}_i = 0. \tag{5.38}$$

Substituting Equation 5.36 for $\Delta\vec{v}_{ij}$ and rearranging yields

$$\sum_{j=1}^{N} \int_0^{\Delta s_j} \frac{\gamma_j(s)ds}{2\pi r_{ij}(s)} \left[\frac{z_i - z_j(s)}{r_{ij}(s)}\hat{i} - \frac{x_i - x_j(s)}{r_{ij}(s)}\hat{k} \right] \cdot \vec{n}_i = -\vec{V}_\infty \cdot \vec{n}_i. \tag{5.39}$$

Using $\gamma_j(s) = \gamma_j + (\gamma_{j+1} - \gamma_j)\frac{s}{\Delta s_j}$, we can write

$$\sum_{j=1}^{N} (a_{ij}\gamma_j + b_{ij}\gamma_{j+1}) = -\vec{V}_\infty \cdot \vec{n}_i, \tag{5.40}$$

where a_{ij} and b_{ij} denote integrals evaluated from the above equation.* General expressions for these integrals can be derived for arbitrarily oriented panels, e.g. see Katz and Plotkin (2001) and Fidkowski (2022). By grouping the unknowns together, we write the flow-tangency condition at control point i as

*These integrals depend only on the airfoil geometry, and not on the freestream velocity, which comes in on the right-hand side.

$$\sum_{j=1}^{N+1} c_{ij}\gamma_j = -\vec{V}_\infty \cdot \vec{n}_i, \tag{5.41}$$

where the c_{ij} coefficients are combinations of a_{ij} and b_{ij},

$$c_{i1} = a_{i1}, \quad c_{i2} = a_{i2} + b_{i1}, \quad \ldots \quad , c_{iN} = a_{iN} + b_{i\,N-1}, \quad c_{i\,N+1} = b_{iN}.$$

5.4.4 Solution and Post-Processing

Flow tangency yields N equations, one at each control point, for $N+1$ unknowns, γ_j. To obtain a solvable system, we need one more equation, which comes from the Kutta condition. Recall that for airfoils, the Kutta condition states that the vortex strength at the trailing edge is zero. In our numerical model, the vortex strength at the trailing edge is $\gamma_1 + \gamma_{N+1}$, since we are double-counting the trailing edge node.* Therefore, the additional equation is

$$\gamma_1 + \gamma_{N+1} = 0. \tag{5.42}$$

* This equation is similar to that for the discrete vortex-panel method, and it likewise assumes that the trailing-edge gap is small so that the two points are almost coincident.

With $N+1$ equations for $N+1$ unknowns, we can form a linear system and solve it for the unknown γ_j. The matrix form of the flow-tangency equations and Kutta condition is

$$\begin{bmatrix} c_{11} & c_{12} & \dots & c_{1N} & c_{1\,N+1} \\ c_{21} & c_{22} & \dots & c_{2N} & c_{2\,N+1} \\ \vdots & \vdots & \ddots & \vdots & \vdots \\ c_{N1} & c_{N2} & \dots & c_{NN} & c_{N\,N+1} \\ 1 & 0 & \dots & 0 & 1 \end{bmatrix} \begin{bmatrix} \gamma_1 \\ \gamma_2 \\ \vdots \\ \gamma_N \\ \gamma_{N+1} \end{bmatrix} = \begin{bmatrix} -\vec{V}_\infty \cdot \vec{n}_1 \\ -\vec{V}_\infty \cdot \vec{n}_2 \\ \vdots \\ -\vec{V}_\infty \cdot \vec{n}_N \\ 0 \end{bmatrix}. \tag{5.43}$$

Each row of this linear system represents one of the equations: the first N rows correspond to the N control points, and last row is the Kutta condition. The columns are the coefficients multiplying the unknowns, γ_j, in each equation. The right-hand side is the constant term from each equation. This system can be constructed and solved numerically using a computer program.

After solving for γ_j, we can post-process the results.

- Velocities at the panel midpoints can be calculated as

$$\vec{v}_i = \vec{V}_\infty + \sum_{j=1}^{N} \int_0^{\Delta s_j} \frac{\gamma_j(s)ds}{2\pi r_{ij}(s)} \left[\frac{z_i - z_j(s)}{r_{ij}(s)}\hat{i} - \frac{x_i - x_j(s)}{r_{ij}(s)}\hat{k} \right]. \tag{5.44}$$

- The pressure and pressure coefficient at the panel midpoints are then given by Bernoulli's equation,

$$p_i = p_\infty + \frac{1}{2}\rho V_\infty^2 - \frac{1}{2}\rho|\vec{v}_i|^2, \tag{5.45}$$

$$c_p = \frac{p - p_\infty}{\frac{1}{2}\rho V_\infty^2} = 1 - \frac{|\vec{v}_i|^2}{V_\infty^2}.$$

- Forces and moments can be calculated by integrating the pressure distribution over the panels.
- Lift can be calculated using the Kutta–Joukowsky theorem,

$$L' = \rho V_\infty \Gamma = \rho V_\infty \oint_{\text{airfoil}} \gamma ds = \rho V_\infty \sum_{j=1}^{N} \frac{\gamma_j + \gamma_{j+1}}{2} \Delta s_j,$$

where we have used the trapezoidal rule for integrating the vortex distribution.†

† Trapezoidal integration is exact in this case because the distribution is linear on each panel.

There exist multiple vortex-panel airfoil programs. A popular one is XFOIL (Drela 1989), and a similar implementation of it in MATLAB is called mfoil (Fidkowski 2022). These codes include a viscous boundary-layer solver for computing shear stresses and hence drag, which is otherwise absent in an inviscid solver due to d'Alembert's paradox.‡

‡ Airfoils with a trailing-edge gap are not closed, and hence will experience a small amount of pressure drag even without viscosity.

Example 5.4 Using a vortex-panel solver

The mfoil code uses the continuous vortex-panel method similar to the one presented in this section. The unknowns are the vortex strengths at the nodes, but the equations are slightly different: instead of setting the normal component of the velocity equal to zero at control points, mfoil sets the value of the stream function equal to a constant on the airfoil nodes. Doing this makes the airfoil a streamline of the flow, which is equivalent to enforcing flow tangency.

Listing 5.4 shows a sample script for creating a 4-digit series NACA airfoil and analyzing it with a variable number of panels at a prescribed angle of attack. Table 5.1 shows the resulting lift coefficients: we see decreasing-magnitude changes as the number of panels increases, indicating convergence to a final, continuous, value.

Table 5.1 mfoil c_ℓ convergence.

# panels, N	c_ℓ
20	.84643
40	.85491
80	.85732
160	.85805
320	.85832

Listing 5.4 Airfoil analysis using the vortex-panel code mfoil.

```
function runmfoil

alpha = 5; % angle of attack [deg]

Nv = [20,40,80,160,320];
for n = 1:length(Nv),
  % create an airfoil with Nv(n) panels
  m = mfoil('naca','2412', 'npanel',Nv(n)); % create airfoil
  m.setoper('alpha',alpha);                  % set angle of attack
  m.param.legpos = [.6, .2];                 % set legend position
  m.solve();                                 % solve
  fprintf(1,' N=%d, cl = %.5f\n', Nv(n), m.post.cl);
end
set(gcf, 'position', [400, 400, 450, 450]);
exportgraphics(gcf, '../figs/n2412a5.pdf');
```

Figure 5.26 shows the pressure coefficient distribution on the airfoil for the case with the finest paneling. Recall that in potential flow, the pressure coefficient is

$$c_p = \frac{p - p_\infty}{\frac{1}{2}\rho V_\infty^2} = 1 - \frac{|\vec{v}|^2}{V_\infty^2}. \tag{5.46}$$

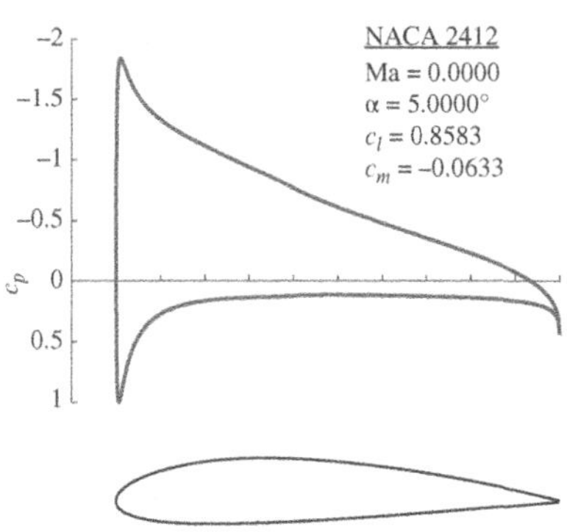

Figure 5.26 Pressure distribution from an inviscid analysis using mfoil.

When the airfoil is producing positive lift, we expect higher pressures on the bottom surface, the "pressure" side, and lower ones on the top surface, the "suction" side. Therefore, it is customary to use an inverted axis or to plot -c_p when plotting c_p on both surfaces concurrently, as in Figure 5.26. Doing so makes the upper curve correspond to the upper surface.

The pressure coefficient is 1 at the leading-edge stagnation point, and approaches 1 at the trailing edge – the gap there prevents it from being a true stagnation point. On the upper surface, the lowest pressure is near the front of the airfoil, and it then rises steadily toward the trailing edge. On the lower surface, the pressure coefficient is close to zero throughout, indicating that $p \approx p_\infty$ there. Finally, the moment coefficient, pitch-up positive, is reported about the quarter-chord location, and a negative value for the moment coefficient indicates that the center of pressure is slightly aft of the quarter-chord.

5.5 The Thin-Airfoil Model

Panel methods provide a numerical approach for computing the forces, moments, and pressure distributions for potential flow over airfoils. With some approximations, an analytical technique, known as thin-airfoil theory, is also possible. If these assumptions are violated, the vortex-panel method provides a recourse. The analytical approach is useful in that it provides closed-form solutions that can be used to estimate forces and moments and spot-check the validity of more sophisticated numerical techniques.

5.5.1 Assumptions

In analyzing thin airfoils, just like in panel methods, we assume two-dimensional potential flow, which means that the flow is incompressible and irrotational. The incompressible assumption is good for low-speed flight, when the Mach number is less than about 0.3. Even for higher Mach numbers, we can use the incompressible results after applying compressibility corrections, as we will discuss in Chapter 8. The irrotational assumption is valid if the flow is inviscid, since an inviscid flow that starts out irrotational (e.g. uniform freestream) will remain irrotational.* When a small amount of viscosity is present, such as in air, the flow does become rotational, but usually only in a thin region near the boundary. The majority of the flow is still irrotational, as we will see in Chapter 6.

*This is a consequence of the vorticity transport equation, presented in Section 3.4, which is obtained by taking the curl of the momentum conservation equation.

In thin-airfoil theory, we also make simplifying assumptions about the shape of the airfoil. These are:

- small thickness, $t(x)/c \lesssim 0.12$
- small camber slope, $dz_c(x)/dx \lesssim 0.5$
- small angle of attack, $\alpha \lesssim 10°$
- small camber, $z_c(x)/c \lesssim 0.1$

We now consider these assumptions in more detail and discuss their implications on the flow model and tangency equations.

Small thickness:

With thickness assumed small, we replace the airfoil with its camber line, as illustrated in Figure 5.27. This zero-thickness approximation of the airfoil gives rise to the name *thin*-airfoil theory. The reason why the approximation is made is that it allows for a closed-form analytical solution.

For a zero-thickness airfoil, the new problem statement in terms of the velocity potential is

$$\nabla^2\phi = 0, \tag{5.47}$$

subject to the boundary conditions

$$\begin{aligned} \nabla\phi &\rightarrow \vec{V}_\infty \quad \text{far from the airfoil,} \\ \nabla\phi \cdot \vec{n} &= 0 \quad \text{on the camber line,} \end{aligned} \tag{5.48}$$

where $\vec{n}$ is the normal vector on the camber line, as shown in Figure 5.27. The second condition enforces flow tangency on the camber line and makes it a streamline of the flow. The small thickness assumption enables this simplified camber-line condition as opposed to requiring flow tangency on the surface.

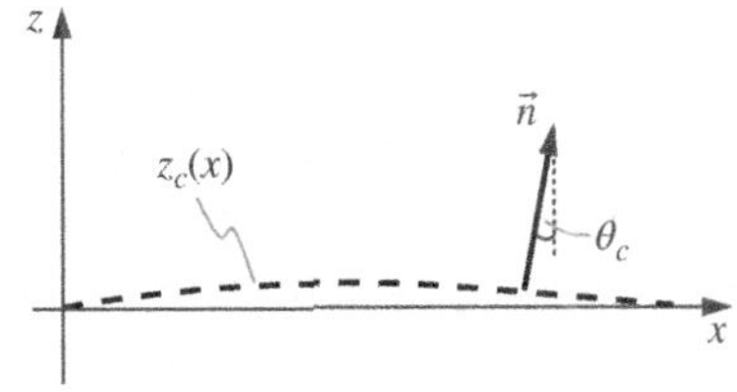

Figure 5.27 Using the small thickness assumption, we replace the airfoil geometry with the airfoil camber line.

We separate ϕ into the potential due to the freestream and the potential due to the airfoil by writing

$$\phi(x,z) = \underbrace{\phi_\infty}_{\text{freestream}} + \underbrace{\phi_a}_{\text{effect of airfoil}}, \tag{5.49}$$

where the freestream potential is simply $\phi_\infty = \vec{V}_\infty \cdot \vec{x}$, so that

$$\nabla\phi_\infty = \vec{V}_\infty = V_\infty \cos\alpha\hat{i} + V_\infty \sin\alpha\hat{k}. \tag{5.50}$$

In this expression, $V_\infty = |\vec{V}_\infty|$ is the velocity magnitude. The normal vector is

$$\vec{n} = \sin\theta_c\hat{i} + \cos\theta_c\hat{k}, \quad \text{where } \theta_c = \tan^{-1}\left(-\frac{dz_c}{dx}\right). \tag{5.51}$$

The normal vector inclination angle, θ_c, is illustrated in Figure 5.27. So the flow-tangency condition, enforced on the camber line, becomes

$$0 = \nabla\phi\cdot\vec{n} = V_\infty \cos\alpha\ \sin\theta_c + V_\infty \sin\alpha\ \cos\theta_c + \nabla\phi_a\cdot\vec{n}. \tag{5.52}$$

Small camber slope and small α:
This assumption states that α and θ_c are small angles. Therefore, we make the following approximations:

$$\begin{aligned} \sin\alpha &\approx \alpha, \\ \cos\alpha &\approx 1, \\ \theta_c = \tan^{-1}\left(-\frac{dz_c}{dx}\right) &\approx -\frac{dz_c}{dx}, \\ \sin\theta_c \approx \theta_c &\approx -\frac{dz_c}{dx}, \\ \cos\theta_c &\approx 1. \end{aligned} \tag{5.53}$$

With these approximations, the camber-line flow-tangency condition, Equation 5.52, becomes

$$0 = V_\infty\left(-\frac{dz_c}{dx}\right) + V_\infty\alpha + \frac{\partial\phi_a}{\partial z}, \tag{5.54}$$

where we have used $\nabla\phi_a\cdot\vec{n} = \frac{\partial\phi_a}{\partial x}\sin\theta_c + \frac{\partial\phi_a}{\partial z}\cos\theta_c \approx \frac{\partial\phi_a}{\partial z}$.† Re-arranging this equation gives

$$\frac{\partial\phi_a}{\partial z} = V_\infty\left(\frac{dz_c}{dx} - \alpha\right) \quad \text{(on the camber line).} \tag{5.55}$$

† This approximation is valid assuming $\frac{\partial\phi_a}{\partial x}$ is not much larger than $\frac{\partial\phi_a}{\partial z}$, which will be the case for our vortex sheet singularity distribution.

Small camber:
The last simplification uses the assumption of small camber. Instead of enforcing the flow-tangency boundary condition on the camber line, we enforce it on the chord line, as illustrated in Figure 5.28. This does not mean that the flow is made to be tangent to the chord line, since we still use the normal vector from the camber line in the flow-tangency equation. This simplification only slightly changes the location at which the camber-line flow tangency is enforced. It is necessary for obtaining analytical solutions for general airfoils, as otherwise the location of flow tangency would change with the geometry.‡

‡ By analogy with the vortex-panel method, we can think of this assumption as keeping the matrix **A** constant and taking into account geometry changes only through the normal vectors in the right-hand side, **b**.

The error introduced by the small-camber approximation is as small as the errors introduced in the previous approximations. The resulting *flow-tangency boundary condition* is

$$\left(\frac{\partial \phi_a}{\partial z}\right)_{z=0} = V_\infty \left(\frac{dz_c}{dx} - \alpha\right) \quad \text{on the chord line.} \tag{5.56}$$

Note that this equation is enforced at all x for $0 \leq x \leq c$. To solve it, we use a specific singularity distribution, a vortex sheet, to model ϕ_a, the potential due to the presence of the airfoil.

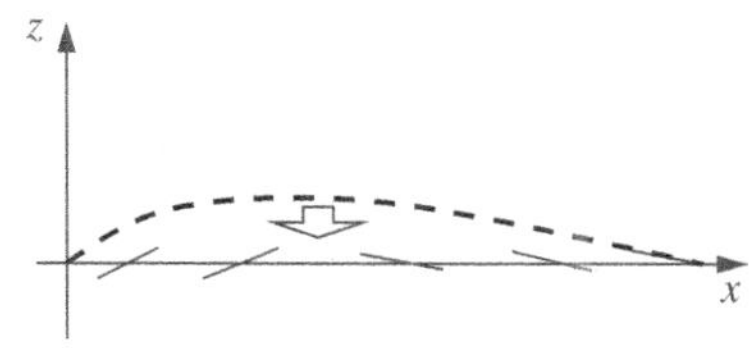

Figure 5.28 The small-camber assumption allows us to apply the flow-tangency boundary condition at the airfoil chord line instead of at the camber line. The line segments on the chord line illustrate the camber slopes at different x locations.

5.5.2 Fundamental Equation of Thin-Airfoil Theory

To produce lift, the potential-flow model for the airfoil needs to have a nonzero net vortex strength, Γ. This is because $L' = \rho V_\infty \Gamma$ by the Kutta–Joukowsky theorem. Before calculating Γ, we need to decide how to introduce the vortex into the model.

One choice that allows for an analytical solution is a continuous distribution of vortices, the vortex sheet introduced in Section 5.4.1. Recall that in the sheet strength, $\gamma(x)$, is a vortex strength per unit length. By Equation 5.33, $\gamma(x)$ is also the horizontal velocity jump between the lower and upper surfaces of the airfoil. When a single vortex sheet is used to model the airfoil, we need to apply the Kutta condition, which requires $u_{\text{upper}} = u_{\text{lower}}$ at the trailing edge, so

$$\gamma(x{=}c) = 0. \tag{5.57}$$

The task is to solve for $\gamma(x)$ using the flow-tangency boundary condition and the Kutta condition. The flow-tangency equation, Equation 5.56, requires the gradient of the potential ϕ_a. This potential is the result of a superposition of all infinitesimal vortices on the sheet. Consider the influence of a piece of the vortex sheet at position x on the vertical velocity at another position, x_0, as shown in Figure 5.29. Recall that the velocity induced by a vortex is inversely proportional to the distance from the vortex, so that

$$dw = -\frac{\gamma dx}{2\pi r} = -\frac{\gamma dx}{2\pi(x_0 - x)}. \tag{5.58}$$

The minus sign in this equation is present because $\gamma > 0$ induces a negative dw at $x_0 > x$. Integrating the infinitesimal velocity contributions from the entire vortex sheet yields the net vertical velocity at x_0,*

$$\left.\frac{\partial \phi_a}{\partial z}\right|_{x_0} = \int_0^c \frac{-\gamma(x)dx}{2\pi(x_0 - x)}. \tag{5.59}$$

* The velocity induced by the vortex sheet is the gradient of the potential, $\nabla\phi_a$, so the vertical component is $\frac{\partial \phi_a}{\partial z}$.

Substituting this expression into the flow-tangency boundary condition, Equation 5.56, yields the *fundamental equation of thin-airfoil theory*,

$$\frac{1}{2\pi}\int_0^c \frac{\gamma(x)dx}{(x_0 - x)} = V_\infty\left(\alpha - \left.\frac{dz_c}{dx}\right|_{x_0}\right), \quad \gamma(c) = 0. \tag{5.60}$$

In the next section, we will solve this integral equation for $\gamma(x)$, using V_∞, α, and dz_c/dx as given quantities.

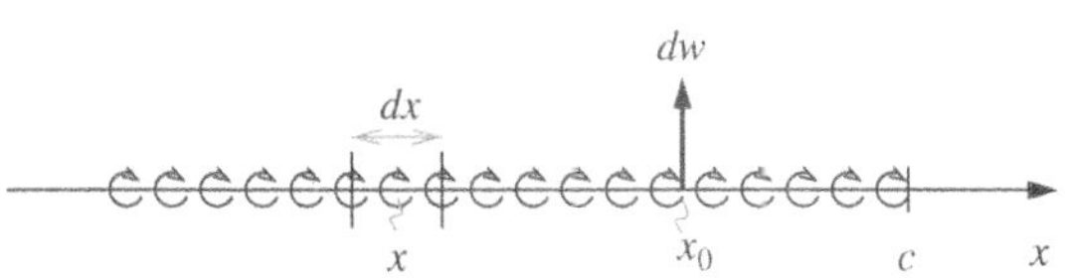

Figure 5.29 A portion of the vortex sheet at x influences the vertical velocity at x_0.

5.6 Thin-Airfoil Solutions

This section presents solutions to the fundamental equation of thin-airfoil theory, Equation 5.60. The solutions for symmetrical airfoils are presented first because of their relative simplicity. Incorporating camber requires a Fourier-series representation of the camber-line slope, making the solution to cambered airfoils more complicated, as expected. Force and moment coefficients are computed in both cases, leading to simple yet quite practical formulas.

5.6.1 Symmetric Airfoils

Symmetric airfoils are ones without camber. Under the assumptions of thin-airfoil theory, these airfoils are treated as *flat plates*. In this case, $dz_c/dx = 0$, which means that Equation 5.60 becomes

$$\frac{1}{2\pi}\int_0^c \frac{\gamma(x)dx}{(x_0 - x)} = V_\infty \alpha. \tag{5.61}$$

One way to solve this equation, a way that sets the stage for the more complicated case of cambered airfoils, is to transform the x variable into a new coordinate, θ, according to

$$x = \frac{1}{2}c(1 - \cos\theta). \tag{5.62}$$

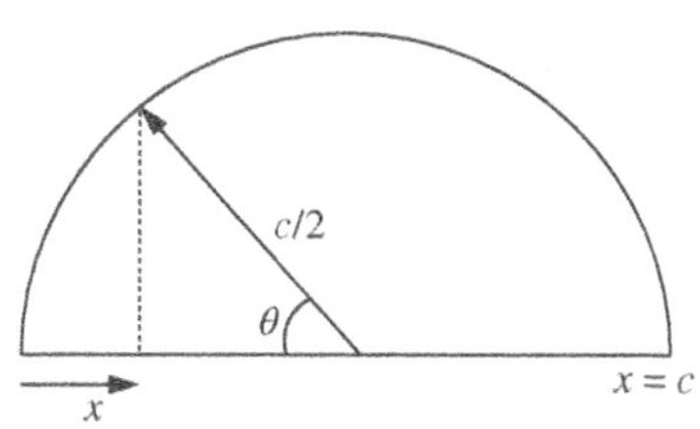

Figure 5.30 Transformation between x and θ.

The coordinate θ ranges from 0 to π as x ranges from 0 to c. However, the transformation is not linear, as illustrated graphically in Figure 5.30. Near the leading and trailing edges, $x = 0$ and $x = c$, a change in θ leads to almost no change in x. In contrast, in the middle of the airfoil, x changes most rapidly with θ. This observation is also evident from the derivative of Equation 5.62,

$$dx = \frac{1}{2}c\sin\theta d\theta. \tag{5.63}$$

Substituting the transformations of x and dx into Equation 5.61, we obtain

$$\frac{1}{2\pi}\int_0^\pi \frac{\gamma(\theta)\sin\theta d\theta}{(\cos\theta - \cos\theta_0)} = V_\infty \alpha, \tag{5.64}$$

where θ_0 corresponds to x_0. Also, the Kutta condition is $\gamma(\theta = \pi) = 0$. This equation has a simple closed-form solution*

*Deriving this solution requires the mathematical theory of integral equations that is beyond the scope of this text.

$$\gamma(\theta) = 2\alpha V_\infty \frac{1 + \cos\theta}{\sin\theta}. \tag{5.65}$$

Note that $\gamma(\pi) = 0$ by l'Hôpital's rule, so the Kutta condition is satisfied.
To convert back to x, the relevant formulas are

$$\cos\theta = 1 - 2\frac{x}{c}, \tag{5.66}$$

$$\sin\theta = \sqrt{1 - \cos^2\theta} = 2\sqrt{\frac{x}{c}\left(1 - \frac{x}{c}\right)}, \tag{5.67}$$

so that

$$\gamma(x) = 2\alpha V_\infty \sqrt{\frac{c - x}{x}}. \tag{5.68}$$

At the leading edge, $\gamma(x)$ goes to infinity.[†] The singularity should also not come as a surprise, since for nonzero angles of attack, the flow must turn an infinitely sharp corner at the leading edge, and by streamline curvature, Equation 3.122, a zero radius turn causes an infinite pressure gradient.

† Fortunately, this does not cause problems when calculating the total vortex strength, since the integral of $\gamma(x)$ is still finite.

With $\gamma(x)$ know, we can calculate the sectional force on the airfoil. Recall that the force, $\vec{F}'$, on a two-dimensional object in potential flow is given by the integral of the pressure acting on the object's surface, which for a thin airfoil becomes an integral of the pressure difference over the chord,

$$\vec{F}' = \oint_{\text{airfoil}} p\vec{n}dl = \int_0^c (p_{\text{lower}} - p_{\text{upper}})\hat{k}dx. \tag{5.69}$$

This force is in the $\hat{k}$ direction, and under the small-angle assumption ($\cos\alpha \approx 1$), its magnitude is precisely the sectional lift, L',

$$L' = \int_0^c (p_{\text{lower}} - p_{\text{upper}})dx. \tag{5.70}$$

We can calculate $p_{\text{lower}} - p_{\text{upper}}$ from Bernoulli's equation,

$$p_{\text{lower}} + \frac{1}{2}\rho(u_{\text{lower}}^2 + w_{\text{lower}}^2) = p_{\text{upper}} + \frac{1}{2}\rho(u_{\text{upper}}^2 + w_{\text{upper}}^2). \tag{5.71}$$

Mass conservation applied to the contour in Figure 5.21 gives $w_{\text{lower}} = w_{\text{upper}}$, which means that[‡]

$$\begin{aligned}\Delta p(x) &\equiv p_{\text{lower}}(x) - p_{\text{upper}}(x) \\ &= \frac{1}{2}\rho\left[u_{\text{upper}}(x)^2 - u_{\text{lower}}(x)^2\right] \\ &= \rho\underbrace{\frac{u_{\text{lower}}(x) + u_{\text{upper}}(x)}{2}}_{V_\infty \cos\alpha \approx V_\infty}\underbrace{(u_{\text{upper}}(x) - u_{\text{lower}}(x))}_{\gamma(x)}.\end{aligned}$$

‡ In the last expression, the average of u_{upper} and u_{lower} is the horizontal component of the freestream velocity because for any x, $\gamma(x)$ only introduces a jump in the tangential velocity, without a net offset. In addition, the rest of the vortex sheet cannot induce a horizontal velocity component at x.

Therefore, the pressure difference between the lower and upper surfaces of the airfoil is

$$\Delta p(x) \equiv p_{\text{lower}}(x) - p_{\text{upper}}(x) = \rho V_\infty \gamma(x). \tag{5.72}$$

Substituting this expression into the sectional lift integral in Equation 5.70 gives

$$L' = \int_0^c \rho V_\infty \gamma(x)dx. \tag{5.73}$$

The integral can be evaluated by transforming from x to θ and using Equation 5.65,

$$\begin{aligned}L' &= \rho V_\infty \int_0^\pi 2\alpha V_\infty \frac{1+\cos\theta}{\sin\theta}\frac{1}{2}c\sin\theta\, d\theta \\ &= \rho V_\infty^2 c\alpha \int_0^\pi (1+\cos\theta)\, d\theta = \rho V_\infty^2 c\alpha\pi.\end{aligned} \tag{5.74}$$

The sectional lift coefficient is then

$$c_\ell = \frac{L'}{\frac{1}{2}\rho V_\infty^2 c} = \frac{\rho V_\infty^2 c\alpha\pi}{\frac{1}{2}\rho V_\infty^2 c} = 2\pi\alpha. \tag{5.75}$$

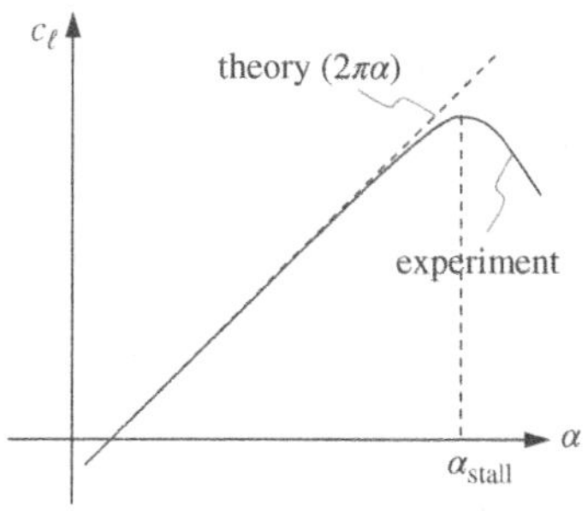

Figure 5.31 Qualitative comparison of the lift coefficient predicted by thin-airfoil theory to that expected from a real experiment, for a symmetric airfoil. Loss of lift is observed in the experiment when the airfoil stalls.

This is an important result that often provides a back-of-the-envelope estimate of the lift coefficient for a thin airfoil. It applies to symmetrical airfoils, although the lift-curve slope, 2π, is also valid for cambered airfoils.

How good is this theory? Figure 5.31 shows qualitatively how a comparison between theory and experiment might look. The theory is good until the boundary layer on the airfoil separates and stall occurs. *Stall* is the loss of lift caused by flow separation that occurs when the angle of attack, α, is sufficiently large in magnitude. The α at which stall occurs depends on the airfoil geometry and on the Reynolds number. For example, for a NACA 0012 airfoil, stall occurs at approximately $\alpha = 12°$ for a Reynolds number $\text{Re} = 3 \times 10^6$.

In addition to the lift force, L', of engineering interest in airfoil analysis is also the sectional moment on the airfoil. For an airfoil, this is the *pitch moment*, and its definition relies on the choice of a location about which the moment is calculated. This location will generally be an (x, z) coordinate, but for thin airfoils, since thickness is neglected, the location will be just the x coordinate. We begin by calculating the moment about the leading edge, which is $x = 0$, as illustrated in Figure 5.32.

We will use the convention that a clockwise (CW) moment is positive. In other words, a positive moment tends to *pitch the airfoil up*. Note that even if the moment about the leading edge is negative, the airfoil may still tend to pitch up about its center of gravity, which will be somewhere between the leading and trailing edge. The leading-edge moment is given by including x in the integral of the pressure difference,§

§ In the last step, $1 - \cos^2\theta$ is $\sin^2\theta$, and the integral evaluates to

$$\int_0^\pi \sin^2\theta = \frac{\pi}{2}.$$

$$\begin{aligned} M'_{LE} &= -\int_0^c x(p_{\text{lower}} - p_{\text{upper}})dx \\ &= -\int_0^\pi \frac{c}{2}(1-\cos\theta)\rho V_\infty 2\alpha V_\infty \frac{1+\cos\theta}{\sin\theta}\frac{1}{2}c\sin\theta\, d\theta \\ &= -\frac{1}{2}\rho V_\infty^2 c^2\alpha \int_0^\pi (1-\cos^2\theta)\, d\theta \\ &= -\frac{1}{4}\rho V_\infty^2 c^2 \alpha\pi. \end{aligned} \tag{5.76}$$

The sectional moment can be non-dimensionalized by dividing it by the dynamic pressure and the chord length squared – the extra length is needed because a moment is a force times a moment arm. Therefore, the leading-edge *sectional moment coefficient* is

$$c_{m,LE} = \frac{M'_{LE}}{\frac{1}{2}\rho V_\infty^2 c^2}. \tag{5.77}$$

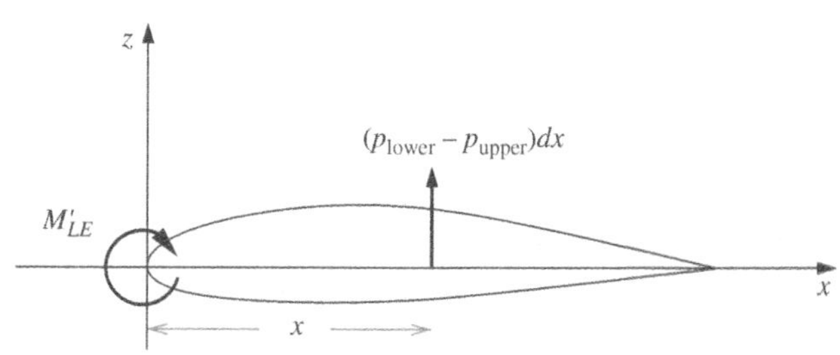

Figure 5.32 Calculation of the moment about the leading edge.

In the case of a symmetric airfoil, this is, using Equation 5.76 and Equation 5.75,

$$c_{m,LE} = -\frac{\pi}{2}\alpha = -\frac{c_\ell}{4}. \tag{5.78}$$

Moments can be defined about other points,¶ and we can consider the moment about another x location, x_p. The equation for M'_p involves a shift of the moment arm,

¶ We consider points on the airfoil chord line for defining the moment, which means that $z = 0$, due to the thin-airfoil assumption, and hence, only x is relevant.

$$\begin{aligned} M'_p &= -\int_0^c (x - x_p)(p_{\text{lower}} - p_{\text{upper}})dx \\ &= -\int_0^c x(p_{\text{lower}} - p_{\text{upper}})dx + \int_0^c x_p(p_{\text{lower}} - p_{\text{upper}})dx \\ &= M'_{LE} + x_p L'. \end{aligned} \tag{5.79}$$

This equation translates the leading-edge moment to another point on the airfoil. The *center of pressure* is the location x_{cp} about which the moment M'_{cp} is zero,

$$0 = M'_{cp} = M'_{LE} + x_{cp}L' = -\frac{1}{4}\rho V_\infty^2 c^2 \alpha\pi + x_{cp}\rho V_\infty^2 c\alpha\pi. \tag{5.80}$$

Solving for x_{cp}, we find that *the center of pressure is at the quarter-chord location* for a symmetric airfoil

$$x_{cp} = \frac{c}{4} \quad \text{for a symmetric airfoil.} \tag{5.81}$$

Symmetric airfoil summary

$$\begin{aligned} c_\ell &= 2\pi\alpha \\ c_{m,LE} &= -\frac{c_\ell}{4} \\ c_{m,c/4} &= 0 \\ x_{cp} = x_{ac} &= \frac{c}{4} \end{aligned}$$

Note, x_{cp} is independent of α. This means that the moment at the quarter-chord is also independent of the angle of attack,‖ and hence, that the quarter-chord location is also the *aerodynamic center*. The aerodynamic center, x_{ac}, is defined as the location at which the moment is independent of angle of attack.**

‖ It will be zero regardless of α for symmetric airfoils.

** For a real airfoil in which the forces and moments are not linear with respect to α, this condition is one on the moment derivative, $dM_{ac}/d\alpha = 0$, at a given angle of attack.

5.6.2 Cambered Airfoils

Cambered airfoils introduce the complication of an arbitrary camber line slope, dz_c/dx, in the thin-airfoil theory equation. Using θ instead of x simplifies the solution approach, with the main effort being a Fourier transform of dz_c/dx. Instead of working with an arbitrary camber-line slope function, we will work with Fourier coefficients that characterize that function. It turns out that only the first few coefficients will be required for force and moment calculations.

Using Equation 5.60 and Equation 5.62, the fundamental equation of thin-airfoil theory in terms of θ is

$$\frac{1}{2\pi}\int_0^\pi \frac{\gamma(\theta)\sin\theta d\theta}{\cos\theta - \cos\theta_0} = V_\infty\left(\alpha - \left.\frac{dz_c}{dx}\right|_{\theta_0}\right). \tag{5.82}$$

One way to solve this equation analytically for $\gamma(\theta)$ is to use a Fourier-series expansion. We start by expanding the camber-line slope as a cosine series,

$$\left.\frac{dz_c}{dx}\right|_{\theta_0} = \sum_{n=0}^{\infty} B_n \cos n\theta_0 = B_0 + \sum_{n=1}^{\infty} B_n \cos n\theta_0. \tag{5.83}$$

Substituting into Equation 5.82 yields

$$\frac{1}{2\pi}\int_0^\pi \frac{\gamma(\theta)\sin\theta d\theta}{\cos\theta - \cos\theta_0} = V_\infty\left(\alpha - B_0 - \sum_{n=1}^{\infty} B_n \cos n\theta_0\right). \tag{5.84}$$

With the help of two trigonometric identities,* we can make the sides equal if we set

$$\gamma(\theta) = 2V_\infty \left[(\alpha - B_0)\frac{1+\cos\theta}{\sin\theta} + \sum_{n=1}^{\infty} B_n \sin n\theta\right]. \tag{5.85}$$

This is a sine-series expansion of $\gamma(\theta)$ with an additional $(\alpha - B_0)$ term that takes care of the symmetric case.† For a general cambered airfoil, we can compute the B_n coefficients from dz_c/dx and then use Equation 5.85 to obtain $\gamma(\theta)$, from which forces and moments will follow. Note, the calculation of B_n from dz_c/dx involves the inverse Fourier transform,

$$B_0 = \frac{1}{\pi}\int_0^\pi \frac{dz_c}{dx}\, d\theta, \tag{5.86}$$

$$B_n = \frac{2}{\pi}\int_0^\pi \frac{dz_c}{dx} \cos n\theta \, d\theta \quad (n > 0), \tag{5.87}$$

which can be derived from Equation 5.83. Given a camber geometry, we can calculate the coefficients B_n, and together with the angle of attack, α, we obtain $\gamma(\theta)$ from Equation 5.85.

Consider the force on a cambered airfoil. The equation for the pressure difference between the lower and upper surfaces, Equation 5.72, still holds,

$$\Delta p(\theta) = \rho V_\infty \gamma(\theta). \tag{5.88}$$

However, $\gamma(\theta)$ changes when considering cambered versus symmetric airfoils. Using Equation 5.85, we can evaluate the sectional lift as‡

$$\begin{aligned} L' &= \int_0^c \Delta p(x)\, dx \\ &= \rho V_\infty \int_0^\pi \gamma(\theta)\frac{c}{2}\sin\theta\, d\theta \\ &= \rho V_\infty^2 c \int_0^\pi \left[(\alpha - B_0)\frac{1+\cos\theta}{\sin\theta} + \sum_{n=1}^{\infty} B_n \sin n\theta\right]\sin\theta\, d\theta \\ &= \rho V_\infty^2 c \int_0^\pi \left[(\alpha - B_0)(1+\cos\theta) + \sum_{n=1}^{\infty} B_n \sin n\theta \sin\theta\right] d\theta \\ &= \rho V_\infty^2 c \left[\alpha\pi - B_0\pi + B_1\frac{\pi}{2}\right]. \end{aligned}$$

Dividing by $\frac{1}{2}\rho V_\infty^2 c$, we obtain the lift coefficient,

$$c_\ell = 2\pi\left(\alpha - B_0 + \frac{1}{2}B_1\right). \tag{5.89}$$

We can write this expression as

$$c_\ell = 2\pi\left[\alpha - \alpha_{L=0}\right], \tag{5.90}$$

*The two relevant trigonometric identities are

$$\frac{1}{\pi}\int_0^\pi \frac{1+\cos\theta}{\cos\theta - \cos\theta_0}\, d\theta = 1,$$

and

$$\frac{1}{\pi}\int_0^\pi \frac{\sin n\theta \sin\theta}{\cos\theta - \cos\theta_0}\, d\theta = -\cos n\theta_0.$$

† If all of the B_n are zero, we obtain the symmetric solution.

‡ The last step uses the identities:

$$\int_0^\pi \cos\theta\, d\theta = 0,$$

$$\int_0^\pi \sin n\theta \sin m\theta\, d\theta = \frac{\pi}{2}\delta_{nm},$$

where $\delta_{nm} = 1$ if $n = m$ and 0 otherwise, with the caveat that the expression evaluates to zero if $n = m = 0$.

where $\alpha_{L=0}$ is the *zero-lift angle of attack*, given by[§]

$$\alpha_{L=0} = B_0 - \frac{1}{2}B_1 = \frac{1}{\pi}\int_0^\pi \frac{dz_c}{dx}(1-\cos\theta). \qquad (5.91)$$

§ In the last step, we substitute for B_0 and B_1 in terms of dz_c/dx.

As the name implies, $\alpha_{L=0}$ is the angle of attack of the airfoil at which the lift is zero. Its value depends on the airfoil geometry, and for symmetric airfoils, it is zero.

Note that the lift slope, $dc_\ell/d\alpha$, is still 2π, just like for symmetric airfoils. Also, a typical airfoil with positive camber produces positive lift at zero angle of attack, which means that $\alpha_{L=0} < 0$.

The calculation of pitch moment about the leading edge is similar to the symmetric case, with the result

$$c_{m,LE} = -\frac{\pi}{2}\left[\alpha - B_0 + B_1 - \frac{B_2}{2}\right]. \qquad (5.92)$$

The center of pressure, x_{cp}, is found from Equation 5.80,

$$0 = M'_{LE} + x_{cp}L' \quad \Rightarrow \quad x_{cp} = -\frac{M'_{LE}}{L'} = -\frac{c_{m,LE}}{c_\ell}c. \qquad (5.93)$$

Substituting the expressions for c_ℓ, Equation 5.89, and $c_{m,LE}$, Equation 5.92, we have[¶]

$$x_{cp} = \frac{\frac{\pi}{2}\left(\alpha - B_0 + B_1 - \frac{1}{2}B_2\right)}{2\pi\left(\alpha - B_0 + \frac{1}{2}B_1\right)}c = \frac{c}{4}\left[1 + \frac{\pi}{c_\ell}(B_1 - B_2)\right]. \qquad (5.94)$$

¶ Unless $B_1 = B_2$, the center of pressure is now not at the quarter-chord, and it moves as c_ℓ varies.

The quarter-chord location, $c/4$, is still special for cambered airfoils. Calculating the moment coefficient there using moment translation gives

$$c_{m,c/4} = c_{m,LE} + \frac{1}{4}c_\ell = \frac{\pi}{4}(B_2 - B_1). \qquad (5.95)$$

Since α is not present in this equation, *$c_{m,c/4}$ is independent of α*, and hence, *the quarter-chord of a cambered airfoil is the aerodynamic center.*[||]

|| This does not mean that the moment about the quarter-chord is necessarily zero; this occurs at the center of pressure, and we saw above that the center of pressure is not necessarily at the quarter-chord for a cambered airfoil. However, the moment at the quarter-chord is special in that it does not vary as the angle of attack is changed. It is only a function of the airfoil geometry.

To recap, the center of pressure, x_{cp}, is where the moment is zero. For symmetric airfoils, it is at the quarter-chord, but for cambered airfoils, it is generally not and moves as α changes. The aerodynamic center, x_{ac}, is where the moment is insensitive to changes in α, $dM'_{ac}/d\alpha = 0$. For both symmetric and cambered airfoils, x_{ac} is at the quarter-chord. Note that these are all results from thin-airfoil theory. Real airfoils with thickness and viscous effects will not have the center of pressure or aerodynamic center exactly as predicted from the theory, but the results will be close as long as the assumptions of thin-airfoil theory are closely satisfied.

In summary, for *cambered airfoils*,

$$c_\ell = 2\pi\left[\alpha - \alpha_{L=0}\right] = 2\pi\left[\alpha - B_0 + \frac{1}{2}B_1\right]$$

$$\alpha_{L=0} = \frac{1}{\pi}\int_0^\pi \frac{dz_c}{dx}(1-\cos\theta)d\theta = \left(B_0 - \frac{1}{2}B_1\right)$$

$$x_{cp} = \frac{c}{4}\left[1 + \frac{\pi}{c_\ell}(B_1 - B_2)\right]$$

$$x_{ac} = \frac{c}{4}$$

$$c_{m,ac} = \frac{\pi}{4}(B_2 - B_1) \text{ independent of } \alpha$$

Example 5.5 Lift coefficient of a cambered airfoil
Consider an airfoil that has a camber line given by

$$z_c(x) = \frac{\epsilon_1}{2c}x^2 + \frac{\epsilon_2}{3c^2}x^3,$$

where ϵ_1 and ϵ_2 are constants. We can determine the lift and moment coefficients on this airfoil at a particular angle of attack α using the results of cambered thin-airfoil theory. These results depend on the B_n coefficients, which come from the camber slope,

$$\frac{dz_c}{dx} = \epsilon_1\frac{x}{c} + \epsilon_2\left(\frac{x}{c}\right)^2. \tag{5.96}$$

The Fourier expansion is in terms of θ, so we re-write dz_c/dx in terms of θ using Equation 5.62,**

** In the last step, we use the trigonometric identity

$$\cos^2\theta = \frac{1}{2}(1 + \cos 2\theta).$$

$$\begin{aligned}\frac{dz_c}{dx} &= \frac{\epsilon_1}{2}(1 - \cos\theta) + \frac{\epsilon_2}{4}(1 - \cos\theta)^2 \\ &= \frac{\epsilon_1}{2} + \frac{\epsilon_2}{4} + \left(-\frac{\epsilon_1}{2} - \frac{\epsilon_2}{2}\right)\cos\theta + \frac{\epsilon_2}{4}\cos^2\theta \\ &= \left(\frac{\epsilon_1}{2} + \frac{3\epsilon_2}{8}\right) - \left(\frac{\epsilon_1 + \epsilon_2}{2}\right)\cos\theta + \frac{\epsilon_2}{8}\cos 2\theta.\end{aligned} \tag{5.97}$$

Comparing this expression to the Fourier-series expansion in Equation 5.83,

$$\frac{dz_c}{dx} = \sum_{n=0}^{\infty} B_n \cos n\theta,$$

we see that, by matching terms,

$$B_0 = \frac{\epsilon_1}{2} + \frac{3\epsilon_2}{8}, \quad B_1 = -\frac{\epsilon_1 + \epsilon_2}{2}, \quad B_2 = \frac{\epsilon_2}{8}. \tag{5.98}$$

All of the other coefficients, $B_n, n > 2$, are zero since B_0, B_1, and B_2 fully describe the camber line slope. Therefore, we have all the information that we need to calculate our force and moment coefficients.†† The lift coefficient is

†† Even if the $n > 2$ B_n coefficients are nonzero, c_ℓ and $c_{m,c/4}$ depend only on B_0, B_1, and B_2.

$$c_\ell = 2\pi\left[\alpha - B_0 + \frac{1}{2}B_1\right] = 2\pi\left[\alpha - \frac{3\epsilon_1}{4} - \frac{5\epsilon_2}{8}\right]. \tag{5.99}$$

The quarter-chord (aerodynamic center) moment coefficient is

$$c_{m,c/4} = \frac{\pi}{4}(B_2 - B_1) = \frac{\pi}{4}\left(\frac{1}{2}\epsilon_1 + \frac{5}{8}\epsilon_2\right). \tag{5.100}$$

Example 5.6 Analysis of a piecewise-cambered airfoil
Consider the piecewise-linear camber line of a thin airfoil shown in Figure 5.33. Assuming thin-airfoil theory, we wish to calculate the following at $\alpha = 2°$:

1. The lift coefficient.
2. The moment coefficient about the quarter-chord.
3. The center of pressure.
4. The stability margin, $SM = (x_{ac} - x_{cp})/c$.‡‡

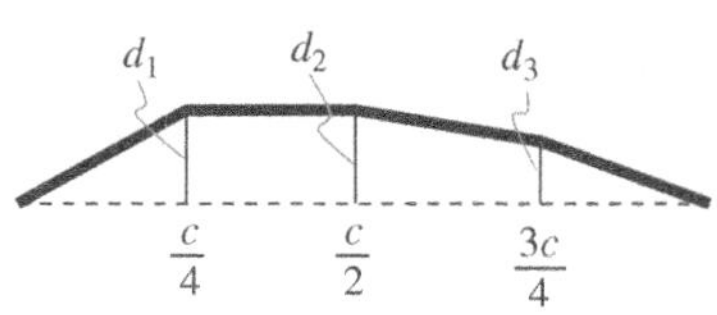

Figure 5.33 Piecewise-linear camber line.

‡‡ The stability margin will be discussed in the next section.

The starting point for this example is the calculation of the first few B_n coefficients. The B_n formula requires the camber-line slope, which is a piecewise-constant function,

$$\frac{dz_c}{dx}(x) = \begin{cases} \dfrac{d_1}{c/4} & x < c/4 \\ \dfrac{d_2 - d_1}{c/4} & c/4 \leq x < c/2 \\ \dfrac{d_3 - d_2}{c/4} & c/2 \leq x < 3c/4 \\ \dfrac{0 - d_3}{c/4} & 3c/4 \leq x \leq c \end{cases} . \tag{5.101}$$

In converting to θ, the constant values of dz_c/dx do not change. Instead, only the ranges do. $x = c/4$ maps to $\theta = \pi/3$, $x = c/2$ maps to $\theta = \pi/2$, and $x = 3c/4$ maps to $\theta = 2\pi/3$. So the camber-line slope as a function of θ is

$$\frac{dz_c}{dx}(\theta) = \begin{cases} \dfrac{d_1}{c/4} & \theta < \pi/3 \\ \dfrac{d_2 - d_1}{c/4} & \pi/3 \leq \theta < \pi/2 \\ \dfrac{d_3 - d_2}{c/4} & \pi/2 \leq \theta < 2\pi/3 \\ \dfrac{0 - d_3}{c/4} & 2\pi/3 \leq \theta \leq \pi \end{cases} .$$

We can now calculate the first few B_n coefficients. Starting with B_0,[§§]

§§ Note that the integrals are with respect to θ, but the camber-line slope is the derivative of z_c with respect to x.

$$\begin{aligned} B_0 &= \frac{1}{\pi} \int_0^{\pi} \frac{dz_c}{dx} d\theta \\ &= \frac{1}{\pi} \left[\int_0^{\pi/3} \frac{4d_1}{c} d\theta + \int_{\pi/3}^{\pi/2} \frac{4(d_2 - d_1)}{c} d\theta + \right. \\ &\quad \left. \int_{\pi/2}^{2\pi/3} \frac{4(d_3 - d_2)}{c} d\theta + \int_{2\pi/3}^{\pi} \frac{4(0 - d_3)}{c} d\theta \right] \\ &= \frac{2}{3c} (d_1 - d_3) . \end{aligned}$$

Next, B_1 is evaluated similarly, but with $\cos\theta$ inside the integral and $2/\pi$ out front,

$$\begin{aligned} B_1 &= \frac{2}{\pi} \int_0^{\pi} \frac{dz_c}{dx} \cos\theta d\theta \\ &= \frac{2}{\pi} \frac{4}{c} [d_1 \sin(\pi/3) + (d_2 - d_1)(\sin(\pi/2) - \sin(\pi/3)) \\ &\quad + (d_3 - d_2)(\sin(2\pi/3) - \sin(\pi/2)) + (-d_3)(\sin(\pi) - \sin(2\pi/3))] \\ &= \frac{8}{\pi c} \left[(\sqrt{3} - 1)d_1 + (2 - \sqrt{3})d_2 + (\sqrt{3} - 1)d_3 \right]. \end{aligned}$$

Finally, B_2 is similar to B_1, but with $\cos 2\theta$ instead of $\cos\theta$,¶¶

$$B_2 = \frac{2}{\pi}\int_0^{\pi} \frac{dz_c}{dx}\cos 2\theta d\theta$$
$$= \frac{2}{\pi}\frac{4}{c}\frac{1}{2}[d_1\sin(2\pi/3) + (d_2 - d_1)(\sin(\pi) - \sin(2\pi/3))$$
$$+(d_3 - d_2)(\sin(4\pi/3) - \sin(\pi)) + (-d_3)(\sin(2\pi) - \sin(4\pi/3))]$$
$$= \frac{4\sqrt{3}}{\pi c}[d_1 - d_3].$$

¶¶ Note that the expressions for B_0 and B_2 both involve $d_1 - d_3$, so that

$$B_0 = \frac{\pi}{6\sqrt{3}}B_2.$$

Choosing a particular geometry, $d_1 = d_2 = 0.1c$, $d_3 = 0.06c$, these expressions yield

$$B_0 = 0.02667, \quad B_1 = 0.3665, \quad B_2 = 0.088213.$$

Using the formulas for the desired quantities, we obtain

$$c_\ell = 2\pi\left[\alpha - B_0 + \frac{1}{2}B_1\right] = 1.203,$$
$$c_{m,c/4} = \frac{\pi}{4}(B_2 - B_1) = -0.21856,$$
$$x_{cp} = \frac{c}{4}\left[1 + \frac{\pi}{c_\ell}(B_1 - B_2)\right] = 0.43166c,$$
$$x_{ac} = 0.25c,$$
$$SM = (x_{ac} - x_{cp})/c = -18.2\%.$$

5.7 Airfoil Design

The goal of airfoil analysis is to determine the performance of a given airfoil geometry at a given operating condition.* In contrast, the goal of airfoil design is to determine the airfoil geometry and possibly the operating condition, given performance requirements and possibly geometry constraints.†

*In inviscid, incompressible flow, the angle of attack, α, is the only operating condition.

†The angle of attack need not be specified in an airfoil design problem, in which case it is a variable that must also be determined.

Three typical airfoil performance requirements pertain to the lift coefficient, the presence of a suction peak, and stability. The simplest is c_ℓ, which, when given, is the lift coefficient that the airfoil must attain. A typical cruise value for an airfoil is $c_\ell \approx 0.5$, whereas landing may demand values of 1 or larger, as indicated in Figure 5.34. Airfoils with high-lift devices such as flaps and slats can attain even higher lift coefficients of 2 or more. The other two requirements deserve more discussion, which is given in the next sections.

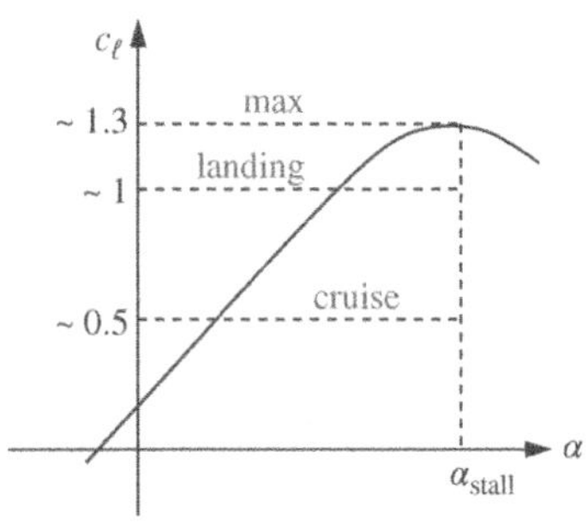

Figure 5.34 Approximate cruise, landing, and maximum c_ℓ values for an airfoil, without high-lift devices.

5.7.1 Leading-Edge Suction Peak

Another design requirement is the mitigation of a leading-edge suction peak. A *suction peak* is a spike in the pressure distribution caused by a sharp turning of the flow. Recall from streamline curvature that the pressure gradient across turning streamlines is inversely proportional to the turning radius. When this radius is very small, the pressure spikes, and such spikes are not desirable once viscous effects are considered.*

*A pressure spike creates a large adverse pressure gradient for fluid elements moving close to the airfoil surface, which can cause flow separation, discussed in Chapter 6.

To avoid suction peaks, the airfoil geometry should be as smooth as possible, which means no corners. Piecewise-linear camber lines and flap hinges ostensibly violate this requirement, but we still use such approximations with the justification that any corners on the geometry could be smoothed out without a large difference in performance. The airfoil thickness also plays a role in smoothing out these changes, as shown in Figure 5.35.

Figure 5.35 An airfoil with an approximate piecewise-linear camber line can have smooth lower and upper surfaces once thickness is included.

In addition to the airfoil geometry, an important location at which a suction peak may form on an airfoil is the leading edge. There, the flow may have to turn a corner if there is a difference in the local flow angle and the airfoil geometry. For a very thin airfoil, the geometry is effectively the camber line, which means that slope of the camber line at the leading edge must align with the local flow angle. Figure 5.36 illustrates this matching: the camber line becomes a streamline of the flow without a slope discontinuity in its connection to the leading-edge stagnation streamline. Note that the camber-line slope at the leading edge is generally not the angle of attack because the flow turns upward prior to meeting the leading edge.

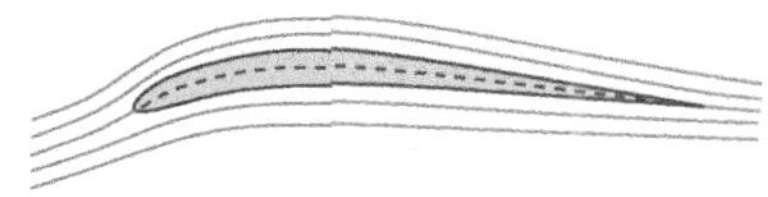

Figure 5.36 Streamlines for a thin airfoil at on-design conditions, without a leading-edge suction peak.

Enforcing this requirement is not as challenging as it may seem, since from Equation 5.72, a pressure spike shows up as a spike in $\gamma(\theta)$, and we have the general form of $\gamma(\theta)$ in Equation 5.85. Considering the leading edge, which is at $\theta = 0$, a spike can occur due to the first term, since

$$\gamma(0) = 2V_\infty \left[(\alpha - B_0)\frac{1+\cos 0}{\sin 0} + 0\right]. \tag{5.102}$$

The $\sin 0$ term in the denominator causes $\gamma(0) \to \infty$ if $\alpha - B_0 \neq 0$.† Hence, we need

$$B_0 = \alpha \text{ for no leading-edge suction peak.} \tag{5.103}$$

† This is for an airfoil with exactly zero thickness. Any thickness or rounding at the leading edge will prevent an infinite spike, even for inviscid flow.

A given airfoil geometry sets B_0, so that $\alpha = \alpha_{des} = B_0$ is called the *design angle of attack*, as this is the angle at which no suction peak occurs. If the airfoil is perfectly thin, there is only one value of α for which the leading-edge suction peak is avoided. However, thickness and rounding of the leading edge help by preventing a suction peak from forming when α is close to, but not exactly, α_{des}, as shown in Figure 5.37. Even though the flow hits the leading edge at an angle higher than that in on-design conditions, the moderate turning radius R prevents a leading-edge suction peak.

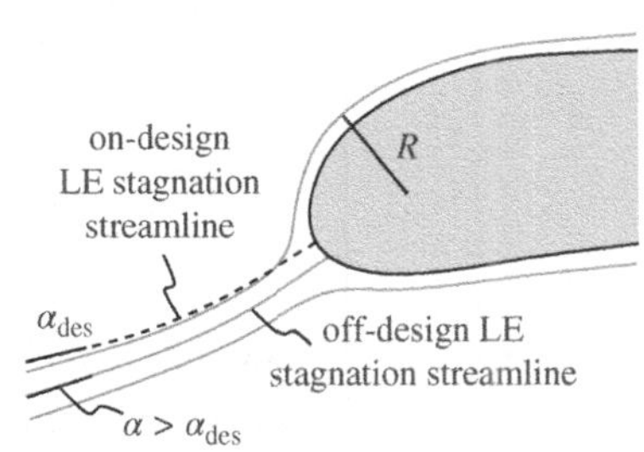

Figure 5.37 A rounded leading edge mitigates the suction peak over a range of angles of attack near α_{des}.

5.7.2 Trimmed Static Stability

A third requirement that can be set in an airfoil design problem pertains to stability. In the case of an airfoil, this is pitch stability, which is the response of the airfoil to perturbations in the angle of attack. A stable response is one in which the perturbations decay, whereas an unstable one occurs when disturbances grow.

For a single flying airfoil, the requirement of stability can be expressed succinctly as

$$\frac{dc_{m,cg}}{d\alpha} < 0, \tag{5.104}$$

where $c_{m,cg}$ is the moment coefficient about the center of gravity of the airfoil. This requirement states that if the angle of attack increases, e.g. through a disturbance, the airfoil should pitch down (a negative, restoring moment) about the center of gravity. Such a response will reduce the angle of attack and mitigate the disturbance. On the other hand, if the sign of this derivative were positive, the perturbation in angle of attack would grow due to a pitch-up moment, leading to an instability.

In trimmed flight, the pitch moment about the center of gravity is zero, which means that $x_{cg} = x_{cp}$.* Also, since Equation 5.104 involves the derivative of a moment with respect to the angle of attack, the aerodynamic center is a good reference point, as the moment there is constant. Translating the moment coefficient from the aerodynamic center to the center of pressure, we have

* This is trimmed flight of an airfoil. For an aircraft, all lifting surfaces would affect x_{cp}.

$$c_{m,cp} = c_{m,ac} + \frac{(x_{cp} - x_{ac})}{c} c_\ell. \tag{5.105}$$

Differentiating with respect to α gives†

† For a first-order analysis, we assume that x_{cp} and x_{ac} remain constant with respect to changes in α.

$$\frac{dc_{m,cp}}{d\alpha} = 0 + \frac{(x_{cp} - x_{ac})}{c} \frac{dc_\ell}{d\alpha}. \tag{5.106}$$

Since $\frac{dc_\ell}{d\alpha} > 0$ for an airfoil in standard operating conditions (e.g. not stalled), the static stability requirement reduces to

$$\frac{dc_{m,cp}}{d\alpha} < 0 \quad \Rightarrow \quad \frac{(x_{cp} - x_{ac})}{c} < 0. \tag{5.107}$$

The *static stability margin* is then defined as

$$SM \equiv \frac{(x_{ac} - x_{cp})}{c}. \tag{5.108}$$

For stability, SM should be positive.‡ It is an indicator of how the aircraft will respond to pitch disturbances. A negative static stability margin is undesirable for a flying wing without active control, as it corresponds to an unstable response: input disturbances in pitch are magnified. However, even though an airfoil may have a negative static stability margin,§ an aircraft with such an airfoil for the main wing can be made stable in pitch by the addition of a horizontal tail.

‡ Note the sign change: $x_{ac} - x_{cp}$ versus $x_{cp} - x_{ac}$.

§ Most cambered airfoils are statically unstable, as we saw in Example 5.6.

Example 5.7 Design of a piecewise-linear camber

Consider again the thin airfoil from Example 5.6, but now in the context of design instead of analysis. The design goal is to determine d_1, d_2, d_3, and α such that the airfoil meets the following geometry and performance requirements:

- Lift coefficient of 0.25
- No leading-edge suction peak ($B_0 = \alpha$)
- 5% static stability, $(x_{ac} - x_{cp})/c = 0.05$
- $d_1 = d_2$¶

¶ This is a geometry constraint: with four unknowns, we need four design requirements.

The general formulas derived in Example 5.6, in terms of d_1, d_2, and d_3, apply to this example as well. The task is reversed, however, as we are given several design requirements and asked to solve for the geometry and angle of attack. We proceed by writing down the equations for the design requirements.

First, the no leading-edge suction peak is straightforward: $B_0 = \alpha$. Using the formula for B_0 from Example 5.6, we have

$$\alpha = \frac{2}{3c}(d_1 - d_3). \tag{5.109}$$

Next, the lift coefficient requirement is

$$c_\ell = 2\pi\left[\alpha - B_0 + \frac{1}{2}B_1\right] \quad \Rightarrow \quad B_1 = \frac{c_\ell}{\pi}. \tag{5.110}$$

The static stability margin requirement is

$$\frac{x_{ac} - x_{cp}}{c} = SM \quad \Rightarrow \quad B_2 - B_1 = \frac{4SM}{\pi} c_\ell. \tag{5.111}$$

Writing these requirements, along with $d_1 = d_2$, in terms of d_1, d_2, and d_3, we obtain the following three equations:

$$d_1 - d_2 = 0,$$
$$\frac{8}{\pi c}\left[(\sqrt{3}-1)d_1 + (2-\sqrt{3})d_2 + (\sqrt{3}-1)d_3\right] = \frac{c_\ell}{\pi},$$
$$\frac{4}{\pi c}\left[(2-\sqrt{3})d_1 + (-4+2\sqrt{3})d_2 + (2-3\sqrt{3})d_3\right] = \frac{4SM}{\pi}c_\ell.$$

This is a 3×3 linear system of equations. Listing 5.5 shows a code that builds the system, solves it, and plots the camber line.

Listing 5.5 Code for solving for a piecewise-linear camber line.

```
function piecelindes
% solves for and plots a piecewise-linear camber line

cl = .25; % lift coefficient
SM = .05; % stability margin
c = 1;    % chord (arbitrary)

% build system for d1, d2, and d3
A = zeros(3,3);   b = zeros(3,1);
s = sqrt(3);
A(1,:) = [1, -1, 0];                        b(1) = 0;
A(2,:) = 8/(pi*c)*[s-1, 2-s, s-1];          b(2) = cl/pi;
A(3,:) = 4/(pi*c)*[2-s, -4+2*s, 2-3*s];  b(3) = 4*SM*cl/pi;

% solve the system
d = A\b
d1 = d(1); d2 = d(2); d3 = d(3);
% angle of attack in degrees
alpha = 2*(d1-d3)/(3*c) * 180/pi

% camber-line breakpoints for plotting
x = [0, c/4, c/2, 3*c/4, c];
z = [0, d1, d2, d3, 0];

% make the plot
figure(1); clf;
plot(x,z, 'ks-','linewidth', 2);
axis equal; axis([0, 1, -.06, .06]); grid on; box off
set(gcf, 'position', [400, 400, 800, 250]);
writeplot('$x/c$', '$z_c/c$', '../figs/piecelindes.pdf', 2);
```

Code notes:

- The choice $c = 1$ is arbitrary but makes plotting easier, since all dimensions are already normalized by c.
- The 3×3 system $\mathbf{Ad} = \mathbf{b}$ is for the 3 unknowns d_1, d_2, and d_3.
- The angle of attack is determined separately after d_1 and d_3 are known.

Running the code gives

$$d_1 = d_2 = 0.036344c, \quad d_3 = -0.0069578c, \quad \alpha = 1.654°. \tag{5.112}$$

Figure 5.38 shows a plot of the airfoil camber line. The negative value of d_3 creates a negative camber near the airfoil trailing edge, which is called a "reflexed" camber.

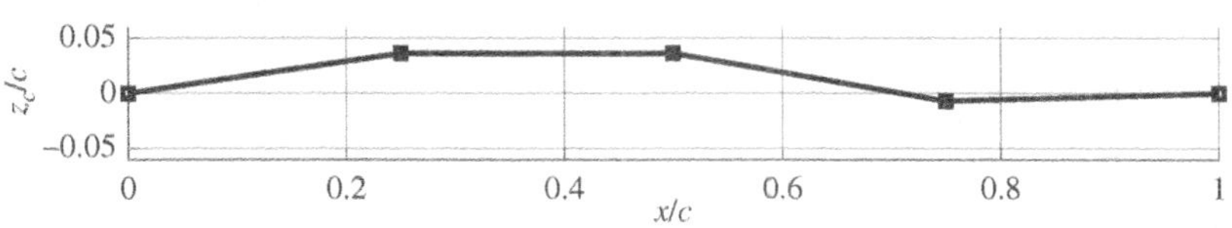

Figure 5.38 Piecewise-linear camber line designed to meet specified lift, suction peak, and stability requirements.

The airfoil design in the previous example had a piecewise-linear camber line, which would require some smoothing in order to avoid sharp corners. The next example presents a design with a smooth camber, made possible by a polynomial

$z_c(x)$. The process remains the same: we calculate the B_n coefficients from the design requirements and then determine the parameters that govern the camber-line shape in order to obtain the desired B_n coefficients.

Example 5.8 Design of a smooth cambered airfoil

In this example, we use thin-airfoil theory to design a smooth camber line, of the form

$$\frac{z_c}{c} = a\frac{x}{c}\left(1 - \frac{x}{c}\right)\left(1 + b\frac{x}{c}\right), \tag{5.113}$$

where the unknowns are the geometry parameters a and b, along with the angle of attack, α. We assume three design requirements,

- Lift coefficient of 0.5
- No leading-edge suction peak
- 10% static stability

As in the previous example, the first step is to convert the design requirements into conditions on the B_n coefficients. First, to achieve no suction peak, we need $B_0 = \alpha$. Next, the c_ℓ requirement is

$$c_\ell = 2\pi\left(\alpha - B_0 + \frac{1}{2}B_1\right) = \pi B_1 \quad \Rightarrow \quad B_1 = \frac{c_\ell}{\pi}. \tag{5.114}$$

To achieve the desired static stability, we need

$$SM = \frac{x_{ac} - x_{cp}}{c} = \frac{\pi}{4c_\ell}(B_2 - B_1) \quad \Rightarrow \quad B_2 = \frac{4c_\ell}{\pi}SM + B_1. \tag{5.115}$$

We can now calculate a and b by computing the camber-line slope and comparing it to the Fourier expansion in Equation 5.83. The camber line slope is

$$\frac{dz_c}{dx} = a\left(1 - \frac{x}{c}\right)\left(1 + b\frac{x}{c}\right) - a\frac{x}{c}\left(1 + b\frac{x}{c}\right) + ab\frac{x}{c}\left(1 - \frac{x}{c}\right). \tag{5.116}$$

Using Equation 5.62, x/c transforms to θ, and after simplification,‖

‖ This simplification uses the identity

$$2\cos^2\theta = \cos 2\theta + 1.$$

$$\frac{dz_c}{dx} = -\frac{ab}{8} + \left(a + \frac{ab}{2}\right)\cos\theta - \frac{3ab}{8}\cos 2\theta. \tag{5.117}$$

Matching terms in the Fourier expansion, Equation 5.83, we identify

$$B_0 = -\frac{ab}{8}, \quad B_1 = \left(a + \frac{ab}{2}\right), \quad B_2 = -\frac{3ab}{8}. \tag{5.118}$$

The last equation gives $ab = -8B_2/3$, and substituting into the second equation,

$$a = B_1 - ab/2 = B_1 + 4B_2/3, \quad b = -8B_2/(3a). \tag{5.119}$$

These are the desired geometry parameters, and the angle of attack is $\alpha = B_0 = -ab/8$. Listing 5.6 shows a code that calculates this solution from given c_ℓ and SM values and plots the camber line.

Code notes:

- a and b only depend on B_1 and B_2, so these are solved for first.
- The design angle of attack (no suction peak) is set to $B_0 = -ab/8$ after obtaining a and b.
- The cubic formula for the camber line is plotted with a large number of x points.

Listing 5.6 Code for solving for a smooth camber line.

```
function smoothdes
% solves for and plots a smooth camber line

cl = .5; % lift coefficient
```

```
SM = .1; % stability margin
c = 1;   % chord (arbitrary)

B1 = cl/pi;, B2 = 4*cl/pi*SM + B1; % B coefficients
a = B1 + 4*B2/3, b = -8*B2/(3*a)    % geometry parameters
alpha = -a*b/8*180/pi               % angle of attack [deg]

% camber line for plotting
x = linspace(0,1,101); z = a*x.*(1-x).*(1+b*x);
figure(1); clf;
plot(x,z, 'k-','linewidth', 2);
axis equal; axis([0, 1, -.06, .06]); grid on; box off
set(gcf, 'position', [400, 400, 800, 250]);
writeplot('$x/c$', '$z_c/c$', '../figs/smoothdes.pdf', 2);
```

Running the code gives

$$a = 0.456, \quad b = -1.302, \quad \alpha = 4.26^\circ. \tag{5.120}$$

Figure 5.39 shows the camber line. Note the reflexed shape, as in Example 5.7: positive camber in the front, and negative camber in the aft portion. This reflex is responsible for the positive static stability of the airfoil combined with no suction peak at positive lift.

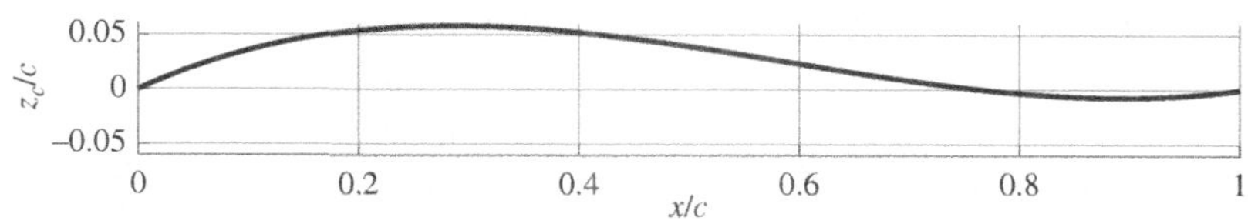

Figure 5.39 Smooth camber line designed to meet specified lift, suction peak, and stability requirements.

This design could now be tested in a panel solver, such as XFOIL or mfoil, in order to determine how closely the design requirements are met. Doing so requires making an airfoil with thickness, and a reasonable approach is to start with a thin, symmetrical airfoil, such as the NACA 0006, and to add the desired camber. Problem 4 exercises this approach.

5.8 Joukowsky Airfoils

So far we have covered both numerical techniques, in the form of discrete and continuous vortex panels, and an analytical technique, in the form of thin-airfoil theory, for analyzing airfoils. Another analytical technique, but one that benefits from numerical post-processing, makes use of the result for potential flow over a cylinder, and a mapping of the complex-number plane (Batchelor 2000). This technique relies on a coordinate transformation that maps the cylinder geometry to an airfoil, as illustrated in Figure 5.40, in such a way that the potential flow equations remain satisfied for the flow around the geometry. Various mappings are possible, and this section presents a simple one that produces a family of airfoils with cusped trailing edges.

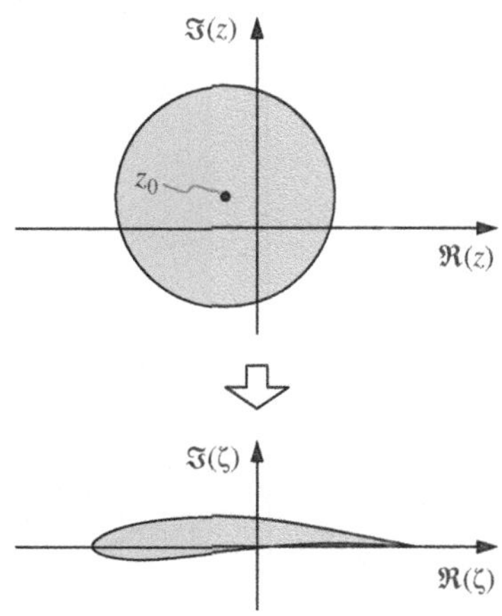

Figure 5.40 Complex-coordinate mapping of a cylinder in the z plane to an airfoil in the ζ plane.

5.8.1 Complex-Coordinate Mapping

Recall from Chapter 4 that a potential flow can be described by a complex potential, $F(z) = \phi + i\psi$, where $z = x + iy$ is the complex coordinate. As long as $F(z)$ is analytic, it describes some potential flow. If the coordinate transformation is also analytic, then the mapped function F also describes a potential flow.

Let $\zeta(z)$ be an invertible mapping from a complex coordinate, z, containing the cylinder geometry to another complex coordinate, ζ, in which the geometry transforms into an airfoil. Define the complex potential $\tilde{F}$ in the ζ plane by

$$\tilde{F}(\zeta) \equiv F(z(\zeta)). \tag{5.121}$$

If $z(\zeta)$ and $F(z)$ are both analytic, then so is $\tilde{F}(\zeta)$. In addition, for any set of points z_{stream} forming a streamline of the flow in the z plane, the corresponding set of mapped points $\zeta(z_{\text{stream}})$ also forms a streamline in the ζ plane, because*

* Recall that the stream function $\psi = \Im(F)$ is constant on a streamline.

$$\text{const} = \Im\left[F(z_{\text{stream}})\right] = \Im\left[\tilde{F}(\zeta(z_{\text{stream}}))\right]. \tag{5.122}$$

5.8.2 Joukowsky Map

One mapping that takes a cylinder into an airfoil shape is the *Joukowsky transformation*, introduced by Joukowsky (1910)*

* The name Zhukovsky has been transliterated in the literature as Joukowsky, and sometimes Joukowski.

$$\zeta = z + \frac{1}{z}. \tag{5.123}$$

The cylinder in the z plane must pass through the point $z = 1 + 0i$, which is a singular point of the derivative of the mapping in Equation 5.123.† The image of this point, at $\zeta = 2 + 0i$, becomes the airfoil trailing edge. In addition, for positive thickness, the cylinder must enclose $z = -1 + 0i$, which maps to the origin in the ζ plane. Other than these two restrictions, the center of the cylinder is arbitrary. If z_0 denotes the cylinder center, -$\Re(z_0)$ dictates the airfoil's thickness, whereas $\Im(z_0)$ dictates its camber.

† At this point, $\zeta' = \frac{d\zeta}{dz} = 0$, which means that $z' = \frac{dz}{d\zeta}$ is undefined, making $\tilde{F}'$ undefined by the chain rule in Equation 5.121.

Example 5.9 Joukowsky airfoils

We can visualize Joukowsky airfoils obtained for different z_0 values by plotting the transformation in Equation 5.123 at many points. The code in Listing 5.7 does this by first placing points on the z-plane cylinder, using the fact that the radius of the cylinder is the distance between z_0 and the point $1 + 0i$. It then applies the Joukowsky transformation to obtain ζ and plots the real and imaginary parts.

Listing 5.7 Joukowsky transformation code.

```
function joukowsky_map(z0)
% Plots a cylinder in the z plane, centered at z0, and the airfoil
% in the zeta plane resulting from the mapping zeta = z + 1/z.

a = norm(1-z0);                % cylinder radius (must hit 1+0i)
t = linspace(0,2*pi,101);  % theta values for cylinder
z = z0 + a*exp(i*t);          % z points on cylinder
zeta = z + 1./z;               % zeta points on airfoil

figure(1); clf; plotcomplex(z);
figure(2); clf; plotcomplex(zeta);

%------------------------------
function plotcomplex(z)
plot(real(z), imag(z), 'k-', 'linewidth', 2);
axis equal; grid on; box off;
```

Figure 5.41 shows the points on the cylinder in the z plane for one choice of z_0, and Figure 5.42 shows airfoils in the ζ plane for different z_0 values. These

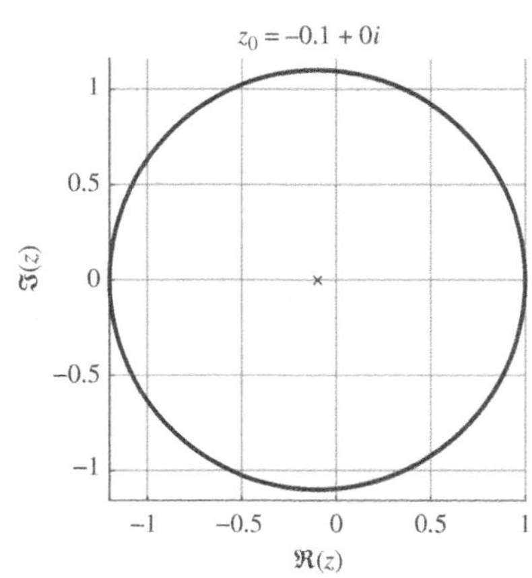

Figure 5.41 z-plane cylinder.

correspond to a symmetric airfoil, a cambered airfoil, and a thick cambered airfoil.

5.8.3 Kutta Condition

From Section 4.3.3, we know how to solve for potential flow over a lifting cylinder, which is one with a vortex of strength Γ. Depending on the value of circulation, Γ, the trailing-edge stagnation point can be anywhere on the aft part of the cylinder. Similarly, after the mapping, this stagnation point can lie on a wide range of locations on the airfoil. However, to enforce the Kutta condition, the stagnation point needs to be at the trailing edge of the airfoil, which is the image of the point $z = 1 + 0i$. Therefore, Γ must be selected to place the stagnation point at this value for the flow over the cylinder, at an angle of attack α. Satisfying the Kutta condition then becomes a geometry problem that uses the formula for the location of stagnation points on the cylinder, Equation 4.75.

As shown in Figure 5.43, which is in the z plane, the freestream flow makes an angle of attack of α relative to the x axis. The vortex of strength Γ turns the flow so that a stagnation point ends up at $1 + 0i$. This means that the trailing-edge stagnation point angle is

$$\theta_s = 2\pi - \alpha - \beta, \quad \beta = \sin^{-1}\left(\frac{\Im(z_0)}{a}\right), \tag{5.124}$$

where $a \equiv |1 - z_0|$ is the radius of the cylinder. From Equation 4.75, the stagnation point angle and Γ are related through

$$\sin\theta_s = -\frac{\Gamma}{4\pi V_\infty a} \quad \Rightarrow \quad \Gamma = -4\pi V_\infty a \sin\theta_s. \tag{5.125}$$

The lift coefficient is then

$$c_\ell = \frac{L'}{\frac{1}{2}\rho V_\infty^2 c} = \frac{\rho U_\infty \Gamma}{\frac{1}{2}\rho V_\infty^2 c} = \frac{2\Gamma}{V_\infty c} = 8\pi\frac{a}{c}\sin(\alpha - \beta), \tag{5.126}$$

where we used $\sin\theta_s = \sin(2\pi - \alpha - \beta) = -\sin(\alpha + \beta)$. Note that if the angles α and β are small, the $c_\ell(\alpha)$ relationship becomes linear with a lift-curve slope of $8\pi a/c$. This value is approximately 2π because $a \approx 1$ and $c \approx 4$,* consistent with thin-airfoil theory.

The lift coefficient in Equation 5.126 is that of both the cylinder *and* the airfoil, because the value of circulation, Γ, does not change under the mapping. This invariance can be shown by considering an enclosing contour far away from the origin of the z plane, since far away from the origin the transformation becomes the identity mapping. By a similar analysis, the angle of attack of the uniform freestream remains constant under the mapping, so that the airfoil sees the same angle of attack as the cylinder. Hence, Equation 5.126 gives the $c_\ell(\alpha)$ equation for both the cylinder in the z plane and the airfoil in the ζ plane.

Example 5.10 Lift curve for Joukowsky airfoils
Equation 5.126 gives the lift coefficient at an arbitrary angle of attack, for a Joukowsky airfoil with a given z_0.† This lift coefficient is consistent with the Kutta condition, in that the trailing edge is a stagnation point. Listing 5.8 presents a code that plots the lift curves for the same three airfoils shown in Figure 5.42, and Figure 5.44 shows the resulting plot. As expected, adding camber by increasing $\Im(z_0)$ shifts the lift curve up, and increasing thickness by making $\Re(z_0)$ more negative increases the lift curve slope.

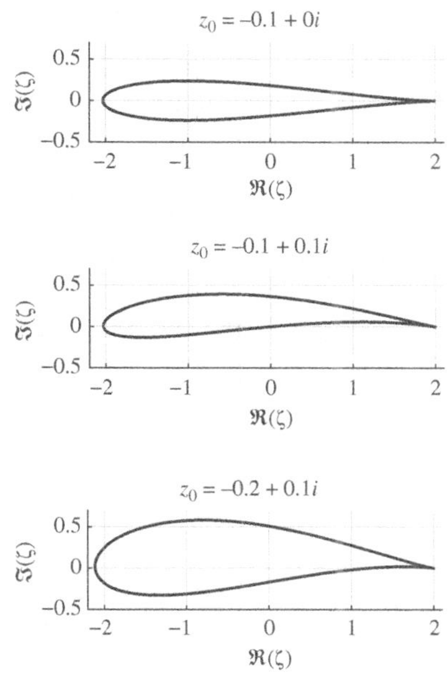

Figure 5.42 ζ-plane airfoils for various values of z_0.

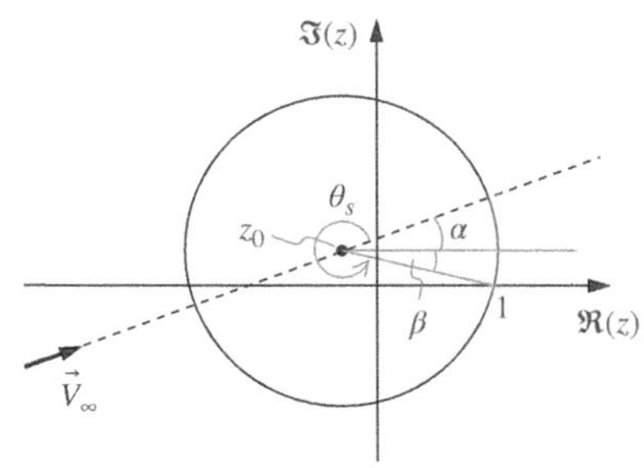

Figure 5.43 Geometry for positioning the stagnation point in the z plane. The radius of the cylinder is a.

* c is exactly 4 for a flat plate, when z_0 is at the origin, since the leading edge is at $z = -1$, which maps to $\zeta = -2$. For other values of z_0, c will be slightly larger than 4, but a will also be slightly larger than 1.

† Note that β depends only on z_0.

Listing 5.8 Code for plotting the lift curve of Joukowsky airfoils.

```
function joukowsky_cl
% plots the lift curve for Joukowsky airfoils

z0v = [-0.1+0i, -0.1+0.1i, -0.2+0.1i];
z0lab = {'$z_0 = -0.1 + 0i$', '$z_0 = -0.1 + 0.1i$', '$z_0 = -0.2+0.1
    i$'};
ctype = {'k--', 'k-', 'k-.'};

figure(1); clf;
for k = 1:length(z0v),
  z0 = z0v(k);                        % cylinder center
  a = norm(1-z0);                     % cylinder radius
  beta = asin(imag(z0)/a);            % TE stag point angle
  av = linspace(-5,15,21);            % angles of attack to run

  t = linspace(0,2*pi,1001); % theta values for cylinder
  z = z0 + a*exp(i*t);              % z points on cylinder
  zeta = z + 1./z;                  % zeta points on airfoil
  c = max(real(zeta)) - min(real(zeta)); % chord

  cl = 8*pi*a/c*sin(av*pi/180+beta);

  plot(av, cl, ctype{k}, 'linewidth', 2, 'displayname', z0lab{k});
  hold on;
end

legend('location', 'southeast', 'fontsize', 14, ...
    'interpreter', 'latex', 'box', 'off');
grid on; box off; axis([-5,15,-1,2.5]);
set(gcf, 'position', [400, 400, 400, 400]);
writeplot('$\alpha$ [deg]', '$c_{\ell}$', '../figs/joukowskycl.pdf'
    ,4);
```

Code notes:

- The `norm` function computes the cylinder radius as the distance between the complex numbers z_0 and 1.
- The chord length is approximated by applying the Joukowsky mapping to many points on the cylinder, and then taking the difference between the largest and smallest x values.
- The lift coefficient calculation uses Equation 5.126.

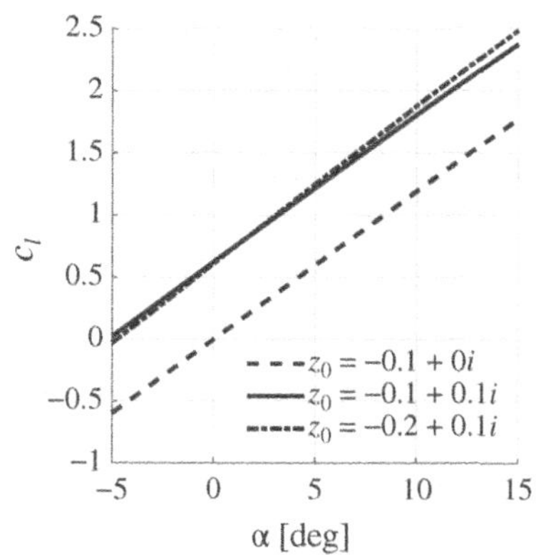

Figure 5.44 c_ℓ versus α for the three Joukowsky airfoils shown in Figure 5.42.

5.8.4 Streamlines

In addition to computing the lift coefficient, we can also use the Joukowsky transformation to visualize the entire flow field around the airfoil in the ζ plane. In particular, plotting contours of the stream function gives the streamlines. To obtain the stream function in the ζ plane, we take the imaginary part of Equation 5.121,

$$\tilde{\psi}(\zeta) = \Im\left(\tilde{F}(\zeta)\right) = \Im\left(F(z(\zeta))\right) = \psi(z(\zeta)). \tag{5.127}$$

For plotting, we use a grid in ζ space, but we need to know z to obtain the stream function. This necessitates an *inverse* transformation from ζ to z. Solving the quadratic that results from $\zeta = 1/z + z$ gives

$$z = \frac{1}{2}\left(\zeta \pm \sqrt{\zeta^2 - 4}\right). \tag{5.128}$$

In this equation, we always take the root lying outside the cylinder. The flow inside the cylinder is fictitious, and we know that the outside of the airfoil maps to the outside of the cylinder, so always taking the root outside the cylinder guarantees that we obtain the correct flow around the airfoil. Knowing the z value, we compute the stream function for flow over a cylinder with circulation, which consists of a superposition of freestream, doublet, and vortex.*

*The freestream speed is arbitrary, the doublet strength is set by the cylinder radius, and the vortex strength is set by Equation 5.125 to satisfy the Kutta condition.

Example 5.11 Streamlines around Joukowsky airfoils
Following the steps outlined in this section, Listing 5.9 shows a code that plots the streamlines around a Joukowsky airfoil given its geometry-defining z_0 value and the angle of attack, α.

Listing 5.9 Code for plotting the streamlines around a Joukowsky airfoil.

```
function joukowsky_stream(z0, alpha, psilevels)
% Plots streamlines for a Joukowsky airfoil defined by z0
% at angle of attack alpha (rad); psilevels defines the
% streamline levels -- use [] for default number.

a = norm(1-z0);                  % cylinder radius
beta = asin(imag(z0)/a);         % Kutta TE stag point angle
Vinf = 1;                        % freestream, arbitrary
Gamma = 4*pi*Vinf*a*sin(alpha+beta);  % vortex strength

% plotting resolution
Nx = 301; Ny = 301;
[X,Y] = ndgrid(linspace(-3,3,Nx), linspace(-1.5,1.5,Ny));

% loop over the points and calculate the stream function
Psi = zeros(size(X));  % initialize to zero
for ix=1:Nx,
  for iy=1:Ny,
    zeta = X(ix,iy) + i*Y(ix,iy); % coord in airfoil plane
    z1 = 0.5*(zeta - sqrt(zeta*zeta-4)); % inverse map - root
    z2 = 0.5*(zeta + sqrt(zeta*zeta-4)); % inverse map + root
    if (norm(z1-z0) < norm(z2-z0)) z = z2; else z = z1; end
    zx = real(z-z0); zy = imag(z-z0); % shifted coordinates
    r   = sqrt(zx^2 + zy^2); % cylinder polar r
    theta = atan2(zy,zx); % cylinder polar theta

    % compute and store stream function for plotting
    Psi(ix,iy) = Vinf*r*sin(theta-alpha)*(1-a^2/r^2) + Gamma/(2*pi)*
     log(r/a);
  end
end

% Plotting: zeta plane streamlines and airfoil
figure(1); clf; if isempty(psilevels), psilevels = 21; end;
contour(X,Y,Psi, psilevels, '-', 'color', '#888888');
hold on; contour(X,Y,Psi, [0, 0], 'k-', 'linewidth', 2);
axis off, axis equal;
```

Code notes:

- The stream function calculation uses Equation 4.70 for lifting flow over a cylinder, with the angle θ decremented by α due to the rotated freestream direction.
- The larger magnitude root of Equation 5.128 is taken as the outside root.

Figure 5.45 shows the streamlines generated by this code for the three Joukowsky airfoils in Figure 5.42. Note that the flow is everywhere tangent to the airfoil surface, and that it leaves the trailing-edge smoothly, indicating that the Kutta condition is satisfied.

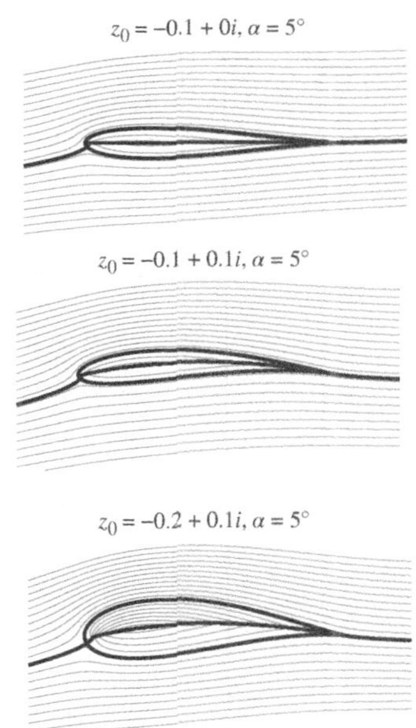

Figure 5.45 Streamlines around three Joukowsky airfoils.

5.9 Summary and Concepts

This chapter applied potential flow to airfoils, one of the most important shapes in aerodynamics. The study of airfoils requires understanding how geometry changes impact the performance, as measured in terms of forces and moments, and this in turn requires accurate analysis and design tools. Several of these were developed in this chapter, all based on potential-flow assumptions, which means that viscous effects were not captured. In particular, drag was not predicted, as it is zero in two-dimensional potential flow. Including viscous effects is left to the material in the next chapter.

Some of the key concepts covered in this chapter are as follows:

- A net vortex strength is required to model the flow turning that creates lift in potential flow.
- The airfoil analysis problem is to determine this vortex strength and how it is distributed, given an airfoil geometry and angle of attack.
- Numerically, the problem is discretized by using a finite number of vortices, or by parameterizing a continuous distribution of vortex strength per unit length by a finite number of coefficients.
- In thin-airfoil theory, the model is a single vortex sheet, and the unknown is the vortex sheet strength function.
- For both approaches, the governing equations rely on the same physical principle: flow tangency, which states that the flow cannot cross a solid wall.
- Applying flow tangency discretely or continuously, with approximations, lets us solve for the unknown distribution of vortex strength.
- Lifting potential flow solutions around objects are mathematically not unique, but the Kutta condition picks out the physically relevant one, in which the flow leaves the trailing edge smoothly.
- The Kutta–Joukowsky theorem lets us quickly calculate the lift from the net vortex strength.
- Flow over select airfoils can also be produced by a conformal mapping of the complex coordinate: this is a mathematically-elegant analysis shortcut, albeit with less control on the resulting geometry.
- Thin-airfoil theory can be used to design airfoils that meet given performance requirements, through the relationship of forces and moments to geometry Fourier coefficients, B_n.

5.10 Problems

1. Justify each answer to the below questions with an explanation. Assume thin-airfoil theory unless otherwise noted.
   a True or false: to produce positive lift, an airfoil must fly at a positive angle of attack.
   b True or false: in the vortex-panel method, the matrix entries depend on the angle of attack.
   c True or false: the center of pressure of an airfoil producing positive lift must be on the airfoil.
   d True or false: a symmetric airfoil can have a positive static stability margin.
   e True or false: deflecting a leading-edge slat down decreases the lift coefficient.
   f True or false: the quarter-chord moment of an airfoil is zero if the lift is zero.
   g True or false: the vortex-panel method can predict the stall angle of an airfoil.

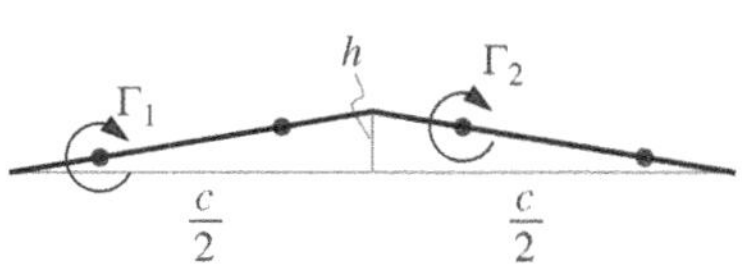

Figure 5.46 A two-panel airfoil model.

2. Consider a two-panel model of a cambered airfoil of chord length c as shown in figure 5.46, in which each panel has a vortex and control point located at the panel quarter and three-quarter-chord locations, respectively. By enforcing flow tangency at the two control points, determine the lift and quarter-chord moment coefficients of the airfoil as a function of the angle of attack. Assume small α and $h \ll c$.

3. Extend the multiple-vortex model of an airfoil presented in Section 5.3.2 to include camber. Still keep the vortices and control points on the chord line, but now use the normal vectors from the camber line, evaluated at the control point x locations, when enforcing flow tangency. Run your code with the camber line from Example 5.8 and compare the on-design lift coefficient values.

4. Verify the airfoil design obtained in Example 5.8 by running a panel code such as mfoil, starting with a thin, symmetric airfoil like the NACA 0006 and modifying the camber to be the cubic function derived in the example. How well does this airfoil satisfy the given requirements?

5. A thin, initially symmetric airfoil can deflect a slat and flap up or down. The airfoil needs to produce a lift coefficient of $c_\ell = 0.6$ without a suction peak and with neutral static stability. Determine the angle of attack and the flap and slat deflection angles, assumed small, that meet the design requirements. Figure 5.47

Figure 5.47 Thin airfoil with a slat and flap.

6. Recall that in the 4-digit NACA series, the first digit times $c/100$ is the maximum camber – denote this value by a, and the second digit times $c/10$ is the location of the maximum camber – denote this value by b. The last two digits give the thickness, which we ignore for a thin airfoil. The camber line of such airfoils is given by

$$\frac{z}{c} = \begin{cases} \left[2b\frac{x}{c} - \left(\frac{x}{c}\right)^2\right]\frac{a}{b^2} & \text{for } 0 \leq \frac{x}{c} \leq b \\ \left[1 - 2b + 2b\frac{x}{c} - \left(\frac{x}{c}\right)^2\right]\frac{a}{(1-b)^2} & \text{for } b < \frac{x}{c} \leq 1 \end{cases}.$$

Consider the `52**`, `55**`, and `58**` airfoils (i.e. $a = .05, b = .2, .5, .8$). Determine the design angle of attack, the design lift coefficient, and the quarter-chord moment coefficient of each airfoil. Discuss the differences.

7. Calculate the trimmed static-stability margin of a NACA 4416 airfoil at $c_\ell = 0.5$ using a panel solver such as mfoil. Do this first with no flap and then again with a flap hinged at 80% chord, deflected 15° down and up. Plot pressure coefficient distributions and discuss the results.

8. Modify the code given in Example 5.11 to plot streamlines in the z-plane, for flow around a cylinder with a given z_0 value. Show contours for the same three airfoils and angles of attack analyzed in Example 5.11 and verify that the trailing stagnation point is at $z = 1 + 0i$.

6

Boundary Layers

The previous chapters considered primarily inviscid flows: that is, flows in which the viscosity, μ, is zero. However, all fluids have some amount of viscosity, which introduces a new transport mechanism of diffusion, and a new boundary condition of no slip past a solid wall. Viscous shear forces acting on fluid elements also cause rotation, which invalidates the potential-flow assumption of irrotational flow and makes the solution process more difficult. The governing Navier–Stokes equations capture all of these effects, but these equations are difficult to solve: only a few simple analytical solutions exist, and numerical approaches are typically expensive and require expert user oversight. Fortunately, when viscous effects are not strong, the governing equations change slightly, yet enough to admit simplified numerical and analytical treatments. It is this setting that frames the material of the current chapter on boundary layers.

6.1 Effect of Viscosity

The role of viscosity in the Navier–Stokes equations, e.g. Equation 3.61, is twofold. First, the equations themselves contain additional terms related to viscous shear transport of momentum. Second, the boundary conditions are augmented to constrain not just the wall-normal but also the wall-tangential velocity. Recall that for inviscid flows, the physical boundary condition at walls was no flow-through, $\vec{v} \cdot \vec{n} = 0$, and as a consequence, walls became streamlines of the flow. However, in the presence of viscosity, a flow must satisfy a *no-slip* condition at the wall,*

$$\vec{v}|_{\text{wall}} = \vec{0}. \tag{6.1}$$

*This is the condition for a stationary wall. For a wall that is moving, the no-slip condition states that the fluid velocity at the wall must equal the wall velocity.

If the viscosity is in some sense "small," we expect the flow to be similar to the inviscid case, which has zero viscosity. That is, the flow over most of the domain should be similar for $\mu \to 0$ and $\mu = 0$. However, the boundary conditions in the two cases are different: in the case of zero viscosity, the flow slips past the wall, whereas with nonzero viscosity, the flow must satisfy the no-slip condition. Therefore, as $\mu \to 0$, there are two competing requirements for the near-wall velocity: it should resemble the nonzero, slipping velocity of the inviscid case, but it also must go to zero at the wall.† The resolution to both requirements is the development of a *boundary layer*, which is a thin region near the wall where the velocity changes rapidly from zero at the wall to an "inviscid" slip velocity close to it.

†This is known as a singular-perturbation problem in mathematics, and the split outer/inner solution procedure for such problems is similar to the split inviscid/viscous approach we take here.

Figure 6.1 illustrates the velocity in a boundary layer, in the form of a qualitative zoomed-in sketch. In many aerodynamic flows of interest, the thickness of the boundary layer is so small that it is nearly imperceptible relative to the rest of the geometry.

Aerodynamics: A Computational Introduction, First Edition. Krzysztof J. Fidkowski.
© 2026 John Wiley & Sons, Inc. Published 2026 by John Wiley & Sons, Inc.

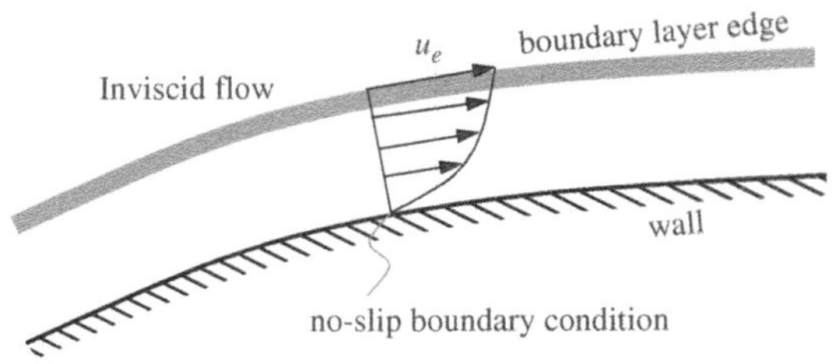

Figure 6.1 Sample velocity profile within a boundary layer. The "edge" of the boundary layer is an imprecisely defined line below which the effects of viscosity start to be important. Outside of the boundary layer, the flow can be assumed inviscid.

We typically measure the effect of viscosity by the *Reynolds number*,

$$\mathrm{Re} = \frac{\rho U_\infty L}{\mu}, \tag{6.2}$$

where U_∞ is the magnitude of the freestream velocity,‡ and L is a characteristic length scale for the object, such as a cylinder diameter or an airfoil chord length. When specifying the Reynolds number for non-conventional geometries, the length scale used should also be specified. Sometimes, this length scale is denoted by a subscript, for example Re_L to denote the Reynolds number based on L. For airfoils, we will use the chord length as the length scale, and Re without a subscript will denote the Reynolds number based on the chord length.

‡ Most of the flows in this chapter will be horizontal in the freestream, and hence the use of U_∞ for the velocity magnitude.

The Reynolds number is a non-dimensional measure of the importance of convective transport relative to viscous transport. When Re is low, viscous effects are relatively important. An example is a small cylinder in molasses. For such flows, boundary layers are not present as viscosity affects a significant portion of the flow, and the region where viscous effects are important is not confined to a thin layer near the object. When Re is high, convective transport dominates, and viscosity is important only in a small region near the boundary, the boundary layer. An example is a large aircraft wing at cruise speed, for which Re can be 10^7 or more. Outside the boundary layer, viscous effects can generally be neglected.

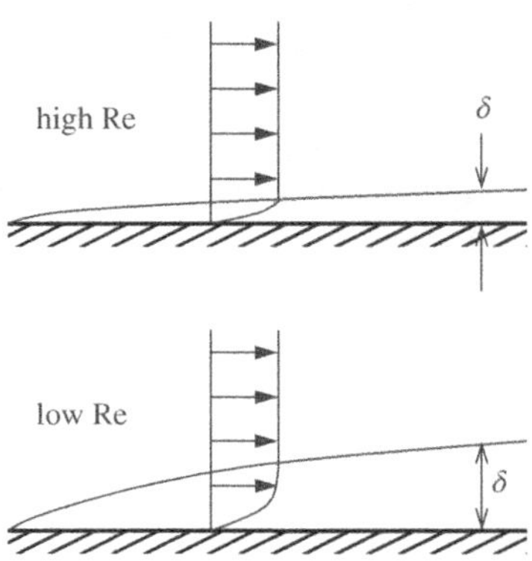

Figure 6.2 Qualitative behavior of boundary-layer thickness, δ, with Reynolds number, Re.

The boundary-layer thickness is a measure of the distance over which viscous effects are important. It is usually determined by analyzing the velocity profile close to the wall. However, there is no single exact point in the velocity profile at which viscous effects stop being important, and so several definitions of boundary-layer thickness exist, as discussed later in the chapter. For now, we just call the representative thickness $\delta(x)$, where x measures distance along the wall, and where the flow can be well-approximated as inviscid for $y \gtrsim \delta(x)$. As illustrated in Figure 6.2, for a given x location, *the boundary-layer thickness generally decreases with increasing Reynolds number.*

In a boundary layer, the velocity gradient is high, and this results in shear stresses on the wall. The setup for calculating this shear stress is given in Figure 6.3. Recall from Equation 3.44 that the shear stress *on the fluid* is given by

$$\begin{aligned} \text{stress on fluid} &= \vec{n} \cdot \tau \\ &= (n_x \tau_{xx} + n_y \tau_{yx})\hat{\imath} + (n_x \tau_{xy} + n_y \tau_{yy})\hat{\jmath} \\ &= -\tau_{yx}\hat{\imath} - \tau_{yy}\hat{\jmath}, \end{aligned} \tag{6.3}$$

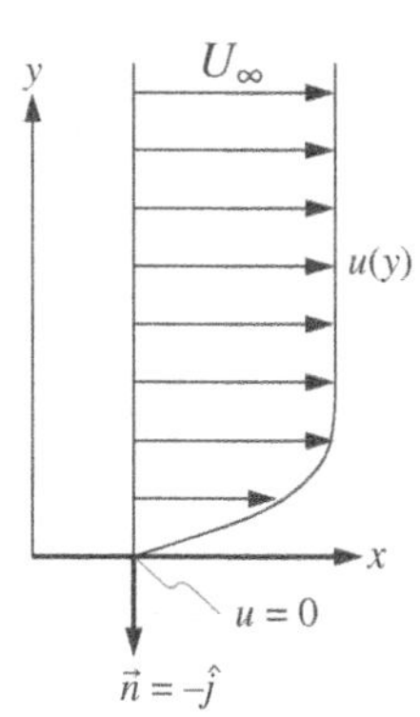

Figure 6.3 Calculation of the shear stress on the wall in a boundary-layer flow.

where $\vec{n}$ is the normal vector pointing outward from the fluid, equal to $-\hat{\jmath}$ in the case of fluid above a horizontal wall. The stress on the wall is equal in magnitude but opposite in direction,

$$\text{stress on wall} = -\vec{n} \cdot \tau = \tau_{yx}\hat{\imath} + \tau_{yy}\hat{\jmath}. \tag{6.4}$$

When the fluid is incompressible,

$$\tau_{ij} = \mu\left(\frac{\partial v_i}{\partial x_j} + \frac{\partial v_j}{\partial x_i}\right), \tag{6.5}$$

and the stress on the wall is§

$$\text{stress on wall} = \mu\left(\frac{\partial u}{\partial y} + \frac{\partial v}{\partial x}\right)\hat{i} + 2\mu\frac{\partial v}{\partial y}\hat{j} = \mu\frac{\partial u}{\partial y}\hat{i}. \tag{6.6}$$

§ At the wall, $u = 0$ for all x, so $\frac{\partial u}{\partial x} = 0$, and by continuity, $\frac{\partial u}{\partial x} + \frac{\partial v}{\partial y} = 0 \Rightarrow \frac{\partial v}{\partial y} = 0.$

This stress is in the x direction, and it is positive when $\frac{\partial u}{\partial y} > 0$. We call the magnitude of this stress the *wall shear stress*,

$$\tau_w = \mu\frac{\partial u}{\partial y}\bigg|_{\text{wall}}. \tag{6.7}$$

The non-dimensional form of the wall shear stress is the *skin-friction coefficient*,

$$c_f = \frac{\tau_w}{\frac{1}{2}\rho U_\infty^2}. \tag{6.8}$$

The terms wall shear stress and skin friction are used interchangeably. The effect of this friction on drag can be very important. For example, on a typical subsonic transport aircraft, skin friction can account for almost half of the total drag!

Before proceeding further, we note that boundary layers do not always have smooth or monotonic velocity profiles as shown in Figure 6.3, and these profiles can vary in time.¶ In such situations, the definitions of wall shear stress and skin-friction coefficient can be made in a time-averaged sense.

¶ For example, we will see later in this chapter that this occurs when boundary layers separate or become turbulent.

6.2 Boundary-Layer Equations

Although we can rely on the Navier–Stokes equations to obtain solutions to viscous flow problems, the Navier–Stokes system is not easy to solve and often requires an expensive numerical approach. It turns out that under assumptions that are satisfied in thin boundary layers, the Navier–Stokes simplify and admit more direct numerical approaches and mathematically elegant similarity solutions. This section presents the simplifying assumptions and the resulting equations.

We consider two-dimensional, steady, and incompressible flow without body forces. The Navier–Stokes equations in this case are

$$\begin{aligned}
&\frac{\partial u}{\partial x} + \frac{\partial v}{\partial y} = 0 && \text{(continuity)}\\
&u\frac{\partial u}{\partial x} + v\frac{\partial u}{\partial y} = -\frac{1}{\rho}\frac{\partial p}{\partial x} + \nu\left(\frac{\partial^2 u}{\partial x^2} + \frac{\partial^2 u}{\partial y^2}\right) && (x\text{ momentum})\\
&u\frac{\partial v}{\partial x} + v\frac{\partial v}{\partial y} = -\frac{1}{\rho}\frac{\partial p}{\partial y} + \nu\left(\frac{\partial^2 v}{\partial x^2} + \frac{\partial^2 v}{\partial y^2}\right) && (y\text{ momentum})
\end{aligned} \tag{6.9}$$

where $\nu \equiv \mu/\rho$ is the *kinematic viscosity*, which has dimensions of length^2/time. Although ν will be assumed small, we cannot neglect it completely, since it is viscosity that creates boundary layers.

Not all of the terms in the Navier–Stokes equations are significant for boundary layers. Under a few assumptions, we can perform an order-of-magnitude analysis* to see which terms to retain.

* In an order-of-magnitude analysis, when we write $u = \mathcal{O}(U_\infty)$, we mean that u is "on the order of" U_∞, i.e. that $u = cU_\infty$, where c is "close to 1". For example, c could be 0.5, 2, 5, but not 1000.

The setup for the boundary-layer analysis is illustrated in Figure 6.4. This is effectively a flat plate of length c, in which the x axis is aligned with the plate and the y axis is perpendicular to it. The boundary layer starts at $x = 0$, and δ is some order-of-magnitude measure of the boundary-layer thickness at $x = c$.

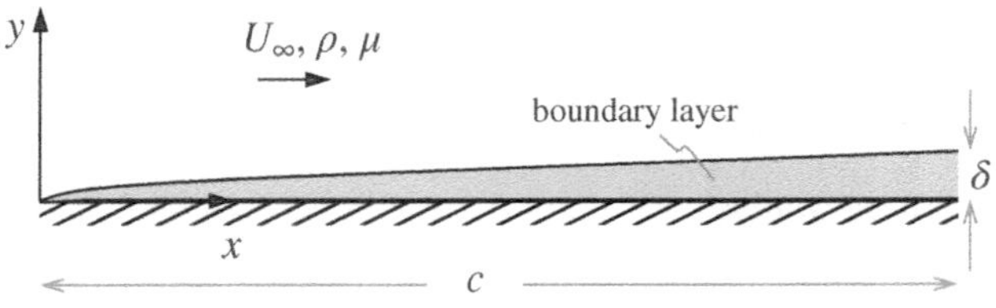

Figure 6.4 Setup for boundary layer analysis and definition of key terms used in the derivation of the boundary-layer equations.

The derivation of the boundary-layer equations relies on the following assumptions:

- Changes in the x direction occur over a distance c,

$$\frac{\partial(\cdot)}{\partial x} = \mathcal{O}\left(\frac{\cdot}{c}\right). \tag{6.10}$$

- Changes in the y direction occur over a distance δ,

$$\frac{\partial(\cdot)}{\partial y} = \mathcal{O}\left(\frac{\cdot}{\delta}\right). \tag{6.11}$$

- The boundary layer is thin,†

$$\delta \ll c. \tag{6.12}$$

- The Reynolds number based on c, Re_c, is large,

$$\mathrm{Re}_c = \frac{U_\infty c}{\nu} \gg 1. \tag{6.13}$$

- Changes in the x velocity are on the order of U_∞,‡

$$\Delta u = \mathcal{O}(U_\infty). \tag{6.14}$$

- Changes in pressure are on the order of ρU_∞^2,§

$$\Delta p = \mathcal{O}(\rho U_\infty^2). \tag{6.15}$$

† This and the next assumption can be taken as defining properties of a boundary layer.

‡ The x velocity is zero at the wall and U_∞ outside the boundary layer, so changes on the order of U_∞ are observed.

§ This assumption comes from Bernoulli's equation, which holds outside the boundary layer in the inviscid flow, if the velocity change there is on the order of U_∞.

We apply these assumptions to the Navier–Stokes equations, starting with the conservation of mass, or continuity equation,

$$\underbrace{\frac{\partial u}{\partial x}}_{\mathcal{O}\left(\frac{U_\infty}{c}\right)} + \underbrace{\frac{\partial v}{\partial y}}_{\mathcal{O}\left(\frac{\Delta v}{\delta}\right)} = 0. \tag{6.16}$$

Since the two terms must add to zero, they must have the same magnitude, albeit an opposite sign. So an order-of-magnitude estimate for changes in the vertical velocity is

$$\Delta v = \mathcal{O}\left(U_\infty \frac{\delta}{c}\right). \tag{6.17}$$

Note that we did not assume any such estimate yet – it comes out of the continuity equation after assuming an order-of-magnitude estimate for U_∞. Since $v = 0$ at the wall, $\Delta v = \mathcal{O}(v)$, and so we can write

$$\frac{v}{U_\infty} = \mathcal{O}\left(\frac{\delta}{c}\right). \tag{6.18}$$

Since $\delta \ll c$, this equation states that the vertical component of the velocity is small relative to the freestream speed, which is consistent with the concept of a thin boundary layer.

Next, we consider x momentum,¶

$$\underbrace{u\frac{\partial u}{\partial x}}_{\mathcal{O}\left(U_\infty \frac{U_\infty}{c}\right)} + \underbrace{v\frac{\partial u}{\partial y}}_{\mathcal{O}\left(U_\infty \frac{\delta}{c}\frac{U_\infty}{\delta}\right)} = \underbrace{-\frac{1}{\rho}\frac{\partial p}{\partial x}}_{\mathcal{O}\left(\frac{1}{\rho}\frac{\rho U_\infty^2}{c}\right)} + \underbrace{\nu\frac{\partial^2 u}{\partial x^2}}_{\mathcal{O}\left(\nu\frac{U_\infty}{c^2}\right)} + \underbrace{\nu\frac{\partial^2 u}{\partial y^2}}_{\mathcal{O}\left(\nu\frac{U_\infty}{\delta^2}\right)}. \tag{6.19}$$

¶ This equation requires an estimate of the order of magnitude of second derivatives, which proceeds as follows:

$$\frac{\partial^2 u}{\partial x^2} = \frac{\partial}{\partial x}\left(\frac{\partial u}{\partial x}\right) = \mathcal{O}\left(\frac{1}{c}\frac{\partial u}{\partial x}\right) = \mathcal{O}\left(\frac{1}{c}\frac{U_\infty}{c}\right) = \mathcal{O}\left(\frac{U_\infty}{c^2}\right).$$

We first look at the last two terms. Since $\delta \ll c$,

$$\frac{\nu U_\infty}{c^2} \ll \frac{\nu U_\infty}{\delta^2}. \tag{6.20}$$

We can therefore neglect $\nu\frac{\partial^2 u}{\partial x^2}$ relative to $\nu\frac{\partial^2 u}{\partial y^2}$.||

|| Dropping this term is significant! It enables the upcoming simplified numerical and analytical solution approaches.

Next, note that all of the other terms in Equation 6.19 are $\mathcal{O}(U_\infty^2/c)$. These are the "inviscid terms" because they represent convective transport. A defining characteristic of boundary layers is that in a boundary layer, the viscous terms balance the inviscid terms. This means

$$\mathcal{O}\left(\frac{U_\infty^2}{c}\right) = \mathcal{O}\left(\frac{\nu U_\infty}{\delta^2}\right) \Rightarrow \left(\frac{\delta}{c}\right)^2 = \mathcal{O}\left(\frac{\nu}{U_\infty c}\right). \tag{6.21}$$

Taking the square root and recognizing the appearance of the Reynolds number on the right, we have**

$$\frac{\delta}{c} = \mathcal{O}\left(\frac{1}{\sqrt{\mathrm{Re}_c}}\right). \tag{6.22}$$

** Although not a quantitative formula, this order-of-magnitude estimate of the thickness of the boundary layer is an important result, and we will see it again in the analytical solutions.

Finally, we make estimates of the various terms that appear in the y-momentum equation,

$$\underbrace{u\frac{\partial v}{\partial x}}_{\mathcal{O}\left(U_\infty^2 \frac{\delta}{c^2}\right)} + \underbrace{v\frac{\partial v}{\partial y}}_{\mathcal{O}\left(U_\infty^2 \frac{\delta}{c^2}\right)} = -\frac{1}{\rho}\frac{\partial p}{\partial y} + \underbrace{\nu\frac{\partial^2 v}{\partial x^2}}_{\mathcal{O}\left(\nu U_\infty \frac{\delta}{c^3}\right)} + \underbrace{\nu\frac{\partial^2 v}{\partial y^2}}_{\mathcal{O}\left(\nu U_\infty \frac{1}{\delta c}\right)}. \tag{6.23}$$

In the order-of-magnitude estimates, multiplying both sides by c/U_∞^2 and using Equation 6.22 yield

$$\mathcal{O}\left(\frac{1}{\sqrt{\mathrm{Re}_c}}\right) + \mathcal{O}\left(\frac{1}{\sqrt{\mathrm{Re}_c}}\right) = -\frac{c}{\rho U_\infty^2}\frac{\partial p}{\partial y} + \mathcal{O}\left(\frac{1}{\mathrm{Re}_c^{3/2}}\right) + \mathcal{O}\left(\frac{1}{\sqrt{\mathrm{Re}_c}}\right). \tag{6.24}$$

For $\mathrm{Re}_c \gg 1$, the $1/\mathrm{Re}_c^{3/2}$ term is negligible compared to the $1/\mathrm{Re}_c^{1/2}$ terms. The remaining terms must balance. One of these terms involves $\frac{\partial p}{\partial y}$, and all of the other ones are $\mathcal{O}(1/\mathrm{Re}_c^{1/2})$. This means that the term with $\frac{\partial p}{\partial y}$ cannot be of magnitude greater than $\mathcal{O}(1/\mathrm{Re}_c^{1/2})$,

$$\frac{c}{\rho U_\infty^2}\frac{\partial p}{\partial y} = \mathcal{O}\left(\frac{1}{\mathrm{Re}_c^{1/2}}\right) \quad \text{(at most)}. \tag{6.25}$$

Using the boundary-layer assumption that changes in y occur over δ, and Equation 6.22,††

$$\frac{c}{\rho U_\infty^2}\frac{(\Delta p)_{\text{in y}}}{\delta} = \mathcal{O}\left(\frac{1}{\mathrm{Re}_c^{1/2}}\right) \Rightarrow \frac{(\Delta p)_{\text{in y}}}{\rho U_\infty^2} = \mathcal{O}\left(\frac{1}{\mathrm{Re}_c}\right) \ll 1. \tag{6.26}$$

†† The notation $(\Delta p)_{\text{in y}}$ refers to the pressure change within the boundary layer, in the y direction.

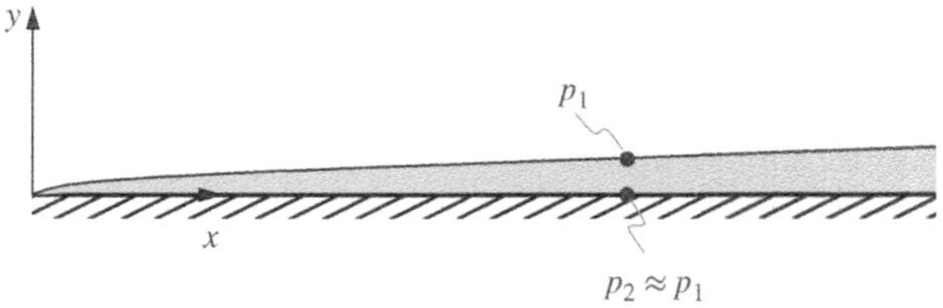

Figure 6.5 An order-of-magnitude analysis of the *y*-momentum equation tells us that pressure is transmitted through a thin boundary layer.

This equation shows that pressure changes in the *y* direction through the boundary layer are small relative to the dynamic pressure, and we therefore neglect them. The *y*-momentum equation therefore reduces to

$$\frac{\partial p}{\partial y} \approx 0 \quad \text{in the boundary layer.} \tag{6.27}$$

In other words, the pressure is transmitted vertically through the boundary layer, as illustrated in Figure 6.5.‡‡

In summary, the two-dimensional, steady, incompressible *boundary-layer equations* are§§

$$\begin{aligned} \frac{\partial u}{\partial x} + \frac{\partial v}{\partial y} &= 0 && \text{(continuity)} \\ u\frac{\partial u}{\partial x} + v\frac{\partial u}{\partial y} &= -\frac{1}{\rho}\frac{\partial p}{\partial x} + \nu\frac{\partial^2 u}{\partial y^2} && (x \text{ momentum}) \\ \frac{\partial p}{\partial y} &= 0 && (y \text{ momentum}) \end{aligned} \tag{6.28}$$

‡‡ If this were not the case, the pressure-based force and moment calculations for airfoils in the previous chapter would not hold in the real case of viscous flow and boundary layers.

§§ Recall that we made assumptions in deriving these equations from the Navier–Stokes system. As a result, these equations apply only within the boundary layer.

Since the *y*-momentum equation states that $\frac{\partial p}{\partial y} = 0$, the pressure inside the boundary layer only depends on *x*, and we can write

$$p = p(x) = p_e(x), \tag{6.29}$$

where $p_e(x)$ is the pressure on the edge of the boundary layer. This pressure corresponds to that from the inviscid solution. We will also denote by $u_e(x)$ the *x* velocity of the inviscid solution at the edge of the boundary layer. Since we can apply Bernoulli's equation in the inviscid region,¶¶

$$\begin{aligned} p_e(x) + \frac{1}{2}\rho u_e(x)^2 &= \text{constant} \\ \Rightarrow \quad \frac{dp_e}{dx} &= -\rho u_e \frac{du_e}{dx}. \end{aligned} \tag{6.30}$$

¶¶ Bernoulli's equation does not apply inside the boundary layer because the flow there is rotational.

This equation relates the pressure gradient*** to the edge velocity gradient, both properties of the inviscid region.

*** We write a total derivative, $\frac{dp_e}{dx}$, because $\frac{\partial p}{\partial y} = 0$.

Example 6.1 Startup Couette flow

The boundary-layer equations derived in this section assumed steady flow, and subsequent sections deal with steady boundary layers. However, there are also unsteady problems that exhibit boundary layer behavior. One simple example is that of startup Couette flow, in which a boundary layer develops next to a flat plate suddenly set into motion.

Consider an initially stationary, infinite flat plate aligned with the *x* direction, as shown in Figure 6.6. At time $t = 0$, the plate is set in motion at velocity U_{wall}. For $t > 0$, the fluid above the plate will begin to move as the viscous

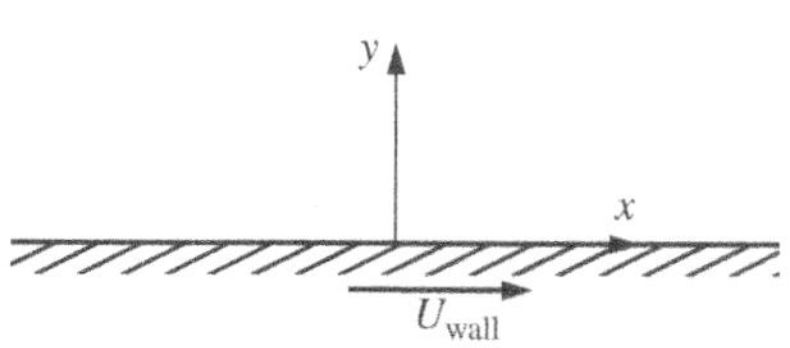

Figure 6.6 Geometry for startup Couette flow.

effects diffuse the momentum away from the plate. We assume that the flow is incompressible and that there are no body forces or pressure gradients.

Since the flat plate is infinite, we treat the flow as fully developed,††† $\frac{\partial \vec{v}}{\partial x} = 0$. The continuity equation then reduces to

$$\frac{\partial u}{\partial x} + \frac{\partial v}{\partial y} = 0 \quad \Rightarrow \quad \frac{\partial v}{\partial y} = 0 \quad \Rightarrow \quad v = \text{constant}. \tag{6.31}$$

Since $v|_{y=0} = 0$ by the no flow-through condition, we have $v = 0$ everywhere. The x-momentum equation then reduces to

$$\frac{\partial u}{\partial t} + u\underbrace{\frac{\partial u}{\partial x}}_{0} + \underbrace{v}_{0}\frac{\partial u}{\partial y} = \nu\underbrace{\frac{\partial^2 u}{\partial x^2}}_{0} + \nu\frac{\partial^2 u}{\partial y^2}, \tag{6.32}$$

$$\frac{\partial u}{\partial t} = \nu\frac{\partial^2 u}{\partial y^2}. \tag{6.33}$$

This is a partial differential equation for u in terms of t and y. However, we can simplify the equation into one in terms of only one variable, η, which is a combination of y and t. We define this *similarity variable* as,‡‡‡

$$\eta = \frac{y}{\sqrt{4\nu t}}. \tag{6.34}$$

To see how the reduced x-momentum equation simplifies with this definition, we write the velocity derivatives in terms of η using the chain rule,

$$\frac{\partial u}{\partial t} = \frac{\partial u}{\partial \eta}\frac{\partial \eta}{\partial t} = \frac{\partial u}{\partial \eta}\frac{(-\eta)}{2t}, \tag{6.35}$$

$$\frac{\partial u}{\partial y} = \frac{\partial u}{\partial \eta}\frac{\partial \eta}{\partial y} = \frac{\partial u}{\partial \eta}\frac{1}{\sqrt{4\nu t}}, \tag{6.36}$$

$$\frac{\partial^2 u}{\partial y^2} = \frac{\partial^2 u}{\partial \eta^2}\frac{1}{4\nu t}. \tag{6.37}$$

Substituting these expressions into Equation 6.33, we obtain

$$\frac{-\eta}{2t}\frac{\partial u}{\partial \eta} = \nu\frac{\partial^2 u}{\partial \eta^2}\frac{1}{4\nu t} \quad \Rightarrow \quad \frac{\partial^2 u}{\partial \eta^2} = -2\eta\frac{\partial u}{\partial \eta}. \tag{6.38}$$

Therefore, the equation did simplify, since y and t no longer appear. We are left with an ordinary differential equation for η! Integrating twice gives

$$u(\eta) = C_1\int_0^{\eta} e^{-(\eta')^2}\,d\eta' + C_2, \tag{6.39}$$

where η' is an integration variable. The boundary conditions are that $u(\eta = 0) = U_{\text{wall}}$ and $u(\eta \to \infty) = 0$. To apply the boundary condition at $\eta \to \infty$, we use§§§

$$\int_0^{\infty} e^{-\eta'^2}\,d\eta' = \frac{\sqrt{\pi}}{2}. \tag{6.40}$$

††† The x coordinate has no significance for an infinite flat plate: observers at different x locations must see the same flow.

‡‡‡ η is non-dimensional, and in fact, its definition can be motivated by dimensional analysis using y, t, and ν. The factor of 4 makes the final expressions nicer but is otherwise not necessary.

§§§ This result can be derived as follows. Let I denote the desired integral. Using the two-dimensional Cartesian coordinates x and y,

$$\begin{aligned} I^2 &= \int_0^{\infty} e^{-x^2}dx\int_0^{\infty} e^{-y^2}dy \\ &= \int_0^{\infty}\int_0^{\infty} e^{-(x^2+y^2)}dx\,dy \\ &= \int_0^{\infty}\int_0^{\pi/2} e^{-r^2}r\,dr\,d\theta \\ &= \frac{\pi}{2}\left[-\frac{1}{2}e^{-r^2}\right]_0^{\infty} = \frac{\pi}{4}. \end{aligned}$$

Taking a square root, $I = \sqrt{\pi}/2$.

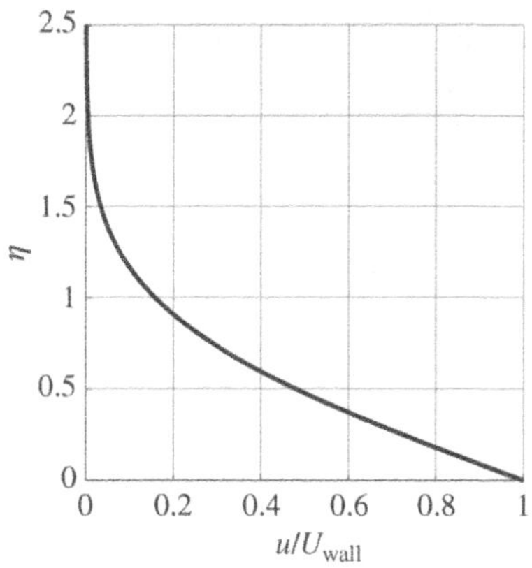

Figure 6.7 Similarity solution velocity profile for startup Couette flow.

Evaluating Equation 6.39 at $\eta = 0$ and $\eta \to \infty$ yields

$$C_1 = -U_{\text{wall}} \frac{2}{\sqrt{\pi}}, \quad C_2 = U_{\text{wall}}, \tag{6.41}$$

so that the full solution is

$$u(\eta) = U_{\text{wall}} \left(1 - \frac{2}{\sqrt{\pi}} \int_0^{\eta} e^{-\eta'^2} d\eta' \right) = U_{\text{wall}} \left(1 - \text{erf}(\eta)\right), \tag{6.42}$$

where $\text{erf}(\eta)$ is called the *error function*, defined by the given normalized integral of the exponential. Figure 6.7 plots this solution as u/U_{wall} versus η. Since $\eta = y/\sqrt{4\nu t}$, we can easily rewrite the solution in terms of y and t. The solution in terms of η gives us the velocity profile at any time, since for a fixed time t, $u(\eta)$ becomes a function only of y. For this reason, the solution we found in this example is a *similarity solution*: at different times, the velocity profiles are similar in that they are scaled versions of each other. We will use the same idea for solving the boundary layer equations analytically.

6.3 A Finite-Difference Boundary-Layer Solver

The parameters in the boundary-layer equations given in Equation 6.28 are the edge velocity, u_e, which affects the pressure gradient, and the kinematic viscosity, ν. Before considering analytical solution approaches, which are restricted to special $u_e(x)$, we consider a numerical solver based on finite-difference approximations. Key to the numerical approach is the fact that the x-momentum boundary-layer equation is missing the $\frac{\partial^2 u}{\partial x^2}$ term, which enables a spatial marching procedure.

The non-trivial boundary-layer equations in Equation 6.28 are continuity and x-momentum, since the y-momentum equation just tells us that pressure is transmitted through the boundary layer. These two equations let us solve for the two unknown fields, $u(x, y)$ and $v(x, y)$. Various discretizations are possible, e.g. see Cebeci and Smith (1974) and Anderson (2016), and here, we present a simple finite-difference approach. Figure 6.8 shows the lattice in the domain, which is assumed rectangular. Curvature of the surface is neglected, and the nodes are assumed to be equally spaced in x and y (although Δx can be different from Δy).

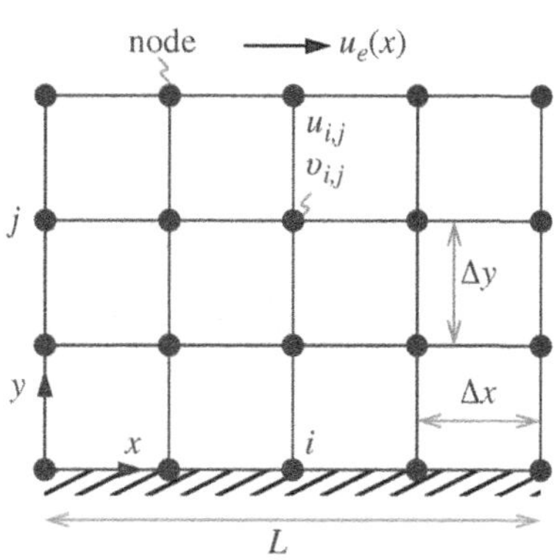

Figure 6.8 Boundary-layer finite-difference discretization.

The two indices that identify a node are i and j, which respectively give the horizontal and vertical position in the grid. A *column* of nodes consists of the vertical stack of nodes at a given i. Marching starts on the left, at $i = 1$, where we assume that the velocity is given in the form of $u_0(y)$ and $v_0(y)$. We will assume that $v_0(y) = 0$, since it is usually much smaller than u, although setting a nonzero v is straightforward. Given the states up to column i, we calculate the column $i + 1$ states in two steps:

1. Calculate u in column $i + 1$ by using the x-momentum equation and the velocities in column i.
2. Calculate v in column $i + 1$ by using continuity and all available velocities, including u at $i + 1$.

Let $u_{i,j}$ and $v_{i,j}$ store u and v at the nodes. For the first step, we discretize the x-momentum equation,

$$u \frac{\partial u}{\partial x} + v \frac{\partial u}{\partial y} = u_e \frac{du_e}{dx} + \nu \frac{\partial^2 u}{\partial y^2}, \tag{6.43}$$

at node (i, j) by using a simple forward difference* in x and central differences in y,

$$u_{i,j}\frac{u_{i+1,j}-u_{i,j}}{\Delta x}+v_{i,j}\frac{u_{i,j+1}-u_{i,j-1}}{2\Delta y}= \\ u_{e,i,j}\left.\frac{du_e}{dx}\right|_{i,j}+\nu\frac{u_{i,j-1}-2u_{i,j}+u_{i,j+1}}{\Delta y^2}. \tag{6.44}$$

We solve this equation for $u_{i+1,j}$, the only unknown from column $i+1$ that appears in this equation.† This unknown appears only in the first term, the discretized x derivative, so the solution for $u_{i+1,j}$ is straightforward.

For the second step, we use the continuity equation, enforced at point $\left(i+\frac{1}{2}, j+\frac{1}{2}\right)$, where the half index denotes an interval midpoint. Referring to Figure 6.9, the discretized continuity equation is

$$\frac{\partial u}{\partial x}+\frac{\partial v}{\partial y}=0 \quad\Rightarrow\quad \frac{u_R-u_L}{\Delta x}+\frac{v_T-v_B}{\Delta y}=0, \tag{6.45}$$

where $u_R=(u_{i+1,j}+u_{i+1,j+1})/2$, $u_L=(u_{i,j}+u_{i,j+1})/2$, $v_T=(v_{i,j+1}+v_{i+1,j})/2$, and $v_B=(v_{i,j}+v_{i+1,j})/2$. We solve this equation for $v_{i+1,j+1}$ and march upward through the column of unknowns at $i+1$, starting with $v=0$ at the wall, where $j=1$.

From these two equations, we are ready to write a solver. Listing 6.1 presents a code that implements the discretization. It takes as input vectors `x` and `y` that contain the node coordinates along the horizontal and vertical directions of the rectangle, a vector `ue` that contains u_e at each x station, a vector `u0` that contains u at the first column, $i=1$, and the kinematic viscosity, `nu`. It returns matrices `U` and `V` containing u and v values at each grid node.

Listing 6.1 A finite-difference boundary-layer solver.

```
function [U,V] = blfd(x, ue, y, u0, nu)
% solves BL equations given initial profile and edge velocity

Nx = length(x)-1; dx = x(2)-x(1); % x spacing, assumed uniform
Ny = length(y)-1; dy = y(2)-y(1); % y spacing, assumed uniform
U = zeros(Nx+1,Ny+1); V = U; % will store x, y velocities
U(1,:) = u0; % set initial u profile at first x station

% start marching along x stations
for i = 1:Nx,

  % use the x-momentum equation to update u
  ueuex = 0.5*(ue(i)+ue(i+1))*(ue(i+1)-ue(i))/dx; % pressure force
  for j = 2:Ny,
    u_yy = (U(i,j+1) - 2*U(i,j) + U(i,j-1))/(dy*dy);
    u_y = (U(i,j+1) - U(i,j-1))/(2*dy);
    U(i+1,j) = U(i,j) + (dx/U(i,j))*(ueuex + nu*u_yy - V(i,j)*u_y);
  end
  U(i+1,Ny+1) = ue(i+1); % edge velocity

  % use the continuity equation to update v
  for j = 2:(Ny+1),
    V(i+1,j) = -V(i,j) + V(i+1,j-1)+V(i,j-1) ...
        - (dy/dx)*(U(i+1,j-1)+U(i+1,j)-U(i,j-1)-U(i,j));
  end
end
```

Given this solver, how do we use it? First, we need to decide on the edge velocity, $u_e(x)$. This velocity defines the problem, and a simple choice to start is $u_e=U_\infty=$ const., which corresponds to a zero-pressure gradient. Second, we need to choose an initial velocity profile, $u_0(y)$. Later in this chapter, we will test the code with certain known profiles, but for now, we just choose a function that behaves as we

*Alternatively, the first-order accurate forward difference at node i can be viewed as a second-order accurate central difference at node $i+1/2$, half-way between nodes i and $i+1$.

†This equation is used for all values of j except the first ($j=1$) and last ($j=N_y$), where we do not have nodes at $j-1$ or $j+1$. No slip is assumed at the wall, $j=1$, and at the top of the boundary layer, u is set to u_e.

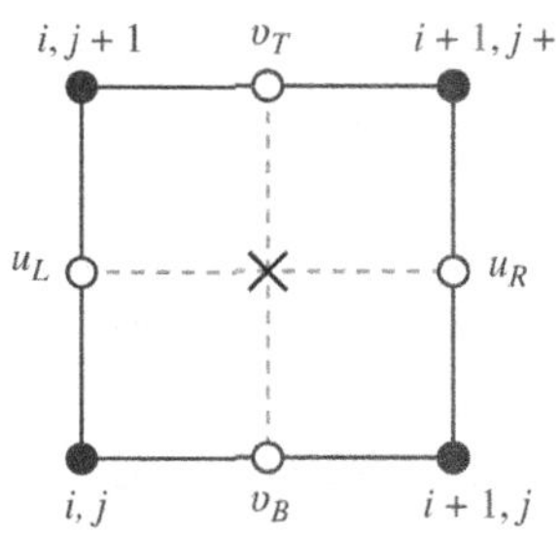

Figure 6.9 Enforcing continuity at lattice midpoints. The $\left(i+\frac{1}{2}, j+\frac{1}{2}\right)$ point is the "x" in the middle.

Code notes:

- The edge velocity, u_e, and initial profile, $u_0(y)$, are inputs.
- The given x and y vectors must be spaced equally.
- The solver marches from left-to-right in space, filling in `U` and `V`.
- In the pressure force term, $\frac{du_e}{dx}$ is discretized with a forward difference, which is a central difference at $i+\frac{1}{2}$, and the average value of u_e from nodes i and $i+1$ is used.
- Stability requires a sufficiently large number of points in x.

would expect in the boundary layer: $u = 0$ at the wall, $u = u_e$ at the top, and smooth variation in between. One possibility is an exponential form,

$$u_0(y) = 1 - e^{-Cy/L}, \tag{6.46}$$

where C is a constant that dictates the boundary-layer thickness, and this is what we will use in an example. Later, we will initialize with known analytical or numerical solutions to test the ability of the solver to evolve the state in the boundary layer.‡

‡ Starting at the leading edge of a boundary layer is not an option because the boundary-layer equations break down there. In particular, the $\frac{\partial^2 u}{\partial x^2}$ term becomes important.

What can we learn from a finite-difference solution? Knowing the velocity at many points, visualization of the entire flowfield is possible. Velocity vectors are one option, although a better choice is streamlines. Recall that since the flow is two-dimensional and incompressible, a stream function, $\psi(x, y)$, exists. We can calculate it numerically using the fact that the stream function is the path integral of the normal velocity, as shown in Figure 2.25. In the discrete setting, we assign $\psi = 0$ to nodes on the wall and then proceed upward through the columns of nodes for a particular i, calculating the stream function with a midpoint approximation to the velocity integral,

$$\psi_{i,j+1} = \psi_{i,j} + \frac{1}{2}\left(u_{i,j} + u_{i,j+1}\right)\Delta y. \tag{6.47}$$

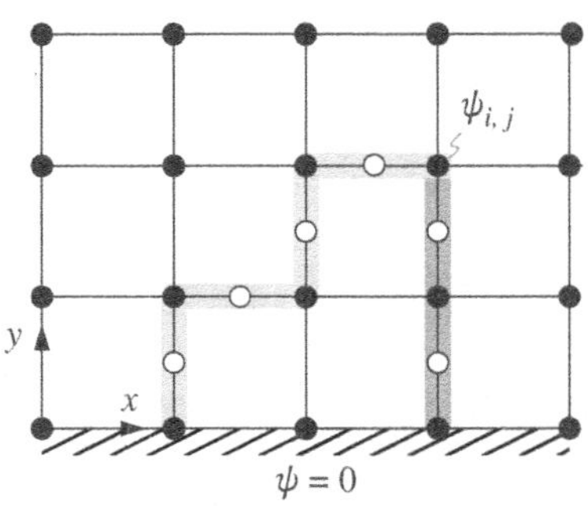

Figure 6.10 Calculation of the stream function by integrating velocities along paths (two shown). The white circles denote locations at which average velocities, normal to the path, are calculated.

This vertical march through column $i+1$ is just one path through the unknowns, but other paths exist, including ones that step right, e.g. from $i-1$ to i, as shown in Figure 6.10. Note, the above equation is based on a central-difference discretization of $u = \psi_y$. To move horizontally, we can write a similar equation for $v = -\psi_x$.

The calculation of the stream function should be independent of the path in the analytical case, where $\nabla \cdot \vec{v} = 0$ everywhere. Fortunately, by our choice of enforcing continuity at lattice cell centers, through Equation 6.45, the calculation of ψ will also be independent of the discrete path through the nodes. Specifically, multiplying Equation 6.45 by $\Delta x \Delta y$ shows that the same change in ψ is observed when moving right and up versus up and right in one cell of the lattice. Hence, up to numerical precision, the ψ values at the nodes will be unique with respect to the path taken.

Listing 6.2 shows a code that calculates the stream function this way and makes streamline plots. It also plots velocity profiles at a given number of x stations.

Code notes:

- The initialization of ψ to zero takes care of the values on the wall.
- The paths for calculating ψ are vertically upward through j values at each i.
- Input vectors x and y allow us to form matrices `X` and `Y` for plotting contours.
- The velocity profiles are scaled by an input factor, `fac`, to make them fit on the plot. The horizontal axis is the x coordinate, but velocity profiles show the x velocity horizontally.

Listing 6.2 Plots streamlines and velocity profiles from a boundary-layer solution.

```
function plotblstream(x,y,U,V,n,fac)
% plots BL stream function and n velocity profiles

% build the stream function from the velocity
Psi = zeros(size(U));
for i = 1:size(U,1)
  for j = 2:size(U,2)
    Psi(i,j) = Psi(i,j-1) + 0.5*(U(i,j-1) + U(i,j));
  end
end

% plot contours of the stream function
[X,Y] = ndgrid(x,y);
contour(X,Y,Psi,11, '-', 'color', '#888888'); hold on;
plot(X(:,1), Y(:,1), 'k-', 'linewidth', 2);
axis equal; axis off; box off;

% plot n velocity profiles scaled by fac
I = round(linspace(1,length(x),n));
for i = 1:length(I)-1,
  plot(x(I(i))+fac*U(I(i),:), y, 'k-', 'linewidth',1);
end
```

Another useful quantity we can obtain from the solution is the skin-friction coefficient, given in Equation 6.8. Its calculation requires the wall shear stress, given in Equation 6.7, which in turn requires the derivative of u with respect to y. A finite-difference approximation of this derivative could be a simple forward difference, but more accurate is a second-order one-sided difference,

$$\left.\frac{\partial u}{\partial y}\right|_{i,1} \approx \frac{-3u_{i,1} + 4u_{i,2} - u_{i,3}}{2\Delta y}. \tag{6.48}$$

In terms of this derivative, the skin-friction coefficient is§

$$c_f|_i = \frac{\tau_w}{\frac{1}{2}\rho U_\infty^2} = \frac{2\nu}{U_\infty^2}\left.\frac{\partial u}{\partial y}\right|_{i,1}. \tag{6.49}$$

§ When the edge velocity varies, we have the option of calculating a *local* skin-friction coefficient, where u_e replaces U_∞.

Example 6.2 Testing the finite-difference boundary-layer solver
This example shows how to use the boundary-layer solver to analyze a boundary layer with some arbitrary parameters. We pick $L = 1$ for the horizontal extent, a constant edge velocity $u_e = U_\infty = 1$, and $\nu = 0.001$, for a Reynolds number of $\text{Re}_L = 1000$. To initialize the velocity at the first station, we use Equation 6.46 with $C = 50$. The choice of C determines how thick the boundary layer is at the start, and the extent of the domain in the y direction should be adequate to accommodate the solution and its growth.¶ Listing 6.3 shows a code that sets up the problem and calls the boundary-layer solver.

¶ If the domain is too small in y, the solution will be inaccurate because the edge velocity will be enforced too close to the wall.

Listing 6.3 Driver code for testing the boundary-layer solver.

```
function blfd_test
% tests the FD BL solver

Uinf = 1;                      % freestream speed
nu = 0.001;                    % viscosity
L = 1;                         % flat plate length
x = linspace(0,L,501);         % x stations
ue = Uinf*ones(size(x));       % ue at each x
y = linspace(0,0.2*L,31);      % y values
u0 = 1 - exp(-50*y/L);         % u0 = initial profile
[U,V] = blfd(x, ue, y, u0, nu); % run the solver

% streamlines and velocity profiles
figure(1); clf;
plotblstream(x, y, U, V, 7, .1);
set(gcf, 'position', 600*[1,1,1,.4]);
writeplot('x', 'y', '../figs/blfd_test_stream.pdf', 2);

% skin friction coefficient distribution
figure(2); clf;
uy = (-U(:,3)+4*U(:,2)-3*U(:,1))/(2*(y(2)-y(1)));
cf = nu*uy/(0.5*Uinf^2);
plot(x,cf, 'k-'); grid on;
set(gcf, 'position', 400*[1,1,1,1]);
writeplot('$x$', '$c_f$', '../figs/blfd_test_cf.pdf', 4);
```

Code notes:

- The number of points in the x direction is much larger than in the y direction, not only because the domain is larger in x, but also to avoid an instability in the marching solver.
- u_e is constant, but a vector needs to be passed into `blfd`, so a constant one $u_e = U_\infty$ is constructed.
- No units are assumed for any of the quantities, but they could be ascribed if necessary. The equations solved are dimensional, but the outputs are either qualitative (streamlines) or dimensionless (skin-friction coefficient).

Once the velocities are calculated, the code plots the streamlines and velocity profiles, shown in Figure 6.11.

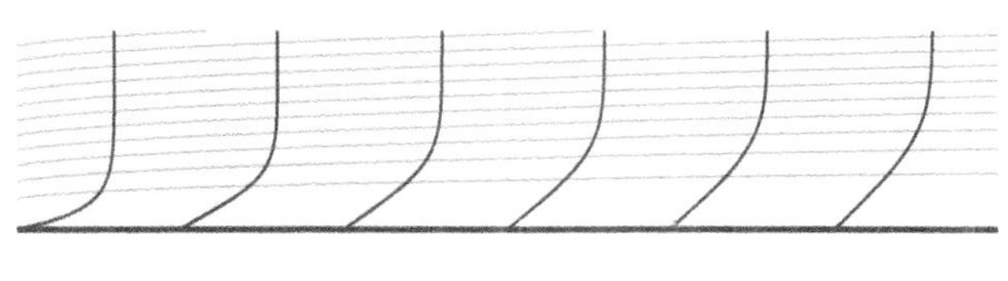

Figure 6.11 Streamlines and velocity profiles for a test run of the boundary-layer solver.

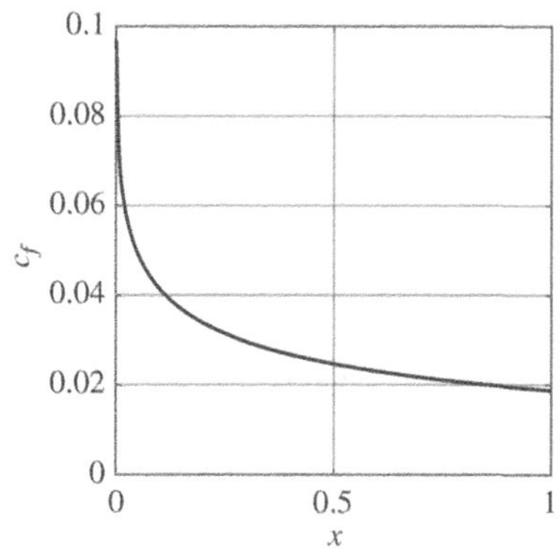

Figure 6.12 Skin-friction distribution for a test run of the boundary-layer solver.

The boundary layer grows in thickness with position downstream, as shown by the increasing vertical extent of the slowdown in the velocity profiles, and by the streamlines deflecting upward.

The test code also computes and plots the skin-friction coefficient distribution, $c_f(x)$, shown in Figure 6.12. The skin friction decreases with x because, as evident from Figure 6.11, the derivative $\frac{\partial u}{\partial y}$ decreases as the boundary layer grows. Note that a low value of $\frac{\partial u}{\partial y}$ in a velocity profile plot looks like a steep slope because of the swap in axes: u is horizontal and y is vertical. Based on these results, the solution makes sense qualitatively for this relatively benign constant edge velocity test. In later sections, we will use the solver for more quantitative comparisons to analytical and other numerical solutions.

6.4 Laminar Flow over a Flat Plate

We can obtain an analytical solution to the boundary-layer equations when the edge pressure, $p_e(x)$, and the edge velocity, $u_e(x)$, are constant. This situation corresponds to a uniform freestream, which would be observed, for example, for flow over a horizontal flat plate, away from any other objects. For a constant pressure, $\frac{\partial p}{\partial x} = \frac{dp_e}{dx} = 0$, so the x-momentum boundary-layer equation is

$$u\frac{\partial u}{\partial x} + v\frac{\partial u}{\partial y} = \nu\frac{\partial^2 u}{\partial y^2}. \tag{6.50}$$

6.4.1 The Blasius Equation

In 1908, Paul Richard Heinrich Blasius* discovered a *similarity solution* to the boundary-layer equations with $\frac{dp_e}{dx} = 0$. A similarity solution for the boundary-layer velocity profile, $u(y)$, is one in which $u(y)$ at different x locations is a scaled version of one universal velocity profile. For example, velocity profiles at two x locations, x_1 and x_2, shown in Figure 6.13, are similar if

*Blasius was a student of Ludwig Prandtl. A translation of his 1908 PhD thesis appears as a 1950 NACA technical memorandum (Blasius 1950).

$$u\,(x_1, \eta\delta(x_1)) = u\,(x_2, \eta\delta(x_2)), \tag{6.51}$$

where $\delta(x)$ is a measure of boundary-layer thickness and η is a non-dimensional measure of the vertical distance along the velocity profile. Such a scaling of the y coordinate accounts for a different vertical extent of the actual velocity profiles at x_1 and x_2. No scaling of the velocity magnitude is needed in this case, since the edge velocity is constant, but one could be applied if u_e depended on x.

A consequence of Equation 6.51 is that we can apply a variable transformation to simplify the boundary-layer equations and solution. In this case, the transformation consists of dividing the y coordinate by the boundary-layer thickness,

$$\eta(x, y) \equiv \frac{y}{\delta(x)}. \tag{6.52}$$

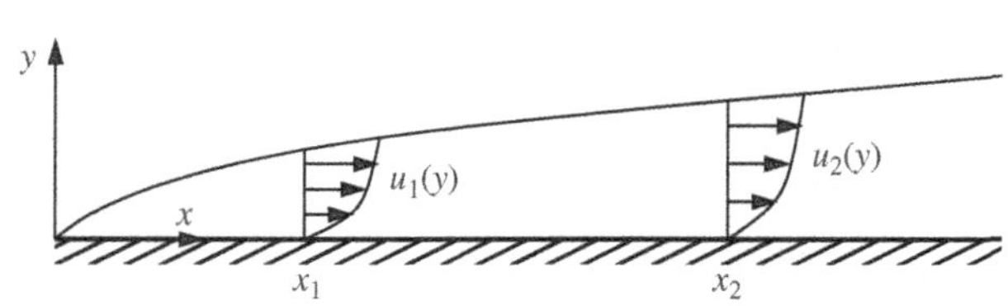

Figure 6.13 In the Blasius solution for laminar flow over a flat plate, velocity profiles at different x locations are scaled versions of each other.

η is the *similarity variable*, and we will soon see that the universal form of the velocity profile is a function only of η.

Recall from the order-of-magnitude analysis, Equation 6.22, that the non-dimensional boundary layer thickness scales inversely with the square root of the Reynolds number. Using x in place of c,† we have

$$\frac{\delta(x)}{x} = \mathcal{O}\left(\frac{1}{\sqrt{\mathrm{Re}_x}}\right), \tag{6.53}$$

† c denotes a horizontal position on the flat plate, which we can just as well call x.

where Re_x is the Reynolds number based on x,

$$\mathrm{Re}_x \equiv \frac{U_\infty x}{\nu}. \tag{6.54}$$

For deriving a similarity solution, the constant in this order of magnitude estimate is irrelevant, since all that matters mathematically is how δ changes with x. Using 1 as the constant in Equation 6.53, we have

$$\delta(x) = \sqrt{\frac{\nu x}{U_\infty}}. \tag{6.55}$$

Substituting this $\delta(x)$ into Equation 6.52 gives an equation for the similarity variable,

$$\eta = \sqrt{\frac{U_\infty}{\nu x}}\,y. \tag{6.56}$$

To complete the variable transformation, we leave the x coordinate alone, so that the new coordinate pair (ξ, η) is

$$\xi = x, \quad \eta = \sqrt{\frac{U_\infty}{\nu x}}\,y. \tag{6.57}$$

In applying this variable transformation to the boundary-layer equations, we use the chain rule to evaluate the derivatives,

$$\frac{\partial}{\partial x} = \frac{\partial \xi}{\partial x}\frac{\partial}{\partial \xi} + \frac{\partial \eta}{\partial x}\frac{\partial}{\partial \eta}, \tag{6.58}$$

$$\frac{\partial}{\partial y} = \frac{\partial \xi}{\partial y}\frac{\partial}{\partial \xi} + \frac{\partial \eta}{\partial y}\frac{\partial}{\partial \eta}, \tag{6.59}$$

where

$$\frac{\partial \xi}{\partial x} = 1, \qquad \frac{\partial \xi}{\partial y} = 0, \qquad \frac{\partial \eta}{\partial x} = -\frac{\eta}{2\xi}, \qquad \frac{\partial \eta}{\partial y} = \sqrt{\frac{U_\infty}{\nu \xi}}. \tag{6.60}$$

At this point, we have two equations, continuity and x-momentum, and two velocity components, u and v. We can simplify this system to one equation for one unknown by using a stream function, ψ.‡ Defining a stream function automatically ensures that the continuity equation is satisfied. The single equation that remains is x-momentum, in which all instances of the velocity components x and y can be replaced by derivatives of the stream function.

‡ We presented the stream function in the context of potential flow. Although the viscous flow inside a boundary layer is not potential flow because it is rotational, the stream function definition only requires $\nabla \cdot \vec{v} = 0$, so it can be used in rotational incompressible flows.

To enable a similarity solution, we need to relate the stream function to the similarity variable, η. One way to do this is through pure dimensional analysis, by noting that ψ has units of velocity times length, and by using the relevant velocity, U_∞, and length, $\delta(x)$, scales. Another way is through the physical interpretation of the stream function as the area flow rate between points in the flow, as introduced in Figure 2.25. Focusing on one x location, we choose the first point on the wall $y = 0$, and the second point at a distance y above the wall, as shown in Figure 6.14. The value of the stream function at (x, y) is then

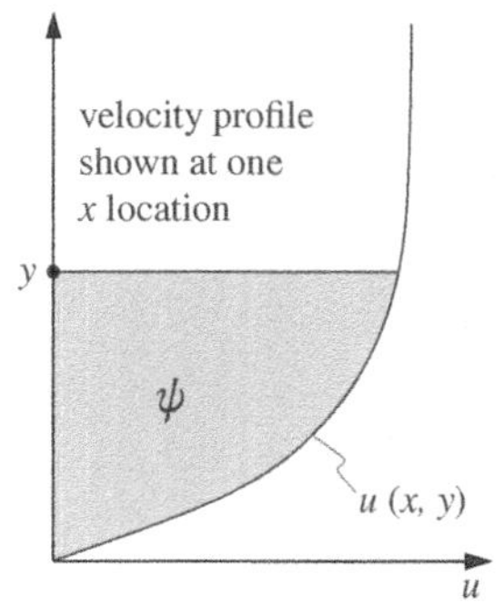

Figure 6.14 The stream function ψ at a point (x, y) in the flow is the area flow rate between that point and the wall.

$$\psi(x, y) = \int_0^y u(x, y)\, dy = U_\infty \delta(x) \underbrace{\int_0^\eta \frac{u(x, y)}{U_\infty}\, d\eta}_{f(\eta)}, \tag{6.61}$$

where we used $y = \eta(x, y)\delta(x)$ and $dy = \delta(x)d\eta$. For a self-similar velocity profile, per Equation 6.51, the integrand $u(x, y)/U_\infty$ should be a function only of η. Hence, we assert that the integral above depends only on η and call it $f(\eta)$. Therefore, the stream function is

$$\psi(x, y) = \delta(x)U_\infty f(\eta) = \sqrt{\nu \xi U_\infty}\, f(\eta), \tag{6.62}$$

where ξ replaced x so that the expression is in terms of the transformed coordinates. To verify that a similarity solution exists, we substitute the above definition of ψ into the x-momentum boundary-layer equation and check if the equation simplifies. Using the definition of the stream function and the chain rule, the velocity components are

$$\begin{aligned} u &= \frac{\partial \psi}{\partial y} = \frac{\partial \xi}{\partial y}\frac{\partial \psi}{\partial \xi} + \frac{\partial \eta}{\partial y}\frac{\partial \psi}{\partial \eta} \\ &= \sqrt{\frac{U_\infty}{\nu \xi}}\sqrt{\nu \xi U_\infty} f'(\eta) = U_\infty f'(\eta), \end{aligned} \tag{6.63}$$

$$\begin{aligned} v &= -\frac{\partial \psi}{\partial x} = -\left(\frac{\partial \xi}{\partial x}\frac{\partial \psi}{\partial \xi} + \frac{\partial \eta}{\partial x}\frac{\partial \psi}{\partial \eta}\right) \\ &= -\frac{1}{2}\sqrt{\frac{\nu U_\infty}{\xi}} f(\eta) + \frac{\eta}{2\xi}\sqrt{\nu \xi U_\infty} f'(\eta) \\ &= \frac{1}{2}\sqrt{\frac{\nu U_\infty}{\xi}}(\eta f' - f). \end{aligned} \tag{6.64}$$

The derivatives of the velocity components are calculated by the chain rule.§ The x derivative of u is

§ In these and the previous calculations, note that ξ does not depend on y, so $\frac{\partial \xi}{\partial y} = 0$, but η depends on both x and y.

$$\frac{\partial u}{\partial x} = \frac{\partial u}{\partial \xi} - \frac{\eta}{2\xi}\frac{\partial u}{\partial \eta} = -\frac{\eta}{2\xi} U_\infty f''(\eta). \tag{6.65}$$

Similarly, the derivative of u with respect to y is

$$\frac{\partial u}{\partial y} = \sqrt{\frac{U_\infty}{\nu \xi}}\frac{\partial u}{\partial \eta} = \sqrt{\frac{U_\infty}{\nu \xi}} U_\infty f''(\eta), \tag{6.66}$$

and the second derivative is

$$\frac{\partial^2 u}{\partial y^2} = \frac{U_\infty}{\nu \xi}\frac{\partial^2 u}{\partial \eta^2} = \frac{U_\infty}{\nu \xi} U_\infty f'''(\eta). \tag{6.67}$$

Substituting these expressions into the x-momentum equation for a zero-pressure gradient flat plate, Equation 6.50, yields, after a few manipulations,

$$\begin{aligned} u\frac{\partial u}{\partial x} + v\frac{\partial u}{\partial y} &= \nu \frac{\partial^2 u}{\partial y^2} \\ -\frac{1}{2} f f'' \frac{U_\infty^2}{\xi} &= \frac{U_\infty^2}{\xi} f'''(\eta). \end{aligned} \tag{6.68}$$

Rearranging gives the *Blasius equation*,¶

$$2f''' + ff'' = 0. \tag{6.69}$$

This is an ordinary differential equation! We converted two partial differential equations for u and v in terms of x and y into one ordinary differential equation for f in terms of η. More importantly, this equation is only in terms of f and η – the original coordinates x and y no longer appear. This elimination of x and y indicates that a similarity solution indeed exists, justifying the assumption about the form of the stream function in Equation 6.62.

To solve the Blasius equation, we also need boundary conditions. In terms of the x and y coordinates, the boundary conditions are

- $y = 0 \quad \Rightarrow \quad u = 0$ (no slip) and $v = 0$ (no flow through)
- $y \to \infty \quad \Rightarrow \quad u \to U_\infty$

In terms of the transformed variables, these conditions become

- $\eta = 0 \quad \Rightarrow \quad f' = 0$ (no slip) and $f = 0$ (no flow through)
- $\eta \to \infty \quad \Rightarrow \quad f' \to 1$

Requiring $f = 0$ at $\eta = 0$ makes the wall a streamline, $\psi = 0 =$ const., which is equivalent to the no flow-through condition. The Blasius equation is a nonlinear ordinary differential equation. It has no analytical solution, but we can solve it numerically, using a numerical method, as shown in the next example. The result is the solution shown in Figure 6.15.

¶ In this equation, $f(\eta)$ is a non-dimensional stream function, and the derivatives denoted by $'$ are with respect to η.

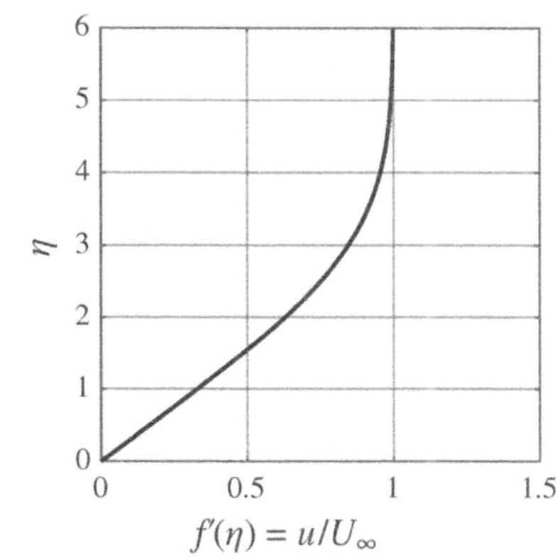

Figure 6.15 Solution of the Blasius equation, obtained numerically. Shown is $f'(\eta)$ versus η, which is a scaled version of the velocity profile, since $u = U_\infty f'(\eta)$.

Example 6.3 Solution of the Blasius equation
One way of obtaining a numerical solution to the Blasius equation is to cast Equation 6.69 as a first-order differential equation, which we do by defining a three-component state vector **F**,

$$\mathbf{F} = \begin{bmatrix} f \\ f' \\ f'' \end{bmatrix} \quad \Rightarrow \quad \mathbf{F}' = \begin{bmatrix} f' \\ f'' \\ f''' \end{bmatrix} = \begin{bmatrix} f' \\ f'' \\ -ff''/2 \end{bmatrix}. \tag{6.70}$$

The third entry in $\mathbf{F}'$ uses the Blasius equation to write $f''' = -ff''/2$. In this form, the state can be integrated in η using a variety of ordinary differential equation solvers. The code in Listing 6.4 uses MATLAB's `ode45` function, which is an explicit Runge–Kutta method, and it solves the equation from 0 to a prescribed η value.

Listing 6.4 Code for solving the Blasius equation.

```
function [eta,F] = blasius
% solves the Blasius equation over a fixed eta range

% initial conditions
f0   = 0;
fp0  = 0;
fpp0 = 0.33206; % can determine by guess or bisection

F0 = [f0, fp0, fpp0];

[eta, F] = ode45(@Fprime, [0, 6], F0);

%------------------------
```

Code notes:

- The wall boundary conditions provide f and f' at $\eta = 0$, two out of the three components for the initial state.
- The third component, $f''(0)$, needs to be set appropriately so that when the integration is finished at large η, $f' = 1$.
- Determining $f''(0)$ can be done with trial and error or bisection, with the result that $f''(0) \approx 0.33206$.
- `Fprime` computes $\mathbf{F}'$ using Equation 6.70. The first line initializes `Fp` to be the same size as `F`.

```
function Fp = Fprime(eta,F);
Fp = F;
Fp(1) = F(2);
Fp(2) = F(3);
Fp(3) = -0.5*F(1)*F(3);
```

Running this code produces the data plotted in Figure 6.15. Since $f' = u/U_\infty$ is a velocity profile, we plot it with axes flipped so that η is on the vertical axis. The terminal value of $\eta = 6$ can be adjusted to obtain data for more or less of the profile. This velocity profile is *universal* for the flat plate. That is, the velocity profile at any x, and for any U_∞, ν, can be scaled to match the one shown in Figure 6.15! This is the useful property of similarity solutions: a one-dimensional function $f(\eta)$ gives the solution $u(x, y)$ to a two-dimensional problem.

Example 6.4 Comparison of the Blasius solution to the finite-difference solver
We can verify the boundary-layer finite-difference solver developed in Section 6.3 for flow over a flat plate by using the Blasius solution. To do this, we initialize the velocity profile on the left side of the domain, $u_0(y)$ to the Blasius solution corresponding to some distance L from the leading edge of a flat plate. The initial y velocity is assumed close to zero but could be passed into the solver for improved accuracy. We then run the solver for another length L to let the boundary layer grow and compare quantities like skin-friction coefficient and the velocity profile at the end of the domain. Listing 6.5 shows the driver code that performs the setup and calls the boundary-layer solver.

Code notes:

- The Blasius solver from Example 6.3 is used to compute the non-dimensional solution for initialization.
- The boundary-layer solver starts at $x = L$ and computes the solution up to $x = 2L$.
- The non-dimensional Blasius solution must be scaled appropriately, η by δ from Equation 6.55 to obtain y, and f' by U_∞ to obtain u.
- The initial velocity profile is interpolated to a set of points that is uniform in y, since there is no guarantee that the η values coming out of the `blasius` function are equally spaced.
- The entire code includes post-processing that is not shown in this listing.

Listing 6.5 Compares the boundary-layer solver with the Blasius solution.

```
function blfd_blasius
% compares FD BL solver to Blasius

Uinf = 1;                        % freestream speed
nu = 0.001;                      % viscosity
L = 1;                           % flat plate length
[eta,F] = blasius;               % non-dim solution for initialization
delta = sqrt(nu*L/Uinf); % delta at x = L
yb = eta*delta;                  % Blasius y values
ub = F(:,2)*Uinf;                % Blasius x-velocity at L
y = linspace(0,2*max(yb),21); % equal spacing in y
u0 = interp1(yb,ub,y, 'linear', Uinf); % interpolate u from ub
x = linspace(L,2*L,501); % x stations to simulate
ue = Uinf*ones(size(x)); % edge velocity at x
[U,V] = blfd(x, ue, y, u0, nu); % simulate another length L
```

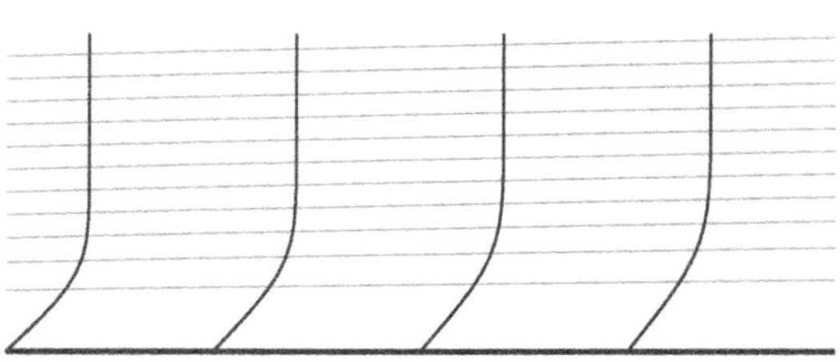

Figure 6.16 Streamlines and velocity profiles for the boundary-layer solver with the Blasius initial profile.

Running the solver produces the streamlines and velocity profiles shown in Figure 6.16. Qualitatively, the boundary layer grows as expected, and the velocity profiles appear self-similar.

Figure 6.17 compares the velocity profile at the right side of the domain, $x = 2L$, from the solver to the Blasius solution. The latter is the universal profile scaled to the correct x location. We see very good agreement, visually identical, between the results.

Finally, Figure 6.18 compares the skin-friction coefficient computed from the solver to the Blasius result, which will be given shortly, in Equation 6.73. Again, we see excellent agreement. This quantitative comparison verifies the boundary-layer solver against a known analytical solution, for a zero-pressure gradient flow. We will later also verify it for a certain special form of the pressure gradient, prior to demonstrating it in the general case.

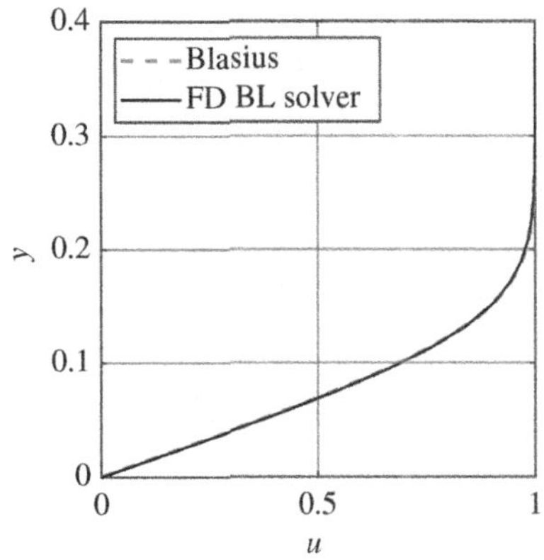

Figure 6.17 Velocity profile at $x = 2L$ as predicted by the finite-difference solver and the exact Blasius solution.

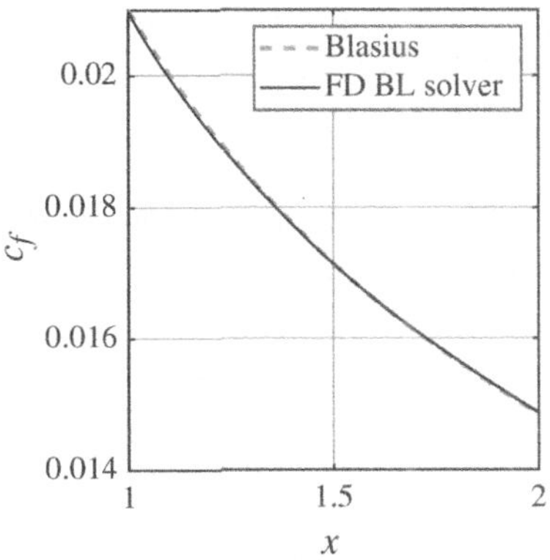

Figure 6.18 Skin-friction distribution for the boundary-layer solver with the Blasius initial profile.

6.4.2 Properties of the Blasius Solution

An engineering quantity of interest in a boundary-layer flow is the wall shear stress, which contributes to the drag force. The first derivative of f gives the velocity profile, while the second derivative gives the velocity gradient $\frac{\partial u}{\partial y}$, by Equation 6.66. We can use the value of f'' at the wall, $\eta = 0$, to calculate the wall shear stress,

$$\begin{aligned} \tau_w &= \mu \left(\frac{\partial u}{\partial y}\right)_{y=0} = \mu U_\infty \left(\frac{\partial f'}{\partial y}\right)_{y=0} \\ &= \mu U_\infty \sqrt{\frac{U_\infty}{\nu x}} \left(\frac{\partial f'}{\partial \eta}\right)_{\eta=0} = \mu U_\infty \sqrt{\frac{U_\infty}{\nu x}} f''(0). \end{aligned} \tag{6.71}$$

Non-dimensionalizing this stress gives the "skin-friction coefficient,"

$$c_f = \frac{\tau_w}{\frac{1}{2}\rho U_\infty^2} = 2\sqrt{\frac{\nu}{U_\infty x}} f''(0) = \frac{2f''(0)}{\sqrt{\mathrm{Re}_x}}. \tag{6.72}$$

Using the numerical solution, $f''(0) \approx 0.332$, the *skin-friction coefficient for the Blasius solution* is

$$c_f = \frac{0.664}{\sqrt{\mathrm{Re}_x}}. \tag{6.73}$$

This equation tells us how the skin-friction coefficient changes with x.* At the leading edge, $x = 0$, the Blasius solution gives $c_f \to \infty$, which is non-physical. The problem with extending the Blasius result to $x = 0$ is that near $x = 0$ the assumptions that went into deriving the boundary layer equations break down.† In reality, c_f will remain finite as $x \to 0$, although it can be very large. To obtain the precise value, we need to solve the full Navier–Stokes equations. Fortunately, the integral of $c_f(x)$ is still finite for the Blasius solution, so we can compute meaningful drag values.

Another quantity of interest in a boundary-layer flow is the thickness of the boundary layer. So far, we have been referring to $\delta(x)$ as the boundary-layer thickness, but it is only the result of an order-of-magnitude analysis, and its definition is arbitrary up to a multiplicative constant.‡

It turns out that there are several practical definitions of the thickness of a boundary layer, all computed from the velocity profile. One is the *99% thickness*, $\delta_{99}(x)$, which is the value of y where $u(x, y) = 0.99u_e(x)$. Recall that $u_e(x)$ is the inviscid-region velocity, at the edge of the boundary layer. For the Blasius solution, $u_e = U_\infty$, and, by inspecting the numerical solution, we can determine

$$\frac{u}{U_\infty} = f' = 0.99 \text{ at } \eta \approx 5.0. \tag{6.74}$$

* c_f is bigger at smaller x, since Re_x is directly proportional to x. This should make sense because the boundary layer is thinner at smaller x, and for self-similar velocity profiles, a thinner boundary layer means a bigger $\partial u/\partial y$.

† For example, the boundary layer is no longer thin relative to x as $x \to 0$, and we cannot drop $\frac{\partial^2 u}{\partial x^2}$ in the x-momentum equation.

‡ Only the functional form of $\delta(x) \sim \sqrt{x}$ was important for the derivation. Scaling δ by a constant changes the Blasius equation and the solution $f(\eta)$, but it does not change the physical solution $u(x, y)$.

§ Note that $\delta_{99}(x) \approx 5\delta(x)$.

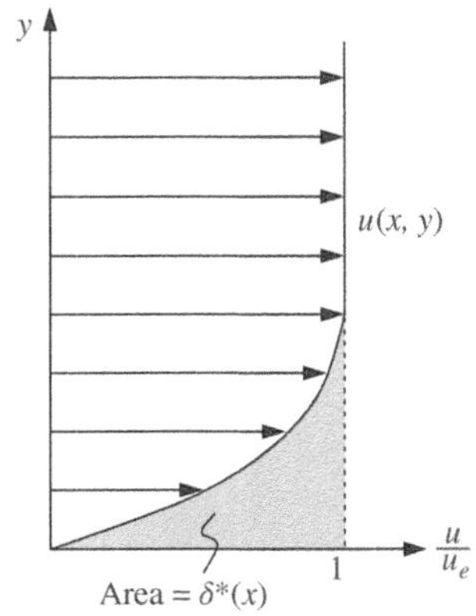

Figure 6.19 The displacement thickness, $\delta^*(x)$, is proportional to the *missing mass flow* due to the presence of the boundary layer compared to the inviscid solution.

Using the definition of η in Equation 6.56,§

$$\eta = \sqrt{\frac{U_\infty}{\nu x}} y \quad \Rightarrow \quad 5.0 = \sqrt{\frac{U_\infty}{\nu x}} \delta_{99} \quad \Rightarrow \quad \frac{\delta_{99}}{x} = \frac{5.0}{\sqrt{\mathrm{Re}_x}}. \tag{6.75}$$

There are other ways to define the boundary-layer thickness, some of which turn out to be more useful for analysis and interpretation of the boundary layer. One of these is the *displacement thickness*,

$$\delta^* \equiv \int_0^{y_1} \left(1 - \frac{u}{u_e}\right) dy, \tag{6.76}$$

where y_1 is a point sufficiently far from the wall that it is completely outside the boundary layer, i.e. where $u \approx u_e$. As illustrated in Figure 6.19, one interpretation of δ^* is that it is proportional to the missing mass flow. In particular, $\rho u_e \delta^*$ is the difference between the mass flow if the boundary layer were not present and the mass flow with the boundary layer was present.

Another interpretation is that $\delta^*(x)$ is equal to the deflection distance of a streamline in the inviscid flow above the boundary layer. To visualize this interpretation, consider a comparison between pure inviscid flow, with no boundary layer, and flow with the boundary layer present, as shown in Figure 6.20. The density ρ is assumed constant, and y_1 is outside the boundary layer. In the first case, the mass flow exiting through the right boundary of the control volume is¶

¶ Without a boundary layer, the velocity is $u = u_e$ for all y.

$$\dot{m} = \int_0^{y_1} \rho u_e dy. \tag{6.77}$$

In the case with the boundary layer, the mass flow exiting the control volume through the right boundary is

$$\dot{m} = \int_0^{y_1} \rho u(y) dy + \int_{y_1}^{y_1+\delta^*} \rho u_e dy = \int_0^{y_1} \rho u(y) dy + \rho u_e \delta^*. \tag{6.78}$$

These two mass flows must be equal because the mass flow entering the two control volumes is the same at $x = 0$.‖ Assuming also that u_e is the same in both cases, we have

‖ Note that no flow crosses the wall or streamline, which make up the bottom and top boundaries of the control volume.

$$\int_0^{y_1} \rho u_e dy = \int_0^{y_1} \rho u(y) dy + \rho u_e \delta^*. \tag{6.79}$$

Solving for δ^* gives the expression in Equation 6.76. Note that by this interpretation, the distance $\delta^*(x)$ away from the wall can be thought of as the "effective body" seen by the inviscid flow.

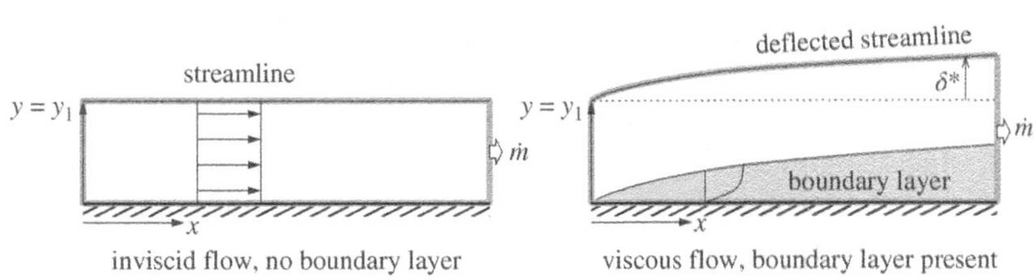

Figure 6.20 Interpretation of δ^* as the deflection of a streamline in inviscid flow. y_1 is a y location outside the boundary layer.

For the Blasius solution, $u_e = U_\infty$ and $u = f'U_\infty$. So the displacement thickness is

$$\delta^* = \int_0^{y_1} (1 - f'(\eta))\, dy = \sqrt{\frac{\nu x}{U_\infty}}\, (\eta_1 - f(\eta_1)), \tag{6.80}$$

where η_1 corresponds to y_1, i.e. some point above the boundary layer. For $\eta > 5.0$, $f'(\eta) \approx 1$ so that any $\eta > 5.0$ can be reasonably taken to be outside the Blasius boundary-layer profile. The limit of $\eta_1 - f(\eta_1)$ as $\eta \to \infty$ is approximately 1.72, a result that can be obtained from the numerical solution using the code in Example 6.3. Dividing by x gives the normalized *displacement thickness for the Blasius solution,***

$$\frac{\delta^*}{x} = \frac{1.72}{\sqrt{\mathrm{Re}_x}}. \tag{6.81}$$

** Note that for Blasius, $\delta^* \approx 1.72\delta \approx 0.34\delta_{99}$, but all measures grow with $\sqrt{x}$.

Another measure of boundary-layer thickness is the *momentum thickness*, defined as

$$\theta \equiv \int_0^{y_1} \left(1 - \frac{u}{u_e}\right) \frac{u}{u_e} dy. \tag{6.82}$$

Physically, θ is proportional to the "missing momentum flow." In particular, $\rho u_e^2 \theta$ is the difference between a hypothetical inviscid momentum flow and the viscous momentum flow in the boundary layer, when the comparison is made at the same mass flow. We will use θ in our study of the integral boundary-layer equation, in Section 6.6. The *momentum thickness for the Blasius solution* is

$$\frac{\theta}{x} = \frac{0.664}{\sqrt{\mathrm{Re}_x}}. \tag{6.83}$$

To recap, the Blasius solution is a solution to the boundary-layer equations for the case $dp_e/dx = 0$, i.e. for laminar flow over a flat plate at zero angle of attack.†† It is a similarity solution in $\eta = \sqrt{U_\infty/(\nu x)}y$, written in the form of the stream function, $\psi = \sqrt{\nu x U_\infty}\, f(\eta)$, where $u = U_\infty f'(\eta)$. The important Blasius results are

$$c_f = \frac{0.664}{\sqrt{\mathrm{Re}_x}}, \quad \frac{\delta_{99}}{x} = \frac{5.0}{\sqrt{\mathrm{Re}_x}}, \quad \frac{\delta^*}{x} = \frac{1.72}{\sqrt{\mathrm{Re}_x}}, \quad \frac{\theta}{x} = \frac{0.664}{\sqrt{\mathrm{Re}_x}}, \tag{6.84}$$

†† A zero angle of attack is equivalent to a zero pressure gradient and constant edge velocity. In addition, *laminar* means not turbulent, a topic that will be addressed later in the chapter.

where $\mathrm{Re}_x = U_\infty x/\nu$. Note, the fact that c_f and θ/x are the same quantities is not a coincidence. We can show that this is the case by evaluating the drag on a section of the flat plate from 0 to x. Using the fact that $\rho U_\infty^2 \theta$ is the missing momentum, and hence the drag, and that the drag on a plate is the integral of the wall shear stress, we have

$$D' = \rho U_\infty^2 \theta(x) = \int_0^x \tau_w(x')\, dx' = \frac{1}{2}\rho U_\infty^2 \int_0^x c_f(x')\, dx'. \tag{6.85}$$

Now, the integral of c_f, evaluated using the above formulas, is

$$\int_0^x c_f(x')\, dx' = \int_0^x \frac{0.664}{\sqrt{\mathrm{Re}_{x'}}}\, dx' = 2(0.664)\frac{x}{\sqrt{\mathrm{Re}_x}} = 2xc_f(x). \tag{6.86}$$

Substituting into Equation 6.85, we find that

$$\rho U_\infty^2 \theta(x) = \frac{1}{2}\rho U_\infty^2 2xc_f(x) \quad \Rightarrow \quad \frac{\theta(x)}{x} = c_f(x). \tag{6.87}$$

Example 6.5 Drag on a thin symmetric airfoil at $\alpha = 0$
A thin airfoil at a zero angle of attack can be reasonably approximated as a flat plate, so that the Blasius results apply. The drag on the airfoil is caused by skin friction on the upper and lower surfaces. Accounting for both, the two-dimensional drag on the airfoil is

$$D' = 2\int_0^c \tau_w(x)dx = 2\left(\frac{1}{2}\rho U_\infty^2\right)\int_0^c c_f(x)dx, \tag{6.88}$$

where c is the chord of the airfoil. Using the definition of the drag coefficient,

$$c_d = \frac{D'}{\frac{1}{2}\rho U_\infty^2 c} = \frac{2}{c}\int_0^c c_f(x)dx. \tag{6.89}$$

Substituting the Blasius result that $c_f(x) = 0.664/\sqrt{\text{Re}_x}$, we have,

$$c_d = \frac{2}{c}\int_0^c \frac{.664}{\sqrt{\text{Re}_x}}dx = \frac{2.656}{\sqrt{\text{Re}_c}}. \tag{6.90}$$

where $\text{Re}_c = U_\infty c/\nu$. Therefore, the airfoil's drag coefficient depends inversely on the square root of the chord-based Reynolds number.[‡‡]

‡‡ It is important to distinguish local versus integrated coefficients. $c_f(x)$ is a local coefficient that has a distribution over a surface. On the other hand, c_d is an integrated coefficient that indicates the amount of drag on the entire airfoil. In addition, $\Re_x$ varies with location on the airfoil (the x value), whereas $\Re_c$ is fixed for an airfoil of a given chord length, c.

The previous example calculated the sectional drag, D', and the corresponding coefficient, c_d, for a two-dimensional thin airfoil. We can extend the calculation to an object in three dimensions, such as a plate with a non-rectangular shape, as shown in the next example.

Example 6.6 Drag on a triangular plate at $\alpha = 0$
A flat plate with an isosceles right-triangle planform[§§] is flying at a zero angle of attack, as shown in Figure 6.21. To evaluate the drag on the flat plate, we integrate the skin friction over the area,

$$D = 2\int_0^a\int_0^{a-y} \tau_w(x)dxdy, \tag{6.91}$$

where a is the side length of the right triangle, the origin is at the right-angle vertex, and the leading factor of 2 accounts for the fact that there are two sides to the flat plate (top and bottom). Using the Blasius solution for c_f, Equation 6.73, and the fact that $\tau_w = \frac{1}{2}\rho U_\infty^2 c_f$, the expression for the drag becomes

$$\begin{aligned} D &= 2\frac{1}{2}\rho U_\infty^2 \int_0^a\int_0^{a-y} \frac{0.664}{\sqrt{\text{Re}_x}}dxdy \\ &= \rho U_\infty^2 \int_0^a\int_0^{a-y} 0.664\sqrt{\frac{\nu}{U_\infty x}}dxdy. \end{aligned} \tag{6.92}$$

§§ The planform is the view of a wing shape from above. In this example, the triangle could represent half of a wing.

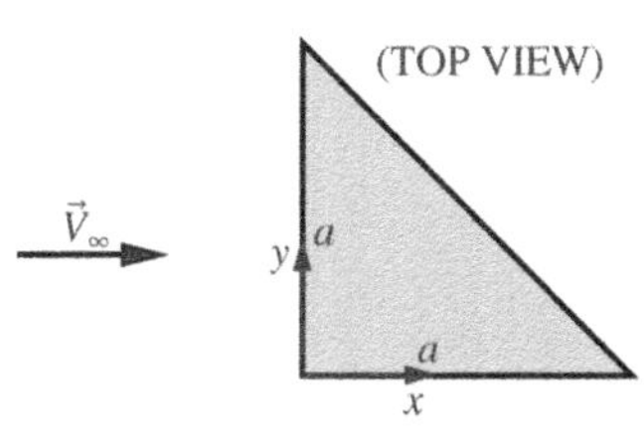

Figure 6.21 A right-triangle flat plate.

Evaluating the inside integral over x gives

$$\begin{aligned} D &= \rho U_\infty^2 \sqrt{\frac{\nu}{U_\infty}} \int_0^a \left[1.328\sqrt{x}\right]_0^{a-y} dxdy \\ &= 1.328\rho U_\infty^2 \sqrt{\frac{\nu}{U_\infty}} \int_0^a \sqrt{a-y}\,dy. \end{aligned} \tag{6.93}$$

Finally, evaluating the remaining integral in y results in

$$\begin{aligned} D &= 1.328\rho U_\infty^2 \sqrt{\frac{\nu}{U_\infty}} \left[-\frac{2}{3}(a-y)^{\frac{3}{2}}\right]_0^a \\ &= \frac{2.656}{3}\rho U_\infty^2 \sqrt{\frac{\nu}{U_\infty}} a^{\frac{3}{2}} \\ &= \frac{2.656}{3}\rho U_\infty^2 a^2 \sqrt{\frac{\nu}{U_\infty a}} \\ &= \frac{2.656}{3}\frac{\rho U_\infty^2 a^2}{\sqrt{\mathrm{Re}_a}}, \end{aligned} \tag{6.94}$$

where $\mathrm{Re}_a = U_\infty a/\nu$ is the Reynolds number based on a. The drag coefficient for a flat plate such as this, which could be a portion of a wing, is obtained by non-dimensionalizing D using the freestream dynamic pressure and the planform area, which is the area projected onto the x–y plane from the top view. For this geometry, the planform area is $S = \frac{1}{2}a^2$, so the drag coefficient is

$$C_D = \frac{D}{\frac{1}{2}\rho U_\infty^2 S} = \frac{D}{\frac{1}{2}\rho U_\infty^2 \frac{1}{2}a^2} = \frac{2.656}{3}\frac{4}{\sqrt{\mathrm{Re}_a}} \approx \frac{3.541}{\sqrt{\mathrm{Re}_a}}. \tag{6.95}$$

6.5 Falkner–Skan Flows

The Blasius self-similar velocity profile is an elegant solution to the boundary-layer equations. However, it only applies to a zero-pressure gradient flow, i.e. one in which the edge velocity is constant. A flat plate or thin airfoil at zero angle of attack can approximate these conditions, but any more interesting flows, such as airfoils with some thickness and non-zero angle of attack, will have edge velocities, and hence pressure gradients, that vary with distance along the surface.* This section presents a technique for analyzing certain types of such flows, ones with a specific form of the edge velocity. Although not a general solution for arbitrary $u_e(x)$, the Falkner–Skan family of solutions illustrates how a boundary layer behaves in both favorable and adverse pressure gradients. Furthermore, the solutions are, like in the Blasius case, self-similar, which greatly simplifies the calculations.

*For example, consider the airfoil c_p plots shown in the last chapter, which indicate significant variation in u_e, since $c_p = 1 - u_e^2/U_\infty^2$.

6.5.1 The Falkner–Skan Equation

The Blasius solution for a zero pressure gradient flow, $dp_e/dx = 0$, is a special case of a general family of flows for which the edge velocity is of a power law form,

$$u_e = Kx^m, \tag{6.96}$$

*The Blasius solution corresponds to $m = 0$ and $K = U_\infty$, so that u_e is constant.

where K and m are constants.* Using Equation 6.30, the corresponding pressure gradient is

$$\frac{dp_e}{dx} = -\rho u_e \frac{du_e}{dx} = -\rho(Kx^m)Kmx^{m-1} = -\rho K^2 m x^{2m-1}. \tag{6.97}$$

When $m > 0$, $\frac{dp_e}{dx} < 0$, and the pressure gradient is said to be *favorable*. Conversely, when $m < 0$, $\frac{dp_e}{dx} > 0$, and the pressure gradient is said to be *adverse*. These two cases are illustrated in Figure 6.22. In a favorable pressure gradient, the pressure force acts in the flow direction, pushing fluid elements along as they move from a high-pressure to a low-pressure region. The opposite occurs in an adverse pressure gradient, for which the pressure force opposes the direction of the flow and slows fluid elements down.

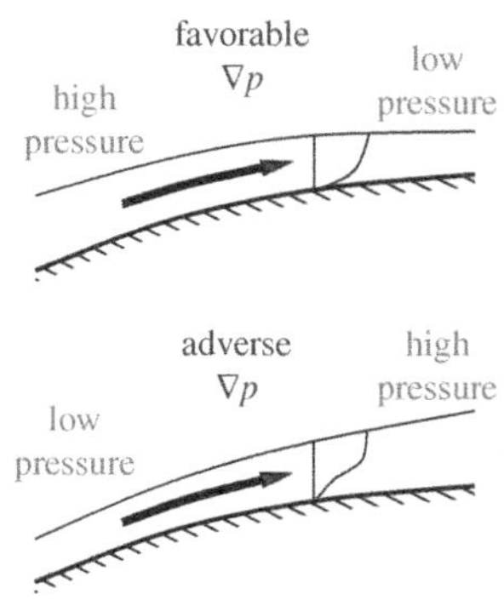

Figure 6.22 Favorable and adverse pressure gradients in a boundary layer. The arrows indicate the direction of the pressure force, which is along -∇p.

These power-law edge velocities result from inviscid "wedge flows" or "corner flows," for which the turning angle is $\beta\pi/2$, where

$$\beta = \frac{2m}{1+m}. \tag{6.98}$$

Figure 6.23 illustrates the case of negative β, which corresponds to corner flows, and positive β, which corresponds to wedge flows. The corner case results in an adverse pressure gradient because the downward slope of the wall means that the inviscid flow encounters more area to flow into and hence slows down. A reduction in the speed means an increase in the pressure, and hence, the pressure force pushes fluid elements back. Conversely, the wedge flow case results in a favorable pressure gradient because the upward slope of the wall constricts the flow and accelerates it, leading to increasing velocity and decreasing pressure.

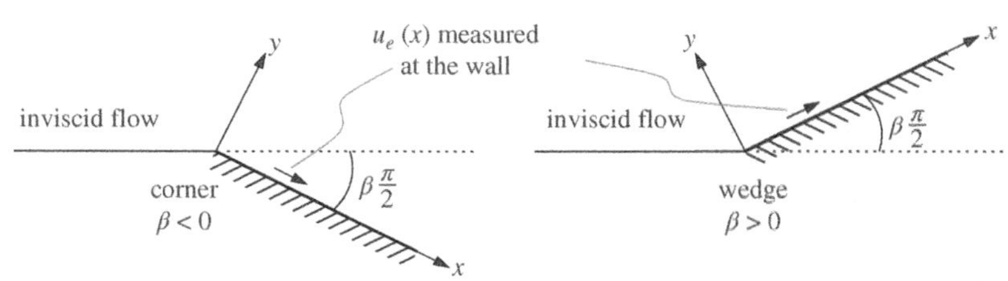

Figure 6.23 Falkner–Skan edge velocities arise from inviscid corner and wedge flows.

Two special cases are $m = 0 \Rightarrow \beta = 0$, for which we obtain the Blasius solution, and $m = 1 \Rightarrow \beta = 1$, for which we obtain *stagnation point flow*. As shown in Figure 6.24, stagnation point flow corresponds to horizontal flow encountering a vertical wall.

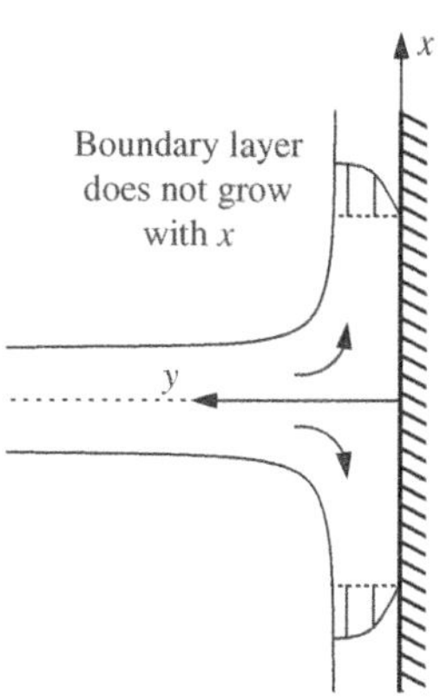

Figure 6.24 Stagnation point flow in two dimensions: $m = 1 \Rightarrow \beta = 1$.

A similarity solution exists for these flows, and the general forms of the boundary-layer thickness scaling and similarity variable are

$$\delta(x) = \sqrt{\frac{2}{m+1}\frac{\nu x}{u_e(x)}}, \qquad \eta = \frac{y}{\delta(x)}. \tag{6.99}$$

The stream function is still given by the same general expression that we derived for the Blasius solution,

$$\psi = \delta(x) u_e(x) f(\eta), \tag{6.100}$$

and the velocity profile is

$$u = u_e(x) f'(\eta). \tag{6.101}$$

Note that $u_e(x)$ now replaces the constant U_∞ edge velocity from the Blasius case. The special case of stagnation point flow is interesting, since in this case, $m = 1 \Rightarrow u_e = Kx$ so that the boundary-layer thickness is

$$\delta(x) = \sqrt{\frac{2}{1+1}\frac{\nu x}{Kx}} = \sqrt{\frac{\nu}{K}} \quad \text{(for stagnation point flow).} \tag{6.102}$$

In this case, δ is independent of x! This means that the boundary layer for a stagnation point flow does not grow or shrink with x. Figure 6.25 illustrates such a boundary layer at the leading edge of an airfoil, where the flow resembles stagnation-point flow due to the airfoil surface being nearly perpendicular to oncoming flow. The zoomed-in velocity magnitude contours show an approximately constant-thickness boundary layer.

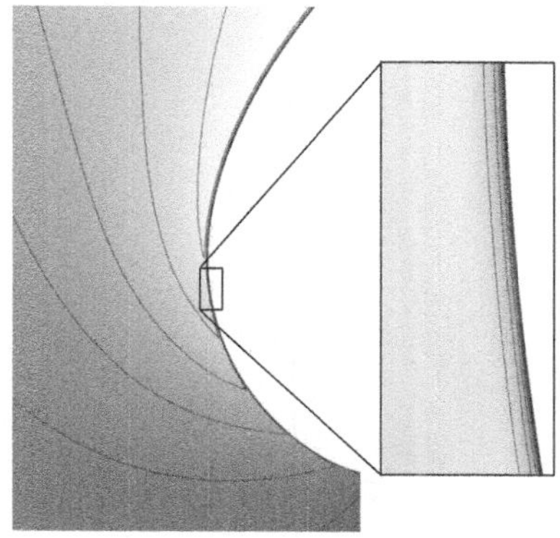

Figure 6.25 Velocity magnitude contours at the leading edge of an airfoil.

After substituting the transformed variables into the x-momentum boundary-layer equation, we obtain the *Falkner–Skan equation* (Schlichting and Gersten 2016),†

$$f''' + ff'' + \beta(1 - f'^2) = 0. \tag{6.103}$$

† $\beta = 0$ corresponds to the Blasius solution, but this equation reduces to Equation 6.69 without the factor of 2 because we defined δ differently in the Falkner–Skan case, with an extra factor of 1/2 in the square root (compare Equation 6.55 to Equation 6.99).

The boundary conditions for the Falkner–Skan equations are the same as for the Blasius equation,

- $\eta = 0 \quad \Rightarrow \quad f' = 0$ (no slip) and $f = 0$ (flow tangency)
- $\eta \to \infty \quad \Rightarrow \quad f' \to 1$

Also, the wall shear stress can be calculated as

$$\tau_w = \mu \frac{\partial u}{\partial y}\bigg|_{y=0} = \mu u_e(x)\sqrt{\frac{m+1}{2}}\sqrt{\frac{u_e(x)}{\nu x}} f''(0), \tag{6.104}$$

where $f''(0)$ is a tabulated quantity obtained from numerical solutions. The *skin-friction coefficient for Falkner–Skan flows* is then

$$c_f = \frac{\tau_w}{\frac{1}{2}\rho u_e^2} = 2\sqrt{\frac{m+1}{2}}\frac{1}{\sqrt{\mathrm{Re}_x}} f''(0), \tag{6.105}$$

where $\mathrm{Re}_x = u_e(x)x/\nu$ for Falkner–Skan flows. Note that we use $u_e(x)$ to non-dimensionalize τ_w in the case when the edge velocity is not constant. A special case is that when $c_f = 0$, i.e. when $f''(0) = 0$. This occurs at $\beta = -0.19884$ and corresponds to a boundary layer that is on the *verge of separation*, which we will discuss in the next section.

A comparison of the boundary-layer profiles for different β is shown in Figure 6.26. Note that a favorable pressure gradient, $m > 0$, which corresponds to $\beta > 0$, has a boundary-layer *thinning* effect compared to the Blasius solution, which is $\beta = 0$. The boundary-layer thickness is visibly smaller as the velocity profiles approach u_e at lower η values with increasing β. These profiles also have a higher skin friction as a result. On the other hand, an adverse pressure gradient, $m < 0$ and $\beta < 0$, has a boundary-layer *thickening* effect compared to the Blasius solution.

Example 6.7 Solution of the Falkner–Skan equation
Just like the Blasius equation, the Falkner–Skan equation is nonlinear and does not have a general analytical solution. We can obtain a numerical one in a

similar way as in Example 6.3, by integrating a three-component state vector from $\eta = 0$ to some terminal value. The state and derivative are now

$$\mathbf{F} = \begin{bmatrix} f \\ f' \\ f'' \end{bmatrix} \quad \Rightarrow \quad \mathbf{F}' = \begin{bmatrix} f' \\ f'' \\ f''' \end{bmatrix} = \begin{bmatrix} f' \\ f'' \\ -ff'' - \beta(1 - f'^2) \end{bmatrix}. \tag{6.106}$$

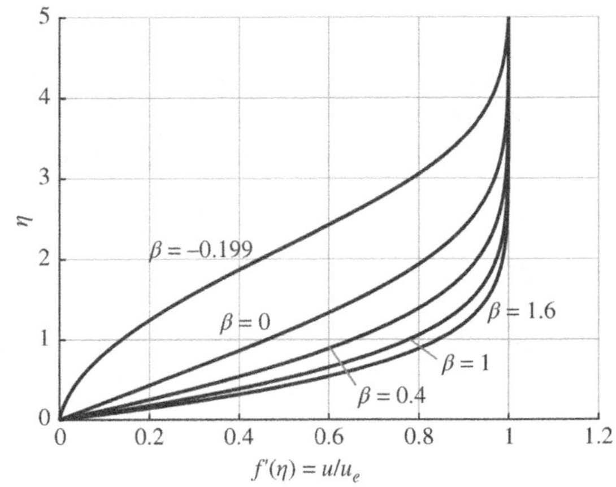

Figure 6.26 Velocity profiles for Falkner–Skan flows with various β (wedge/corner angles). $\beta = 0$ gives the Blasius flat plate profile, and $\beta = 1$ is stagnation point flow. $\beta > 0$ corresponds to a favorable pressure gradient, while $\beta < 0$ corresponds to an adverse pressure gradient.

The code in Listing 6.6 uses MATLAB's `ode45` function to solve the Falkner–Skan equation for a particular β. It determines the appropriate initial value for f'' by using the `fzero` function to determine which value of f'' results in $f' = 1$ for a large η. The auxiliary functions `fz` and `fs` are given in Listing 6.7. This code was used to make Figure 6.26.

Code notes:

- $\eta = 5$ serves as the "large" terminal value of η. This can be increased for more accuracy in determining $f''(0)$.
- β is an extra input into the functions; an alternative approach is to make `beta` a global variable.
- `fz` computes $f' - 1$ at a large η, for a particular choice of $f''(0)$, by integrating the Falkner–Skan equation.

Listing 6.6 Code snippet for solving the Falkner–Skan equation.

```
bv = [0, 0.4, 1, 1.6, -0.199]; % betas
fpp0v = [0.5, 0.8, 1.2, 1.6, 0.05]; % f''(0) guesses
figure(1); clf;

for i = 1:length(bv),
  beta = bv(i)
  fpp0 = fzero(@(z) fz(z,beta),fpp0v(i));
  F0 = [0 0 fpp0]      % initial condition
  [t,F]=ode45(@(t,F) fs(t,F,beta),[0 5],F0);
  plot(F(:,2), t,'k', 'linewidth', 2); hold on
end
```

Listing 6.7 Falkner–Skan code functions.

```
%-----------------------------------------
function z = fz(fpp, beta)
F0 = [0 0 fpp];
[t,F] = ode45(@(t,F) fs(t,F,beta),[0 5], F0);
z = F(end,2)-1;

%---------------------------
function Fp = fs(t,F, beta)
Fp = F;
Fp(1)=F(2);
Fp(2)=F(3);
Fp(3)=-F(1)*F(3)-beta*(1-F(2)^2);
```

Example 6.8 Falkner–Skan derivation
In this example, we fill in some of the details of the Falkner–Skan flow derivation. Consider a symmetric wedge at zero angle of attack, as shown in Figure 6.27.

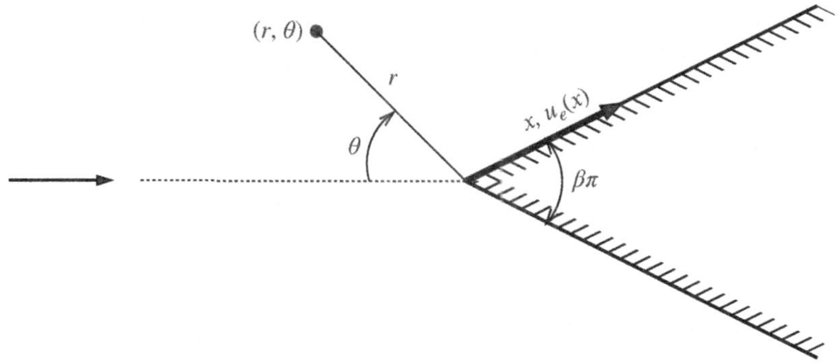

Figure 6.27 Symmetric wedge at zero angle of attack.

In order to estimate the behavior of the boundary layer on the wedge, the edge velocity, $u_e(x)$, is needed. Assuming the boundary layer is thin, we can approximate the edge velocity from the potential flow velocity on the body surface. The potential for wedge flow has the form‡

‡ The form that the velocity potential should take is not obvious, but the next steps show that it allows for flow tangency on the wedge.

$$\phi(r,\theta) = -\frac{K}{m+1} r^{m+1} \cos\left[(m+1)\theta\right], \tag{6.107}$$

where K, $m > 0$ are positive constants. By applying the flow-tangency condition on the wedge, we can relate the wedge angle, $\beta\pi$, to m. The velocity components in cylindrical coordinates are

$$v_r = \frac{\partial \phi}{\partial r} = -Kr^m \cos\left[(m+1)\theta\right], \tag{6.108}$$

$$v_\theta = \frac{1}{r}\frac{\partial \phi}{\partial \theta} = Kr^m \sin\left[(m+1)\theta\right]. \tag{6.109}$$

Flow tangency on the wedge means that $v_\theta = 0$ when $\theta = \pi - \frac{1}{2}\beta\pi$,§

§ Note that $\theta = 0$ corresponds to points upstream of the wedge.

$$v_\theta = Kr^m \sin\left[(m+1)(\pi - \frac{1}{2}\beta\pi)\right] = 0. \tag{6.110}$$

To satisfy this equation, we need the sin term to be zero,

$$\sin\left[(m+1)(\pi - \frac{1}{2}\beta\pi)\right] = 0 \tag{6.111}$$

$$\Rightarrow \quad (m+1)\left(\pi - \frac{1}{2}\beta\pi\right) = \pi$$

$$(m+1)(2-\beta) = 2$$

$$\beta = \frac{2m}{m+1}. \tag{6.112}$$

Next, we show that the edge velocity is $u_e(x) = Kx^m$. The edge velocity is $v_r(r)$ at $\theta = \pi - \frac{1}{2}\beta\pi$. Using Equation 6.108,

$$u_e(x) = v_r(r) = -Kx^m \cos(\pi) = Kx^m. \tag{6.113}$$

Recall that the Falkner–Skan similarity solution is obtained in terms of a non-dimensional η by defining

$$\eta = \frac{y}{\delta(x)}, \quad \delta(x) = \sqrt{\frac{2}{m+1}}\sqrt{\frac{\nu x}{u_e(x)}}. \tag{6.114}$$

These definitions, together with the stream function form, Equation 6.100, can be substituted into the x-momentum equation,

$$u\frac{\partial u}{\partial x} + v\frac{\partial u}{\partial y} = -\frac{1}{\rho}\frac{dp_e}{dx} + \nu\frac{\partial^2 u}{\partial y^2}, \tag{6.115}$$

to obtain the Falkner–Skan equation

$$f''' + ff'' + \beta\left(1 - f'^2\right) = 0. \tag{6.116}$$

The details of this substitution are left to Problem 6.

Example 6.9 Comparison of Falkner–Skan to the finite-difference solver
We return to the boundary-layer solver developed in Section 6.3 and this time verify it for flow with a nonzero-pressure gradient by using the Falkner–Skan family. We choose stagnation-point flow, $m = 1$, which corresponds to a favorable pressure gradient through an accelerating edge velocity, $u_e = Kx$, and which should show a constant-thickness boundary layer. The steps in the comparison are the same as in the Blasius case, Example 6.4: we initialize the velocity profile on the left side of the domain, $u_0(y)$ to the Falkner–Skan solution corresponding to $x = L$, away from the boundary-layer start at $x = 0$, and then run the solver for another length L. Listing 6.8 shows the driver code that performs the setup and calls the boundary-layer solver.

Code notes:

- A Falkner–Skan solver similar to the one used to generate Figure 6.26 was used to obtain the initial condition, $u_0(y)$.
- The boundary-layer solver computes the solution from $x = L$ to $x = 2L$.
- The scaling of the non-dimensional solution uses the Falkner–Skan definition of $\delta(x)$, Equation 6.99.
- The entire code includes post-processing routines that are not shown.

Listing 6.8 Driver code for comparing the boundary-layer solver with the Falkner–Skan solution.

```
function blfd_fs
% compares FD BL solver to Falkner-Skan

m = 1;                        % FS m value (stag point flow)
K = 1;                        % ue = K*x^m
nu = 0.001;                   % viscosity
L = 1;                        % reference length
[eta,F] = falknerskan(m);% non-dim solution for initialization
delta = sqrt(2/(m+1)*nu*L/(K*L^m)); % delta at L
yf = eta*delta;               % FS y values
uf = F(:,2)*K*L^m;            % FS x-velocity at L
y = linspace(0,max(yf),31); % equal spacing in y
u0 = interp1(yf,uf,y, 'linear', K*L^m); % interpolate u from uf
x = linspace(L,2*L,501); % x stations to simulate
ue = K*x.^m;                  % edge velocity at x
[U,V] = blfd(x, ue, y, u0, nu); % simulate another length L
```

Running the solver produces the streamlines and velocity profiles shown in Figure 6.28. Qualitatively, the boundary-layer thickness appears to be constant, which is expected for stagnation-point flow. Note that u_e increases with x, since for $m = 1$, $u_e = Kx$.

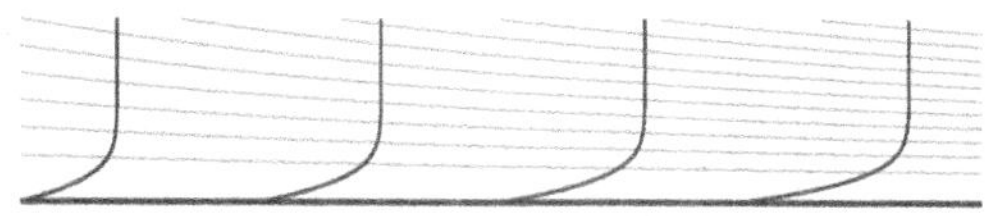

Figure 6.28 Streamlines and velocity profiles for the boundary-layer solver applied to a stagnation-point edge velocity.

Figure 6.29 compares the velocity profile at the right side of the domain, $x = 2L$, from the solver to the Falkner–Skan solution. We see very good agreement, visually identical, between the results. Figure 6.29 also compares the skin-friction coefficient computed from the solver to the Falkner–Skan result in Equation 6.105.¶ We again observe excellent agreement.

¶ This is *local* c_f calculation, non-dimensionalized by $u_e(x)$. Hence, even though from the velocity profiles we see that $\frac{\partial u}{\partial y}$ increases, the skin friction coefficient decreases because $u_e = Kx$, and τ_w is non-dimensionalized by u_e^2.

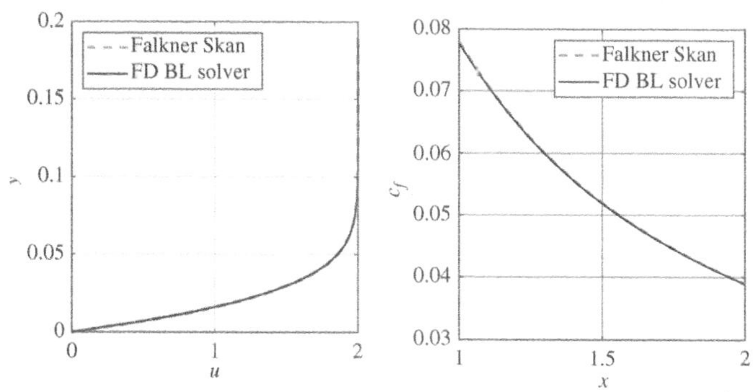

Figure 6.29 Velocity profile at $x = 2L$ and skin friction distribution for the boundary-layer solver with the stagnation-point initial profile.

6.5.2 Flow Separation

So far we have considered *attached* flows, in which the boundary layer is thin and fluid elements move in the positive x direction in the boundary layer. Such flows are consistent with the boundary-layer concept, that most of the flow is inviscid, and that viscous effects are confined to a small region near the wall. However, we know from experience that this description is not always accurate, especially for bluff bodies or airfoils at high angles of attack. For these flows, large, often unsteady, wakes occur as the inviscid flow cannot follow the shape of the object. Such situations constitute separated flow, and in this section, we define it and discuss the flow characteristics that contribute to its onset.

A flow *separates* when the local skin friction, c_f, becomes negative, as illustrated in Figure 6.30. Recall from Equation 6.8 that the skin-friction coefficient is related to the wall shear stress by

$$c_f = \frac{\tau_w}{\frac{1}{2}\rho u_e^2} = \frac{\mu}{\frac{1}{2}\rho u_e^2} \left.\frac{\partial u}{\partial y}\right|_{y=0}, \tag{6.117}$$

where we have non-dimensionalized by the dynamic pressure based on the boundary-layer edge velocity, $u_e(x)$, for the general case when $u_e(x)$ is not constant.* Thus, at the separation point,

*The non-dimensionalization does not matter for determining separation, which is based on the sign of $\left.\frac{\partial u}{\partial y}\right|_{y=0}$.

$$c_f = 0 \quad \Rightarrow \quad \left.\frac{\partial u}{\partial y}\right|_{y=0} = 0, \tag{6.118}$$

which gives rise to the third velocity profile shown in Figure 6.30. Also, once separated,

$$c_f < 0 \quad \Rightarrow \quad \left.\frac{\partial u}{\partial y}\right|_{y=0} < 0. \tag{6.119}$$

This negative value of $\frac{\partial u}{\partial y}\big|_{y=0}$ means that u immediately above the wall is negative, which indicates reversed flow, as shown on the right in Figure 6.30.

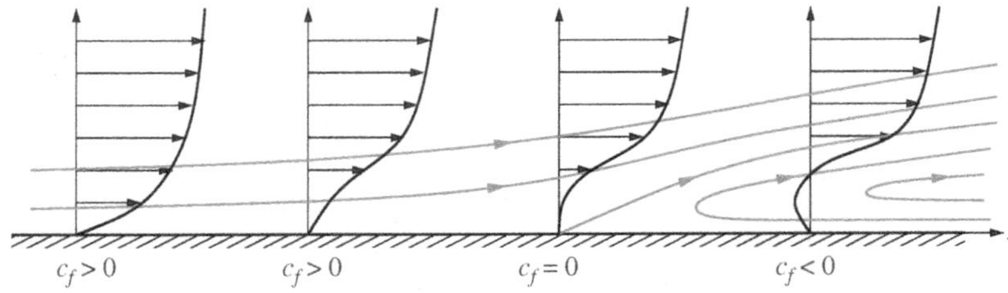

Figure 6.30 At the point of separation, the skin friction is zero. After separation, the flow reverses direction and the skin friction is negative. The gray lines are streamlines.

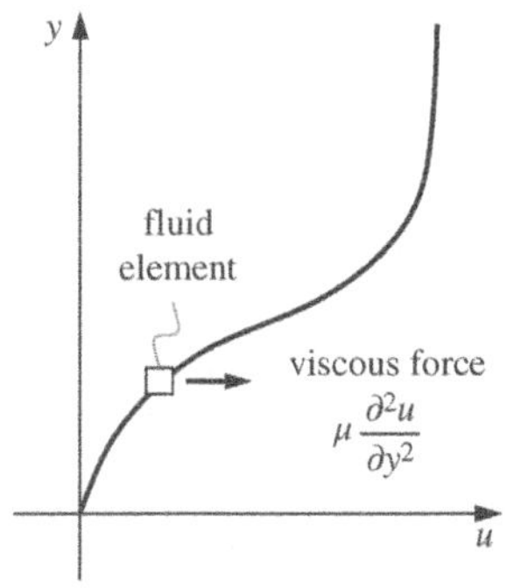

Figure 6.31 Viscous force on a fluid element in a boundary-layer velocity profile close to separation.

For a flat plate with $dp_e/dx = 0$, the Blasius solution, we obtained a self-similar velocity profile for which $\left.\frac{\partial u}{\partial y}\right|_{y=0}$ was always positive. Therefore, this flow will not separate, since the skin friction can never be zero. To make a flow separate, the fluid elements near the wall need to be decelerated to zero velocity, which means that an extra force, beyond the viscous force, is needed. Without body forces, the only possibility is a pressure force, which, to cause separation, must act to slow the fluid elements down. This is an adverse pressure gradient, as illustrated in Figure 6.22.

Hence, we conclude that for a boundary layer to separate, *an adverse pressure gradient must be present*. Since pressure is transmitted through a boundary layer, an adverse pressure gradient decelerates the entire velocity profile, and the slow-moving fluid close to the wall will generally be the first to reverse direction as a result.

For separation to occur, an adverse pressure gradient must not only be present, but it also must be sufficiently strong. As shown in Figure 6.31, for a boundary-layer velocity profile close to separation, fluid elements near the wall actually experience a viscous force that accelerates them in the $+x$ direction. This acceleration is due to the change in the concavity of the velocity profile, brought about by the adverse pressure gradient. Thus, to decelerate the fluid element to zero velocity, the adverse pressure gradient must be strong enough to overcome this viscous force. In addition, the adverse pressure gradient must also act on the boundary layer over a sufficiently-long distance/time. Fluid elements start with positive x momentum when the boundary layer begins, at a leading edge or stagnation point, and this momentum must be brought to zero for separation to occur. A certain distance is required for this deceleration to occur due to the adverse pressure gradient.†

† This is true for separation over smooth bodies. Sharp geometry changes, e.g. corners (Figure 6.35), can also cause separation, due to both a strong adverse pressure gradient and a rapid change in the velocity direction required to keep the boundary layer attached.

On an airfoil, adverse pressure gradients become stronger as the angle of the attack of the airfoil increases. This is illustrated in Figure 6.32, which shows pressure coefficient distributions for inviscid flow over a NACA 0012 airfoil at two angles of attack. Note that a downward slope in the c_p versus x curve, when c_p is increasing on the inverted vertical axis, corresponds to an adverse pressure gradient. Therefore, separation is more likely at higher angles of attack. This is consistent with the discussion of stall in Chapter 5, for example as indicated in Figure 5.31, which was defined as the loss of lift when the flow separates at high angles of attack.

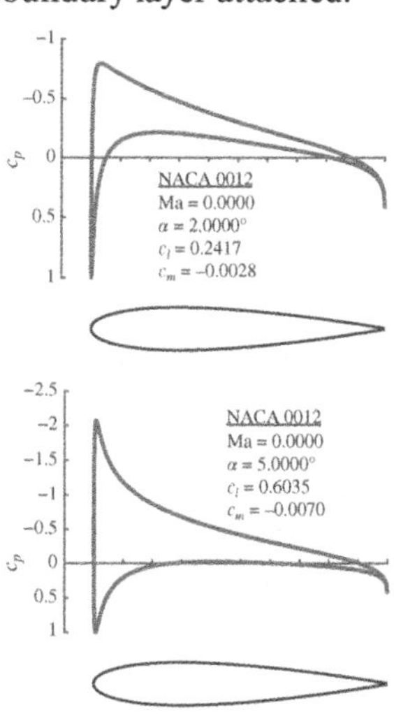

Figure 6.32 Adverse pressure gradients, prominent on the upper surface of the airfoil, become stronger with increasing angle of attack. These c_p plots are produced using mfoil, in inviscid mode.

When a flow separates, the pressure distributions generally do not match the inviscid results because the viscous effects are no longer confined to a thin boundary-layer region near the surface. Figure 6.33 shows the effect of separation for a NACA 0012 airfoil, through pressure and skin-friction coefficient distributions at two angles of attack in viscous flow at Reynolds number, Re = 20,000. For this Reynolds number, separation occurs on the top surface at both angles of attack. The point of separation can be identified by inspecting the c_f plot and noting where c_f first becomes negative. This occurs at $x/c \approx 0.5$ for $\alpha = 2°$ and at $x/c \approx 0.2$

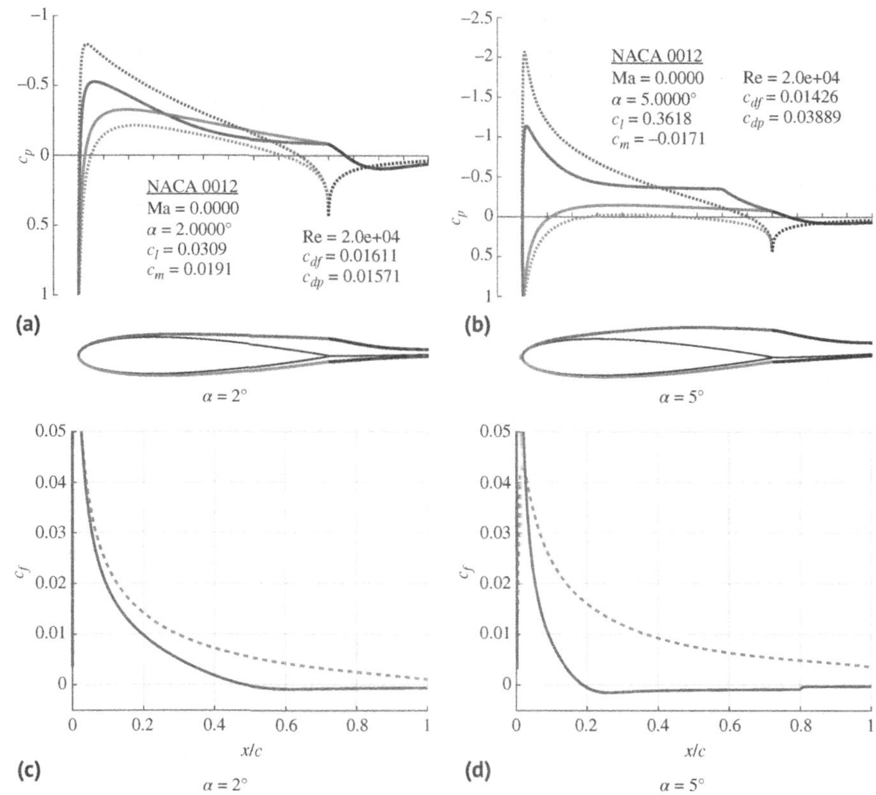

Figure 6.33 Viscous results for the NACA 0012 airfoil at Re = 20,000, produced using the mfoil program. Separation occurs in both cases on the upper surface, earlier at the higher angle of attack. The skin-friction plot is useful for identifying the point of separation. Note that the two curves on each plot correspond to the upper/lower surfaces of the airfoil.

for $\alpha = 5°$. Note that at the higher angle of attack, the larger adverse pressure gradient results in earlier separation.

The c_p distributions in Figure 6.33 show both the viscous and inviscid results, with the latter as dotted lines. The effect of a thick boundary layer and upper-surface separation is quite pronounced on the pressure distribution: the viscous results are not very close to the inviscid ones. Notably, it is not just the region near separation that is different, but the entire c_p distribution changes due to viscous effects. Boundary-layer thickening and separation create a large displacement thickness, which changes the shape of the effective body seen by the flow. The inviscid flow over the upper surface experiences less flow turning, and a suction pressure that is correspondingly not as low as in the inviscid case. The resulting viscous lift coefficients, as measured by the area between the upper and lower c_p curves, are lower than their inviscid counterparts.‡

‡ In addition, the trailing-edge behavior is different: in the viscous case, the effective body is not closed, more so than the inviscid case with a small gap, due to the displacement thickness propagating into the wake, so that the trailing edge is not a stagnation point.

6.5.3 Separation Control

On blunt bodies, such a cylinder, separation can occur very quickly because of high adverse pressure gradients. For example, note the strong downward slopes in the pressure coefficient distribution for a cylinder shown in Figure 4.26. On smooth shapes, the point of separation is highly dependent on the Reynolds number, Re. In particular, *for laminar flows, separation will occur sooner at higher Reynolds numbers.* For low Reynolds numbers, the larger size and faster growth of the boundary-layer displacement thickness attenuates geometry changes that produce adverse pressure gradients. As the Reynolds number increases, the displacement thickness of the boundary layer decreases and grows more slowly, so that the surface pressure distribution approaches that of the inviscid solution – with strong adverse pressure gradients present. This is true for *laminar* flows. For turbulent flows, which are likely at high Reynolds numbers, separation can actually be delayed, as we will soon see.

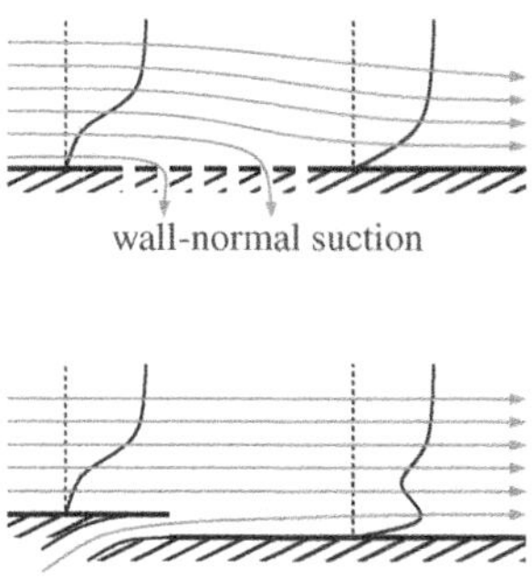

Figure 6.34 Separation control strategies.

Separation is not desirable for aerodynamic performance, due to reduced lift and increased drag. Therefore, various measures exist to control or delay separation. When the geometry is smooth, two common strategies for delaying separation, shown in Figure 6.34, are

1. Wall- normal suction, in which slow-moving fluid is removed from near the wall, and faster-moving fluid from above the wall takes its place.
2. Tangential blowing, in which kinetic energy is added to the boundary layer, accelerating the fluid near the wall and preventing it from reversing direction.

There are limits to these control measures. For example, in the presence of sharp corners, the geometry change and very large adverse pressure gradient just past the corner usually set the separation point there for any Reynolds number, as shown in Figure 6.35.

Figure 6.35 Flow is likely to separate immediately at sharp corners.

Separation control can be done passively by taking advantage of pressure differences between the lower, "pressure," side and the upper, "suction," side of an airfoil. In particular, gaps between airfoil elements, such as flaps and slats, can be used in an advantageous manner to re-energize the boundary layer on the upper surface of an airfoil, where adverse pressure gradients are strongest. Figure 6.36 shows the pressure distribution from a computational fluid dynamics simulation of flow over a three-element airfoil, one with a slat, a main element, and a flap. In this high-lift configuration, the pressure is relatively high underneath the airfoil, and the gaps between the elements allow air to bleed to the lower-pressure suction side. The placement of the elements is such that the stream of bled air blows tangentially along the upper surface, thereby re-energizing the boundary layer. As a result, this airfoil can achieve a higher lift coefficient than if the gaps were closed. Note, the effect on the lift of the loss of pressure underneath the airfoil due to the bleed is small in comparison to the benefit of keeping the upper surface flow attached.

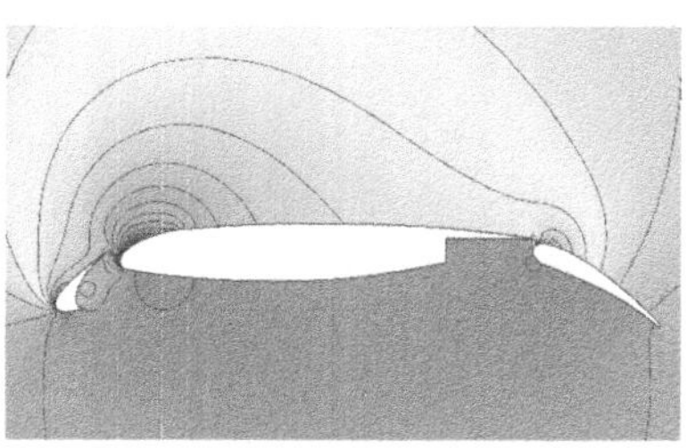

Figure 6.36 Pressure distribution for a multi-element airfoil.

6.6 The Integral Boundary-Layer Equation

The Blasius and Falkner–Skan solutions are special cases when the edge velocity takes on a simple form (constant or a power law). How do we solve the boundary-layer equations for arbitrary edge velocities and pressure gradients? One approach is to solve the equations numerically, using, for example, the finite difference method presented in Section 6.3. However, in this section, we will focus on a less accurate, but simpler, approach that involves first integrating the boundary-layer equations through the boundary-layer thickness (Drela 2014).

6.6.1 Derivation

The motivation for this method is that of engineering interest from a boundary-layer solution are typically only a few quantities: $c_f(x)$, $\delta^*(x)$, and $\theta(x)$, not the entire velocity profile $u(y)$. To "wash out" the details of the $u(y)$ profile, we will integrate the boundary-layer equations in y, starting from the wall and ending at the edge of the boundary layer.* Integrating out one variable converts a partial differential equation in x, y to an ordinary differential equation in x.

Before integrating in y, we first combine the continuity and x-momentum equations in order to obtain a single equation that will produce the desired engineering quantities after integration. To do this, we multiply the continuity equation by $(u_e - u)$, and subtract the momentum equation,

$$(u_e - u)\left[\frac{\partial u}{\partial x} + \frac{\partial v}{\partial y} = 0\right]$$
$$-\left[u\frac{\partial u}{\partial x} + v\frac{\partial u}{\partial y} = u_e\frac{du_e}{dx} + \nu\frac{\partial^2 u}{\partial y^2}\right].$$

The result is†

$$(u_e - u)\frac{\partial u}{\partial x} - u\frac{\partial u}{\partial x} + (u_e - u)\frac{\partial v}{\partial y} - v\frac{\partial u}{\partial y} = -u_e\frac{du_e}{dx} - \nu\frac{\partial^2 u}{\partial y^2}. \tag{6.120}$$

Combining terms,

$$\frac{\partial}{\partial x}\left[(u_e - u)u\right] + \frac{\partial}{\partial y}\left[(u_e - u)v\right] = (-u_e + u)\frac{du_e}{dx} - \nu\frac{\partial^2 u}{\partial y^2}, \tag{6.121}$$

where we used the product rule for the x and y derivatives,

$$\frac{\partial}{\partial x}\left[(u_e - u)u\right] = (u_e - u)\frac{\partial u}{\partial x} + u\frac{du_e}{dx} - u\frac{\partial u}{\partial x}, \tag{6.122}$$

$$\frac{\partial}{\partial y}\left[(u_e - u)v\right] = (u_e - u)\frac{\partial v}{\partial y} - v\frac{\partial u}{\partial y}. \tag{6.123}$$

We now integrate Equation 6.121 term by term through the boundary layer, from $y = 0$ to $y = y_1$,‡

$$\frac{d}{dx}\int_0^{y_1}(u_e - u)u\,dy + \int_0^{y_1}\frac{\partial}{\partial y}\left[(u_e - u)v\right]dy = -\int_0^{y_1}(u_e - u)\frac{du_e}{dx}dy - \nu\int_0^{y_1}\frac{\partial^2 u}{\partial y^2}dy. \tag{6.124}$$

These integrals are evaluated as follows:

$$\int_0^{y_1}\frac{\partial}{\partial y}\left[(u_e - u)v\right]dy = (u_e - u)v\Big|_0^{y_1} = 0,$$

$$-\nu\int_0^{y_1}\frac{\partial^2 u}{\partial y^2}dy = -\nu\frac{\partial u}{\partial y}\Big|_0^{y_1} = -\left(0 - \frac{\tau_w}{\rho}\right) = \frac{\tau_w}{\rho},$$

$$\int_0^{y_1}(u_e - u)u\,dy = u_e^2\int_0^{y_1}\left(1 - \frac{u}{u_e}\right)\frac{u}{u_e}\,dy = u_e^2\theta,$$

$$\int_0^{y_1}(u_e - u)\frac{du_e}{dx}\,dy = u_e\frac{du_e}{dx}\int_0^{y_1}\left(1 - \frac{u}{u_e}\right)dy = u_e\frac{du_e}{dx}\delta^*.$$

Substituting these integral evaluations into Equation 6.124 gives the dimensional form of the *von Karman integral equation*

$$\frac{d}{dx}\left(u_e^2\theta\right) + u_e\frac{du_e}{dx}\delta^* = \frac{\tau_w}{\rho}. \tag{6.125}$$

* The edge of the boundary layer, denoted by $y = y_1$, does not need to be a precise location, because integrating a distance into the "inviscid" region, where $u = u_e$, will not change the results.

† In the momentum equation, Bernoulli's equation is used to write the pressure gradient in terms of u_e, as in Equation 6.30,

$$\frac{-1}{\rho}\frac{dp_e}{dx} = u_e\frac{du_e}{dx}.$$

‡ The x derivative on the first term commutes with the y integral.

This equation expresses a relationship between quantities that characterize the boundary-layer velocity profile in an integral sense. These are the wall shear stress, τ_w, the momentum thickness, θ, and the displacement thickness, δ^*. All of these quantities are of potential engineering relevance.

We can non-dimensionalize Equation 6.125 by expanding the first term using the product rule, dividing by u_e^2, and defining the *shape factor* as the ratio of displacement to momentum thickness,

$$H \equiv \frac{\delta^*}{\theta}. \tag{6.126}$$

The resulting non-dimensional form is

$$\frac{d\theta}{dx} + (2 + H)\frac{\theta}{u_e}\frac{du_e}{dx} = \frac{c_f}{2}. \tag{6.127}$$

This is one ordinary differential equation (ODE) for three unknowns, $\theta(x)$, $\delta^*(x)$, and $c_f(x)$. It is exact up to the approximations of the boundary-layer equations. However, as it contains three unknown distributions, it is not closed: we cannot solve it without additional information or assumptions. In the next sections, we consider two approximate solution approaches.

6.6.2 Assumed Profiles

One approach to solving Equation 6.127 is to assume a single-parameter velocity profile,

$$u(x, y) = u_e(x)h\left(\frac{y}{\delta(x)}\right), \tag{6.128}$$

where $h(\cdot)$ is some assumed function that governs the profile shape, and $\delta(x)$ is the single unknown parameter that is a measure of the vertical stretching of the profile, i.e. the boundary-layer thickness. This parameter does not have to equal any existing measure of boundary-layer thickness, such as δ^*, θ, δ_{99}. Instead, it is a parameter that is used to determine how these thicknesses scale with x, which means that the existing thickness measures will be directly proportional to $\delta(x)$.

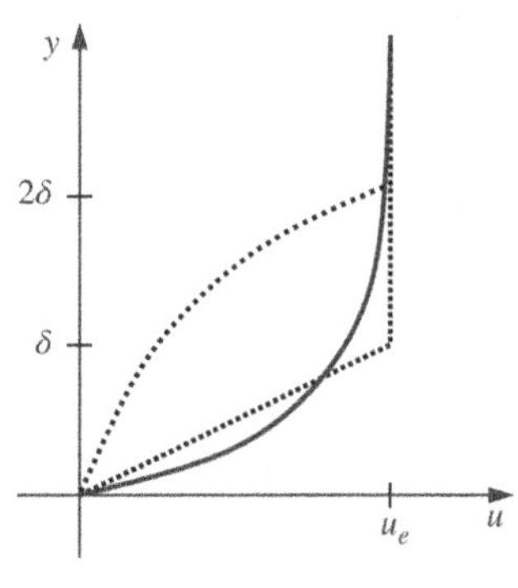

Figure 6.37 Various assumed profile shapes.

The function $h(\cdot)$ in Equation 6.128 determines the profile shape, and sample ones are shown in Figure 6.37. It affects the solution by prescribing the relationship between $\delta(x)$ and the quantities appearing in Equation 6.127. How do we choose it? This is where the approximate nature of the method comes in. Unless we know the exact velocity profile, we have to estimate its shape. This estimate may rely on some knowledge of the problem: for example, whether the flow is in a favorable or adverse pressure gradient, which will determine whether we opt for a filled-out profile with high velocities close to the wall, or a profile with an inflection point, resembling one close to separation. Fortunately, the assumed profile does not need to be exact to obtain decent results for the distribution of skin friction and displacement/momentum thickness.*

* A drawback of the method, however, is that the profile does not change with x, which is not realistic over a long stretch of x values in a changing pressure gradient.

Using the assumed profile in the definitions of $\theta(x)$, $\delta^*(x)$, and $c_f(x)$ yields expressions for these variables in terms of $\delta(x)$. The first two require integrating the profile, whereas c_f requires evaluating the derivative at the wall. Substituting these expressions into the Von Karman integral equation, Equation 6.127 yields an ODE for $\delta(x)$. The solution to the ODE can then be used in the expressions for $\theta(x)$, $\delta^*(x)$, and $c_f(x)$, as shown in the following example.

Example 6.10 Flat-plate solution using the method of assumed profiles
In this example, we will solve for $\theta(x)$, $\delta^*(x)$, and $c_f(x)$ on the flat plate, for which $u_e = U_\infty$ = constant, using the method of assumed profiles. This is not a practical application, since we have the Blasius solution for the flat plate, but the exercise serves to verify the method and to assess the impact of the assumed function $h(\cdot)$ on the accuracy of the results.

For this example, we assume that the velocity profile takes the following piecewise form,

$$u(x,y) = \begin{cases} U_\infty \sin\left(\dfrac{\pi y}{2\delta(x)}\right) & \text{for} \quad 0 \le y \le \delta(x) \\ U_\infty & \text{for} \quad y > \delta(x) \end{cases}, \tag{6.129}$$

where $\delta(x)$ is the single parameter that describes the velocity profile. Figure 6.38 illustrates this profile: the sine variation between $y = 0$ and $y = \delta$ has the qualitatively correct appearance of a boundary-layer profile.

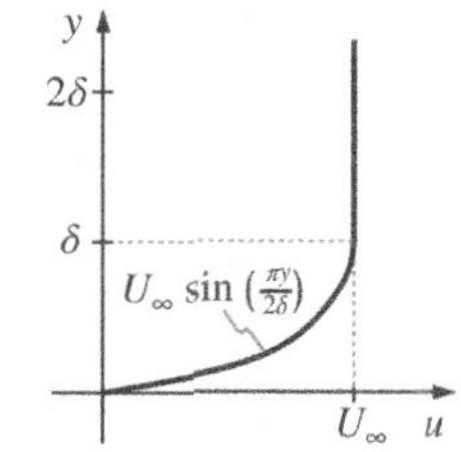

Figure 6.38 Piecewise-sine assumed velocity profile.

Using the assumed velocity profile, we can calculate δ^*, θ, H, and c_f as functions of $\delta(x)$. Substituting the assumed velocity profile into the respective definitions yields

$$\begin{aligned}
\delta^*(x) &= \int_0^{\delta(x)} \left(1 - \frac{u}{u_e}\right) dy = \int_0^{\delta(x)} \left(1 - \sin\left(\frac{\pi y}{2\delta(x)}\right)\right) dy \\
&= \left(1 - \frac{2}{\pi}\right) \delta(x) \approx .3634\delta(x),
\end{aligned} \tag{6.130}$$

$$\begin{aligned}
\theta(x) &= \int_0^{\delta(x)} \frac{u}{u_e}\left(1 - \frac{u}{u_e}\right) dy \\
&= \int_0^{\delta(x)} \sin\left(\frac{\pi y}{2\delta(x)}\right)\left(1 - \sin\left(\frac{\pi y}{2\delta(x)}\right)\right) dy \\
&= \left(\frac{2}{\pi} - \frac{1}{2}\right) \delta(x) \approx .1366\delta(x),
\end{aligned} \tag{6.131}$$

$$H(x) = \frac{\delta^*(x)}{\theta(x)} \approx 2.6598, \tag{6.132}$$

$$c_f(x) = \frac{\tau_w}{\frac{1}{2}\rho u_e^2} = \frac{2\nu \frac{\partial u}{\partial y}\big|_{y=0}}{U_\infty^2} = \frac{2\nu}{U_\infty^2} \frac{U_\infty \pi}{2\delta(x)} = \frac{\pi \nu}{U_\infty \delta(x)}. \tag{6.133}$$

In the integrals, the upper limit was taken as $y_1 = \delta(x)$, since the velocity is just uniform freestream for $y > \delta(x)$.†

We now substitute the above results into the integral momentum equation and derive an ordinary differential equation (ODE) for $\delta(x)$. The boundary-layer integral momentum equation is given by Equation 6.127, which in this case reduces to‡

$$\frac{d\theta}{dx} = \frac{c_f}{2}, \tag{6.134}$$

† There is no danger of using y_1 larger then δ in this case, since the integrands in δ^* and θ contain $(1 - u/u_e)$, which is zero when $u = u_e = U_\infty$.

‡ Since $u_e = U_\infty$ = constant, $\frac{du_e}{dx} = 0$.

Table 6.1 Comparison of the assumed sine profile to Blasius.

	Sine	Blasius	error
$\frac{\delta^*}{x}$	$\frac{1.7425}{\sqrt{\mathrm{Re}_x}}$	$\frac{1.7208}{\sqrt{\mathrm{Re}_x}}$	1.26%
$\frac{\theta}{x}$	$\frac{0.6551}{\sqrt{\mathrm{Re}_x}}$	$\frac{0.6640}{\sqrt{\mathrm{Re}_x}}$	1.34%
H	2.6598	2.5916	2.63%
c_f	$\frac{.6551}{\sqrt{\mathrm{Re}_x}}$	$\frac{.6640}{\sqrt{\mathrm{Re}_x}}$	1.34%

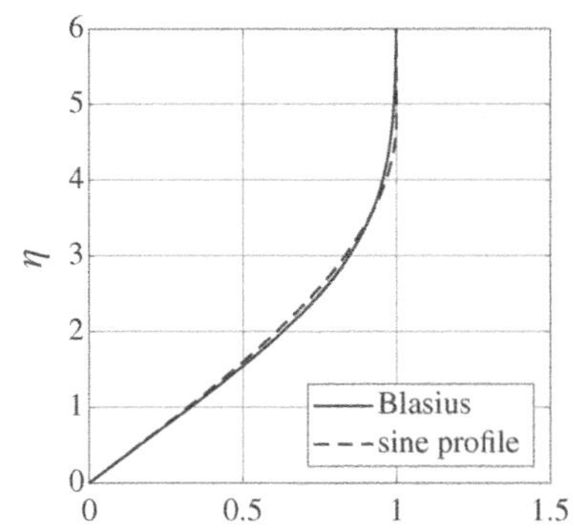

Figure 6.39 Comparison of the assumed sine profile in Example 6.10, Equation 6.129, to the Blasius solution. η is taken from the definition used in the Blasius solution, Equation 6.56.

Substituting the above results for θ and c_f gives the ordinary differential equation

$$\left(\frac{2}{\pi} - \frac{1}{2}\right)\frac{d\delta}{dx} = \frac{\pi\nu}{2U_\infty\delta(x)}. \tag{6.135}$$

We can solve this equation for $\delta(x)$ by assuming that the leading edge is at $x = 0$, so that $\delta(0) = 0$. Re-arranging the equation gives

$$\frac{\pi^2\nu}{U_\infty(4-\pi)}dx = \delta d\delta. \tag{6.136}$$

Integrating both sides then yields

$$\frac{\pi^2\nu x}{U_\infty(4-\pi)} + C = \frac{1}{2}\delta^2. \tag{6.137}$$

Since $\delta(x = 0) = 0$, $C = 0$. Therefore,

$$\delta = \pi\sqrt{\frac{2\nu x}{U_\infty(4-\pi)}}, \qquad \frac{\delta}{x} = \pi\sqrt{\frac{2}{\mathrm{Re}_x(4-\pi)}} \approx \frac{4.7953}{\sqrt{\mathrm{Re}_x}}, \tag{6.138}$$

where $\mathrm{Re}_x = U_\infty x/\nu$. Using this $\delta(x)$ in the expressions for $\delta^*, \theta, H,$ and c_f yields the results shown in Table 6.1. Each result is compared to the analytical Blasius solution, and the agreement is quite good: the percentage errors are all 1–3%. This agreement can be explained by the fact that the sine profile approximates the Blasius profile very well, as shown in Figure 6.39. Note, however, that in assuming the sine profile, we did not assume how the profile stretched in the y direction as x changed. This information came from the solution of the integral boundary-layer equation for $\delta(x)$.

Example 6.11 Finite-difference solver for a general $u_e(x.)$

Previous examples demonstrated the boundary-layer solver developed in Section 6.3 for a constant and power-law $u_e(x)$. In this section, we demonstrate the solver in a more practical setting, for flow over an airfoil. The case is a NACA 2410 airfoil at Re = 30,000, $\alpha = 2°$. To obtain the edge velocity, we run the mfoil code, which gives $u_e(x)$ starting at stagnation. We initialize the boundary-layer solver using the stagnation-point velocity profile from the Falkner–Skan family. Doing so requires the correct value of $K = \frac{du_e}{dx}$, which we estimate using a simple forward difference, and the kinematic viscosity, set from the Reynolds number and unit freestream speed and chord length. Note that in the case of stagnation-point flow, $\delta(x)$ is constant and equal to $\sqrt{\nu/K}$, per Equation 6.102. Listing 6.9 shows the code that runs this case.

Code notes:

- mfoil is run in viscous mode to allow for an accurate comparison of skin friction.
- mfoil uses calibrated correlations to close the IBL equations, making a c_f comparison possible.
- mfoil does not give velocity profiles.
- For stability at this Re, the boundary-layer solver requires more x stations.
- The entire code includes post-processing routines that are not shown.

Listing 6.9 Driver code for comparing the boundary-layer solver with mfoil.

```
function blfd_mfoil
% tests boundary layer solver using mfoil ue data

% set parameters, run mfoil, get u_e
Uinf = 1; Re = 30000; c = 1;
nu = Uinf*c/Re; % kinematic viscosity
m = mfoil(naca='2410',npanel=199); % construct airfoil
m.setoper(alpha=2, Re=30000); % set conditions
m.solve(); % obtain the solution
Is = m.vsol.Is{2}; % upper surface indices
```

```
uem = m.glob.U(4,Is); % ue values on the upper surface nodes
xm = m.foil.s(Is) - m.isol.sstag; % arc length on upper surface
K = (uem(2)-uem(1))/(xm(2)-xm(1)); % ue = Kx near stagnation

% set equal spacing in x, interpolate ue
x = linspace(min(xm), 0.2*max(xm), 8001); % domain to simulate
ue = interp1(xm, uem, x,'spline');

% run BL FD solver with FS stag point initialization
[eta,F] = falknerskan(1);% non-dim solution, m=1
delta = sqrt(nu/K);        % stag point thickness
yf = eta*delta;            % FS y values
uf = F(:,2)*ue(1);         % FS x-velocity at first point
y = linspace(0,4*max(yf),61); % equal spacing in y
u0 = interp1(yf,uf,y, 'linear', ue(1)); % interpolate u from uf
```

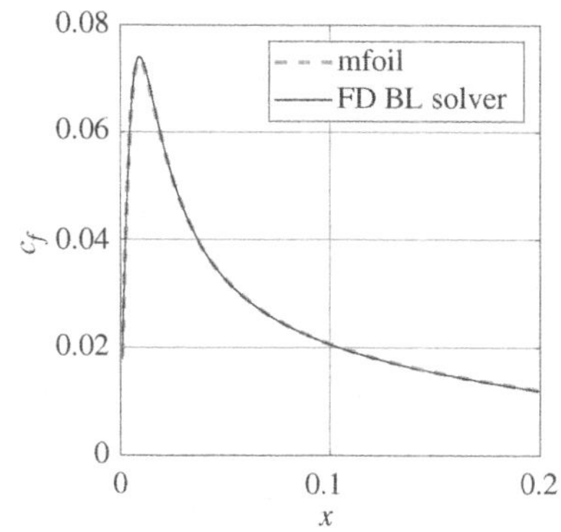

Figure 6.40 Skin-friction distribution comparison between the boundary-layer solver and mfoil.

Running the solver produces the streamlines and velocity profiles shown in Figure 6.41, which represents the first 20% of the airfoil upper surface. The boundary-layer thickness increases rapidly after a short length as the flow encounters an adverse pressure gradient after rounding the leading-edge of the airfoil. The boundary-layer solver assumes a rectangular domain, but the boundary layer on an airfoil curves to follow the geometry. This approximation will yield accurate results as long as the curvature is not overly high, i.e. when the boundary-layer thickness is much smaller than the radius of curvature.

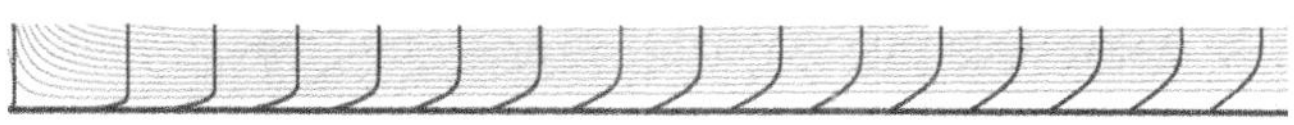

Figure 6.41 Streamlines and velocity profiles for the boundary-layer solver applied to $u_e(x)$ data from mfoil.

Figure 6.40 compares the skin-friction coefficient computed from the solver to the mfoil result. This c_f is non-dimensionalized by U_∞, not $u_e(x)$. We observe excellent agreement with the c_f computed from mfoil.

6.6.3 Thwaites's Method

In the method of assumed profiles presented in the last section, choosing a single velocity profile requires some guess work. A profile that is not representative of the actual boundary-layer flow may produce large errors. In addition, with a single profile, one cannot model situations in which the profile changes as the boundary layer develops. Such a capability is important in the prediction of separation, since the shape of the velocity profile changes significantly between attached and separated flow.

One solution method that allows for the boundary-layer profile to change in x is due to Thwaites (1949). The starting point for this method is the integrated boundary-layer equation. Instead of assuming a single velocity profile, however, Thwaites's method employs a family of profiles. An empirically based correlation is possible when the velocity profile is non-dimensionalized as follows:

- The x velocity is non-dimensionalized by $u_e(x)$, $U \equiv u/u_e$.
- The y coordinate is non-dimensionalized by the momentum thickness, $\theta(x)$, yielding $\eta \equiv y/\theta$.*

Thwaites conjectured that, for a wide range of boundary-layer flows, the important aspects of a velocity profile are the skin-friction coefficient, c_f, and the shape factor, $H = \delta^*/\theta$, and that both of these correlate only to the profile curvature at

* This is a different η definition from the ones used previously in self-similar flows such as Blasius and Falkner–Skan.

the wall. For the purpose of a correlation to experiment, this curvature is a non-dimensional measure calculated on the normalized velocity profile,†

$$\lambda \equiv -\left.\frac{d^2U}{d\eta^2}\right|_{\eta=0} = -\left.\frac{\theta^2}{u_e}\frac{\partial^2 u}{\partial y^2}\right|_{y=0} = -\left.\frac{\theta^2}{\nu u_e}\nu\frac{\partial^2 u}{\partial y^2}\right|_{y=0} = \frac{\theta^2}{\nu}\frac{du_e}{dx}. \tag{6.139}$$

† In the last step, the x-momentum boundary layer equation is used, evaluated at the wall, $y = 0$, where $u = v = 0$. The left-hand side of the equation is thus zero in steady flow, leaving

$$0 = u_e\frac{du_e}{dx} + \nu\left.\frac{\partial^2 u}{\partial y^2}\right|_{y=0},$$

from which the equality follows.

λ is called *Thwaites's parameter*. The shear stress used for the correlation is also measured on the normalized velocity profile, and it is defined as

$$l \equiv \left.\frac{dU}{d\eta}\right|_{\eta=0} = \left.\frac{\theta}{u_e}\frac{\partial u}{\partial y}\right|_{y=0} = \frac{u_e\theta}{\nu}\frac{\rho\nu\left.\frac{\partial u}{\partial y}\right|_{y=0}}{\rho u_e^2} = \frac{u_e\theta}{\nu}\frac{c_f}{2}. \tag{6.140}$$

Thwaites conjectured that both l and H depend only on λ; that is, $l = l(\lambda)$, $H = H(\lambda)$. The non-dimensional integral boundary-layer equation in Equation 6.127 then becomes

$$\frac{d\theta}{dx} = \frac{c_f}{2} - (2 + H)\frac{\theta}{u_e}\frac{du_e}{dx} = \frac{\nu}{u_e\theta}l(\lambda) - (H(\lambda) + 2)\frac{\nu}{u_e\theta}\lambda. \tag{6.141}$$

Multiplying both sides by $2u_e\theta/\nu$, we obtain

$$\frac{u_e}{\nu}\frac{d\theta^2}{dx} = \underbrace{2\,[l(\lambda) - (H(\lambda) + 2)\lambda]}_{\equiv F(\lambda)}, \tag{6.142}$$

Where $F(\lambda)$ is defined by the above combination of $l(\lambda)$ and $H(\lambda)$. Thwaites compared $F(\lambda)$ for several experimental data sets and found that $F(\lambda)$ for all of the data collapsed onto a nearly linear function,‡

‡ This step, in addition to the assumption that l and H depend only on λ, is where the approximation comes in for Thwaites's method.

$$F(\lambda) \approx 0.45 - 6\lambda. \tag{6.143}$$

Multiplying Equation 6.142 by u_e^5 and using the definition of Thwaites's parameter from Equation 6.139 give

$$\frac{u_e^6}{\nu}\frac{d\theta^2}{dx} = 0.45u_e^5 - 6u_e^5\frac{\theta^2}{\nu}\frac{du_e}{dx}. \tag{6.144}$$

Moving the last term to the left-hand side and recognizing the product rule yield

$$\frac{1}{\nu}\frac{d}{dx}\left(\theta^2 u_e^6\right) = 0.45u_e^5. \tag{6.145}$$

This equation can be integrated in x. If the integration begins at the leading edge, where $x = 0$ and $\theta = 0$, the result is§

§ In this equation, x' is an integration variable, whereas x is the location at which θ is desired. Thus, θ at x depends on the edge velocities in the range from 0 to x.

$$\theta(x)^2 = \frac{0.45\nu}{u_e(x)^6}\int_0^x u_e(x')^5 dx'. \tag{6.146}$$

This expression lets us calculate the boundary-layer momentum thickness as a function of x. Once $\theta(x)$ is known, Thwaites's parameter is found from Equation 6.139, $\lambda(x) = \frac{\theta(x)^2}{\nu}\frac{du_e}{dx}$. The non-dimensional skin friction $l(\lambda(x))$ and the shape factor $H(\lambda(x))$ are then calculated from $\lambda(x)$ using curve-fit relations of the experimental data. Displacement thickness can then be calculated from $\delta^*(x) = H(x)\theta(x)$, and the skin-friction coefficient from $l(x)$, using Equation 6.140, as $c_f(x) = 2\nu l(x)/(u_e(x)\theta(x))$.

The curve-fit relations for $l(\lambda)$ and $H(\lambda)$ are given as piecewise functions,¶

¶ Note that H and l are not defined outside of these ranges.

$$\begin{aligned}
\text{For } -0.1 \le \lambda \le 0: \quad & l(\lambda) = 0.22 + 1.402\lambda + \frac{0.018\lambda}{0.107 + \lambda} \\
& H(\lambda) = 2.088 + \frac{0.0731}{0.14 + \lambda} \\
\text{For } 0 < \lambda \le 0.1: \quad & l(\lambda) = 0.22 + 1.57\lambda - 1.8\lambda^2 \\
& H(\lambda) = 2.61 - 3.75\lambda + 5.24\lambda^2
\end{aligned} \tag{6.147}$$

Example 6.12 Thwaites's method for a flat plate

Thwaites's method is used to predict boundary-layer properties given a known inviscid solution, in the form of an edge velocity, $u_e(x)$. A simple check of the method is for the case of a constant edge velocity, $u_e(x) = U_\infty$. We already know the solution to this case, the Blasius solution, and hence, this is a verification exercise rather than a prediction.

For a constant u_e, Thwaites's parameter is

$$\lambda = \frac{\theta^2}{\nu}\frac{du_e}{dx} = 0. \tag{6.148}$$

The squared momentum thickness is obtained from Equation 6.146,

$$\theta(x)^2 = \frac{0.45\nu}{u_e(x)^6}\int_0^x u_e(x')^5 dx' = \frac{0.45\nu}{U_\infty^6}U_\infty^5 x = \frac{0.45\nu x}{U_\infty}. \tag{6.149}$$

Taking the square root,

$$\theta = \sqrt{\frac{0.45\nu x}{U_\infty}} = \sqrt{0.45}\sqrt{\frac{\nu}{U_\infty x}}x = 0.671\mathrm{Re}_x^{-0.5}x. \tag{6.150}$$

Compared to the Blasius solution, the dependence on Re_x and x is the same, and the coefficient 0.671 is only 1% off from the expected value of 0.664. We evaluate the shape factor from the curve fit in Equation 6.147 and obtain $H = 2.61$, so the displacement thickness is

$$\delta^* = H\theta = 1.751\mathrm{Re}_x^{-1/2}x. \tag{6.151}$$

This expression is also consistent with the Blasius solution, and the coefficient 1.751 is very close to the Blasius-solution value of 1.721. The normalized shear stress follows from Equation 6.147, $l = 0.22$, so that the skin-friction coefficient is

$$c_f = \frac{2\nu l}{u_e\theta} = \frac{0.44\nu}{U_\infty 0.671\mathrm{Re}_x^{-1/2}x} = \frac{0.656}{\mathrm{Re}_x^{1/2}}. \tag{6.152}$$

Compared to Blasius, the dependence of c_f on $\mathrm{Re}_x^{-1/2}$ is the same, and the factor 0.656 is about 1% off from the Blasius value of 0.664.||

|| This analysis shows that Thwaites's method reproduces the Blasius solution quite well. However, the power of Thwaites's method lies in its ability to predict boundary-layer properties for non-constant edge velocities. Problem 7 addresses such a case.

6.7 Turbulent Boundary Layers

The boundary-layer solutions considered so far have been steady state and stable. Such *laminar* flows exist and can be generated experimentally, as long as the Reynolds number is not overly high. When the Reynolds number based on some measure of boundary-layer thickness exceeds ~1000, the flow is likely to remain neither steady nor stable. Whereas in a stable flow, small perturbations are damped and the system returns to its equilibrium, in an unstable flow, perturbations are amplified. The unstable flow that results generally leads to *turbulence*, which is flow that exhibits random fluctuations in space and time. This section discusses both stability and turbulence, in the context of boundary-layer flows.*

*Turbulent flow can occur outside of boundary layers, including in pipes, wakes behind bluff bodies, and on a large scale in atmospheric flows, with the latter being the familiar case in aviation: an aircraft encountering "turbulence" is flying through an atmospheric state of turbulent flow.

6.7.1 Instability and Transition

In a stable flow, small perturbations are damped and the system returns to its equilibrium. On the other hand, in an unstable flow, perturbations are amplified. Boundary layers can be stable or unstable, depending on several parameters. To determine stability, we can perform a *perturbation analysis*. The basic steps of such an analysis are:

1. Linearize the Navier–Stokes equations about a steady-state solution.
2. Assume sinusoidal disturbances in space and time for the flow state, velocity, and pressure.
3. Substitute the perturbations into the linearized equations.
4. Solve an eigenvalue problem for the complex temporal or spatial frequencies that govern the growth/decay of disturbances.

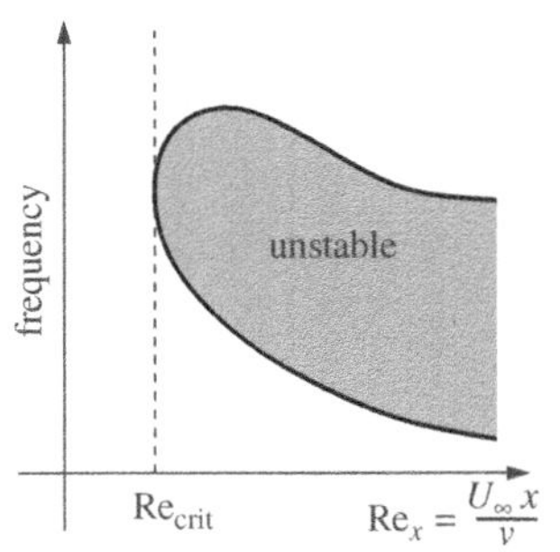

Figure 6.42 Qualitative result from stability analysis of a boundary layer. Frequency refers to the spatial or temporal frequency of the assumed perturbations.

The equation that arises in the last step of this analysis is called the *Orr-Sommerfeld* equation (White 1974). The result of solving this equation is a "thumb" plot of the form qualitatively illustrated in Figure 6.42. Note that for Re_x above some number, $\mathrm{Re}_{\mathrm{crit}}$, the boundary layer will be unstable to perturbations at certain frequencies.

When a boundary layer is unstable, even the smallest disturbance is amplified. Such a disturbance can come from surface roughness, small vibrations of the body, acoustic noise, or variations in the freestream properties. Since these sources introduce disturbances at a broad spectrum of frequencies, it is generally the case that *if a boundary layer can go unstable it will.* $\mathrm{Re}_{\mathrm{crit}}$ is known as the *critical Reynolds number*, above which the boundary layer becomes unstable.

Recall that $\mathrm{Re}_x = U_\infty x/\nu$, which means that a boundary layer will tend to become more unstable with increasing distance from the leading edge. In particular, the boundary layer becomes unstable when

$$x > x_{\mathrm{crit}} = \frac{\nu \mathrm{Re}_{\mathrm{crit}}}{U_\infty}. \tag{6.153}$$

The region over which instabilities grow is called the *transition region*. The word *transition* is used to denote the change in the flow character from stable/orderly *laminar* flow, to random/chaotic *turbulent* flow. The extent of the transition region depends on the Reynolds number and the amplitude of the perturbations present. In our analysis, we will often assume that transition from laminar to turbulent flow occurs instantaneously.*

*This is a good approximation at high Reynolds numbers. At low or moderate Reynolds numbers, the distinction between the location of onset of instabilities, x_{crit}, and the location of the start of turbulent flow, x_{tr}, will be more pronounced. The latter is generally considered the "transition location," although turbulent features, such as chaotic behavior and turbulent spots, do appear earlier (White 1974).

Figure 6.43 illustrates the qualitative behavior of a boundary layer in transition from laminar to turbulent flow. During transition, instabilities grow until the perturbations to the flow are no longer small. At this point, the linear analysis breaks down, regions of *spanwise vorticity* form and break down, and areas of turbulence, which are random spatial and temporal fluctuations, appear and coalesce into fully turbulent flow. The development of spanwise vorticity, which is three-dimensional, is not predicted by the linearized two-dimensional theory and associated *Tollmien–Schlichting waves*, which are the unstable traveling waves that result from the stability analysis.†

† Nonlinear analysis is necessary to identify how the three-dimensional perturbations grow, but qualitatively, their development accelerates the growth of the instability.

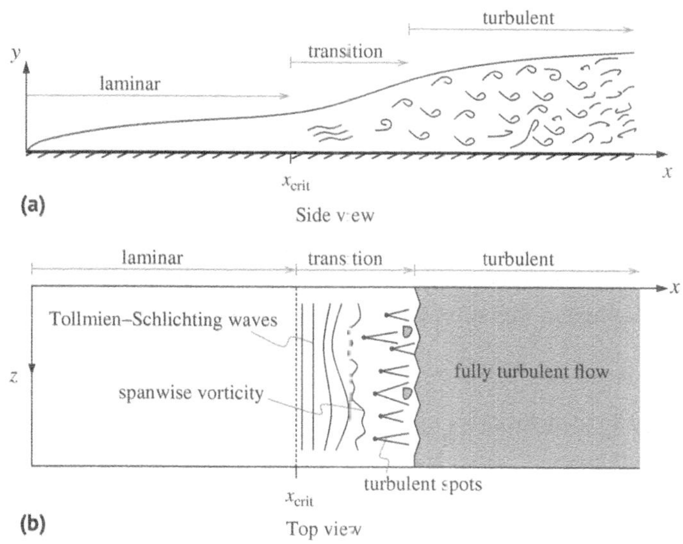

Figure 6.43 Side and top views of transition from laminar to turbulent flow.

After the development of spanwise vorticity, the final process before turbulence is the appearance of *turbulent spots*, which are intense local regions of turbulent flow, set off seemingly randomly by the state of the nonlinear breakdown and external factors, such as surface imperfections. These spots are intermittent, in that they appear and disappear in time. Downstream, the frequency of their appearance increases both in space and time, and eventually, they grow and coalesce into fully turbulent flow.

The transition process described here is called *natural transition*. It is a gradual process starting with the growth of unstable Tollmien–Schlichting waves that eventually break down and develop turbulent spots, which then lead to the turbulent flow. Faster transition can be observed if the flow is subject to larger disturbances, such as from rough walls, a noisy freestream, vibrations, or acoustic waves, as illustrated in Figure 6.44 schematically for flow over an airfoil. These disturbances can seed waves before the linearized stability analysis predicts their onset, and hence, transition will occur at a lower Reynolds number. The accelerated mechanism is called *bypass transition*, and if it is done intentionally with a wire or roughness strip, the boundary layer is said to be *forced* or *tripped* (Drela 2014). Finally, *separated-flow transition* occurs when a laminar flow separates, which is a large disturbance that is often accompanied by unsteady behavior. The disturbance accelerates the transition to turbulence, and if the separation angle is not excessive, the flow may reattach, leading to a laminar separation bubble (Drela 1989), illustrated in Figure 6.45.

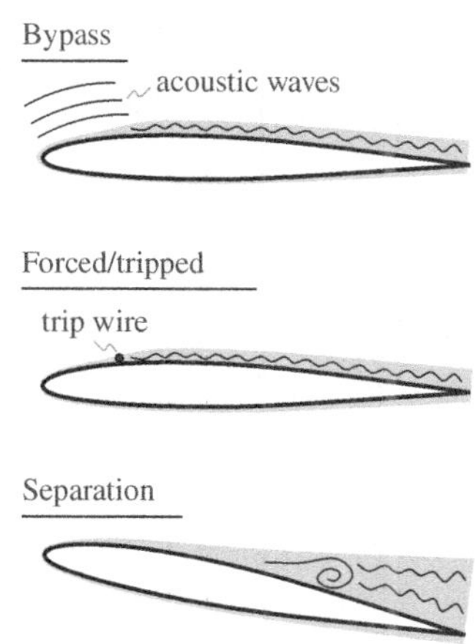

Figure 6.44 Mechanisms of transition from laminar to turbulent flow.

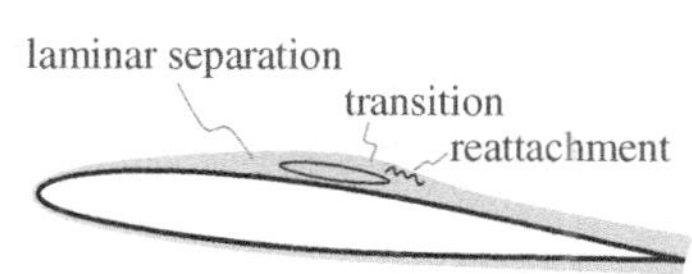

Figure 6.45 Bubble created by laminar separation and reattachment due to transition to turbulence.

6.7.2 Turbulent Flow and Reynolds Averaging

Turbulent flow is characterized by *random spatial and temporal variations* and is inherently more complicated than laminar flow. Fluctuations in the velocity and pressure occur in all three dimensions,* and fluid packets called *eddies*, of many sizes, move, rotate, and stretch throughout the flow. This movement enhances mixing of mass, momentum, and energy. The fluctuations are often self-sustaining, meaning that they draw energy from large bulk motions and release it through viscous dissipation at very small scales. Pope (2000) presents a thorough discussion of the characteristics of turbulent flow.

One way to understand the effects of turbulence is based on *Reynolds averaging*, a process that defines time-averaged quantities (Reynolds 1987; Reynolds 2004).

* The randomness is not white noise but instead contains structure according to a spectral energy distribution that can often be quantified.

For example, for the x component of velocity, we define the time average $\bar{u}$ by

$$\overline{u}(\vec{x}) = \frac{1}{T}\int_0^T u(\vec{x}, t)dt, \tag{6.154}$$

where the time horizon T is large enough to smooth out turbulent variations, but small enough to not affect the longer-time unsteadiness, such as an aircraft maneuver or control surface deflection, in the problem. Here, for simplicity, we assume that aside from the turbulent variations, the flow is steady.† With the average velocity defined in Equation 6.154, the fluctuating part is

† This means that we can make T as large as we want, i.e. $T \to \infty$.

$$u'(\vec{x}, t) = u(\vec{x}, t) - \overline{u}(\vec{x}). \tag{6.155}$$

This averaging is illustrated graphically in Figure 6.46. Note that averaging the fluctuating part gives

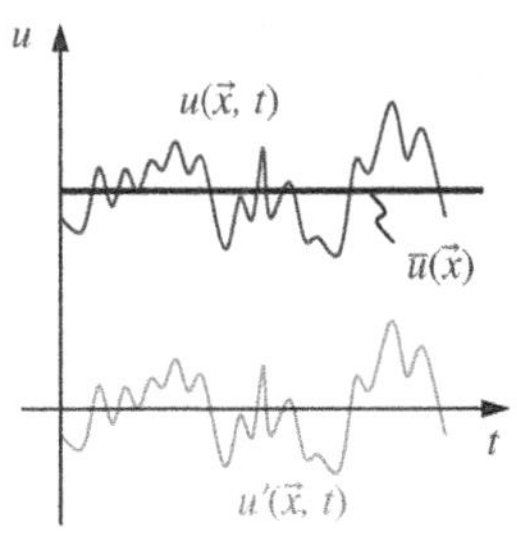

Figure 6.46 Graphical illustration of the definition of the average, $\overline{u}$, and fluctuating part, u', of the x velocity.

$$\begin{aligned}\overline{u'} &= \frac{1}{T}\int_0^T u'(\vec{x}, t)dt \\ &= \frac{1}{T}\int_0^T \left(u(\vec{x}, t) - \overline{u}(\vec{x})\right) dt \\ &= \overline{u}(\vec{x}) - \overline{u}(\vec{x}) = 0.\end{aligned}$$

Therefore, fluctuations have zero mean. Having defined the average and fluctuating part of a state, we now seek equations that govern the averaged, or mean, flow and in the process try to develop insight into how the fluctuations affect the mean flow equations. Using Equation 6.155 and working in index notation, where the velocity components are v_i, the continuity equation becomes

$$\partial_i v_i = 0 \quad \Rightarrow \quad \partial_i(\bar{v}_i + v_i') = 0 \quad \Rightarrow \quad \partial_i\bar{v}_i + \partial_i v_i' = 0. \tag{6.156}$$

Applying the time-average to this equation yields‡

‡ We use the fact that a spatial derivative commutes with a time integral,

$$\overline{\partial_i v_i'} = \partial_i \overline{v_i'} = 0.$$

$$\partial_i\bar{v}_i = 0. \tag{6.157}$$

So the average velocity satisfies the familiar form of the continuity equation. Turning now to the momentum equation, with no body forces for simplicity, we apply the averaging operation to the entire equation to obtain

$$\overline{\frac{Dv_j}{Dt}} = \overline{-\frac{1}{\rho}\partial_j p + \nu\nabla^2 v_j}. \tag{6.158}$$

Averaging the right-hand side is straightforward because the pressure and velocity appear linearly, so the result is

$$-\frac{1}{\rho}\partial_j\overline{p} + \nu\nabla^2\overline{v}_j. \tag{6.159}$$

The averaging of the left-hand side is complicated by the nonlinear term that appears inside the material derivative. Using the continuity equation,§ we write the material derivative of the velocity as

§ This equation lets us add $\partial_i v_i = 0$ in the second step.

$$\begin{aligned}\frac{Dv_j}{Dt} &= \partial_t v_j + v_i\partial_i v_j \\ &= \partial_t v_j + v_i\partial_i v_j + (\partial_i v_i)v_j \\ &= \partial_t v_j + \partial_i(v_i v_j).\end{aligned} \tag{6.160}$$

Averaging then gives

$$\begin{aligned}\overline{\frac{Dv_j}{Dt}} &= \overline{\partial_t v_j} + \overline{\partial_i(v_i v_j)} \\ &= \partial_t \bar{v}_j + \partial_i \overline{v_i v_j} \\ &= \partial_t \bar{v}_j + \partial_i \overline{(\bar{v}_i + v_i')(\bar{v}_j + v_j')} \\ &= \partial_t \bar{v}_j + \partial_i \overline{\bar{v}_i \bar{v}_j + v_i' v_j'} \\ &= \partial_t \bar{v}_j + \partial_i \left(\bar{v}_i \bar{v}_j\right) + \partial_i \overline{v_i' v_j'} \\ &= \frac{\bar{D}\bar{v}_j}{\bar{D}t} + \partial_i \overline{v_i' v_j'}.\end{aligned} \tag{6.161}$$

So we see that the material derivative of the velocity is the material derivative of the mean velocity¶ plus an extra term that is the divergence of the average of the product of fluctuations. Substituting into the averaged momentum equation, Equation 6.158, we obtain

¶ The $\frac{\bar{D}(\cdot)}{\bar{D}t}$ notation indicates that we use the mean velocity in the convective derivative.

$$\frac{\bar{D}\bar{v}_j}{\bar{D}t} = -\frac{1}{\rho}\partial_j \bar{p} + \nu \nabla^2 \bar{v}_j - \underbrace{\partial_i \overline{v_i' v_j'}}_{\text{new term}}. \tag{6.162}$$

This looks like the original momentum equation, written for $\bar{v}_j$ instead of v_j, except that the last term is new. It is the divergence of $\overline{v_i' v_j'}$, which is the *Reynolds stress tensor*. This tensor has a similar effect on the flow as the viscous stress tensor, but it is more difficult to analyze because we need to know something about the fluctuations. Together, the averaged continuity and momentum equations go by the name of the *Reynolds-averaged Navier–Stokes (RANS)* equations. The solution of the RANS equations requires a *closure model* for the Reynolds stresses, i.e. a model that expresses them in terms of the mean quantities. Various such models are available, with a range of rigor, complexity, and accuracy, but we will not discuss these here. For a detailed presentation of many common turbulence closure models, see Wilcox (2006).

6.7.3 Qualitative Observations and Empirical Correlations

Although the solution of the RANS equations is beyond the scope of this book, we can make some general qualitative observations and discuss empirical correlations. One important observation is that turbulent boundary layers experience *more mixing* than laminar boundary layers. In the Reynolds-averaged equations derived above, the Reynolds stresses account for this mixing. In fact, the Reynolds stresses can be thought of as adding extra dissipation to the flow, although the analysis is not this simple in practice.

Based on the observation of increased mixing, we can make some qualitative remarks about differences between turbulent and laminar boundary layers. In particular, increased mixing of x momentum vertically through the boundary layer means that:

- High-momentum fluid in the upper part of the boundary layer makes its way down near the wall.
- Low-momentum fluid near the wall makes its way to the top of the boundary layer.

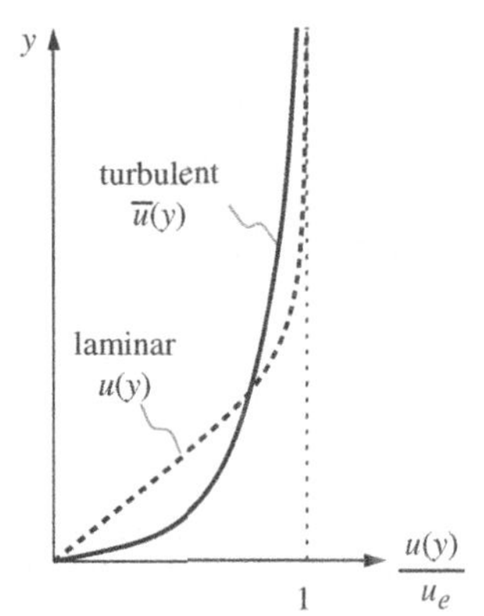

Figure 6.47 Comparison of a turbulent boundary-layer velocity profile (averaged) to a laminar profile at the same Re_x. The turbulent profile is thicker and exhibits higher skin friction.

The resulting boundary-layer profile is shown in Figure 6.47. Note that we are showing the *averaged* turbulent profile, $\bar{u}(y)$, compared to a laminar profile, $u(y)$, at the same Re_x. The increased mixing in the turbulent case results in a *fuller* profile near the wall and a more pronounced velocity deficit in the upper part of the boundary layer. Therefore, compared to the laminar case, turbulent boundary layers will have:

1. a higher skin-friction coefficient because $\left.\frac{\partial u}{\partial y}\right|_{\text{wall}}$ is larger in the turbulent case,
2. a larger thickness because of the greater extent of low-momentum fluid in the vertical direction, away from the wall.

Empirical formulas for a turbulent boundary layer on a flat plate with zero-pressure gradient were derived by Prandtl (Kuethe and Chow 1997). They are

$$\begin{aligned} \frac{\delta^*}{x} &= 0.0460\,\mathrm{Re}_x^{-1/5}, \\ \frac{\theta}{x} &= 0.0360\,\mathrm{Re}_x^{-1/5}, \\ c_f &= 0.0592\,\mathrm{Re}_x^{-1/5}. \end{aligned} \tag{6.163}$$

Note that a turbulent boundary layer grows more rapidly with respect to distance along the flat plate,

$$\delta^* \sim x^{1/2} : \text{Laminar (e.g. Blasius)} \tag{6.164}$$

$$\delta^* \sim x^{4/5} : \text{Turbulent} \tag{6.165}$$

Also, the skin-friction coefficient is larger and decays more slowly with respect to Re_x. This is illustrated in Figure 6.48, which shows the laminar and turbulent skin-friction profiles on a logarithmic plot. If we know in which regime the boundary layer is, laminar versus turbulent, we can read off the appropriate c_f from the plot. Note that there is no Re_x that serves as the absolute dividing line between laminar and turbulent flow for all cases. Some flows, for example over smooth surfaces, will stay laminar up to $\mathrm{Re}_x = 10^6$ or more, while others, for example over rough walls, will transition to turbulence at $\mathrm{Re}_x = 10^5$ or earlier. The transition location therefore depends on characteristics of the flow and the geometry. A reasonable rule of thumb for many airfoils is that transition occurs around $\mathrm{Re}_x = 5 \times 10^5$.

The larger c_f of a turbulent boundary layer means that to reduce the skin-friction drag, we want to maintain laminar flow for as long as possible. For example, sailplanes have smooth and clean wing surfaces to maintain laminar flow and hence minimize drag. This is harder to do with large transport aircraft that fly much faster and hence have a much larger Reynolds number.

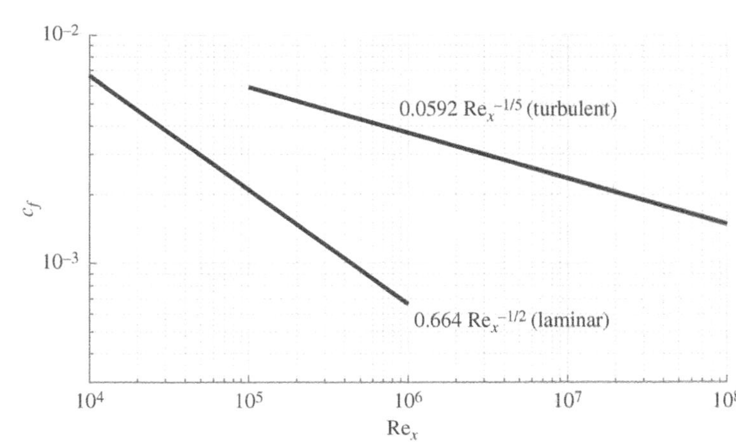

Figure 6.48 Skin-friction coefficient, c_f, versus $\mathrm{Re}_x = U_\infty x/\nu$ for laminar and turbulent flow on a flat plate.

Although reducing skin friction is generally a good idea for reducing drag, *skin friction is not the only source of drag*. If the flow undergoes separation, there is a resulting drag force on the body due to the pressure integral on the body surface. Separation prevents full pressure recovery at the rear (trailing edge) of the body, which means that the pressure there is lower than it would be if separation were not present. This lower pressure then no longer offsets the high pressure at the front (leading edge) of the body, and a resulting drag force is produced. This force is called *pressure drag* or *drag due to separation*. As shown in Figure 6.49, the separation drag on a small cylinder can be comparable to the separation drag on a well-designed large airfoil at the same conditions.

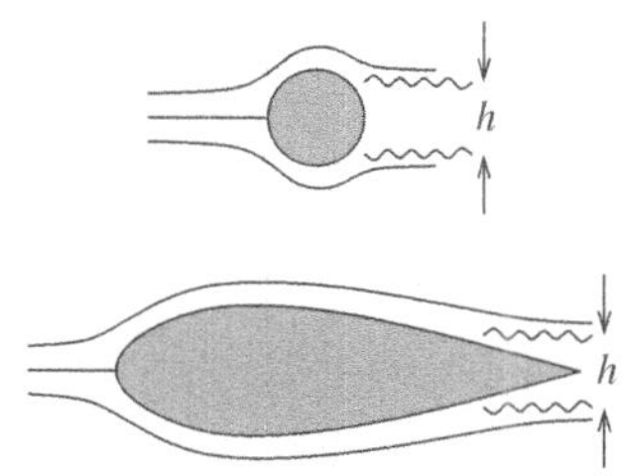

Figure 6.49 Pressure drag on a small bluff body such as a cylinder will be quite high, comparable to that on a much larger but more streamlined shape, such as an airfoil.

To reduce drag due to separation, we need to prevent or delay separation. Here, turbulent boundary layers have an advantage: *turbulent boundary layers are less prone to separation* compared to laminar boundary layers. This is due to the fuller velocity profiles in turbulent boundary layers, caused by increased mixing with high-momentum flow away from the wall. The resulting extra momentum close to the wall makes turbulent boundary layers less likely to separate.

Hence, sometimes in the interest of reducing the *total* drag, the sum of skin friction plus drag due to separation, we actually prefer turbulent flow! Such a situation occurs when the drag is dominated by separation, for example in flow over a sphere, such as a golf ball. In this case, skin friction is relatively low because the exposed surface area is not very large. However, in the laminar case, the flow separates almost immediately on the aft side of the sphere due to the large adverse pressure gradient there. Thus, drag due to separation dominates, and we stand to benefit from delaying the point of separation. If the flow is not already turbulent, and if the Reynolds number is in the transitional regime, we can delay separation by "forcing" the boundary layer to be turbulent. There are several ways to do this: one approach is a trip wire, but a simpler one that does not depend on the freestream direction is through increased roughness everywhere, for example through dimples. Both of these strategies are shown in Figure 6.50. Although the resulting turbulent boundary layer causes more skin-friction drag on the exposed surface, it keeps the boundary layer attached longer, and in cases such as a golf ball, this reduction in drag due to separation more than offsets the increased skin friction.

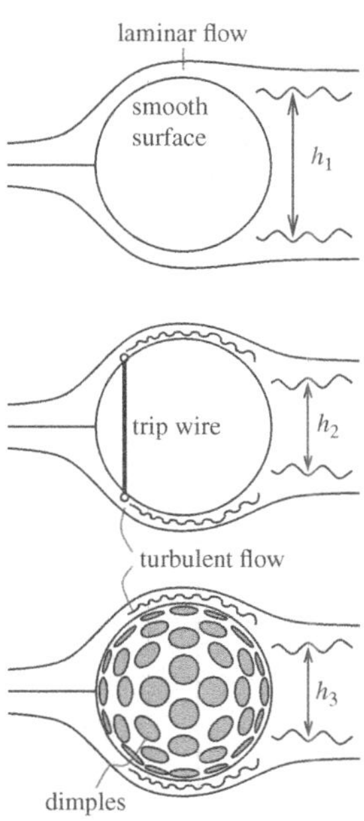

Figure 6.50 We can force a laminar boundary layer in the transitional Reynolds number regime to become turbulent by creating disturbances, for example with a trip wire or surface roughness. Note the reduced size of the separated region in the two cases with turbulent flow – this translates to a reduced drag due to separation in these cases.

Note, however, that many aerospace vehicles are streamlined and do not suffer from large drag due to separation, at least at design conditions. So adding dimples to these vehicles would generally not be a good idea, since it would only increase the skin-friction drag. Moreover, large transport aircraft cruise at Reynolds numbers well above the transitional regime, so that the boundary layers are typically already turbulent for the majority of the wing.

Example 6.13 Drag on an aircraft fuselage

Consider the fuselage of a large transport aircraft, which has a diameter of 4 m and a length of 40 m. We will ignore the effect of the rounded nose and tapered tail and assume a straight cylindrical fuselage, as shown in Figure 6.51. The aircraft is flying at 100 m/s at an altitude where $\rho = 1$ kg/m^3 and $\mu = 1.7 \times 10^{-5}$ kg/(m.s), and the fuselage is aligned with the freestream velocity.

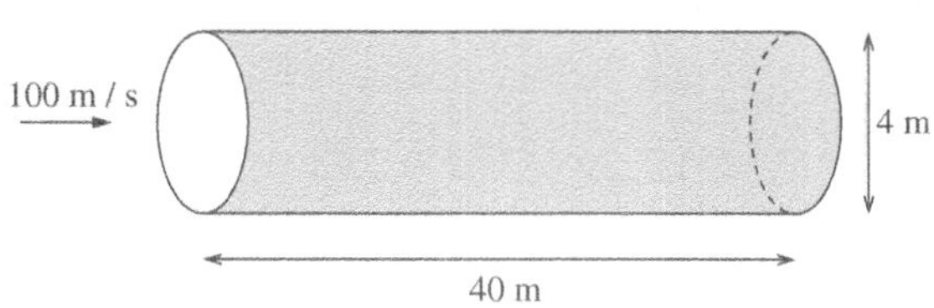

Figure 6.51 Fuselage used in the drag analysis in Example 6.13.

We assume that transition from laminar to turbulent flow occurs instantaneously at $\mathrm{Re}_x = \mathrm{Re}_{\mathrm{crit}} = 10^6$ and the following flat-plate local skin-friction formulas hold,

$$c_f = \frac{0.664}{(\mathrm{Re}_x)^{.5}} \quad \text{(laminar)}, \quad c_f = \frac{0.0592}{(\mathrm{Re}_x)^{.2}} \quad \text{(turbulent)}. \tag{6.166}$$

The skin-friction drag on the fuselage can be estimated by treating it as a rolled-up flat plate. The drag is given by an integral of the wall shear stress over the fuselage surface area, which we can imagine unrolled into the shape of a 4πm × 40 m flat plate.

$$\begin{aligned} D &= \int_S \tau_w dS = 4\pi \int_0^{40} \frac{1}{2}\rho U_\infty^2 c_f(x) dx \\ &= 2\pi\rho U_\infty^2 \left[\int_0^{x_{\mathrm{crit}}} c_{f,\mathrm{lam}}(x) dx + \int_{x_{\mathrm{crit}}}^{40} c_{f,\mathrm{turb}}(x) dx \right], \end{aligned} \tag{6.167}$$

where x_{crit} is the transition location corresponding to $\mathrm{Re}_x = \mathrm{Re}_{\mathrm{crit}} = 10^6$. Using the given expressions for c_f, and changing integration variables from x to Re_x, we have,

$$\int_0^{x_{\mathrm{crit}}} c_{f,\mathrm{lam}}(x) dx = \int_0^{\mathrm{Re}_{\mathrm{crit}}} \frac{.664}{(\mathrm{Re}_x)^{.5}} \frac{\nu}{U_\infty} d\mathrm{Re}_x = \frac{\nu}{U_\infty} 1.328 (\mathrm{Re}_{\mathrm{crit}})^{.5}, \tag{6.168}$$

$$\int_{x_{\mathrm{crit}}}^{40} c_{f,\mathrm{turb}}(x) dx = \int_{\mathrm{Re}_{\mathrm{crit}}}^{\mathrm{Re}_{40}} \frac{.0592}{(\mathrm{Re}_x)^{.2}} \frac{\nu}{U_\infty} d\mathrm{Re}_x = \frac{\nu}{U_\infty} .074 \left[\mathrm{Re}_{40}^{.8} - \mathrm{Re}_{\mathrm{crit}}^{.8} \right], \tag{6.169}$$

where $\mathrm{Re}_{40} = 40\, U_\infty/\nu$. Substituting these expressions into the drag formula gives

$$D = 2\pi\rho\nu U_\infty \left[1.328 (\mathrm{Re}_{\mathrm{crit}})^{.5} + .074 (\mathrm{Re}_{40}^{.8} - \mathrm{Re}_{\mathrm{crit}}^{.8}) \right]. \tag{6.170}$$

Using the given quantities, we have

$$D = D_{\mathrm{lam}} + D_{\mathrm{turb}} = 14.1\mathrm{N} + 3887.1\mathrm{N} = 3901.2\mathrm{N}. \tag{6.171}$$

So the percent of drag due to laminar skin friction is only 0.36%. Note that the transition location is

$$x_{\mathrm{crit}} = \frac{\nu \mathrm{Re}_{\mathrm{crit}}}{U_\infty} = 0.17 \text{ m}. \tag{6.172}$$

Since the region of laminar flow is only the first 17 cm of the 40 m fuselage, the contribution of the laminar skin friction to the total drag is expected to be very small.

Example 6.14 Transition and separation on an airfoil
Locations of transition and separation can often be identified from plots of the skin-friction coefficient. Consider a NACA 2620 airfoil at $\alpha = 4°$, in a transitional Reynolds number of $\text{Re} = 5 \times 10^6$, analyzed using mfoil. The pressure coefficient distribution for this airfoil is shown in Figure 6.52a.

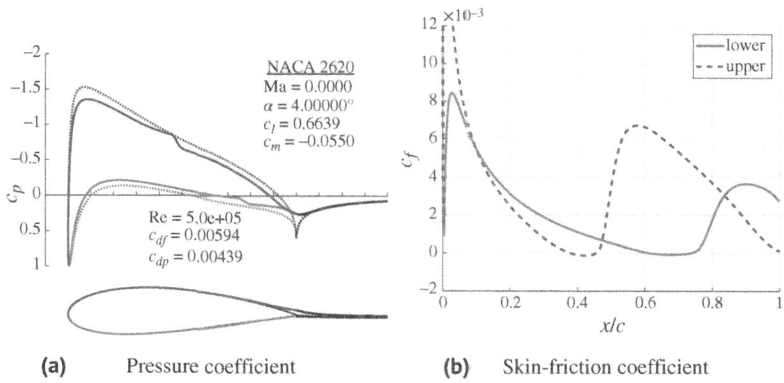

Figure 6.52 Pressure coefficient and skin-friction coefficient distributions for an airfoil at a transitional Reynolds number. Locations of transition and separation can be identified from the c_f plot.

Recall that separation is defined as the point at which the skin-friction coefficient, c_f, drops below zero. From the c_f plot in Figure 6.52b, the flow separates at $x/c \approx 0.38$ on the upper surface, and at $x/c \approx 0.64$ on the lower surface.

Furthermore, the c_f plot lets us discern the approximate locations of transition to turbulence. From Figure 6.48, we see that the skin friction for a turbulent boundary layer is higher than that of a laminar boundary layer for the same Re_x. Therefore, we can identify the location of transition from laminar to turbulent flow as the point where c_f jumps from a low value to a high value. On the upper surface, this occurs at $x/c \approx 0.45$, while on the lower surface transition occurs at $x/c \approx 0.75$.*

*These values are estimates obtained by looking at the start of the c_f rise. mfoil also calculates the transition locations, and for this case, it gives $x/c = 0.46$ and $x/c = 0.75$ for the upper and lower surfaces, respectively.

Example 6.15 Viscous effects on lift
In this example, we use the mfoil program to study the behavior of the lift coefficient on a NACA 0010 airfoil as the Reynolds number is varied. Suppose that the angle of attack is fixed at $\alpha = 2°$ and consider Reynolds numbers in the range $10^4 \leq \text{Re} \leq 10^7$.

First, we create a plot of c_ℓ versus Re, using logarithmic spacing for Re and standard spacing for c_ℓ, including enough points to resolve interesting areas of the plot. Listing 6.10 presents the code that does this, and Figure 6.53 shows the resulting plot.

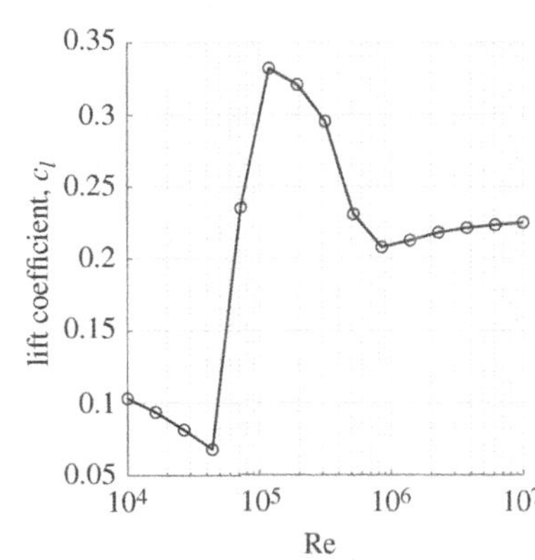

Figure 6.53 Lift coefficient variation with Reynolds number.

Listing 6.10 Code for generating a lift coefficient versus Reynolds number plot.

```
m = mfoil(naca='0010',npanel=399);
adeg = 2; % angle of attack in degrees

% cl plot
Rev = logspace(4,7,15);
for i = 1:length(Rev)
  m.setoper(alpha=adeg, Re=Rev(i), Ma=0.0);
  m.param.doplot = false;
  m.solve();
  clv(i) = m.post.cl;
end
figure(1); clf;
semilogx(Rev, clv, 'ko-');
```

```
grid on; box off;
set(gca, 'xtick', logspace(4,7,4));
set(gcf, 'position', [400, 400, 400, 400]);
writeplot('Re', 'lift coefficient, $c_{\ell}$', '../figs/visclift_cl.
    pdf', 4);
```

For Re $\lesssim 4\times10^4$, we observe that c_ℓ decreases with increasing Re. This occurs because there is a region of separated flow on the upper surface that is increasing in size. In particular, the flow separates sooner at higher Reynolds numbers. This is illustrated in Figure 6.54, which shows $c_f(x)$ plots on the upper surface at Re $=1\times10^4$ and Re $=3\times10^4$. Listing 6.11 shows the code snippet for plotting the c_f and δ^* distributions on only the upper surface of the airfoil.

Listing 6.11 Code snippet for plotting c_f and δ^* distributions on the upper surface.

```
m.setoper(alpha=adeg, Re=1e4, Ma=0.0); m.solve();
xy = [m.foil.x, m.wake.x]; Il = m.vsol.Is{1}; Iu = m.vsol.Is{2};
figure(1); plot(xy(1,Iu), m.post.cf(Iu), 'k-', 'displayname', 'Re =
    $1 \times 10^4$'); hold on;
figure(2); plot(xy(1,Iu), m.post.ds(Iu), 'k-', 'displayname', 'Re =
    $1 \times 10^4$'); hold on;
```

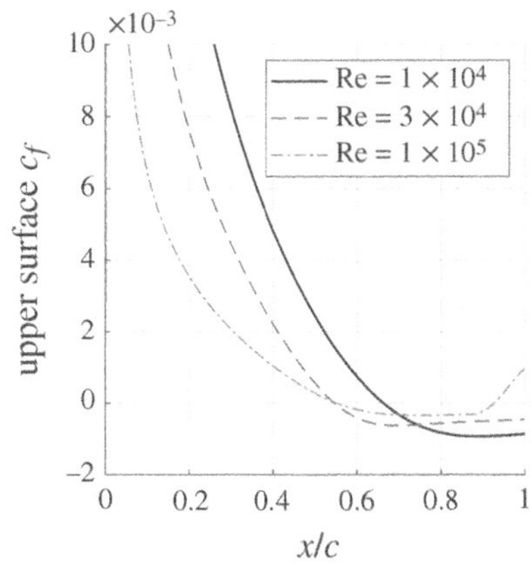

Figure 6.54 Skin-friction distributions on the upper surface.

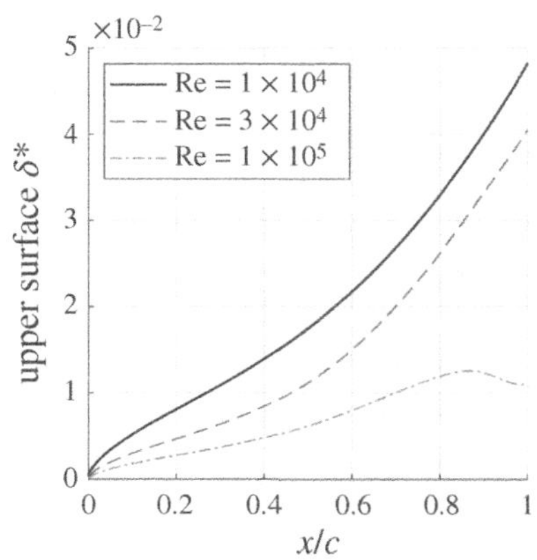

Figure 6.55 Displacement thickness distributions on the upper surface.

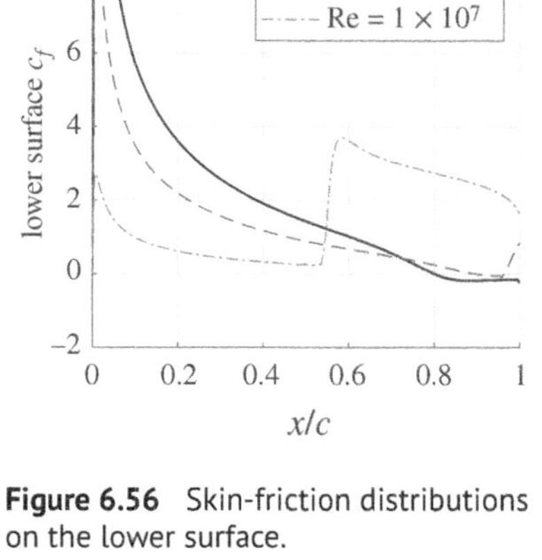

Figure 6.56 Skin-friction distributions on the lower surface.

Note that c_f goes negative earlier for Re $=3\times10^4$. Early separation on the upper surface means that the streamlines are curved less over the upper surface, since the flow does not follow the geometry, causing the lift coefficient to drop.

Just before Re $\approx 10^5$, c_ℓ suddenly increases. The increase in c_ℓ is due to transition to turbulence near the trailing edge of the upper surface, causing reattachment of the separated region and a sudden increase in the lift. The transition is evident in the $c_f(x)$ plot, which is shown in Figure 6.54 for Re $=3\times10^4$ (prior to transition) and Re $=1\times10^5$ (just after transition). Note the rapid rise in c_f near the trailing edge on the upper surface at Re $=1\times10^5$, indicative of transition to turbulence. This transition causes reattachment, as seen by the plot of $\delta^*(x)$ for the same Reynolds numbers in Figure 6.55. For Re $=3\times10^4$, δ^* keeps increasing for the separated flow. On the other hand, for Re $=10^5$, we see that δ^* begins to decrease after $x/c \gtrsim 0.85$ (reattachment), so that the flow experiences more curvature over the top, increasing the lift.

Around Re $\approx 4\times10^5$, we find that c_ℓ drops fairly rapidly. The drop in lift is due to transition to turbulent flow on the *lower* surface of the airfoil. Prior to Re $\lesssim 3\times10^5$, there is a region of separated flow on the lower surface, near the trailing edge, as indicated by the c_f plot in Figure 6.56 (see Re $=3\times10^5$). This separated region actually helps to increase the lift, because it deflects the streamlines on the lower surface down more compared to if the streamlines were following the geometry. This is also indicated on the δ^* plot in Figure 6.57, where for Re $=3\times10^5$, there is a kind of "flap effect": for $x/c \gtrsim 0.7$, the displacement thickness is larger on the lower surface compared to the upper surface, simulating an effective downward flap deflection of the camber line, and thereby increasing the lift. For higher Re, the flow on the lower surface transitions to turbulence, as shown on the c_f plot for Re $=8\times10^5$, and this region of separated flow is eliminated (the flow reattaches). The δ^* plot confirms that the flap effect is gone, and this is responsible for the drop in the lift.

Finally, for Re $\gtrsim 10^6$, c_ℓ increases slowly. The increase in c_ℓ at high Re is due to the boundary layers decreasing in thickness with increasing Re. Most of the flow is turbulent by this point, and the boundary-layer thickness decreases rather slowly with increasing Re. Nevertheless, there is a noticeable change, as indicated by the displacement thickness plots in Figure 6.57. The displacement thickness is smaller for $Re = 1 \times 10^7$ compared to for $Re = 8 \times 10^5$ over the majority of the airfoil, specifically on the upper surface. This decrease in the boundary-layer thickness on the upper surface causes more curvature of the streamlines as the flow follows the geometry more closely, increasing the lift coefficient.

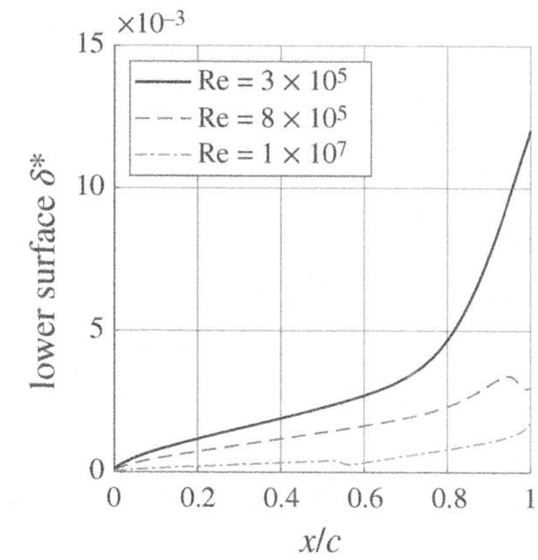

Figure 6.57 Displacement thickness distributions on the lower surface.

6.8 Summary and Concepts

This chapter brought back viscosity, which we first saw in dynamics in Chapter 3 but then neglected in the analysis of potential flow and airfoil theory. The reason why we neglected viscosity is because solving the Navier–Stokes equations for viscous flow is challenging and computationally expensive. For this reason, most of the analysis in this chapter involved the boundary-layer equations, which come from Navier–Stokes with a few simplifications inspired by order-of magnitude analysis at a high Reynolds number. We solved these equations numerically using a finite-difference method and then presented similarity solutions, including Blasius and Falkner–Skan. For flows with general boundary-layer edge velocities, cheaper alternatives to solving the Navier–Stokes equations include the integral boundary layer equations for flows near walls and Reynolds averaging for turbulent flows. Both of these techniques require closures to make the equations solvable, which introduce approximations and modeling errors.

Some of the key concepts covered in this chapter are as follows:

- The Reynolds number measures the relative importance of convective and viscous transport: a high Reynolds number corresponds to a small viscosity.
- Boundary layers form near walls of high Reynolds-number flows.
- The boundary-layer equations govern the flow inside a boundary layer: of primary importance is the wall tangential, or x-momentum equation, as the y-momentum equation only tells us that pressure is transmitted through the boundary layer.
- The nonlinear boundary-layer equations can be solved numerically using a finite-difference spatial marching technique due to the absence of a second derivative of the velocity with respect to x in the x-momentum equation.
- Semi-analytical similarity solutions exist for zero-pressure gradient and power-law edge velocity flows: these are the Blasius and Falkner–Skan solutions.
- A boundary layer separates under a sufficiently strong adverse pressure gradient that acts over a sufficiently large distance.
- Similarity solutions cannot predict separation, but Thwaites's method can, through an empirical correlation method.
- The integral boundary-layer equations combined with assumed profiles yield an inexpensive, albeit approximate, solution technique for boundary layers.
- Most of the flows in this chapter were laminar: turbulent boundary layers require a different solution approach, such as Reynolds averaging.
- Qualitative aspects of turbulent boundary layers include higher skin friction, larger thickness, and a fuller velocity profile that is harder to separate: these are all manifestations of the increased momentum mixing in a turbulent boundary layer.

6.9 Problems

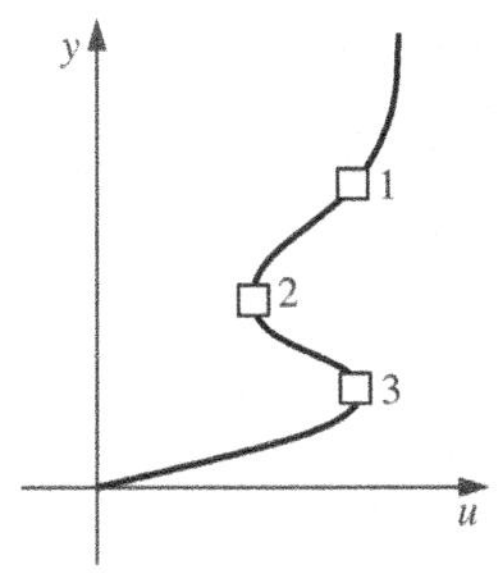

Figure 6.58 Fluid elements in a boundary layer.

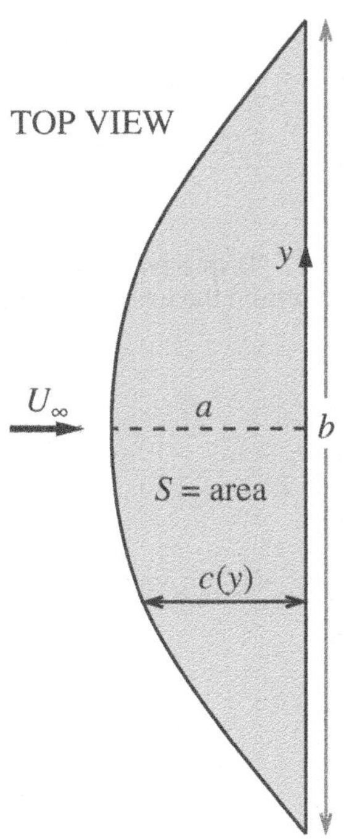

Figure 6.59 Wing with a parabolic chord distribution. The flow is from left to right.

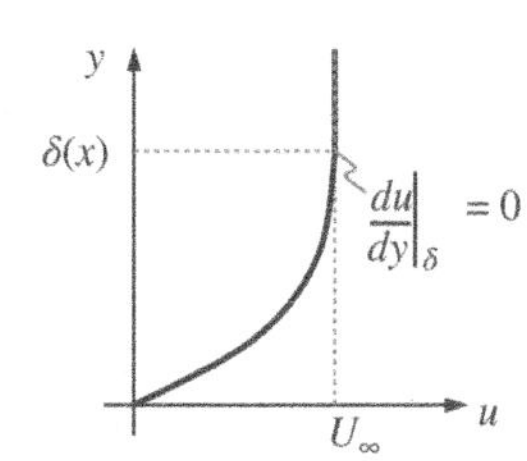

Figure 6.60 A quadratic assumed profile.

1. Justify each answer to the below questions with an explanation.
- a True or false: the viscous force on a fluid element in a boundary layer is always nonzero.
- b True or false: transition to turbulence can happen in a favorable pressure gradient.
- c True or false: the finite-difference boundary layer solver developed in this chapter can simulate separation.
- d In the Blasius solution, what is the limit of f as $\eta \to \infty$?
- e True or false: in a boundary layer, as Re_x increases, the thickness increases.
- f In an adverse pressure gradient, which of the fluid elements shown in Figure 6.58 has the largest magnitude acceleration?
- g True or false: the method of assumed profiles can be used to predict separation.

2. A flat plate at zero angle of attack is placed in a wind tunnel where the airspeed is 15 m/s. The air has density $\rho = 1.225$ kg/m^3 and viscosity $\mu = 1.79 \times 10^{-5}$ kg/(ms). Consider two locations: 10 cm and 40 cm downstream of the leading edge. Assuming laminar flow, calculate the following quantities at both locations and comment on how they change with x: displacement thickness, 99% thickness, momentum thickness, and skin-friction coefficient.

3. In the Blasius boundary-layer solution, the vertical velocity away from the wall is nonzero. Show that v/U_∞ scales inversely with $\sqrt{\mathrm{Re}_x}$, and make a plot of v/U_∞ (horizontal axis) versus η (vertical axis) for $\mathrm{Re}_x = 100$ and $\mathrm{Re}_x = 1000$, using the range $0 \le \eta \le 10$. What is the asymptotic value of v/U_∞ as $\eta \to \infty$?

4. A thin wing with a parabolic chord distribution is flying at speed U_∞ and zero angle of attack, as shown in Figure 6.59. The fluid has density ρ and viscosity μ. Determine the drag force, D, and the drag coefficient, $C_D \equiv D/(\frac{1}{2}\rho U_\infty^2 S)$, where S is the planform area of the wing. Assume laminar flow and express your drag coefficient answer in terms of the Reynolds number based on a, $\mathrm{Re}_a \equiv U_\infty a/\nu$.

5. Repeat the previous problem for the case where the flow starts laminar at the leading edge but then transitions to turbulent at $x_{\mathrm{crit}} = 0.25a$. In addition to calculating D and C_D, sketch the regions of laminar and turbulent flow over the wing, and calculate and label relevant distances.

6. Complete the Falkner–Skan derivation in Example 6.8 by showing that the x-momentum boundary-layer equation reduces to the Falkner–Skan equation for the given similarity variable and boundary-layer thickness expressions.

7. Consider a NACA 2412 airfoil in incompressible flow at $\alpha = 2°$, $\mathrm{Re} = 2 \times 10^5$. From a panel solver such as mfoil, obtain *inviscid* $u_e(x)$ data on the airfoil surface. Using Thwaites's method, compute and plot $\theta(x)$, $H(x)$, $\delta^*(x)$, and $c_f(x)$

on the airfoil's upper surface. Identify the location of separation and compare it to the *viscous* panel solver solution. Discuss how accurate Thwaites's method is in this case.

8. Use the integral boundary-layer equation with a quadratic velocity profile parameterized by a boundary-layer thickness, $\delta(x)$, as shown in figure 6.60 to solve for the boundary-layer properties $\delta^*(x)$, $\theta(x)$, and $c_f(x)$ in the case of flow over a flat plate. Compare your results to the Blasius solution.

7

Finite Wings

This chapter extends the theory and methods developed for airfoils to wings that have a finite span. Flows around these wings have a three-dimensional character that complicates their analysis: wings using the same airfoil but with different three-dimensional shapes will behave differently. In particular, sections of the wing cannot be analyzed in isolation, and modeling their interaction requires information about the wing geometry and operating conditions. The underlying physical model used for the analysis in this chapter is potential flow, which admits solutions via superposition and hence allows for relatively simple numerical and analytical methods. A notable difference compared to airfoils will be the appearance of a drag force even in the absence of viscosity. Minimizing this drag for a given lift will lead to the concept of efficiency of a wing's lift distribution.

7.1 Fundamentals

Key to potential-flow modeling of the lift generated by a finite wing is a vortex line, which is a three-dimensional extension of the point vortex presented in Chapter 4. Indeed, a point vortex appears in the two-dimensional plane that is the perpendicular cross-section of an infinite three-dimensional vortex line. However, in three dimensions, vortex lines need not be straight, and we will need to calculate the velocities induced by vortex segments or curves.

7.1.1 Lumped-Vortex Model

In Chapter 4, we found that under the assumptions of potential flow, the lift per unit span on an object is proportional to the net vortex strength, or circulation, around the object,

$$L' = \rho V_\infty \Gamma. \tag{7.1}$$

This was the result of the Kutta–Joukowsky theorem derived in Section 4.5. Thin-airfoil theory, presented in Chapter 5, related Γ to the geometry of an airfoil, using a vortex sheet model in which Γ was distributed continuously along the chordline, as shown in Figure 5.19, so that

$$\Gamma = \int_0^c \gamma(x)dx. \tag{7.2}$$

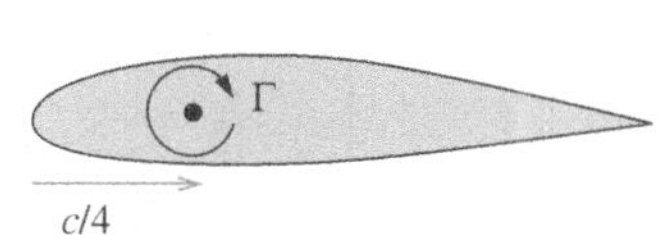

Figure 7.1 Single-vortex, or "lumped," sectional model used in the lifting-line theory of finite wings.

Throughout most of this chapter, we will use an even simpler airfoil model obtained by lumping the vorticity into one point: a vortex of strength Γ located at the airfoil's quarter-chord location, as shown in Figure 7.1. The location at which

Aerodynamics: A Computational Introduction, First Edition. Krzysztof J. Fidkowski.

© 2026 John Wiley & Sons, Inc. Published 2026 by John Wiley & Sons, Inc.

we place this vortex does not affect the lift and drag calculation. Although predicting pitch moment is not possible in this lumped model, choosing the quarter-chord location for the vortex at least recovers the center of pressure for a symmetrical airfoil. This sectional model will be the starting point for our finite-wing theory.

7.1.2 Coordinate System

In our study of finite wings, we will use a *wind-aligned coordinate system*, shown in Figure 7.2, in which

- The x axis, $\hat{i}$, points in the direction of the freestream velocity.
- The y axis, $\hat{j}$, points out the right wing.
- The z axis, $\hat{k}$, points up to complete the coordinate system.

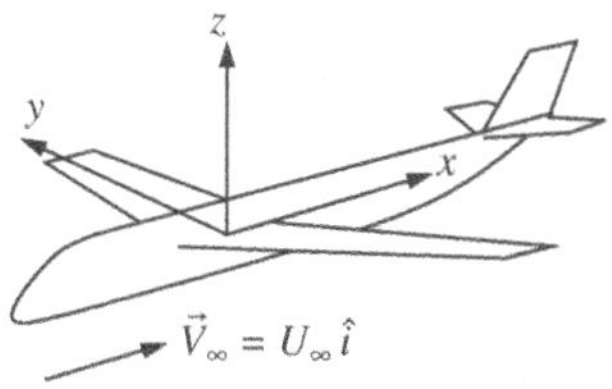

Figure 7.2 Wind-aligned coordinate system used in the aerodynamic analysis of finite wings.

This coordinate system is specific to aerodynamics and differs from other common coordinate systems, for example ones that are used for controls or structural analysis. In the aerodynamic wind-aligned system, drag will be the force on the wing in the x direction, and lift will be the force in the z direction.

7.1.3 The Wake Concept

Consider first a three-dimensional wing of infinite span, as shown in Figure 7.3. Since the span is infinite, each section behaves like a two-dimensional airfoil, and all sections are identical. The vortex at the quarter-chord of each section becomes a *vortex line* or *filament* of strength Γ. A sometimes useful analogy is to think of a vortex line as a wire carrying current, and the velocity field as the induced magnetic field.*

*When the "current" is in the positive y direction, the induced velocity is given by the right-hand rule, rotating as shown in Figure 7.3, turning the freestream flow downwards.

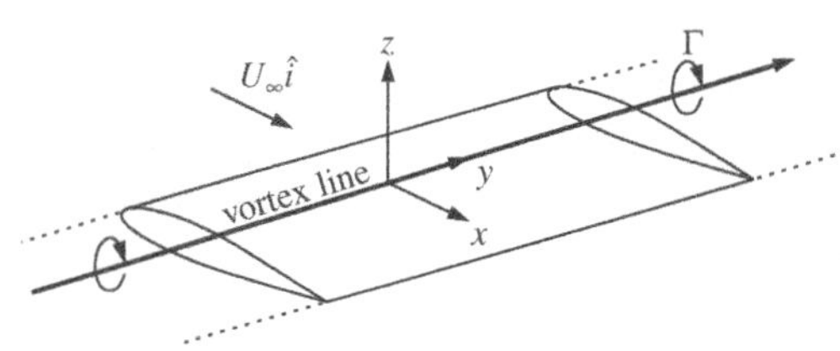

Figure 7.3 A vortex line in a wing of infinite span.

In reality, however, wings are finite. What happens to the vortex line when we truncate the wing at a finite span? *Helmholtz's second theorem*, presented in Helmholtz (1867), Prandtl and Tietjens (1934), and Saffman (1995), among others, states that a vortex line cannot end abruptly. This means that vortex lines must either extend to infinity or form closed loops. Without the wing to hold it in place,† the vortex line will convect downstream, as illustrated in Figure 7.4. The strength of the vortex line will be the same, Γ, everywhere. The trailing legs of this *horseshoe vortex* can extend to infinity, or they can be joined into a loop by a *starting vortex*. For steady-state analysis, the starting vortex can be ignored. The remaining horseshoe vortex is a very simple model for a finite wing.

†The vortex line is a mathematical model for the turning of the flow by the wing. Outside of the span of the wing, there is no geometry to create new flow turning, and hence no vortex line.

However, a single horseshoe vortex model is too simple. It implies that the circulation, $\Gamma(y)$, and hence lift, $L'(y)$, distribution are constant along the span. But a real wing does not have a constant $\Gamma(y)$. First, at wing tips, the pressure between the top and bottom surfaces of the wing matches so that $L' = 0 \Rightarrow \Gamma = 0$ there. Second, a constant or even nearly constant $\Gamma(y)$ is not optimal structurally or aerodynamically.‡ Therefore, to describe a real wing, we need a way to model a variable spanwise circulation.

‡A constant spanwise lift distribution would create a large bending moment at the root, and it would cause too much drag.

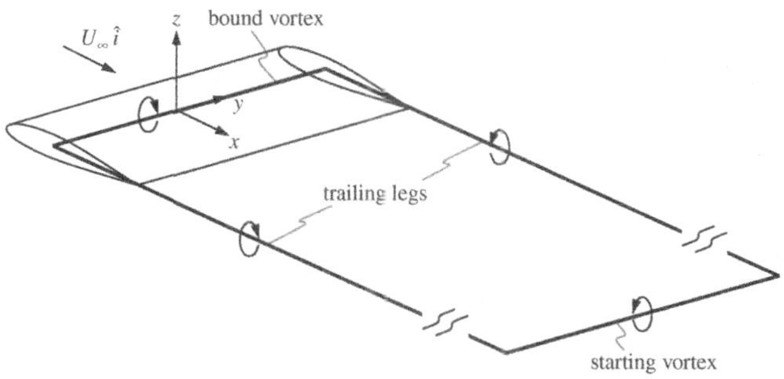

Figure 7.4 A simple "horseshoe" model of a finite wing.

Helmholtz's first theorem, presented in Helmholtz (1867), Prandtl and Tietjens (1934), and Saffman (1995), tells us that the strength of a single vortex line cannot change. So, to model a non-constant $\Gamma(y)$, we need to "shed" vorticity along the span, as illustrated in Figure 7.5. We will assume that the shed vorticity convects downstream, in the direction of the x axis.[§] This assumption will result in a flat wake that will be tractable analytically.

§ The shed vortex lines will generally not be straight because they will be affected by the local velocity, but if we assume that $U_\infty \hat{i}$ is the dominant velocity contribution, the straight vortex line assumption is a good approximation.

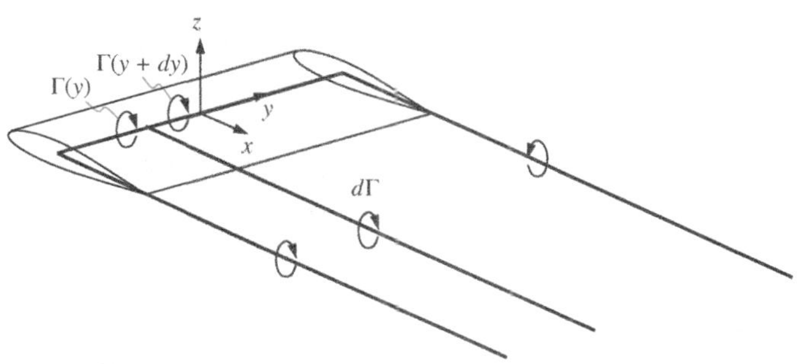

Figure 7.5 To create a non-constant $\Gamma(y)$, vorticity must be shed along the span.

Referring to Figure 7.5, the vorticity shed between spanwise locations y and $y + dy$ is

$$d\Gamma = \frac{d\Gamma}{dy} dy. \tag{7.3}$$

Since $d\Gamma/dy$ is typically negative on the right wing (lift decreases as we move from the root to the tip), the shed vorticity there is in the counter-clockwise direction when viewed from the rear. On the left wing, $d\Gamma/dy$ is typically positive, so that the shed vorticity is in the clockwise direction. All of the shed vorticity put together leads to a *vortex sheet* behind the wing, as shown in Figure 7.6. This vortex sheet is also known as the *wake*. We will think of the wake as a superposition of an infinite number of vortex lines, with infinitesimal strengths $d\Gamma$.

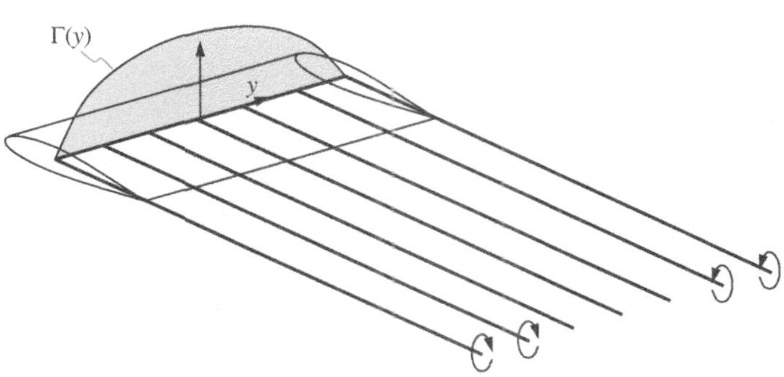

Figure 7.6 A vortex sheet behind the wing caused by shed vorticity.

In reality, the vortex sheet is not flat – it tends to curl up at the tips, as shown in Figure 7.7. This curling up is due to velocities in the z and y directions that the wake induces on itself. Such a wake is called *force-free*, as opposed to the flat wake, which would require some external force to keep it flat.

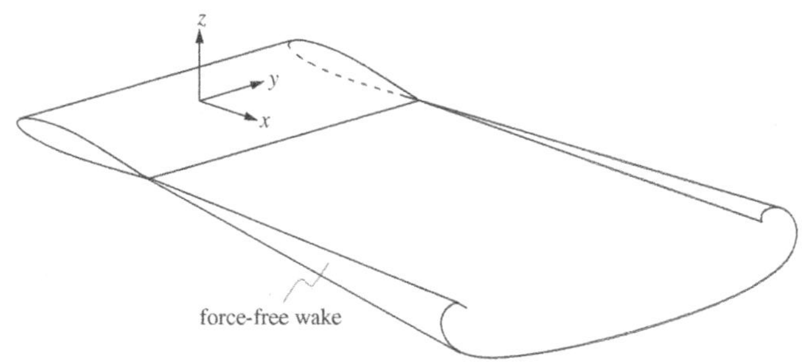

Figure 7.7 A real force-free wake curls up at the tips and does not remain flat.

In our analysis of finite wings, we will assume a flat, instead of a force-free, wake. This assumption greatly simplifies the modeling, because the rollup of the wake depends on the lift distribution, so that using a force-free wake would require calculating the wake geometry separately for each analysis. Moreover, a flat wake still yields accurate answers for lift and drag. This is because if there is no crossflow, the force required to keep a wake flat has no x, or drag, component, and no z, or lift, component, so the calculated lift and drag values will be correct (Kroo and Smith 1991).¶

¶ A real wake will have some crossflow, but if it changes shape slowly, the crossflow will be small relative to the freestream, and errors in drag and lift made by assuming a flat wake will be small.

Example 7.1 Numerical simulation of wake rollup
The trailing vortices in a wake roll up because of their mutual effect on each other's velocities. If the position of the vortices does not change too quickly with distance downstream, a reasonable approximation of their interaction is a two-dimensional one in each cut plane (y, z) downstream. We can then use the moving vortex analysis of Section 4.4.2 to determine where the vortices go.

The code in Listing 7.1 performs such a simulation, by initializing N vortices to equally spaced (in y) positions along the span. The strength of each vortex is the shed vorticity $d\Gamma$ in Equation 7.3. The particular function for $\Gamma(y)$ used in the code is the elliptical lift distribution, which is discussed later in this chapter, in Section 7.5.1. Note that the vortices are not all of the same strength: the tip ones are stronger.

Code notes:

- The span and freestream speed are arbitrary.
- The aspect ratio measures the span relative to the average chord and is defined in Equation 7.20.
- `G` is the distribution of trailing vortex strength, $d\Gamma$.
- We do not place vortices at the tips because the chosen $d\Gamma/dy$ is undefined there.
- `X` stores the (y, z) positions of the vortices.
- `getv` is a helper function that calculates the velocities of each of the vortices.
- The wake is plotted every 100 iterations as the vortices connected together.

Listing 7.1 Wake rollup simulation.

```
function rollup
% simulates wake rollup of vorticity

N = 200; % number of vortices
b = 1; U = 1; % wing span, freestream speed
AR = 10; CL = 0.5; % aspect ratio, lift coefficient
y = linspace(-b/2,b/2,N+2); y = y(2:end-1); dy = y(2)-y(1);
G = -8*U*CL/(pi*AR)*(y/b)./sqrt(1-4*(y/b).^2)*dy; % elliptical
X = [y; zeros(1,N)]; % state = y,z positions

% time stepping: predictor-corrector
T = 8*b/U; Nt = 400; dt = T/Nt;
figure(1); clf; set(gcf, 'position', 400*[1,1,1,.2]);
for n = 0:Nt,
  if (mod(n,100)==0) % plot the wake
    plot(X(1,:),X(2,:),'k.-'); axis equal; axis off; drawnow;
    exportgraphics(gcf, sprintf('../figs/rollup%d.pdf', n));
  end
  V1 = getv(X,G);
  V2 = getv(X + dt*V1,G);
  X = X + 0.5*dt*(V1 + V2);

end
```

```
%------------------------------------
function V = getv(X, G)
% computes velocity given positions and strengths
N = size(X,2); V = zeros(size(X));
for i = 1:N, % loop over vortices
  vi = [0;0]; % velocity of vortex i
  for j = 1:N, % loop over all other vortices
    if (i==j), continue; end
    r2 = (X(1,i)-X(1,j))^2 + (X(2,i)-X(2,j)).^2;
    vi = vi + G(j)/(2*pi*r2)*[X(2,i)-X(2,j); X(1,j)-X(1,i)];
  end
  V(:,i) = vi;
end
```

Running the simulation for the chosen time range results in the plots shown in Figure 7.8. The rollup of the vortices is clear, especially at the tips, where the trailing vorticity is of highest magnitude. For longer times, which effectively means for larger distances downstream, the entire vortex sheet wake is affected. The cores of the rolled-up tips show oscillatory distributions of the vortices, and this is due to the use of discrete vortices. When these get too close together, the inverse distance in the velocity formula leads to very large values. A continuous vortex sheet, presented in Section 5.4.1, would provide a better-behaved model of the wake.

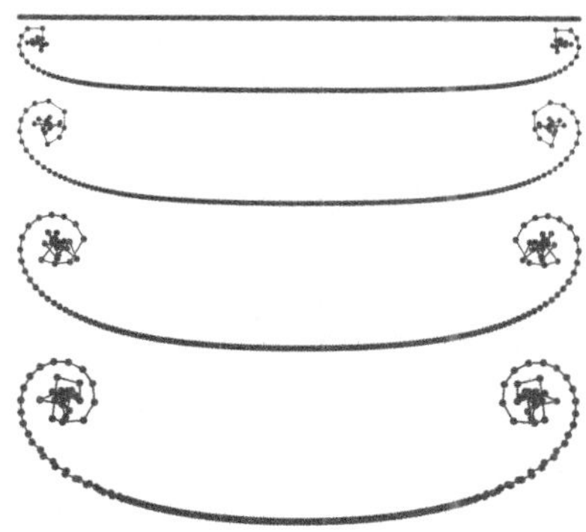

Figure 7.8 Rollup of wake vortices.

7.1.4 Velocity Induced by a Vortex Line

The derivation of finite-wing models requires the calculation of the velocity induced by a single vortex line. Recall that in two dimensions, the velocity induced by a vortex of strength Γ is purely in the tangential direction,

$$v_\theta = -\frac{\Gamma}{2\pi r} \quad \text{(two dimensions).} \tag{7.4}$$

In three dimensions, we have the *Biot–Savart law,** which states (Lamb 1932; Prandtl and Tietjens 1934; Houghton et al. 2016)

$$d\vec{v} = \frac{\Gamma}{4\pi}\frac{d\vec{l} \times \vec{r}}{r^3}. \tag{7.5}$$

*This is the same equation as the one for the magnetic field induced by a wire carrying a current.

As shown in Figure 7.9, this equation gives the effect of the circulation on a differential element, $d\vec{l}$, of the vortex line on the velocity increment, $d\vec{v}$, at a point a distance r away from the differential element. Integrating the Biot–Savart law along the entire vortex line then gives the total velocity induced by the line at a given point in space.

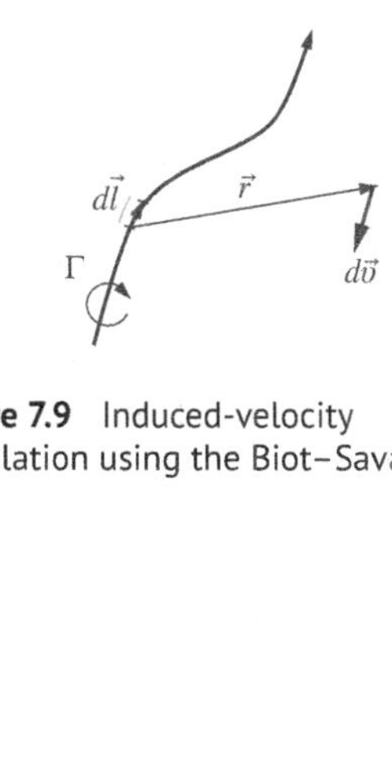

Figure 7.9 Induced-velocity calculation using the Biot–Savart law.

Example 7.2 Velocity induced by an infinite vortex line

Consider a straight, infinite vortex line, as shown in Figure 7.10. Suppose that we are interested in the velocity at a distance h away from the line. A convenient coordinate system is one in which the vortex line and point of interest both lie in the $x-y$ plane. From the cross product in Equation 7.5, the induced velocity will then be out of the page, that is, in the $\hat{k}$ direction.

The total induced velocity is given by an integral of the Biot–Savart formula along the vortex line. In the chosen coordinate system, x varies along the vortex line, so that this integral is

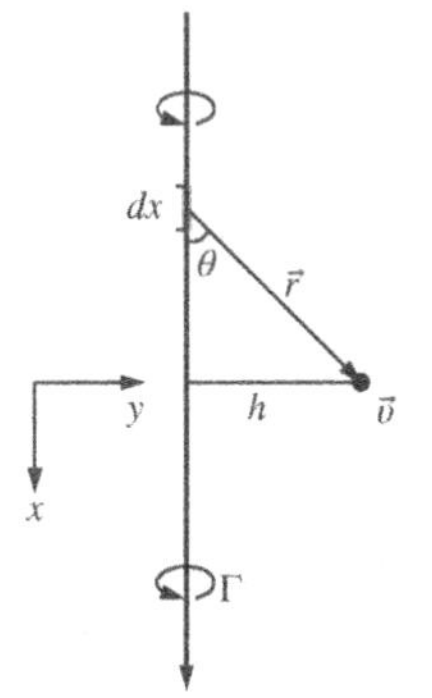

Figure 7.10 Setup for the calculation of velocity induced by a single vortex line. Note that the x coordinate is a negative number at the point where dx is drawn.

$$\vec{v} = \int_{-\infty}^{\infty} \frac{\Gamma}{4\pi} \frac{\vec{dl} \times \vec{r}}{r^3}, \quad \vec{dl} = dx\hat{\imath}. \tag{7.6}$$

A variable transformation simplifies this integral. Instead of using x, we use the angle θ indicated in Figure 7.10. At any point on the vortex line, this is the angle between $\hat{\imath}$ and $\vec{r}$, the vector that points from the vortex line to the point of interest. As x ranges from $-\infty$ to ∞, the angle θ ranges from 0 to π. To relate x and θ, we use the geometry in Figure 7.10, with the result

$$r = \frac{h}{\sin\theta}, \tag{7.7}$$

$$x = -h\cot\theta \quad \Rightarrow \quad dx = \frac{h d\theta}{\sin^2\theta}, \tag{7.8}$$

$$\vec{dl} \times \vec{r} = dxr\sin\theta\hat{k} = \frac{h^2}{\sin^2\theta} d\theta\hat{k}. \tag{7.9}$$

Note that the formula $x = -h\cot\theta$ comes from the fact that $h/(-x) = \tan\theta$. Substituting these expressions into Equation 7.6, we have

$$\vec{v} = \int_0^{\pi} \frac{\Gamma}{4\pi} \frac{1}{r^3} \frac{h^2}{\sin^2\theta} d\theta\hat{k} = \int_0^{\pi} \frac{\Gamma}{4\pi h} \sin(\theta) d\theta\hat{k}. \tag{7.10}$$

The integral of $\sin\theta$ from 0 to π is 2, so the velocity induced by the vortex line is†

† This is the familiar two-dimensional point-vortex formula for the velocity, where the two-dimensional plane contains the point and is perpendicular to the vortex line.

$$\vec{v} = \frac{\Gamma}{2\pi h}\hat{k}. \tag{7.11}$$

Again, we can infer the $\hat{k}$ direction by applying the right-hand rule to the diagram in Figure 7.10. If the circulation of the line were reversed, the induced velocity would be in the opposite direction.

Example 7.3 A vortex ring
Consider potential flow induced by a circular vortex ring of strength Γ and radius R, as shown in Figure 7.11.

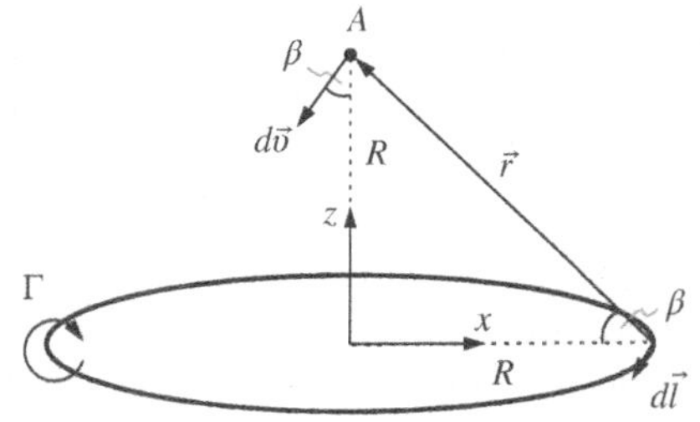

Figure 7.11 Calculation of the velocity induced by a vortex ring.

We are interested in the velocity vector a distance R above the center of the ring, which is point A in the figure. Using the Biot–Savart law and parameterizing the integral along the circular ring by the cylindrical-coordinate angle, θ, we have

$$\vec{v}_A = \oint_C d\vec{v} = \oint_C \frac{\Gamma}{4\pi}\frac{d\vec{l}\times\vec{r}}{r^3} = \int_0^{2\pi} \frac{\Gamma}{4\pi}\frac{Rd\theta r(-\hat{k}R/r)}{r^3}$$
$$= \frac{\Gamma}{4\pi}\frac{R2\pi r(-\hat{k}R/r)}{r^3} = -\frac{\Gamma}{2}\frac{R^2}{r^3}\hat{k} = -\frac{\Gamma}{4\sqrt{2}}\frac{1}{R}\hat{k}, \tag{7.12}$$

where we used $r = \sqrt{2}R$.[‡] This velocity is purely in the $\hat{k}$ direction, and due to the R^2/r^3 term in the result, its magnitude increases as the point A moves closer to the ring. The largest magnitude of $\Gamma/(2R)$ occurs when A is in the center of the ring.

[‡] The key step in this calculation is the evaluation of $\vec{dl}\times\vec{r}$. One could do this explicitly by writing expressions for $\vec{dl}$ and $\vec{r}$ in a global coordinate system. A simpler way is to observe that all velocity vector components in directions other than $\hat{k}$ cancel out during the integration due to symmetry, so we only need to keep track of the $\hat{k}$ component of $\vec{dl}\times\vec{r}$, which is

$$|\vec{dl}||\vec{r}|\cos\beta = Rd\theta r\frac{R}{r}.$$

Example 7.4 Velocity induced by a vortex segment
Suppose that we are interested in the velocity at point A, a distance h from a vortex segment of strength Γ, as shown in Figure 7.12. To compute the induced velocity at point A, $\vec{v}_A$, we parameterize the integral in the Biot–Savart expression by the angle, θ, between $\hat{i}$ and $\vec{r}$, as in Example 7.2. The only difference is in the limits of integration, which are now from $\theta = \beta_1$ to $\theta = \pi - \beta_2$,

$$\begin{aligned}\vec{v}_A &= \int_{\beta_1}^{\pi-\beta_2} \frac{\Gamma}{4\pi}\frac{1}{r^3}\frac{h^2}{\sin^2\theta}d\theta\hat{k} \\ &= \int_{\beta_1}^{\pi-\beta_2} \frac{\Gamma}{4\pi h}\sin(\theta)d\theta\hat{k} \\ &= \frac{\Gamma}{4\pi h}\left(-\cos(\pi-\beta_2)+\cos(\beta_1)\right)\hat{k} \\ &= \frac{\Gamma}{4\pi h}\left(\cos\beta_1+\cos\beta_2\right)\hat{k}.\end{aligned} \tag{7.13}$$

Note that the velocity magnitude only depends on the distance of point A to the segment and the two angles, β_1 and β_2.[§]

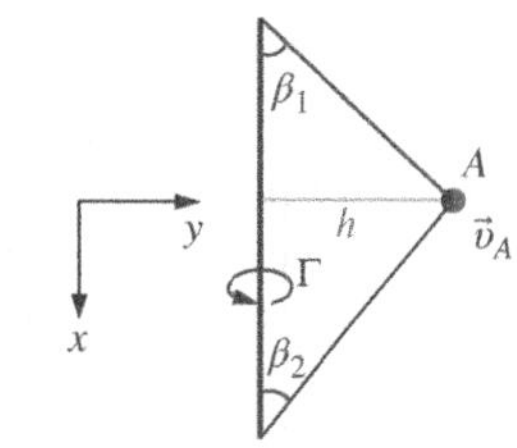

Figure 7.12 Calculation of the velocity induced by a vortex segment.

[§] This general formula reduces to the infinite vortex result, $|\vec{v}_A| = \Gamma/(2\pi h)$, when $\beta_1, \beta_2 \to 0$ and the semi-infinite result, $|\vec{v}_A| = \Gamma/(4\pi h)$, when one angle goes to zero while the other stays at $\pi/2$.

7.2 Horseshoe Vortices

This section presents wing models based on a horseshoe vortex. The simplest of these is just one horseshoe for a wing, and although quite crude, such a model already yields insight into the relationship between lift and drag and the interaction of multiple aircraft in proximity. Stringing together multiple horseshoe vortices improves accuracy, with the limitation of applicability to high aspect ratio wings and no sweep. These models have no chordwise resolution, but they can be made predictive by a combination with a sectional lift model.

7.2.1 A Single Vortex

Recall that a simple model for an aircraft wing is a horseshoe vortex, shown in Figure 7.13 without the starting vortex, which is assumed to be infinitely far away.

The force on the wing can be approximated by the three-dimensional generalization of the Kutta–Joukowsky theorem,

$$\vec{F} = \rho b\Gamma\vec{v}_0\times\hat{j}, \tag{7.14}$$

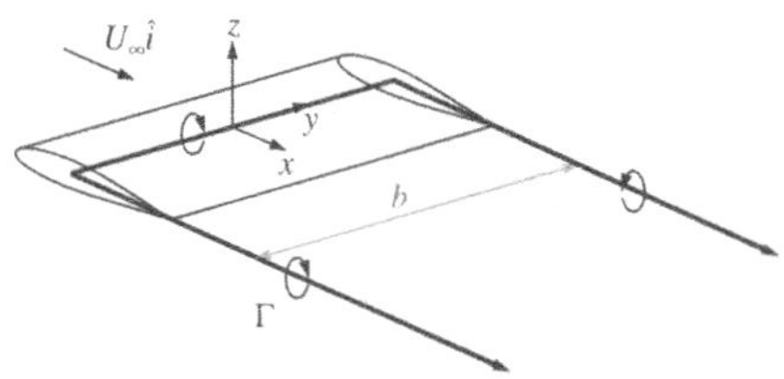

Figure 7.13 An open horseshoe vortex.

where $\vec{v}_0$ is the velocity in the middle of the bound vortex on the wing, which is at the origin of Figure 7.13.* For a freestream aligned with the x direction, the x component of this force is the induced drag and the z component is the lift.

* By symmetry, the bound vortex, which goes through the origin, does not contribute to $\vec{v}_0$.

Using the above formula, we calculate the induced drag, D_i, as a function of lift, L. The velocity at the center of the wing, $\vec{v}_0$, is induced by the two trailing vortices. Each vortex is a distance $h = b/2$ away from the wing's center, and each induces a velocity in the -$\hat{k}$ direction, so using Equation 7.13,

$$\vec{v}_0 = U_\infty \hat{i} + 2\frac{\Gamma}{4\pi(b/2)}(\cos 90^\circ + \cos 0^\circ)(-\hat{k}) = U_\infty \hat{i} - \frac{\Gamma}{\pi b}\hat{k}. \tag{7.15}$$

Note that we include the free-stream velocity in $\vec{v}_0$. Substituting $\vec{v}_0$ into Equation 7.14, the force on the horseshoe vortex is

$$\vec{F} = \rho b \Gamma \vec{v}_0 \times \hat{j} = \underbrace{\rho b \Gamma U_\infty}_{L} \hat{k} + \underbrace{\frac{\rho \Gamma^2}{\pi}}_{D_i} \hat{i}. \tag{7.16}$$

To express the induced drag in terms of the lift, we solve for Γ in terms of L and use this value in the expression for D_i,†

† From this formula, we can see that to reduce the induced drag for a given weight, which equals lift in steady-level flight, the span, b, should be large to keep the span loading, L/b, small. From an operating point of view, increasing dynamic pressure, ρU_∞^2, also reduces induced drag.

$$\Gamma = \frac{L}{\rho b U_\infty} \quad \Rightarrow \quad D_i = \frac{L^2}{\pi \rho b^2 U_\infty^2}. \tag{7.17}$$

We can rewrite the above result using non-dimensional force coefficients, which are different from the two-dimensional counterparts because we now work with three-dimensional forces instead of sectional, or per unit length, forces. The standard normalization of the lift and drag of a wing uses the dynamic pressure, $\frac{1}{2}\rho U_\infty^2$, and planform area, S,‡

‡ The planform area is the area of the wing when viewed from above, i.e. projected into the $x-y$ plane.

$$C_{Di} = \frac{D_i}{\frac{1}{2}\rho U_\infty^2 S}, \quad C_L = \frac{L}{\frac{1}{2}\rho U_\infty^2 S}. \tag{7.18}$$

These are the induced drag and lift coefficients of the wing. Using these formulas to substitute for D_i and L in Equation 7.17, we obtain a relationship between the lift coefficient and the induced drag coefficient.

$$C_{Di}\frac{1}{2}\rho U_\infty^2 S = \frac{C_L^2\left(\frac{1}{2}\rho U_\infty^2 S\right)^2}{\pi \rho b^2 U_\infty^2} \quad \Rightarrow \quad C_{Di} = \frac{C_L^2 S}{2\pi b^2} = \frac{C_L^2}{2\pi AR}, \tag{7.19}$$

where the *aspect ratio* is defined by§

§ The aspect ratio is a measure of the span relative to the average chord. Sailplanes have high AR wings, whereas wings of high-speed jets are usually low AR.

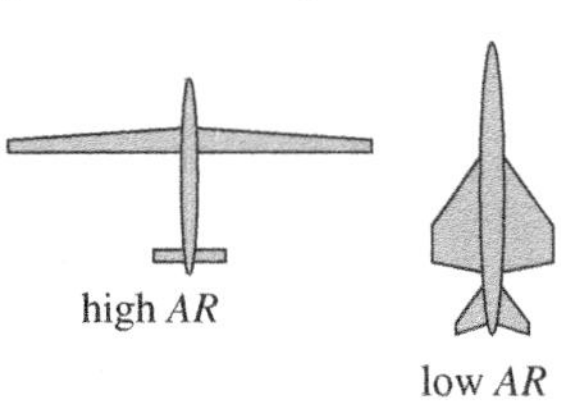

$$AR \equiv \frac{b^2}{S}. \tag{7.20}$$

The result in Equation 7.19 is not that accurate, in that the drag coefficient is under-predicted by about a factor of 2, but it does give the correct quadratic dependence of the induced drag coefficient on the lift coefficient, which we will derive more accurately in Section 7.3.

7.2.2 Formation Flight

The simple horseshoe vortex model from the previous section can be used to analyze the benefit of formation flight, which is a reduction in induced drag when two aircraft fly close to each other.* Consider two identical aircraft flying in the formation shown in Figure 7.14. We assume that each aircraft weight is W and that the aircraft are in steady-level flight. The goal is to calculate the induced drag on each aircraft in terms of $W, \rho U_\infty^2, b, \beta, \gamma$.

* Even though the absolute drag result in the horseshoe vortex model is not that accurate, of interest in formation flight is the relative drag reduction, which is reasonably predicted by the horseshoe vortex model.

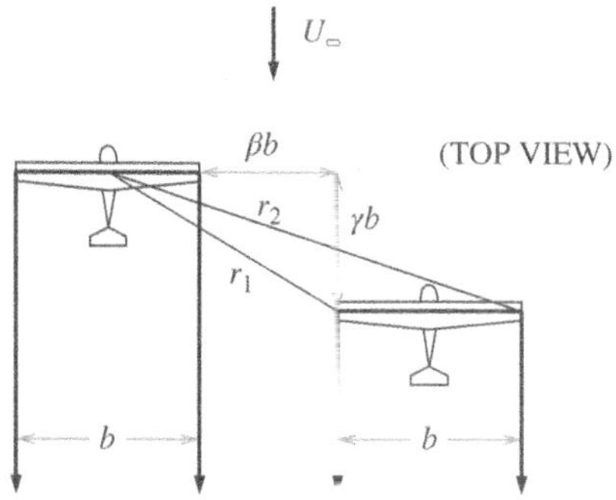

Figure 7.14 Two aircraft in formation.

The lift and induced drag on each aircraft are given by the generalized Kutta–Joukowsky formula for a horseshoe vortex, Equation 7.14,

$$L = \rho b \Gamma U_\infty, \quad D_i = -\rho b \Gamma w, \tag{7.21}$$

where w is the vertical velocity, which we call the *downwash*, at the center of the wing. Since the aircraft are in steady-level flight, lift is equal to weight, W, so that

$$W = L = \rho b \Gamma U_\infty \quad \Rightarrow \quad \Gamma = \frac{W}{\rho b U_\infty}. \tag{7.22}$$

Therefore, to calculate the induced drag on each aircraft, we need to calculate the downwash values at the wing centers. To do this, we use the formula in Equation 7.13 multiple times for each finite or semi-infinite vortex segment. The downwash on the front aircraft, w_1, is caused by all four trailing vortices and the bound vortex of the rear aircraft.†

† In Figure 7.14, we define

$$r_1 = b\sqrt{\left(\beta + \frac{1}{2}\right)^2 + \gamma^2},$$

$$r_2 = b\sqrt{\left(\beta + \frac{3}{2}\right)^2 + \gamma^2}.$$

$$\begin{aligned} w_1 = &-\frac{\Gamma}{\pi b} + \frac{\Gamma}{4\pi b(\beta + 1/2)}\left(1 - \frac{\gamma b}{r_1}\right) - \frac{\Gamma}{4\pi b(\beta + 3/2)}\left(1 - \frac{\gamma b}{r_2}\right) \\ &+ \frac{\Gamma}{4\pi \gamma b}\left(\frac{-(\beta + 1/2)b}{r_1} + \frac{(\beta + 3/2)b}{r_2}\right), \end{aligned} \tag{7.23}$$

where the last term is the bound-vortex contribution. Similarly, the downwash on the rear aircraft, w_2, is caused by the four trailing vortices and the bound vortex of the front aircraft,

$$\begin{aligned} w_2 = &-\frac{\Gamma}{\pi b} + \frac{\Gamma}{4\pi b(\beta + 1/2)}\left(1 + \frac{\gamma b}{r_1}\right) - \frac{\Gamma}{4\pi b(\beta + 3/2)}\left(1 + \frac{\gamma b}{r_2}\right) \\ &- \frac{\Gamma}{4\pi \gamma b}\left(\frac{-(\beta + 1/2)b}{r_1} + \frac{(\beta + 3/2)b}{r_2}\right), \end{aligned} \tag{7.24}$$

where again the last term is the bound-vortex contribution. The drag values are then

$$(D_i)_1 = -\rho b \Gamma w_1, \quad (D_i)_2 = -\rho b \Gamma w_2. \tag{7.25}$$

The reduction of the induced drag on each aircraft, compared to flying alone, depends on their relative position, γ and β, but in general, the rear aircraft will benefit more than the front aircraft. Figure 7.15 shows this by plotting the drags normalized by the flying-alone drag: note that as the separation β increases, the benefit decreases.

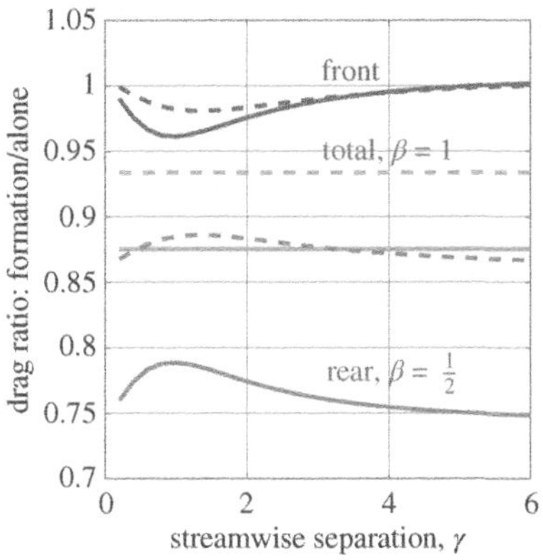

Figure 7.15 Formation drag ratios. Values lower than one indicate a drag benefit compared to flying alone. Dashed lines are for $\beta = 1$, solid are for $\beta = \frac{1}{2}$.

Consider now the sum of the induced drag values on both aircraft. Is there a benefit to the system from flying in formation? Using the above equations, the sum of the induced drag values is

$$\begin{aligned}(D_i)_1 + (D_i)_2 &= -\rho b\Gamma\,(w_1 + w_2)\\ &= -\rho b\Gamma\left(-\frac{2\Gamma}{\pi b} + \frac{2\Gamma}{4\pi b(\beta + 1/2)} - \frac{2\Gamma}{4\pi b(\beta + 3/2)}\right)\\ &= \frac{2\rho\Gamma^2}{\pi} - \frac{\rho\Gamma^2}{2\pi}\frac{1}{(\beta + 1/2)(\beta + 3/2)}\end{aligned}$$

The first term in this result is the induced drag sum for the aircraft flying independently (e.g. $\beta \to \infty$). The second term represents the reduction in drag due to the formation flight. The ratio of drag reduction to the non-formation total drag is

$$\frac{\frac{\rho\Gamma^2}{2\pi}\frac{1}{(\beta+1/2)(\beta+3/2)}}{\frac{2\rho\Gamma^2}{\pi}} = \frac{1}{4}\frac{1}{(\beta + 1/2)(\beta + 3/2)}. \tag{7.26}$$

As β decreases, meaning that the aircraft are closer together spanwise, the benefit in drag reduction increases.‡

‡ Note that the drag savings does not depend on γ. This is a consequence of what's called *Munk's stagger theorem*, presented by Munk (1921) and Glauert (1923): the induced drag of a system does not depend on the streamwise spacing of the individual components (e.g. wings) of the system.

7.2.3 A Predictive Model

The single horseshoe vortex model of Section 7.2.1 is not predictive in that it does not tell us how much lift or drag a wing will produce in a particular flight configuration. That is, we have not related the vortex strength Γ to the geometry of the wing or to the angle of attack and freestream speed.

We can make the horseshoe vortex model predictive by combining it with a two-dimensional sectional model for the lift generated by an airfoil. What distinguishes the three-dimensional model from the two-dimensional one is the effect of the downwash, which is the vertical velocity induced by the trailing vortices. The downwash affects the angle of attack seen by the sectional model, which in turn affects the lift.

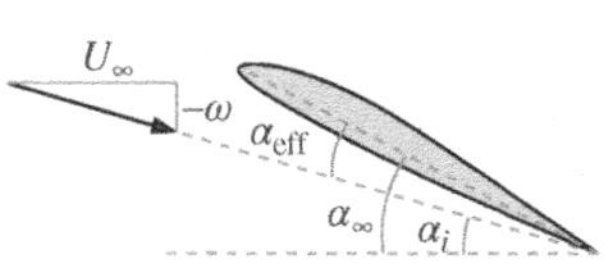

Figure 7.16 Effective angle of attack definition.

For a single horseshoe vortex, we measure the downwash at the root of the wing, and the velocity there is given by Equation 7.15. The downwash is $w = -\Gamma/(\pi b)$, and it affects the angle of attack seen by the root section. Specifically, since w is negative (assuming positive lift), the sectional angle of attack is reduced by the induced angle of attack, as indicated in Figure 7.16, which shows the root section. Assuming small angles, the induced angle of attack is, in radians,

$$\alpha_i \equiv -\frac{w}{U_\infty}. \tag{7.27}$$

So if the wing is flying at an angle of attack of α_∞, the root section sees an angle of

$$\alpha_{\text{eff}} = \alpha_\infty - \alpha_i. \tag{7.28}$$

Using a linear angle-lift relationship from airfoil theory, the sectional lift model is

$$c_\ell = a_0(\alpha_{\text{eff}} - \alpha_{L=0}) = a_0\left(\alpha_\infty + \frac{w}{U_\infty} - \alpha_{L=0}\right), \tag{7.29}$$

where a_0 is the two-dimensional lift-curve slope,* and $\alpha_{L=0}$ is the zero-lift angle of attack.

* Note, $a_0 = 2\pi$ for thin airfoils.

The sectional lift coefficient relates to Γ by the Kutta–Joukowsky theorem,

$$L' = \rho U_\infty \Gamma \quad \Rightarrow \quad c_\ell = \frac{L'}{\frac{1}{2}\rho U_\infty^2 c} = \frac{2\Gamma}{U_\infty c}. \tag{7.30}$$

Substituting this expression into Equation 7.29, we obtain

$$\frac{2\Gamma}{U_\infty c} = a_0 \left(\alpha_\infty + \frac{w}{U_\infty} - \alpha_{L=0} \right). \tag{7.31}$$

The unknown in this equation is Γ, which also appears on the right in $w = -\Gamma/(\pi b)$. Solving for Γ gives

$$\Gamma = \frac{a_0(\alpha_\infty - \alpha_{L=0})}{\frac{2}{U_\infty c} + \frac{a_0}{\pi U_\infty b}}. \tag{7.32}$$

The wing's lift coefficient is then†

† In the horseshoe vortex model, the lift distribution is constant, so $L = L'b$.

$$C_L = \frac{L'b}{\frac{1}{2}\rho U_\infty^2 S} = \frac{2\Gamma b}{U_\infty S} = \left(\frac{a_0}{\frac{S}{bc} + \frac{a_0}{2\pi AR}} \right)(\alpha_\infty - \alpha_{L=0}). \tag{7.33}$$

The factor in front of the angle difference in the last expression is the three-dimensional lift-curve slope. For a high aspect-ratio ($AR \to \infty$) rectangular ($S = bc$) wing, this slope reduces to a_0, which is consistent with the sectional model. For general finite wings, the slope will be less than a_0. Finally, the induced drag coefficient depends quadratically on the lift coefficient according to Equation 7.19.

7.2.4 Multiple Vortices Along a Line

A single horseshoe vortex is not a very accurate model because it does not predict any variation of lift across the span of the wing. A simple way to improve the model is to put together N vortices, spaced equally along the span, as shown in Figure 7.17. This model will still have limitations, including an inability to account for chordwise variations and sweep, but it should show marked improvements over a single vortex.

The unknowns in this model are the strengths of the N horseshoe vortices, Γ_j. If we were given a desired sectional lift distribution over the span, we could choose each Γ_j to match it according to the Kutta–Joukowsky formula, $L'(y) = \rho U_\infty \Gamma(y)$.* Although not predictive, such an approach would still allow us to calculate the corresponding induced drag coefficient by summing the drag forces on each individual horseshoe.

* The result would be a staircase approximation to the given distribution.

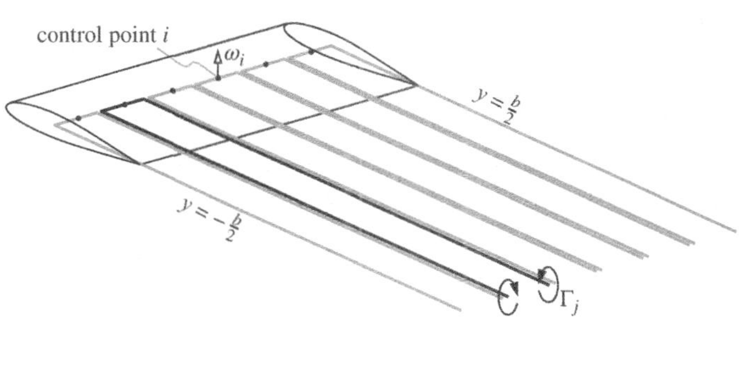

Figure 7.17 A multiple horseshoe vortex model of a wing. One of the $N = 6$ vortices shown, vortex j, is highlighted, as is one of the control points, i.

To make the model predictive, or even to just calculate the induced drag, we need to calculate the downwash at the midpoints of the horseshoe vortices. This downwash is caused by the trailing vortices of all of the horseshoes. The bound portions do not contribute because they are all collinear. Denote by y_i^M, y_i^L, and y_i^R, respectively the y locations of the midpoint, left side, and right side of the ith horseshoe vortex. At y_i^M, the two trailing vortices of horseshoe j each contribute a velocity magnitude of $\Gamma/(4\pi r)$, where r is the distance from the vortex to the midpoint.

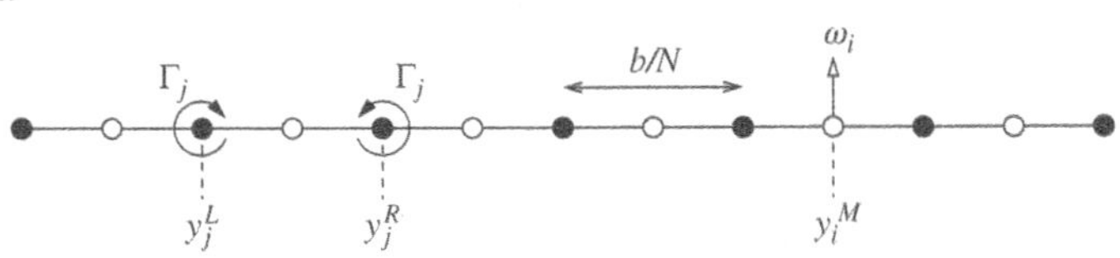

Figure 7.18 Downwash calculation in the multiple horseshoe vortex model of a wing.

Using Figure 7.18 to obtain the correct signs, we sum the contributions from all N vortices at midpoint y_i^M to obtain

$$\begin{aligned} w_i &= \sum_{j=1}^{N}\left[\frac{-\Gamma_j}{4\pi(y_i^M - y_j^L)} + \frac{\Gamma_j}{4\pi(y_i^M - y_j^R)}\right] \\ &= \sum_{j=1}^{N}\frac{1}{4\pi}\left[\frac{-1}{(y_i^M - y_j^L)} + \frac{1}{(y_i^M - y_j^R)}\right]\Gamma_j. \end{aligned} \tag{7.34}$$

This equation shows that the downwash depends linearly on the vortex strengths. We can therefore write the vector of downwash values at all midpoints, $\mathbf{w}$, as a matrix-vector multiplication involving the vortex strengths,

$$\mathbf{w} = \mathbf{W}\Gamma, \tag{7.35}$$

where Γ is the vector of vortex strengths. Comparing to Equation 7.34, the (i, j) entry of the $N \times N$ matrix $\mathbf{W}$ is

$$W_{ij} = \frac{1}{4\pi}\left[\frac{-1}{(y_i^M - y_j^L)} + \frac{1}{(y_i^M - y_j^R)}\right]. \tag{7.36}$$

The unknowns in the predictive model are the N vortex strengths, Γ_j, and the equations are the N sectional lift expressions at the horseshoe midpoints. One side of the equation comes from the sectional lift given by the Kutta–Joukowsky expression. At y_i^M, the sectional lift from Kutta–Joukowsky is $L' = \rho U_\infty \Gamma_i$, so that the sectional c_ℓ is

$$c_\ell = \frac{L'}{\frac{1}{2}\rho U_\infty^2 c_i} = \frac{2\Gamma_i}{U_\infty c_i}, \tag{7.37}$$

where c_i is the chord length at y_i^M.†

† This chord length can vary over the span, so that non-rectangular wings can be analyzed.

The other side of the equation is the airfoil model, which is the same as in Equation 7.29. Setting the two c_ℓ expressions equal gives

$$\frac{2\Gamma_i}{U_\infty c_i} = a_0\left(\alpha_\infty + \frac{w_i}{U_\infty} - \alpha_{L=0,i}\right). \tag{7.38}$$

We can write this set of equations in matrix form, where the unknowns are the vortex strengths. Note that the downwash w_i involves the unknowns, so we put w_i on the left-hand side,

$$\frac{2\Gamma_i}{U_\infty c_i} - \frac{a_0 w_i}{U_\infty} = a_0(\alpha_\infty - \alpha_{L=0,i}). \tag{7.39}$$

Substituting the matrix downwash formula from Equation 7.35, we obtain

$$\frac{2\Gamma_i}{U_\infty c_i} - \sum_{j=1}^{N} \frac{a_0}{U_\infty} W_{ij}\Gamma_j = a_0(\alpha_\infty - \alpha_{L=0,i}). \tag{7.40}$$

This equation is now in the form, $\mathbf{A\Gamma} = \mathbf{F}$, where

$$A_{ij} = \frac{2\delta_{ij}}{U_\infty c_i} - \frac{a_0}{U_\infty} W_{ij}, \quad F_i = a_0(\alpha_\infty - \alpha_{L=0,i}). \tag{7.41}$$

Note, δ_{ij} is 1 if $i = j$ and 0 otherwise.

After solving for the vortex strengths, we can compute the lift and drag coefficients. We proceed by first calculating the force on each individual vortex. By Equation 7.14, the lift and induced drag forces on horseshoe vortex i, of span Δy, are‡

$$L_i = \rho U_\infty \Gamma_i \Delta y, \quad D_i = -\rho w_i \Gamma_i \Delta y. \tag{7.42}$$

‡ We drop the "i" subscript from the next few induced drag equations to not confuse the subscript with the index i.

The total lift and induced drag forces are then sums over the vortices,

$$L = \sum_{i=1}^{N} L_i, \quad D = \sum_{i=1}^{N} D_i. \tag{7.43}$$

The force coefficients are§

$$C_L = \frac{L}{\frac{1}{2}\rho U_\infty^2 S}, \quad C_D = \frac{D}{\frac{1}{2}\rho U_\infty^2 S}. \tag{7.44}$$

§ For a rectangular wing, $S = bc$ is the planform area.

Example 7.5 Modeling a rectangular wing with multiple vortices

This example shows how to implement the multiple horseshoe vortex model for a rectangular wing that consists of symmetric airfoils. Cambered airfoils, flaps, variable chord, and wing twist could all be included but are left out for simplicity in this example. The code given in Listing 7.2 follows the procedure outlined in this section in order to set up and solve the matrix system for the unknown vortex strengths.

Listing 7.2 Multiple vortex wing model code.

```
function [CL, CD] = multiplehv(ainf, N);
% Computes the lift and induced drag coefficients of a wing using N
% horseshoe vortices.  The angle of attack, ainf, is in radians.

% Definitions
a0 = 2*pi;          % 2D lift curve slope
b  = 10;            % span
c  = 1;             % chord
aL0 = 0;            % zero-lift angle of attack
U  = 1;             % freestream speed (arbitrary)
rho = 1;            % density (arbitrary)

dy = b/N;             % span length of each horseshoe
y = linspace(-b/2,b/2, N+1); % horseshoe endpoints
ym = linspace((dy-b)/2, (b-dy)/2, N); % horseshoe midpoints

% set up the system and downwash matrices
W = zeros(N,N); A = W; F = zeros(N,1);
for i = 1:N,  % loop over spanwise midpoints
  for j = 1:N, % loop over unknowns
    W(i,j) = (-1/(ym(i)-y(j)) + 1/(ym(i)-y(j+1)))/(4*pi);
    A(i,j) = -(a0/U)*W(i,j);
  end
```

Code notes:

- The inputs into the code are the angle of attack and the number of vortices.
- The density and freestream speed are arbitrarily set to unity, as they cancel out in the calculation of the force coefficients.
- The main loop that creates the matrix **A** also creates the downwash matrix **W**, which is used in the induced drag calculation.
- After solving for the unknown Γ_j values, the code calculates the lift and drag on the wing by summing together the forces on the individual horseshoes, from Equation 7.14.

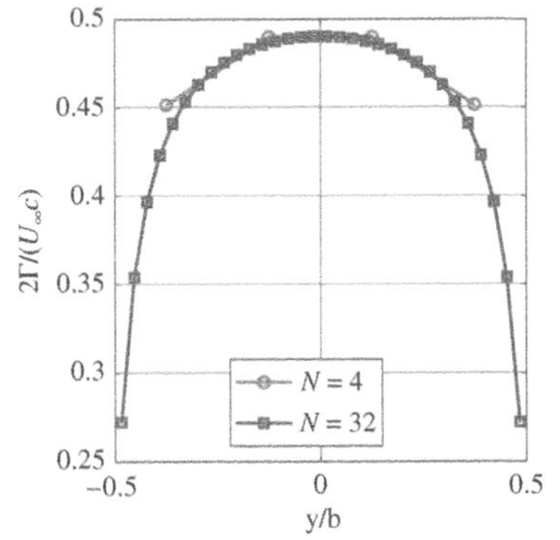

Figure 7.19 Normalized vortex strength distributions.

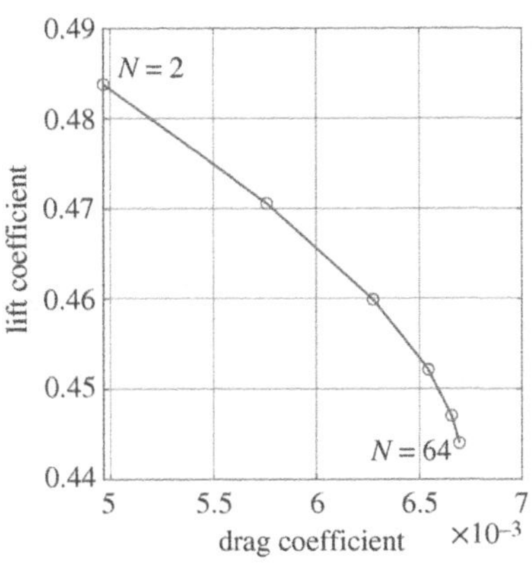

Figure 7.20 Force coefficients for a rectangular wing modeled with N horseshoe vortices.

```
    A(i,i) = A(i,i) + 2/(U*c); % cl effect on eqn
    F(i) = a0*(ainf-aL0); % right-hand side
end

% solve the system for the vortex strengths
G = A\F;

% compute the lift and induced drag coefficients
L = 0; D = 0;
w = W*G; % downwash at each vortex midpoint
L =  rho*dy*U*sum(G); % lift
D = -rho*dy*sum(w.*G); % drag
S = b*c;
CL = L/(0.5*rho*U^2*S);
CD = D/(0.5*rho*U^2*S);
```

Figure 7.19 shows the distributions of Γ_j versus the span for $N = 4$ and $N = 32$. The strongest vortices are near the root, and their strength tapers off near the tips. The sharp variation of the vortex strength near the tips suggests that a non-equal distribution of vortices, with smaller Δy near the tips, could be more efficient in resolving the lift distribution with fewer vortices.

The code assumes a flat-plate ($a_0 = 2\pi$) rectangular wing of span $b = 10$ and chord $c = 1$. At an angle of attack $\alpha_\infty = 5°$, we analyze the same wing using various N, from 2 to 64. Figure 7.20 shows the resulting values of the lift coefficient and drag coefficient, plotted together. Each data point is the (drag, lift) coefficient result from one run. With increasing N, the values of C_L and C_{Di} appear to converge, in that the differences in the coefficients become smaller for each successive doubling of N.

7.3 Lifting-Line Theory

The previous section showed how one or more horseshoe vortices can be used to represent the lift distribution on a wing. Instead of discrete vortices, we can also use smooth sinusoidal functions to represent the lift distribution.* The result is Prandtl's lifting-line theory, which is based on a continuous vortex sheet model for the shed vorticity. This section presents the theory and its implications, after clarifying the applicable assumptions.

*This idea is similar to the camber slope representation in thin-airfoil theory.

7.3.1 Assumptions

Prandtl's lifting-line theory (Bertin and Cummings 2001; Katz and Plotkin 2001; Anderson 2016) requires several assumptions about the flow and the wing geometry. These assumptions are as follows:

- Three-dimensional, steady, potential flow.*
- High aspect ratio ($AR \gtrsim 8$).
- Low/zero sweep ($\Lambda \approx 0$).
- Zero crossflow ($v_y = 0$).†
- A flat wake, as shown in Figure 7.6.

*Here, this means inviscid, irrotational, and incompressible flow.

† This assumption lets us treat each section as two-dimensional.

Figure 7.21 illustrates a typical wing geometry and defines the coordinate system. The sweep angle, Λ, is measured as the angle between the y axis and the quarter-chord line. Note that variations of the sectional properties across the span, y, are allowed, including variable chord length, geometric twist, and sectional camber.

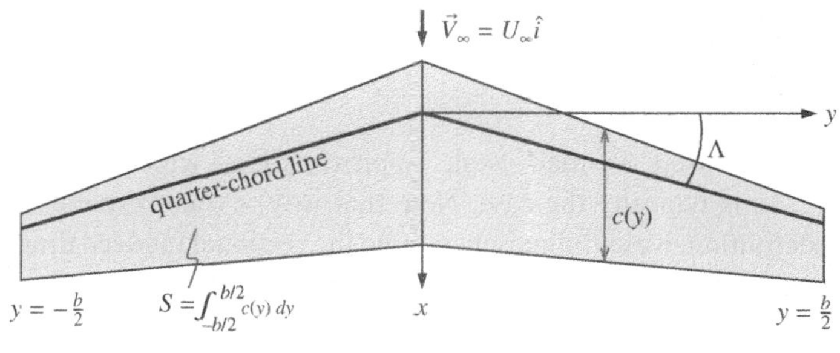

Figure 7.21 Geometry for Prandtl's lifting-line theory. The sectional chord is $c(y)$, the wing span is b, and the planform area of the entire wing is S.

7.3.2 Sectional Forces

An important observation with regard to Figure 7.6 is that shed vorticity in the wake induces a vertical velocity at the wing, changing the force on the wing. A wing producing positive lift, in the $+\hat{k}$ direction, will generally shed vorticity that induces a *downwash*, which is velocity in the $-\hat{k}$ direction, at the wing.* This is illustrated in Figure 7.22.

* It is possible to have local spots of upwash even if producing positive net lift, depending on the lift distribution over the span.

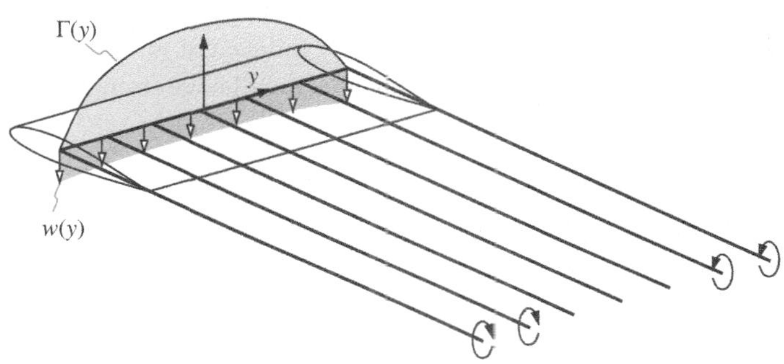

Figure 7.22 Trailing vortices induce a downwash at the wing.

We denote downwash by $w(y)$, the z component of the velocity vector. Note that we usually will have $w(y) < 0$. So, instead of the pure freestream, the wing sees†

† See also Figure 7.16.

$$\vec{v}(y)|_{\text{wing}} = \vec{V}_\infty + w(y)\hat{k} = U_\infty \hat{\imath} + w(y)\hat{k}, \tag{7.45}$$

where we used the fact that in the wind-aligned coordinate system, the freestream is in the $\hat{\imath}$ direction. The sectional force at span location y, is given by the generalized Kutta–Joukowsky theorem,

$$\begin{aligned} \vec{F}' &= \rho\vec{v}(y) \times [\Gamma(y)\hat{\jmath}] \\ &= \rho U_\infty \Gamma(y)\hat{\imath} \times \hat{\jmath} + \rho w(y)\Gamma(y)\hat{k} \times \hat{\jmath} \\ &= \underbrace{\rho U_\infty \Gamma(y)\hat{k}}_{L'(y)} \underbrace{- \rho w(y)\Gamma(y)\hat{\imath}}_{D_i'(y)}. \end{aligned} \tag{7.46}$$

The first term, in the $\hat{k}$ direction, is the sectional lift force, since it is perpendicular to the freestream. The second term, in the $\hat{\imath}$ direction, is parallel to the freestream and hence it is a sectional (induced) drag force. Note that sectional induced drag will generally be a positive number, since typically $w(y) < 0$ and $\Gamma > 0$. The lift result should look familiar, since it is the same as the two-dimensional Kutta–Joukowsky result in Section 4.5, and it is similar to the horseshoe vortex result without the span b. The sectional induced drag result is specific to three dimensions: it states that the drag is directly related to the downwash. In two dimensions, we had no trailing vortices, no downwash, and hence no drag.

Similarly to Equation 7.27, we define the sectional *induced angle of attack* as

$$\alpha_i(y) = \frac{-w(y)}{U_\infty}, \tag{7.47}$$

where we have used a small angle approximation, which is valid when $w(y) \ll U_\infty$, as is typically the case. Note that $w(y) < 0$ corresponds to $\alpha_i > 0$. Using this definition, we can concisely rewrite the sectional induced drag in terms of the sectional lift,‡

‡ This relationship can also be interpreted as the sectional lift vector being "tilted" backward by an angle of α_i, giving rise to the sectional drag which is $L'(y)\sin(\alpha_i) \approx L'(y)\alpha_i$ by the small angle approximation.

$$D_i'(y) = -\rho w(y)\Gamma(y) = \rho\left(\frac{-w(y)}{U_\infty}\right)U_\infty\Gamma(y) = \alpha_i L'(y). \tag{7.48}$$

To compute $D_i'(y)$, we need the induced angle of attack, $\alpha_i(y)$, or equivalently the downwash, $w(y)$.

7.3.3 Downwash Calculation

To obtain the downwash on the wing, we need to integrate the contributions from all of the trailing vortices. Suppose that we are interested in the vertical velocity at span location y_0, as illustrated in Figure 7.23. Trailing vortices at all other span locations will contribute to the vertical velocity induced at y_0. Consider first just one infinitesimal trailing vortex originating at y. It will have strength

$$d\Gamma = \frac{d\Gamma}{dy}(y)dy.$$

At y_0, this trailing vortex induces an infinitesimal velocity in the $\hat{k}$ direction of*

* We use Equation 7.13 for a semi-infinite vortex, i.e. with $\beta_1 = 90°$ and $\beta_2 = 0°$.

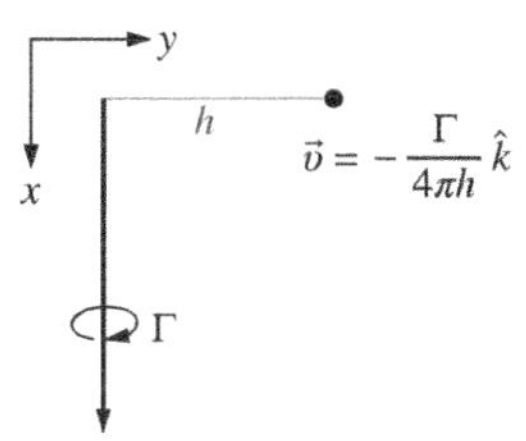

$$dw = -\frac{d\Gamma}{4\pi(y_0 - y)}.$$

Note that when $d\Gamma$ is positive, the trailing vortex induces clockwise velocity when viewed from the rear of the aircraft (i.e. looking along the $-\hat{\imath}$ axis). This vortex induces a downward velocity at $y_0 > y$, and hence the minus sign above.

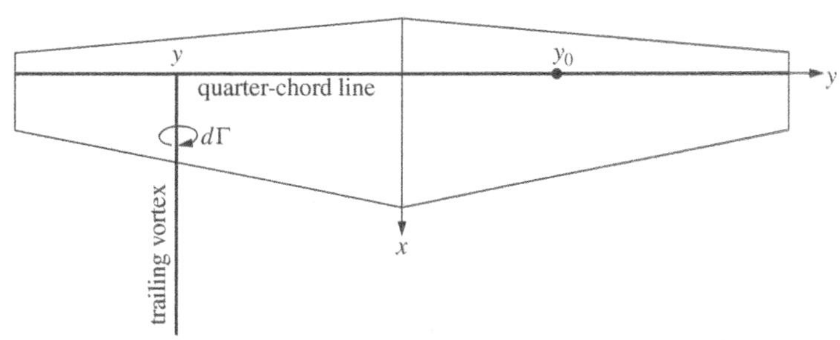

Figure 7.23 Setup for calculating the downwash at span location y_0.

Integrating over the span, we obtain the downwash at point y_0 due to the entire wake,

$$w(y_0) = -\frac{1}{4\pi}\int_{-b/2}^{b/2}\frac{d\Gamma/dy}{y_0 - y}dy. \tag{7.49}$$

Note that the trailing vortices close to y_0 have a larger effect on the induced velocity than those far away, because $y - y_0$ is in the denominator. Also, large $d\Gamma/dy$ means stronger shed vorticity and hence a larger downwash. The induced angle of attack at y_0 is then $-w(y_0)/U_\infty$,

$$\alpha_i(y_0) = \frac{1}{4\pi U_\infty}\int_{-b/2}^{b/2}\frac{d\Gamma/dy}{y_0 - y}dy. \tag{7.50}$$

The strength of vorticity shed in the wake is determined by $\Gamma(y)$, specifically via the slope, $d\Gamma/dy$, but the distribution of $\Gamma(y)$ is not known ahead of time in a wing analysis.†

† Solving for Γ is a coupled problem: the shed vorticity in the wake affects the downwash at all points along the span of the wing, whereas the downwash affects the lift, and hence $\Gamma(y)$, through the induced angle of attack. The details of this interaction give rise to the fundamental equation of Prandtl's lifting-line theory.

7.3.4 Fundamental Equation of Lifting-Line Theory

To understand the relationship between downwash and the lift distribution, we use results from airfoil theory to relate the sectional lift coefficient to an effective sectional angle of attack, which was introduced for the horseshoe vortex in Equation 7.28. We use the same definition here, where the downwash reduces the angle of attack of a section by α_i, the induced angle of attack.

Consider a section of the wing at y_0. Suppose that if downwash were not present, this section would see an angle of attack of*

$$\alpha(y_0) = \underbrace{\alpha_\infty}_{\text{aircraft angle of attack}} + \underbrace{\alpha_g(y_0)}_{\text{geometric twist}} . \tag{7.51}$$

* The decomposition of α into an aircraft angle of attack and a geometric twist angle of the section with respect to the aircraft is somewhat arbitrary. It is done for convenience, especially when the twist of each section is expressed in a body-fixed coordinate system. However, $\alpha(y_0)$ remains well-defined as the angle between the freestream and the chordline of the airfoil at section y_0.

As in the case of the horseshoe vortex, illustrated in Figure 7.16, the effect of downwash is that it superimposes on the freestream and reduces the angle of attack, so that the section actually sees an effective angle of attack of

$$\alpha_{\text{eff}}(y_0) = \alpha(y_0) - \alpha_i(y_0). \tag{7.52}$$

Recall that the induced angle of attack, $\alpha_i(y_0) = -w(y_0)/U_\infty$, is positive when the downwash is indeed downward, i.e. when $w(y_0) < 0$. Combining Equation 7.51 and Equation 7.52, we obtain the following angle summation relationship:

$$\alpha_\infty + \alpha_g(y_0) = \alpha_{\text{eff}}(y_0) + \alpha_i(y_0). \tag{7.53}$$

This relationship is illustrated graphically in Figure 7.24.

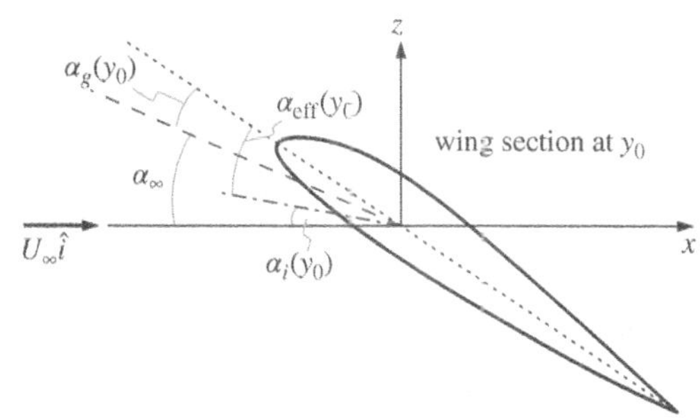

Figure 7.24 Relationship between the angle of attack of the aircraft relative to the freestream, α_∞, the twist angle of a wing section at y_0, $\alpha_g(y_0)$, the induced angle of attack at y_0, $\alpha_i(y_0)$, and the effective angle of attack at y_0, $\alpha_{\text{eff}}(y_0)$.

The effective angle of attack determines the sectional lift coefficient,

$$c_\ell(y_0) = a_0(y_0)\left[\alpha_{\text{eff}}(y_0) - \alpha_{L=0}(y_0)\right], \tag{7.54}$$

where $a_0(y_0)$ is the sectional lift-curve slope,† and $\alpha_{L=0}(y_0)$ is the sectional zero-lift angle of attack. The zero-lift angle of attack is determined by the shape of the airfoil. When a wing uses different airfoils at different sections, $\alpha_{L=0}$ varies with position along the span.

† If the airfoils used in the wing are thin, then we can use $a_0 = 2\pi$. However, Prandtl's lifting-line theory does not require that the lift curve slope be 2π, and this is why we leave a_0 as a variable.

From Equation 7.46, we can express the sectional lift coefficient in terms of the vortex strength

$$L'(y_0) = \rho U_\infty \Gamma(y_0) \quad \Rightarrow \quad c_\ell(y_0) = \frac{2\Gamma(y_0)}{U_\infty c(y_0)}. \tag{7.55}$$

Substituting this result into Equation 7.54 yields

$$\alpha_{\text{eff}}(y_0) = \frac{c_\ell(y_0)}{a_0(y_0)} + \alpha_{L=0}(y_0) = \frac{2\Gamma(y_0)}{a_0(y_0)U_\infty c(y_0)} + \alpha_{L=0}(y_0). \tag{7.56}$$

The angle summation equation in Equation 7.53 now becomes

$$\alpha_\infty + \alpha_g(y_0) = \frac{2\Gamma(y_0)}{a_0(y_0)U_\infty c(y_0)} + \alpha_{L=0}(y_0) + \alpha_i(y_0). \tag{7.57}$$

Finally, we substitute the definition of the induced angle of attack, Equation 7.50, to obtain the *fundamental equation of Prandtl's lifting-line theory*,‡

‡ This equation came from an angle summation statement: the effective angle of attack equals the sectional angle of attack minus the induced angle of attack.

$$\alpha_\infty + \alpha_g(y_0) = \frac{2\Gamma(y_0)}{a_0(y_0)U_\infty c(y_0)} + \alpha_{L=0}(y_0) + \int_{-b/2}^{b/2} \frac{(d\Gamma/dy)dy}{4\pi U_\infty(y_0 - y)}. \tag{7.58}$$

This equation must hold at every y_0 location. The effect of the downwash comes in through the induced angle of attack, which depends on an integral of $d\Gamma/dy$.

In a wing analysis problem, we know:

$$\begin{aligned} \alpha_\infty + \alpha_g(y_0) &= \text{sectional angle of attack} \\ c(y_0) &= \text{sectional chord length} \\ U_\infty &= \text{freestream velocity magnitude} \\ a_0(y_0) &= \text{two-dimensional lift-curve slope (e.g. } 2\pi) \\ \alpha_{L=0}(y_0) &= \text{sectional zero-lift angle of attack} \end{aligned}$$

Although the only unknown in Equation 7.58 is $\Gamma(y_0)$, we need to solve an integro-differential equation to calculate it.

7.3.5 Coordinate Transformation

To simplify Equation 7.58, we use a coordinate transformation for y, the distance along the span,

$$y = \frac{-b}{2}\cos\theta. \tag{7.59}$$

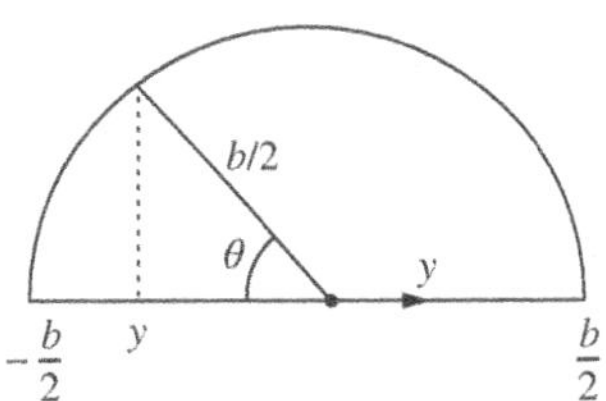

Figure 7.25 Graphical interpretation of the transformation from y to θ.

That is, we transform y, which ranges from $-b/2$ to $b/2$, to the new variable θ, which ranges from 0 to π, as illustrated in Figure 7.25. Note that this transformation is not the same as the one used for x in our discussion of thin-airfoil theory.*

* It is still nonlinear: y changes most slowly with θ at the wing tips.

Now instead of solving for $\Gamma(y)$, we solve for $\Gamma(\theta)$. This is most easily done using a Fourier expansion,

$$\Gamma(\theta) = 2bU_\infty \sum_{n=1}^{\infty} A_n \sin n\theta, \tag{7.60}$$

where the expansion coefficients A_n are dimensionless, since the quantity bU_∞ has units of circulation. We do not use any $\cos(\theta)$ terms because Γ must vanish at the wing tips, which are at $\theta = 0$ and $\theta = \pi$. The expansion of $\Gamma(\theta)$ is illustrated graphically in Figure 7.26.

To obtain an equation in θ, we substitute the expansion for $\Gamma(\theta)$ into Prandtl's lifting-line equation, Equation 7.58. A tricky term is the induced angle of attack,

$$\begin{aligned}\alpha_i(y_0) &= \frac{1}{4\pi U_\infty}\int_{-b/2}^{b/2}\frac{d\Gamma/dy}{y_0-y}dy\\ &= \frac{1}{4\pi U_\infty}\int_0^{\pi}\frac{d\Gamma/d\theta}{-b/2\cos\theta_0+b/2\cos\theta}d\theta\\ &= \frac{1}{2\pi b U_\infty}\int_0^{\pi}\frac{d\Gamma/d\theta}{\cos\theta-\cos\theta_0}d\theta.\end{aligned} \tag{7.61}$$

To obtain $d\Gamma/d\theta$, we differentiate the Fourier expansion in Equation 7.60,

$$\frac{d\Gamma}{d\theta} = 2bU_\infty\sum_{n=1}^{\infty} nA_n\cos n\theta, \tag{7.62}$$

Therefore,

$$\alpha_i(y_0) = \frac{1}{\pi}\int_0^{\pi}\sum_{n=1}^{\infty} nA_n\frac{\cos n\theta}{\cos\theta-\cos\theta_0}d\theta = \sum_{n=1}^{\infty} nA_n\frac{\sin n\theta_0}{\sin\theta_0}, \tag{7.63}$$

where the last step is the result of an integral identity. Substituting the expansion for $\Gamma(\theta)$ and the above expression for α_i into Equation 7.58, we obtain Prandtl's lifting-line equation in terms of θ,

$$\alpha_\infty + \alpha_g(\theta) = \frac{4b}{a_0(\theta)c(\theta)}\sum_{n=1}^{\infty}A_n\sin n\theta + \alpha_{L=0}(\theta) + \sum_{n=1}^{\infty}nA_n\frac{\sin n\theta}{\sin\theta}. \tag{7.64}$$

We dropped the 0 subscript from θ since this equation no longer has an integral with a separate integration variable.

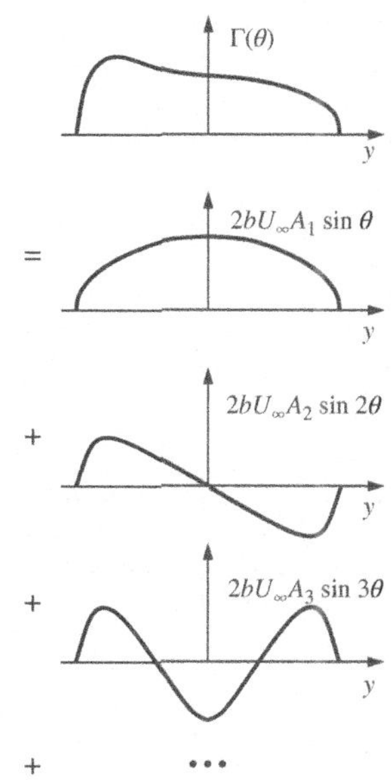

Figure 7.26 Fourier expansion of $\Gamma(\theta)$ into modes.

7.4 Lifting-Line Solutions

This section presents two techniques for solving Equation 7.64. These are both numerical and approximate, with the quality of the approximation improving with increasing cost, as measured by the number of unknowns.

7.4.1 Point Collocation

Given a wing geometry and sectional properties, $c(\theta)$, b, $\alpha_g(\theta)$, $\alpha_{L=0}(\theta)$, $a_0(\theta)$, the unknowns in Equation 7.64 are the A_n coefficients. Since n ranges from 1 to ∞, there are infinitely many unknowns! However, since Equation 7.64 must be satisfied for all $\theta \in (0, \pi)$, we have an infinite number of equations too. One way to make the problem tractable is to consider only a finite number of unknowns. That is, we truncate the infinite series in Equation 7.64 and only consider N coefficients: A_n, $n = 1,2, \ldots, N$. When we do this, the system becomes overdetermined, as we generally can no longer satisfy Equation 7.64 for all locations on the wing, θ. There are various ways to reduce the number of equations, and in this section, we present one approach: *point collocation*.

The idea in point collocation is to *enforce the required equation at a finite number of points*. The steps in this approach applied to Prandtl's lifting-line equation are as follows:

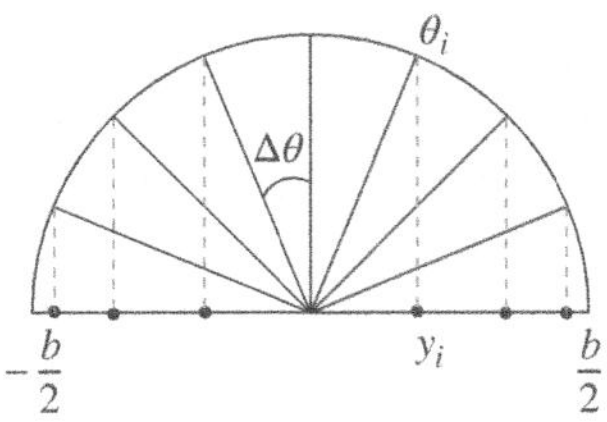

Figure 7.27 Cosine spacing (equal intervals in θ) used in the point collocation solution approach to Prandtl's lifting-line equation.

1. Choose N spanwise locations at which to enforce Equation 7.64. Denote these *collocation points* by θ_i, which correspond to locations y_i. The locations of these points are important. Clearly, they should be spread out over the entire wing to capture any sectional variations well. However, this alone is not enough: for example, an equal spacing in y leads to a numerically ill-conditioned system that results in an oscillating lift distribution for large N. Much better is an equal spacing in θ, which yields a *cosine spacing* in y, as shown in Figure 7.27. See Trefethen (2000) for a thorough discussion of the impact of point spacing on interpolation and other numerical methods.
2. At the collocation points, calculate the sectional properties $c(\theta_i)$, $\alpha_g(\theta_i)$, $\alpha_{L=0}(\theta_i)$, and $a_0(\theta_i)$.
3. Write down N equations by enforcing Equation 7.64 at the N collocation points.
4. From the equations in the previous step, set up a linear $N \times N$ matrix system by putting all terms involving the unknowns A_n on the left-hand side, and all other terms on the right-hand side.* The equations will now look like

$$\frac{4b}{a_0(\theta_i)c(\theta_i)}\sum_{n=1}^{N} A_n \sin n\theta_i + \sum_{n=1}^{N} nA_n \frac{\sin n\theta_i}{\sin \theta_i} = \alpha_\infty + \alpha_g(\theta_i) - \alpha_{L=0}(\theta_i),$$

which in matrix form looks like

$$\begin{bmatrix} M_{11} & M_{12} & \dots & M_{1N} \\ M_{21} & M_{22} & \dots & M_{2N} \\ \vdots & \vdots & \ddots & \vdots \\ M_{N1} & M_{N2} & \dots & M_{NN} \end{bmatrix} \begin{bmatrix} A_1 \\ A_2 \\ \vdots \\ A_N \end{bmatrix} = \begin{bmatrix} F_1 \\ F_2 \\ \vdots \\ F_N \end{bmatrix}. \tag{7.65}$$

where†

$$M_{in} = \frac{4b}{a_0(\theta_i)c(\theta_i)} \sin n\theta_i + n\frac{\sin n\theta_i}{\sin \theta_i} \tag{7.66}$$

$$F_i = \alpha_\infty + \alpha_g(\theta_i) - \alpha_{L=0}(\theta_i) \tag{7.67}$$

5. Solve the matrix system, Equation 7.65, for the A_n coefficients.

* In a lift-constrained analysis, the angle of attack, α_∞ is an unknown, and so it moves to the left-hand side of the equation. In contrast, A_1, which relates to the lift coefficient, is then known, so that terms with A_1 move to the right-hand side.

† In the matrix form, i is the row number, and n is the column number.

The resulting vortex strength distribution,

$$\Gamma(\theta) = 2bU_\infty \sum_{n=1}^{N} A_n \sin n\theta, \tag{7.68}$$

will in general be an *approximation* to the true distribution because of the truncated series expansion. We expect the accuracy to improve with increasing N, and often, only several modes are sufficient to analyze typical wings with smoothly varying lift distributions along the span.

7.4.2 Force Calculation

Having solved for the A_n coefficients, we can now calculate the forces on the wing. Of primary interest is the total lift and the total induced drag on the wing, although knowing the entire distribution of lift across the span lets us calculate moments about the x axis, e.g. rolling and bending moments.*

First, the sectional lift at one span location θ is given by Equation 7.46, with Γ calculated using the truncated Fourier expansion in Equation 7.68,

$$L'(\theta) = \rho U_\infty \Gamma(\theta) = 2b\rho U_\infty^2 \sum_{n=1}^{N} A_n \sin n\theta, \tag{7.69}$$

The *total lift*, L, is the integral of the sectional lift along the span,

$$L = \int_{-b/2}^{b/2} L'(y)dy. \tag{7.70}$$

Transforming to the θ variable and substituting for $L'(\theta)$, we obtain†

$$L = \int_0^\pi \underbrace{\left[2b\rho U_\infty^2 \sum_{n=1}^{N} A_n \sin n\theta,\right]}_{L'(\theta)} \underbrace{\left[\frac{b}{2}\sin\theta d\theta\right]}_{dy} = \frac{\pi}{2}b^2\rho U_\infty^2 A_1. \tag{7.71}$$

This equation says that only the A_1 term contributes to lift! All of the other A_n, $n > 1$, do not affect the total lift produced by the wing.‡ However, all A_n affect the lift *distribution*, $L'(y)$. The corresponding *lift coefficient* of the wing is

$$C_L = \frac{L}{\frac{1}{2}\rho U_\infty^2 S} = \frac{\pi b^2 A_1}{S} = \pi AR A_1, \tag{7.72}$$

where we used the definition of aspect ratio, $AR = b^2/S$.

Turning now to drag, from Equation 7.48, the *sectional induced drag* is

$$D_i'(\theta) = \alpha_i(\theta)L'(\theta) = \left[\sum_{n=1}^{N} nA_n \frac{\sin n\theta}{\sin\theta}\right]\left[2b\rho U_\infty^2 \sum_{m=1}^{N} A_m \sin m\theta\right], \tag{7.73}$$

where we used different indices in the two sums, since the two quantities are being multiplied together, and we need to keep the indices separate. To obtain the *total induced drag*, we integrate along the span,

$$D_i = \int_{-b/2}^{b/2} D_i'(y)dy. \tag{7.74}$$

As in the lift case, we evaluate the integral by transforming to θ,

$$\begin{aligned} D_i &= \int_0^\pi \left[\sum_{n=1}^{N} nA_n \frac{\sin n\theta}{\sin\theta}\right]\left[2b\rho U_\infty^2 \sum_{m=1}^{N} A_m \sin m\theta\right]\left(\frac{b}{2}\sin\theta d\theta\right) \\ &= b^2\rho U_\infty^2 \sum_{n=1}^{N} nA_n \sum_{m=1}^{N} A_m \int_0^\pi \sin n\theta \sin m\theta d\theta \\ &= D_i = \frac{\pi}{2}b^2\rho U_\infty^2 \sum_{n=1}^{N} nA_n^2, \end{aligned} \tag{7.75}$$

where in the last step, we again used orthogonality of the sine functions. Note that while lift depends only on A_1, the induced drag depends on all of the A_n coefficients. The corresponding *induced-drag coefficient* is

* We cannot obtain pitch moments because lifting-line theory does not provide information about the chordwise pressure distribution.

† Since $y = -\frac{b}{2}\cos\theta$, $dy = \frac{b}{2}\sin\theta$. Also, the integral

$$\int_0^\pi \sin n\theta \sin m\theta \, d\theta$$

is 0 if $m \neq n$, and $\pi/2$ otherwise.

‡ This conclusion can also be reached graphically, by looking at the $\sin n\theta$ modes mapped to y: all modes beyond the first one have a net zero contribution to the lift.

$$C_{Di} = \frac{D_i}{\frac{1}{2}\rho U_\infty^2 S} = \pi AR \sum_{n=1}^{N} nA_n^2. \tag{7.76}$$

When the lift is not zero, we can rewrite the induced-drag coefficient using $C_L = \pi ARA_1$,§

$$C_{Di} = \pi ARA_1^2 \sum_{n=1}^{N} n\frac{A_n^2}{A_1^2} = \frac{C_L^2}{\pi AR}\left[1 + \sum_{n=2}^{N} n\frac{A_n^2}{A_1^2}\right]$$
$$= \frac{C_L^2}{\pi AR}(1+\delta). \tag{7.77}$$

§ The quantity δ, defined as

$$\delta = \sum_{n=2}^{N} n\frac{A_n^2}{A_1^2},$$

is always non-negative, and its definition relies on $A_1 \neq 0$, which means that the wing must be producing some lift.

We can also define the *span efficiency factor*, e, as

$$e = \frac{1}{1+\delta}. \tag{7.78}$$

Note that since $\delta \geq 0, e \leq 1$. Substituting into the expression for the induced drag coefficient, we obtain a classical relationship between the lift and induced-drag coefficients,¶

$$C_{Di} = \frac{C_L^2}{\pi ARe}. \tag{7.80}$$

¶ Multiplying both sides by the dynamic pressure and the planform area, we obtain the corresponding dimensional form of this equation,

$$D_i = \frac{1}{\frac{1}{2}\rho U_\infty^2 \pi}\left(\frac{L}{b}\right)^2 \frac{1}{e}. \tag{7.79}$$

This expression is sometimes more useful for design purposes: it states that the induced drag is proportional to the square of the *span loading*, L/b.

This equation tells us that the induced-drag coefficient depends *quadratically* on the lift coefficient.

Example 7.6 Analysis of a rectangular wing

Consider a rectangular wing of aspect ratio AR, as shown in Figure 7.28. Assume that the wing is untwisted, $\alpha_g(\theta) = 0$, and uncambered, $\alpha_{L=0}(\theta) = 0$. Also, suppose that the lift-curve slope is $a_0 = 2\pi$.

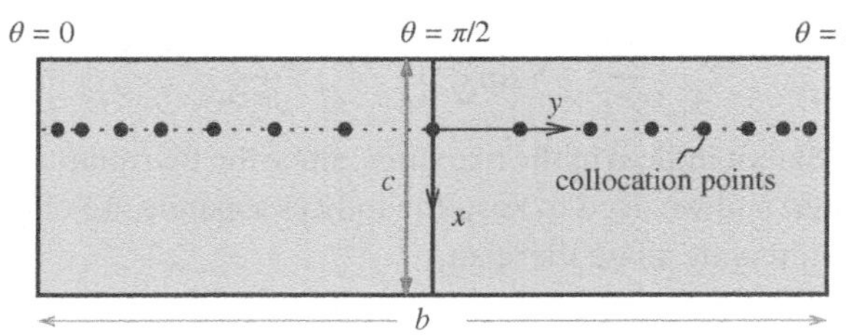

Figure 7.28 A rectangular wing and the collocation points on one half of the wing.

We write a code to solve Equation 7.64 for a finite number of unknowns, A_n, using the point collocation method. Figure 7.28 shows the cosine-spacing placement of the collocation points over the entire wing. The number of unknowns is the same as the number of collocation points, so that we have the same number of equations as unknowns. In this case, we could take advantage of symmetry‖ and only solve for the odd coefficients, $A_1, A_3, A_5, \ldots$ by placing points only on the right half of the wing, but to simplify the presentation, we solve for all A_n and place the collocation points over the entire wing.

‖ Since the wing is left-right symmetric, and since there is no sideslip (freestream is aligned with the x axis), the lift distribution has to be left-right symmetric. This means that $\Gamma(\theta)$ must be symmetric about $y = 0$ which corresponds to $\theta = \pi/2$. The even modes, $\sin(2\theta), \sin(4\theta), \ldots$ are *antisymmetric* about $\theta = \pi/2$, and hence, they cannot be present in the expansion. So $A_2, A_4, A_6, \ldots$ must be zero.

Listing 7.3 presents a code that analyzes this wing numerically using the collocation method. The code takes as input the freestream angle of attack, α_∞, and the wing aspect ratio, AR. It computes the lift coefficient, C_L, and the induced-drag coefficient, C_{Di}, by building and solving the system given in Equation 7.65.

Listing 7.3 A lifting-line code for analysis of a rectangular wing.

```
function [CL, CDi] = rectwing(alpha, AR)
% Computes the lift and induced drag coefficients of a rectangular
% wing of aspect ratio AR at an angle of attack alpha, in radians.

% Definitions
a0 = 2*pi;            % 2D lift curve slope
N  = 40;              % number of unknowns
M  = zeros(N,N);      % coefficient matrix
F  = zeros(N,1);      % right-hand-side vector

t = linspace(pi/(2*N), pi-pi/(2*N), N); % collocation theta points

% set up system
for i = 1:N,   % loop over spanwise locations
  for n = 1:N,   % loop over unknowns
    M(i,n) = (4*AR/a0 + n/sin(t(i)))*sin(n*t(i));
  end
  F(i) = alpha; % fill in right-hand side
end

A = M\F;            % solve the system
CL = pi*AR*A(1); % lift coefficient
CDi = pi*AR*sum([1:N]'.*(A.^2)); % induced drag coefficient
```

Code notes:

- The inputs into the code are the freestream angle of attack and the aspect ratio of the wing.
- No collocation points are placed at the wing tips: they start at $\pi/(2N)$ away from $\theta = 0$ and $\theta = \pi$. This is because the tips do not provide any new information (we know $\Gamma = 0$ there), and including these points causes indeterminate forms in the M_{in} calculation.
- With zero twist and camber, the right-hand side vector is constant and equal to α_∞.
- After solving for the unknown A_n values, the code calculates the lift and drag coefficients using the formulas in terms of A_n.

For the present wing, we have,

$$\frac{4b}{a_0(\theta_i)c(\theta_i)} = \frac{4b}{a_0 c} = \frac{4b^2}{a_0 bc} = \frac{4AR}{a_0}. \tag{7.81}$$

Using Equation 7.66 and Equation 7.67, the entries of the matrix and right-hand side then become

$$M_{in} = \left(\frac{4AR}{a_0} + \frac{n}{\sin\theta_i}\right)\sin(n\theta_i), \qquad F_i = \alpha_\infty. \tag{7.82}$$

Running the code with various angles of attack and plotting the resulting C_L, C_{Di} pairs yield the drag polar. Figure 7.29 shows the polars generated with this code for several aspect ratios.

The quadratic dependence of the induced-drag coefficient on the lift coefficient is evident by the parabolic shape of the polars. As the aspect ratio increases, the parabolas become shallower, which means that for a given lift coefficient, the induced-drag coefficient is smaller. This is expected from Equation 7.80, although the relationship between C_{Di} and AR is not a simple inverse one, because the span efficiency factor, e, changes with AR for a rectangular wing. In fact, e decreases with increasing AR, an effect that can be mitigated by adding taper to the wing.

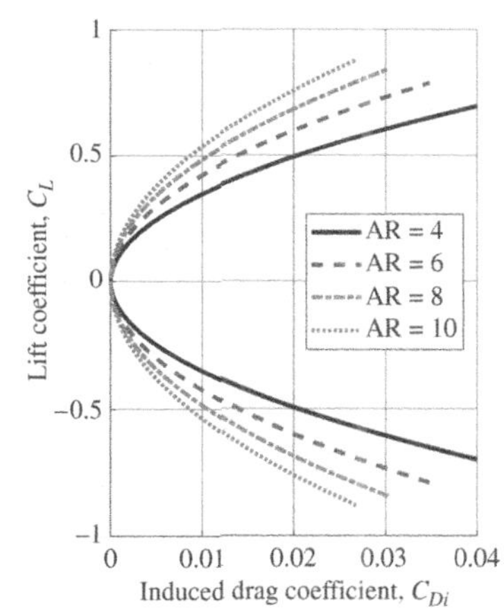

Figure 7.29 Drag polars for an untwisted and uncambered rectangular wing at several aspect ratios.

7.4.3 Weighted Residuals

Collocation is not the only method for solving Prandtl's lifting-line equation. With collocation, we have to decide where to put the collocation points, and not all choices work well.* An alternative solution approach is to not enforce PLL at discrete points, but instead to enforce equations in an integral sense. This is known as the method of weighted residuals.

Both the collocation method and the weighted residual method solve for a finite number of A_n coefficients, but they differ in how the equations are obtained. In collocation, the equations are obtained by requiring Prandtl's lifting-line (PLL) equation to hold at a finite number of points. In the method of weighted residuals,

* For example, points that are equally spaced in y lead to poor conditioning of the resulting system for a large number of points.

the equations are obtained by multiplying Prandtl's lifting-line equation by a finite number of test functions and requiring that for each test function, the weighted integral of the left-hand side matches that of the right-hand side.†

† The extent to which Prandtl's lifting-line equation is not pointwise satisfied is the residual, and each test function is a weight.

In weighted residuals, we choose test functions instead of points. A choice that spares us an additional decision is to use the same functions as those in the approximation of the lift distribution. These are the sine functions, $\sin(m\theta)$, for $m = 1, 2, 3, \ldots$. The weighted residual equations are then‡

‡ Another subtle choice is the variable over which the integral is performed, e.g. y versus θ. We use θ because the test functions are defined in terms of θ. Choosing a different variable amounts to using different test functions, scaled by the Jacobian of the variable transformation.

$$\int_0^\pi (\text{PLL LHS}) \sin(m\theta) d\theta = \int_0^\pi (\text{PLL RHS}) \sin(m\theta) d\theta, \tag{7.83}$$

where LHS/RHS refer to the left/right-hand sides of the lifting-line equation.§ Prandtl's lifting-line equation in terms of θ, with the unknowns on the left-hand side, is

§ Each m corresponds to one equation, so we have one equation per $\sin(m\theta)$ test function.

$$\sum_{n=1}^{N} \left[\frac{4b}{a_0(\theta)c(\theta)} + \frac{n}{\sin\theta} \right] (\sin n\theta) A_n = \alpha_\infty + \alpha_g(\theta) - \alpha_{L=0}(\theta). \tag{7.84}$$

Multiplying this equation by $\sin(m\theta)$¶ and integrating over θ yield

¶ When using only odd Fourier coefficients in the presence of left-right symmetry, we must also use only odd m values, since residuals associated with even m are automatically going to be zero.

$$\sum_{n=1}^{N} M_{mn} A_n = F_m, \tag{7.85}$$

where

$$M_{mn} = \int_0^\pi \left[\frac{4b}{a_0(\theta)c(\theta)} + \frac{n}{\sin\theta} \right] \sin(n\theta)\sin(m\theta) d\theta, \tag{7.86}$$

$$F_m = \int_0^\pi \left[\alpha_\infty + \alpha_g(\theta) - \alpha_{L=0}(\theta) \right] \sin(m\theta) d\theta. \tag{7.87}$$

Some of these integrals can be evaluated analytically depending on the wing geometry and twist distribution. However, for arbitrary wing shapes, numerical integration provides a more general approach.

A simple numerical integration technique is trapezoidal integration, with the caveat that we should avoid including the wingtips ($\theta = 0, \pi$) to prevent indeterminate 0/0 forms in the calculation of M_{mn}. One way to handle this constraint, illustrated in Figure 7.30, is to remove the boundary points and to increase the integration weight on the adjacent points. For improved accuracy, we make the boundary intervals half the size of the regular interval, so that their integration weight is the same as an interior point.

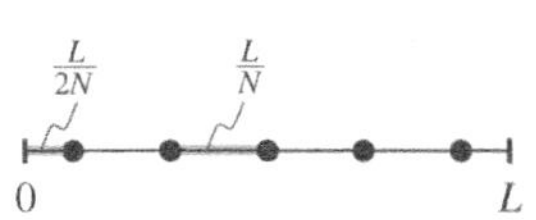

Figure 7.30 Trapezoidal integration without endpoints. The black dots are the integration points, and the weights are all L/N.

Working with the full wing, the range of θ is 0 to π. For n_q points, the interval size using a half interval at the tips is $\Delta\theta = \pi/n_q$. Denote by θ_i the locations of the n_q integration points, starting at $\theta_1 = \Delta\theta/2$ and ending at $\theta_{n_q} = \pi - \Delta\theta/2$. The corresponding weights, w_i, are all $\Delta\theta$. The integrals over θ become sums over integration points,

$$M_{mn} = \sum_{i=1}^{n_q} \left[\frac{4b}{a_0(\theta_i)c(\theta_i)} + \frac{n}{\sin\theta_i} \right] \sin(n\theta_i)\sin(m\theta_i) w_i, \tag{7.88}$$

$$F_m = \sum_{i=1}^{n_q} \left[\alpha_\infty + \alpha_g(\theta_i) - \alpha_{L=0}(\theta_i) \right] \sin(m\theta_i) w_i. \tag{7.89}$$

The following examples show the implementation of this method and some of its advantages over collocation.

Example 7.7 Weighted residuals for a rectangular wing
Consider the same rectangular wing as in Example 7.6, which is left-right symmetric, but for which again we solve for all of the A_n coefficients. Listing 7.4 presents a code that analyzes this wing using the method of weighted residuals. The code takes as input the freestream angle of attack, α_∞, and the wing aspect ratio, AR. It computes the lift coefficient, C_L, and the induced-drag coefficient, C_{Di}, by building and solving the system given in Equation 7.65.

Listing 7.4 An implementation of the method of weighted residuals.

```
function [CL, CDi] = mwr(alpha, AR)
% Computes lift and induced drag coefficients of a
% rectangular wing using weighted residuals.
% alpha = angle of attack [rad], AR = aspect ratio

% Definitions
a0 = 2*pi;              % 2D lift curve slope
N = 40;                 % number of unknowns
M  = zeros(N,N);        % coefficient matrix
F  = zeros(N,1);        % right-hand-side vector

% trapezoidal integration, excluding theta = pi
nq = 500;                          % number of points
dt = pi/nq;                        % interval size
t = linspace(dt/2, pi-dt/2, nq);   % points
w = ones(size(t))*dt;              % weights

% set up system
for m = 1:N,   % loop over test functions
  % loop over unknowns
  for n = 1:N,
    M(m,n) = dot(w,(4*AR/a0 + n./sin(t)).*sin(n*t).*sin(m*t));
  end
  % right-hand side
  F(m) = dot(w, alpha.*sin(m*t));
end

A = M\F;               % solve the system
CL = pi*AR*A(1);  % lift coefficient
CDi = pi*AR*sum([1:N]'.*(A.^2)); % induced drag coefficient
```

Code notes:

- The integration sums are performed using a dot product of the integrand and the vector of weights.
- The number of integration points, n_q, needs to be sufficiently high to ensure accurate trapezoidal integration.
- n_q should be at least as large as the number of unknowns, N, but more improves accuracy, especially when the wing geometry has non-smooth features.
- n_q does not affect the size of the linear system, which remains $N \times N$.
- More accurate numerical integration techniques exist that require fewer points.

After solving for the A_n coefficients, the force calculation in the rest of the code is the same as in the collocation method. Figure 7.31 shows the polars generated with this code for several aspect ratios, and these are nearly identical to those in Figure 7.29.

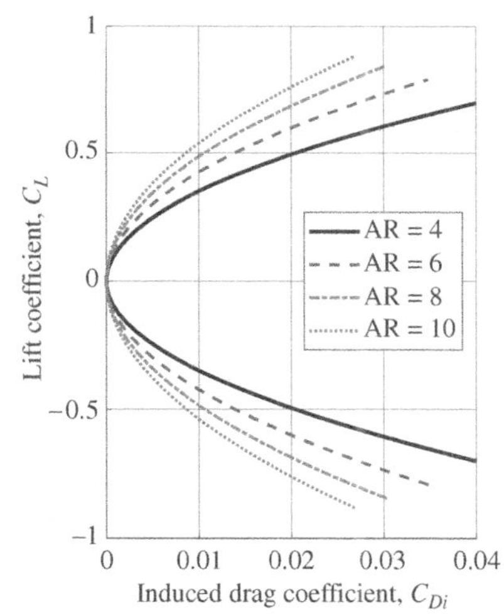

Figure 7.31 Drag polars for an untwisted and uncambered rectangular wing at several aspect ratios, generated using the method of weighted residuals.

The previous example shows that collocation and the method of weighted residuals give very similar results for a plain rectangular wing. However, when we add geometry features with a discontinuity along the span, differences between the two methods arise, and for the same number of unknowns, weighted residuals can be more accurate. The next example illustrates this for the case of a wing with a flap.

Example 7.8 Wing with a flap
Continuing with the same rectangular wing as in the previous examples, we introduce a flap over the inboard half, as shown in Figure 7.32.

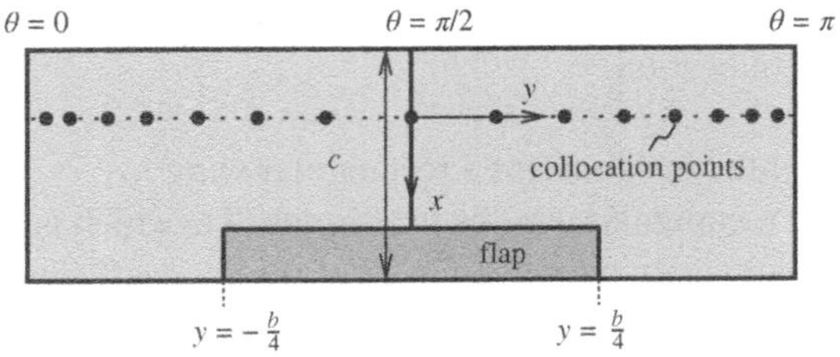

Figure 7.32 A rectangular wing with a flap on the first half of each wing.

The effect of the flap enters Prandtl's lifting-line equation through the sectional zero-lift angle of attack, $\alpha_{L=0}(y)$. This quantity is zero for sections not on the flap, since the wing's airfoil is assumed uncambered in this example. However, when a flap is deflected, the value becomes nonzero. We can compute $\alpha_{L=0}$ for a given flap hinge location and deflection angle using either thin-airfoil theory or a panel method. For this example, we pick a reasonable value of $\alpha_{L=0} = -5°$ for sections with the flap.||

|| A negative value is consistent with a flap-down deflection.

The code modification, given in Listing 7.5, is in the calculation of the right-hand side vector, F_i. For collocation, we compute $\alpha_{L=0}$ at the collocation points and subtract it from F_i, per Equation 7.67. For the method of weighted residuals, we include $\alpha_{L=0}$ at the integration points and subtract it from the integrand, per Equation 7.89.

Code notes:

- `t` is a vector of θ values.
- `tflap` is the θ location of the end of the flap.
- `aL0_noflap` is zero here.
- `aL0_flap` is $\alpha_{L=0}$ for sections on the flap.

Listing 7.5 Sectional zero-lift angle of attack calculation in the presence of a flap.

```
% flap effect: zero lift AOA at collocation points
aL0 = ones(size(t))*aL0_noflap;
I = find(t<tflap);
aL0(I) = aL0_flap;
```

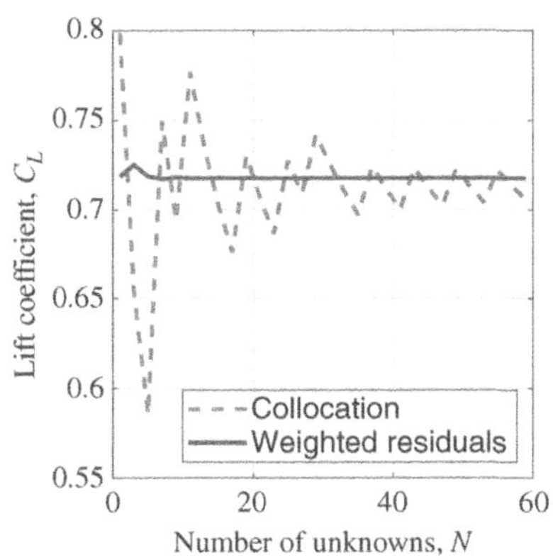

Figure 7.33 Lift convergence versus number of unknowns using collocation and weighted residuals.

We now look at how the lift coefficient behaves as we change the number of unknowns, N. Figure 7.33 shows the results for both collocation and the method of weighted residuals. The collocation method shows oscillations as we increase the number of unknowns, and these are due to the discrete sampling nature of collocation: the fraction of points on the flap, and their location, changes as we increase N. On the other hand, the method of weighted residuals shows a much smoother C_L, with an accurate answer even for small N. This result of course depends on the fact that we are using a large number of integration points (500 in this case), but we could use a more accurate integration method with fewer points. In addition, the size of the linear system remains independent of the number of integration points.

7.5 Wing Design

So far we have presented methods for analyzing high-aspect ratio wings using both horseshoe vortices, in a purely numerical approach, and using a continuous $\Gamma(y)$ distribution in Prandtl's lifting-line theory, a combined analytical and numerical approach. The latter approach yields a connection between the wing performance metrics, lift and induced drag, and parameters governing the vortex distribution, the A_n coefficients. This section presents a design approach that takes advantage of this connection by expressing optimal design requirements in terms of the A_n.

7.5.1 Elliptical Lift Distribution

A common aerodynamic design criterion is to minimize the induced drag for a given amount of lift, and doing this turns out to be quite straightforward in Prandtl's lifting-line theory. Recall the expressions for the lift and induced-drag coefficients from the previous section,

$$C_L = \pi AR A_1, \quad C_{Di} = \pi AR \sum_{n=1}^{N} nA_n^2. \tag{7.90}$$

The lift coefficient sets A_1, but any nonzero A_n, $n > 1$, only serves to increase the induced drag. Therefore, *the minimum induced drag is achieved when $A_n = 0$ for $n > 1$.** With only $A_1 \neq 0$, the sectional lift distribution is, using Equation 7.69,

$$L'(\theta) = 2b\rho U_\infty^2 \sum_{n=1}^{N} A_n \sin n\theta = 2b\rho U_\infty^2 A_1 \sin\theta. \tag{7.91}$$

Using $\sin\theta = \sqrt{1-\cos^2\theta}$ and the transformation from y to θ in Equation 7.59, the equation in terms of y is

$$L'(y) = \underbrace{2b\rho U_\infty^2 A_1}_{L'(y=0)} \sqrt{1-\left(\frac{2y}{b}\right)^2}. \tag{7.92}$$

The expression in the underbrace is the sectional lift at the wing root, $y = 0$, since the square root evaluates to 1 for $y = 0$. We can gain some more insight into this lift distribution by squaring both sides of the equation and rearranging terms,

$$\left(\frac{L'(y)}{L'(0)}\right)^2 + \left(\frac{2y}{b}\right)^2 = 1. \tag{7.93}$$

As shown in Figure 7.34, this is an equation for the upper half of an ellipse! The conclusion is that in Prandtl's lifting-line theory, minimum drag is attained with an *elliptical lift distribution*. The sectional lift, L', when plotted against the span, y, forms the upper half on an ellipse. The elliptical lift distribution has two important properties:

1. *The induced angle of attack is constant.* Recall that α_i is given by,

$$\alpha_i = \sum_{n=1}^{N} nA_n \frac{\sin n\theta}{\sin\theta} = A_1 \text{ when } A_n = 0 \text{ for } n > 1.$$

   So for the elliptical lift distribution, A_1 is equal to the induced angle of attack, and the downwash is constant,

$$w = -\alpha_i U_\infty = -A_1 U_\infty = \text{constant}.$$

   This is illustrated in Figure 7.35.

*Solutions to optimization problems are rarely this simple, but a similar approach is used in the design of slender bodies in supersonic flow, as discussed in Section 8.6.4.

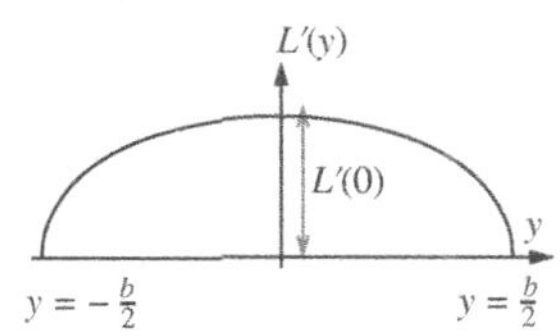

Figure 7.34 An elliptical lift distribution.

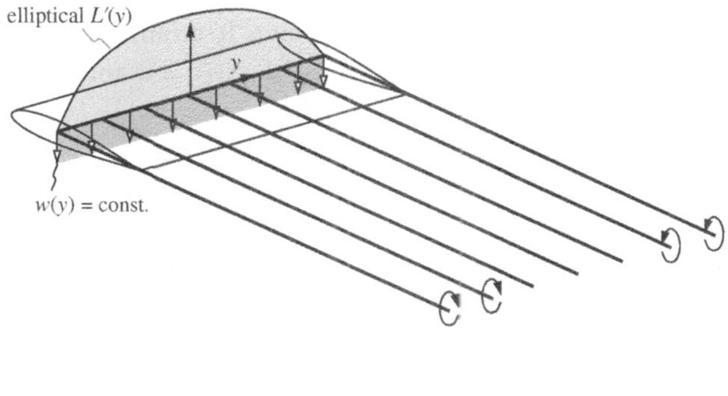

Figure 7.35 The downwash is constant for an elliptical lift distribution.

2. *The span efficiency factor is 1:*†

$$\delta = \sum_{n=2}^{N} n\left(\frac{A_n}{A_1}\right)^2 = 0 \quad \Rightarrow \quad e = (1+\delta)^{-1} = 1.$$

† This means that the induced-drag coefficient for an elliptical lift distribution is

$$C_{Di} = \frac{C_L^2}{\pi AR}.$$

A natural question is: how do we design a wing with an elliptical lift distribution? We must satisfy Prandtl's lifting-line equation, Equation 7.64, which simplifies greatly when $A_n = 0$ for $n \geq 2$,

$$\alpha_\infty + \alpha_g(\theta) = \frac{4b}{a_0(\theta)c(\theta)} A_1 \sin\theta + \alpha_{L=0}(\theta) + A_1. \tag{7.94}$$

In wing design, several terms can be adjusted to make the above equation hold. These include‡

‡ We do not treat $a_0(\theta)$ as a design parameter, since it does not vary much – e.g. it is 2π for thin airfoils.

- $\alpha_g(\theta)$ = sectional geometric twist,
- $\alpha_{L=0}(\theta)$ = sectional zero-lift angle of attack,
- $c(\theta)$ = sectional chord length.

As an example, suppose that we are designing a wing that is flat, $\alpha_g(\theta) = 0$, and has symmetric sections, $\alpha_{L=0}(\theta) = 0$. Also, suppose that $a_0(\theta) = a_0 =$ constant. What does the sectional chord length distribution look like for an elliptical lift distribution? In this case, Equation 7.94 becomes

$$\alpha_\infty = \frac{4b}{a_0 c(\theta)} A_1 \sin\theta + A_1. \tag{7.95}$$

The first term on the right-hand side is the only one that involves θ. For the equation to hold, this term must be constant. Therefore,

$$\frac{\sin(\theta)}{c(\theta)} = \text{constant} \quad \Rightarrow \quad c(\theta) = c_r \sin\theta, \tag{7.96}$$

where c_r is the chord length at the root, $\theta = \pi/2$. In terms of y, this equation is

$$c(y) = c_r\sqrt{1 - \left(\frac{2y}{b}\right)^2}. \tag{7.97}$$

This is an *elliptical chord distribution*, since the plot of the chord versus the span forms the upper part of an ellipse. To satisfy the zero-sweep assumption of Prandtl's lifting-line theory, the planform will appear as shown in Figure 7.36.

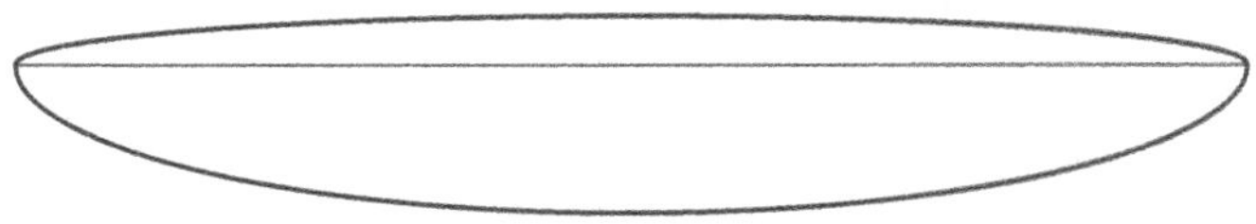

Figure 7.36 A wing with an elliptical chord distribution and zero sweep. The quarter-chord line is parallel to the y axis.

Varying the chord distribution is not the only way to achieve an elliptical lift distribution. We can also vary the geometric twist or the zero-lift angle of attack. For example, a rectangular wing, in which $c(y) = c =$ constant, can produce an elliptical lift distribution if it is twisted such that the wing tips are at a lower angle of attack than the wing root, which is called *washout*. To attain the optimal elliptical lift distribution, the geometric twist, assuming $\alpha_{L=0}(\theta) = 0$ and $a_0 =$ constant, is given by Equation 7.94,§

§ Note that this twist distribution depends on A_1, and hence on the lift coefficient. Therefore, a wing designed in this manner will only produce an elliptical lift distribution for one value of the lift coefficient and will have to be re-twisted when C_L is changed.

$$\alpha_g(\theta) = -\alpha_\infty + \frac{4b}{a_0 c} A_1 \sin\theta + A_1. \tag{7.98}$$

Wing planforms that are flat, meaning untwisted and uncambered, can be *tapered* to approximate an elliptical chord distribution. Wing taper consists of linearly varying the chord distribution from root to tip, as illustrated in Figure 7.37.

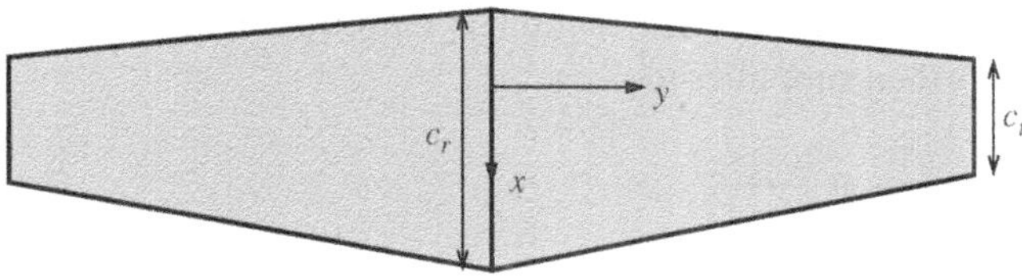

Figure 7.37 A wing with a taper ratio, $\sigma = c_t/c_r < 1$.

The *taper ratio* is defined as the ratio of the tip chord to the root chord:

$$\text{taper ratio} = \sigma = \frac{c_t}{c_r}. \tag{7.99}$$

The effectiveness of taper at approximating an elliptical lift distribution depends on the wing aspect ratio and on the taper ratio. As illustrated in Figure 7.38, a taper ratio of $\sigma \approx 0.3$–0.4 maximizes the span efficiency for a wide range of aspect ratios.

Finally, note that while an elliptical lift distribution will maximize the span efficiency factor, i.e. make $e = 1$, the aspect ratio has an even stronger effect on induced-drag reduction, since

$$C_{Di} = \frac{C_L^2}{\pi ARe}. \tag{7.100}$$

Thus, *high aspect ratios are preferable* for minimizing the induced drag. In dimensional form, e.g. see Equation 7.79, high aspect ratio translates to reduced span loading, L/b. In the limit of $AR \to \infty$ (non-dimensional) or $b \to \infty$ (dimensional), we have an infinite wing, and we recover the two-dimensional result that $C_{Di} \to 0$. As points of reference, subsonic transport aircraft have $AR \approx 10$–15, and some sailplanes have $AR > 30$.

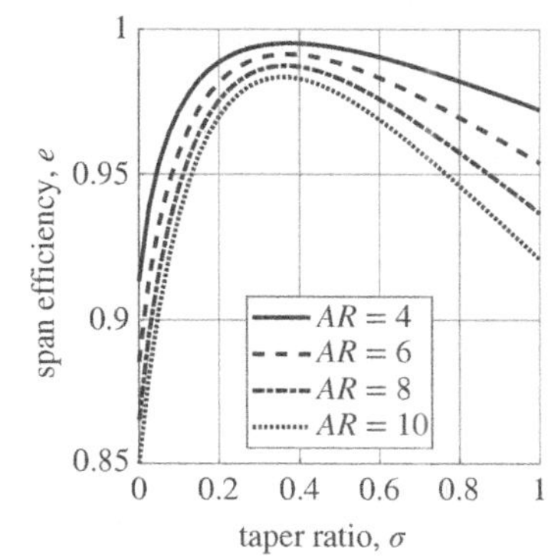

Figure 7.38 Span efficiency factor versus taper ratio, for untwisted and uncambered wings of several aspect ratios.

7.5.2 The Lift-Curve Slope

In Prandtl's lifting-line theory, the lift coefficient, C_L, increases linearly with the angle of attack, α_∞. This can be seen by differentiating Prandtl's lifting-line equation, Equation 7.64,

$$\frac{d}{d\alpha_\infty}\left[\alpha_\infty + \alpha_g(\theta) = \frac{4b}{a_0(\theta)c(\theta)}\sum_{n=1}^{\infty} A_n \sin n\theta + \alpha_{L=0}(\theta) + \sum_{n=1}^{\infty} nA_n \frac{\sin n\theta}{\sin\theta}\right]$$
$$\Rightarrow \quad 1 = \frac{4b}{a_0(\theta)c(\theta)}\sum_{n=1}^{\infty}\frac{dA_n}{d\alpha_\infty}\sin n\theta + \sum_{n=1}^{\infty} n\frac{dA_n}{d\alpha_\infty}\frac{\sin n\theta}{\sin\theta}. \tag{7.101}$$

This is a linear equation for $dA_n/d\alpha_\infty$, and α_∞ does not appear in the equation. Therefore, the derivatives $dA_n/d\alpha_\infty$ *are constant with respect to angle of attack,* α_∞. Since by Equation 7.72 the lift coefficient is directly proportional to A_1,

$$C_L = \pi ARA_1 \quad \Rightarrow \quad \frac{dC_L}{d\alpha_\infty} = \pi AR\frac{dA_1}{d\alpha_\infty} = \text{constant with } \alpha_\infty. \tag{7.102}$$

The derivative $dC_L/d\alpha_\infty$ is called the *lift-curve slope*.

A finite wing will generally have a lift-curve slope that is lower than an infinite (i.e. two-dimensional) wing. For example, consider a wing with an elliptical chord distribution, $c(\theta) = c_r \sin\theta$, and a constant lift-curve slope, $a_0 = \text{constant}$. Also, assume for now that the wing is untwisted, $\alpha_g(\theta) = 0$, and uncambered,

$\alpha_{L=0}(\theta) = 0$, so that the lift distribution is elliptical and only A_1 is nonzero. Equation 7.101 then simplifies to

$$1 = \frac{4b}{a_0 c_r}\frac{dA_1}{d\alpha_\infty} + \frac{dA_1}{d\alpha_\infty}. \tag{7.103}$$

Solving for $dA_1/d\alpha_\infty$ and substituting into Equation 7.102 yield*

$$\frac{dC_L}{d\alpha_\infty} = \frac{\pi AR}{1 + \frac{4b}{a_0 c_r}} = \frac{a_0}{1 + \frac{a_0}{\pi AR}}. \tag{7.104}$$

* For an elliptical chord distribution, the planform area is $S = \pi b c_r/4$, so the aspect ratio is

$$AR = \frac{b^2}{S} = \frac{4b}{\pi c_r}.$$

This result states that the lift-curve slope for a wing with an elliptical chord distribution will be less than the two-dimensional lift-curve slope, a_0. Note that for $AR \to \infty$ the lift-curve slope approaches a_0 and we recover the two-dimensional result as expected. For a general planform shape, we can write,

$$\frac{dC_L}{d\alpha_\infty} = \frac{a_0}{1 + \frac{a_0}{\pi AR}(1+\tau)}, \tag{7.105}$$

where τ depends on the A_n coefficients. For typical wings, a reasonable range for τ is between 0.05 and 0.25. Since $\tau > 0$, *the lift-curve slope for a general planform shape will be less than that for a planform with an elliptical chord distribution.*

Although Equation 7.104 was derived for an uncambered and untwisted wing, the same result holds in the presence of arbitrary twist, $\alpha_g(\theta)$, and camber, $\alpha_{L=0}(\theta)$. This is because $\alpha_g(\theta)$ and $\alpha_{L=0}(\theta)$ do not appear in Equation 7.101, so that the solution for $dA_1/d\alpha_\infty$, and hence $dC_L/d\alpha_\infty$, does not depend on these quantities. Moreover, this argument holds for non-elliptical chord distributions, since Equation 7.101 is true for general $c(\theta)$. Therefore, for arbitrary wings, *geometric twist and sectional camber do not affect the lift-curve slope.*

Looking again at Equation 7.101, we see that the lift-curve slope is affected only by $a_0(\theta)$ and $c(\theta)$. Twist and camber can change C_L, but they will only shift the C_L versus α curve, as illustrated in Figure 7.39.

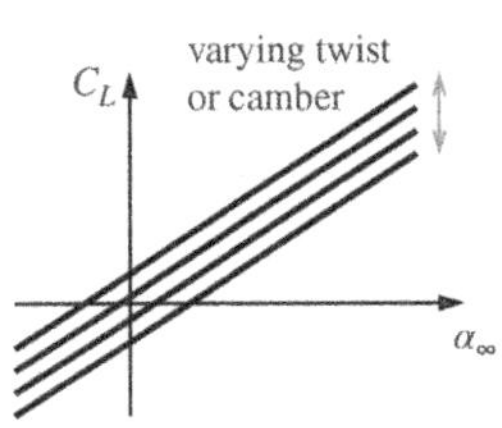

Figure 7.39 In Prandtl's lifting-line theory, twist and camber distributions do not affect the lift-curve slope of a finite wing.

7.5.3 A Special Case

There is a special case that causes some trouble with the above formulas. This is the case when a wing is producing zero total lift, since $L = 0$ means that $A_1 = 0$. If this occurs because the wing is producing no lift distribution, i.e. $L'(y) = 0$, then $\Gamma(y) = 0$ and all of the expansion coefficients $A_1, A_2, A_3, \ldots$ in Equation 7.60 are zero. Therefore in this case, such a wing would be producing zero induced drag, since by Equation 7.75,

$$D_i = \frac{\pi}{2} b^2 \rho U_\infty^2 \sum_{n=1}^{N} n A_n^2 = 0. \tag{7.106}$$

However, suppose that the wing is producing some lift distribution, $L'(y) \neq 0$, but still zero total lift (i.e. the area under the $L'(y)$ curve is zero). Then, we still have $A_1 = 0$, but now some of the A_n coefficients, $n > 1$, must be nonzero according to Equation 7.69. Therefore, by Equation 7.75, we have nonzero induced drag! Induced drag is sometimes called "drag due to lift," but we see from this special case that a more correct characterization of induced drag is "drag due to a *lift distribution.*"*

* In this special case of zero lift but nonzero drag, the span efficiency factor, e, goes to zero, since from Equation 7.80

$$e = \frac{C_L^2}{\pi AR C_{Di}} = 0.$$

Note that in such cases, the span efficiency factor depends on α_∞, since as soon as $C_L \neq 0$, $e > 0$.

To obtain such a special case in reality, we need a wing that produces a lift distribution but zero total lift. That is, some areas of the wing will be producing negative lift while others will be producing positive lift. One way this can happen is if the wing is twisted: at a freestream angle of attack when the wing produces no lift, some sections will be at a positive sectional angle of attack, while other at negative, leading to a lift distribution and induced drag.

Example 7.9 A twisted elliptical wing
Consider an uncambered, $AR = 8$ wing with an elliptical chord distribution. The wing is twisted so that at $\alpha_\infty = 0$, $C_L = 0.5$, $A_3 = 0.1A_1$, and all of the other A_n coefficients are zero. Assume Prandtl's lifting-line theory applies with $a_0(\theta) = 2\pi$.

Given this information, we can calculate the span efficiency factor, e, and the sectional geometric twist, $\alpha_g(\theta)$, at $C_L = 0.5$. The span efficiency factor is $e = 1/(1+\delta)$, where

$$\delta = \sum_{n=2}^{\infty} n\frac{A_n^2}{A_1^2} = 3\frac{A_3^2}{A_1^2} = 0.03. \tag{7.107}$$

Therefore, $e = 1/1.03 \approx 0.971$. The geometric twist can be calculated from Prandtl's lifting-line equation. Since the wing has an elliptical chord distribution, $c(\theta) = c_r \sin\theta$. Also, the wing is uncambered, $\alpha_{L=0}(\theta) = 0$, and the specified condition is for $\alpha_\infty = 0$. Therefore, Equation 7.64 gives

$$\alpha_g(\theta) = \frac{AR}{2}\left(A_1 + A_3\frac{\sin 3\theta}{\sin\theta}\right) + A_1 + 3A_3\frac{\sin 3\theta}{\sin\theta}, \tag{7.108}$$

where we used $4b/c_r = \pi AR$ for an elliptical chord distribution. Since, $C_L = \pi AR A_1$, $A_1 = 0.5/(\pi 8) = 1/(16\pi)$. Substituting the given numbers,

$$\alpha_g(\theta) = \frac{1}{4\pi}\left(1 + 0.1\frac{\sin 3\theta}{\sin\theta}\right) + \frac{1}{16\pi}\left(1 + 0.3\frac{\sin 3\theta}{\sin\theta}\right). \tag{7.109}$$

This twist distribution is plotted in Figure 7.40. Note that the wing tips are twisted up.

We can obtain the sectional lift coefficient by normalizing the sectional lift from Equation 7.69 by the dynamic pressure and local chord,

$$c_\ell = \frac{L'}{\frac{1}{2}\rho U_\infty^2 c} = \frac{2b\rho U_\infty^2 \sum_n A_n \sin n\theta}{\frac{1}{2}\rho U_\infty^2 c_r \sin\theta} = \frac{\pi AR}{\sin\theta}\sum_n A_n \sin n\theta. \tag{7.110}$$

Figure 7.41 plots the sectional lift coefficient versus the normalized span for the given wing lift coefficient of $C_L = 0.5$. The highest sectional lift coefficient occurs at the wing tips, so we expect the tips to be more prone to stall.

Since the chord is not constant, the c_ℓ plot does not represent the lift distribution, which we might want to see if comparing to an elliptical distribution. Hence, another way to normalize the sectional lift is by using the average chord, $\bar{c} \equiv S/b$, which is a constant. For an elliptical wing, $\bar{c} = \pi c_r/4$, so that the normalized lift distribution is

$$c_\ell\frac{c}{\bar{c}} = 4AR\sum_n A_n \sin n\theta. \tag{7.111}$$

Figure 7.42 shows this normalized distribution, which behaves as expected for a typical finite wing: the root produces the highest sectional lift, which then goes to zero at the tips.

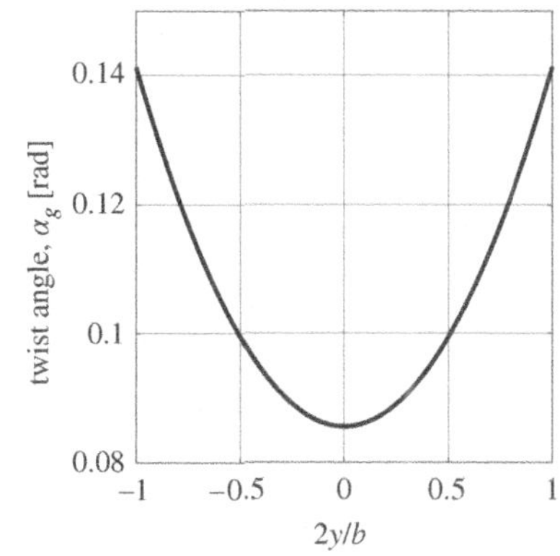

Figure 7.40 Twist distribution.

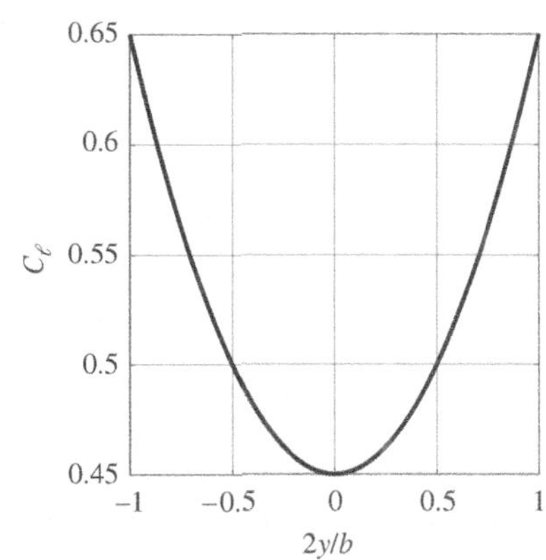

Figure 7.41 Lift coefficient distribution.

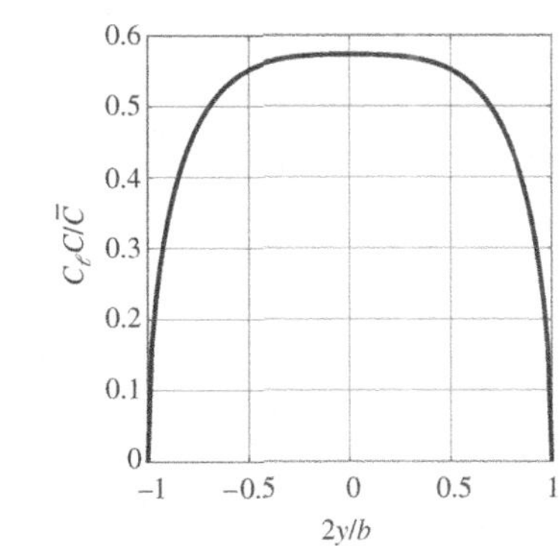

Figure 7.42 Normalized lift distribution.

Next, we plot C_L versus α_∞ and C_L versus C_{Di}. The lift coefficient is a linear function of α_∞,† so we can write

$$C_L = C_L(\alpha_\infty = 0) + \frac{dC_L}{d\alpha_\infty}\alpha_\infty = 0.5 + \frac{\pi AR}{1 + AR/2}\alpha_\infty, \tag{7.112}$$

† Prandtl's lifting-line equation is linear in the A_n coefficients and α_∞, so in the solution, A_1, and hence C_L, depend linearly on α_∞.

where the lift-curve slope is given by Equation 7.104. From C_L, we obtain $A_1 = C_L/(\pi AR)$. For an elliptical wing, A_3 is constant with angle of attack (see Problem 4), and from the given numbers at $\alpha_\infty = 0$, $A_3 = (0.1)(0.5)/(\pi AR) \approx 0.001989$. The induced-drag coefficient is then

$$C_{Di} = \pi AR\left(A_1^2 + 3A_3^2\right). \tag{7.113}$$

Figure 7.43 shows the two relevant plots. The minimum induced drag occurs when the wing is not producing any lift, since by the equation above, A_3 is constant, but A_1 is directly proportional to C_L. This minimum induced drag coefficient is

$$C_{Di,\min} = \pi AR 3A_3^2 = 2.983 \times 10^{-4}. \tag{7.114}$$

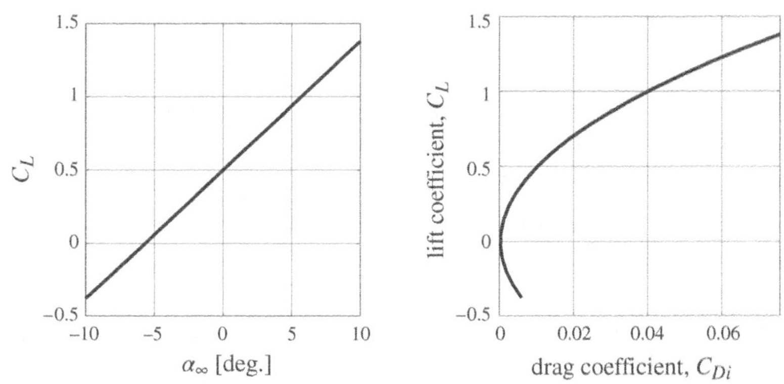

Figure 7.43 Lift and drag coefficient variation with angle of attack for a twisted elliptical lift.

7.6 The Vortex-Lattice Method

As discussed in Section 7.3, Prandtl's lifting-line theory is restricted to wings that have a high aspect ratio and a low sweep. However, many wings exist that are low aspect ratio and/or high sweep. We can analyze such wings numerically using the *vortex-lattice method*. In this section, we outline the steps in this method and give more details for the specific case of a rectangular wing.

7.6.1 Steps

The steps in implementing a vortex-lattice method are similar to those of the vortex-panel method for airfoils. They consist of creating a model for the flowfield using vortices, enforcing flow tangency at points to obtain equations, solving for the unknown vortex strengths, and post-processing the solution to calculate force coefficients. We present these steps in general terms here.

Step 1 Divide the wing into panels, as shown in Figure 7.44. The panels are quadrilaterals, and the number of panels is the product of the number of panels chordwise and spanwise. Let N be the total number of panels.

Step 2 Place a horseshoe vortex on each panel such that,*

- the bound portion is on the quarter-chord line of the panel,
- the trailing vortices are in the x direction.

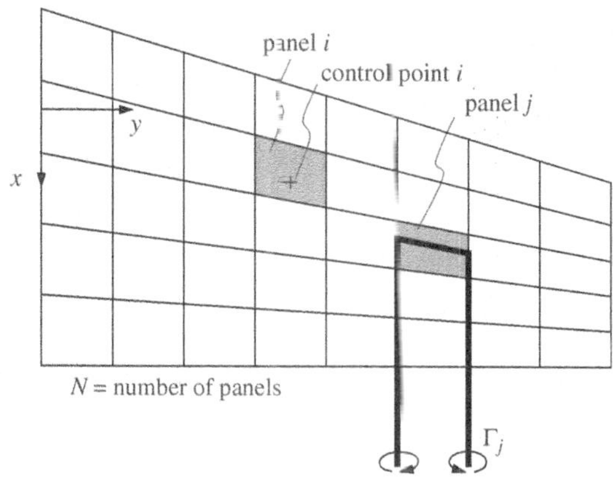

Figure 7.44 Paneling and horseshoe vortex placement used in the vortex-lattice method. Shown is a paneling of the a right wing.

* The choice of the quarter-chord vortex placement, and the three-quarter chord control point placement in the next step, is motivated by the single-vortex flat-plate airfoil model from Section 5.3.1.

This placement is illustrated in Figure 7.44, where Γ_j is the strength of the horseshoe vortex on panel j. The values of Γ_j over all of the panels are the unknowns in the vortex-lattice method.

Step 3 Define control points at the three quarter-chord chordwise, midpoint spanwise, location on each panel, as shown in Figure 7.44. At each control point, calculate the normal vector of the panel and the associated local angle of attack of the panel, $\alpha_{\text{panel } i}$. This angle includes the effects of geometric twist and camber. Note, for this simple model, we do not account for thickness of the airfoil, so $\alpha_{\text{panel } i}$ is calculated from the airfoil camber line.†

† Furthermore, the camber and twist variations are typically assumed small, so the panels are placed on the x, y plane at $z = 0$, and only the normal vector varies from panel to panel.

Step 4 Enforce flow tangency at each control point. Indexing the control points by i, we require that

$$w_i + U_\infty(\alpha_\infty + \alpha_{\text{panel } i}) = 0 \tag{7.115}$$

where w_i is the downwash at control point i induced by all of the horseshoe vortices. This flow-tangency equation is analogous to Equation 5.19 from thin-airfoil theory. Indeed, we make many of the same assumptions in the vortex-lattice method, including small thickness, small camber, small camber slope, and small angle of attack.

Step 5 Set up a linear system based on the flow-tangency equation. To do this, we express the downwash on control point i in terms of the strengths of all of the horseshoe vortices, indexed by j,

$$w_i = \sum_{j=1}^{N} A_{ij}\Gamma_j, \tag{7.116}$$

where $A_{ij}\Gamma_j$ is the downwash on control point i due to the vortex on panel j, and A_{ij} is the influence coefficient of panel j on control point i. The flow-tangency condition then becomes

$$\begin{bmatrix} A_{11} & A_{12} & \dots & A_{1N} \\ A_{21} & A_{22} & \dots & A_{2N} \\ \vdots & \vdots & \ddots & \vdots \\ A_{N1} & A_{N2} & \dots & A_{NN} \end{bmatrix} \begin{bmatrix} \Gamma_1 \\ \Gamma_2 \\ \vdots \\ \Gamma_N \end{bmatrix} = - \begin{bmatrix} U_\infty(\alpha_\infty + \alpha_{\text{panel } 1}) \\ U_\infty(\alpha_\infty + \alpha_{\text{panel } 2}) \\ \vdots \\ U_\infty(\alpha_\infty + \alpha_{\text{panel } N}) \end{bmatrix}.$$

Step 6 Solve the system for the unknowns, Γ_j. Note that the matrix in this linear system does not depend on α_∞ or U_∞: it only depends on the geometry. It can therefore be constructed and factored once for each geometry and used in analysis of different flight conditions.

Step 7 Use the Γ_j to calculate the forces and moments on the wing. To do this, we can use the three-dimensional form of the Kutta–Joukowsky theorem, Equation 7.14, on each panel. We can also analyze the Trefftz plane, as described in Section 7.7, and this approach sometimes yields more accurate results.

Although the vortex-lattice method is applicable to general wing shapes, it still assumes three-dimensional potential flow. If we have thick boundary layers, separation, shocks, or unsteadiness, the vortex-lattice method does not apply. In such cases, we need more sophisticated numerical techniques, such as ones that incorporate boundary layer effects or solve the Navier–Stokes equations.

7.6.2 Application to a Rectangular Wing

The vortex-lattice method can incorporate sweep, twist, dihedral, and cambered airfoils. For simplicity, we look at the case of a flat, untwisted, rectangular wing. Additional geometry features make the downwash calculation more complicated but do not change the overall approach.*

* Problem 7 extends the analysis in this section to a tapered and swept wing.

Consider a rectangular wing of span b and chord c. We place N_b vortices spanwise and N_c vortices chordwise, for a total of $N = N_b N_c$ vortices, as shown in Figure 7.45.

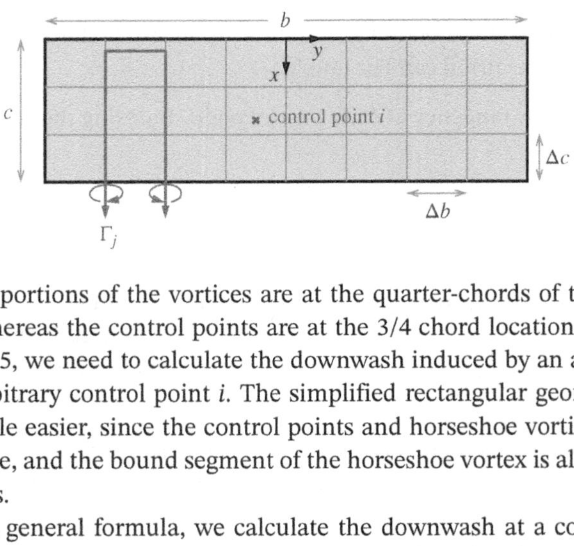

Figure 7.45 $N_b = 8$ spanwise and $N_c = 3$ chordwise paneling of a wing for the vortex-lattice method.

The bound portions of the vortices are at the quarter-chords of the rectangular panels, whereas the control points are at the 3/4 chord locations. To enforce Equation 7.115, we need to calculate the downwash induced by an arbitrary vortex j at an arbitrary control point i. The simplified rectangular geometry makes this task a little easier, since the control points and horseshoe vortices are all in the same plane, and the bound segment of the horseshoe vortex is always aligned with the y axis.

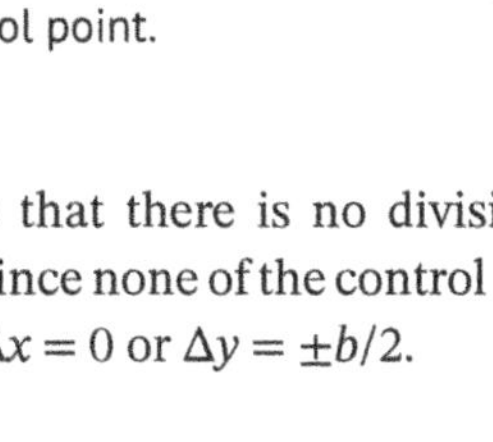

Figure 7.46 Influence of one vortex on a control point.

To obtain a general formula, we calculate the downwash at a control point i displaced by Δx in the x direction and Δy in the y direction from the root of a horseshoe vortex of span Δb and strength Γ_j, as shown in Figure 7.46.

Using Equation 7.13, the z velocity at the control point contains three contributions from the three segments of horseshoe vortex j,†

† Note that there is no division by zero, since none of the control points have $\Delta x = 0$ or $\Delta y = \pm b/2$.

$$
\begin{aligned}
w_i = {} & \frac{\Gamma_j}{4\pi(\Delta y - \Delta b/2)}\left(1 + \frac{\Delta x}{r_2}\right) && \text{(right)} \\
& - \frac{\Gamma_j}{4\pi(\Delta y + \Delta b/2)}\left(1 + \frac{\Delta x}{r_1}\right) && \text{(left)} \\
& - \frac{\Gamma_j}{4\pi\Delta x}\left(\frac{\Delta y + \Delta b/2}{r_1} - \frac{\Delta y - \Delta b/2}{r_2}\right) && \text{(bound),}
\end{aligned}
$$

where r_1 and r_2 are defined in Figure 7.46. Comparing this equation to Equation 7.116, we identify A_{ij} as everything on the right-hand side without Γ_j. Listing 7.6 shows a code that builds the matrix system and solves it for all of the vortex strengths, Γ_j.

Listing 7.6 A vortex-lattice code for the analysis of a rectangular wing.

```
function CL = vortexlattice(alpha, AR)
% Uses vortex lattice to compute the lift coefficient of a
% rectangular wing of aspect ratio AR at angle alpha [rad].

U = 1;       % freestream speed
rho = 1;     % density
c = 1;       % chord
b = AR;      % span
Nb = 40;     % spanwise panels
Nc = 20;     % chordwise panels
N = Nb*Nc; % total number of unknowns

db = b/Nb;  dc = c/Nc;  % panel sizes
A = zeros(N,N);   F = zeros(N,1); % matrix and RHS
for i = 1:N, % loop over control points
  xi = [(floor((i-1)/Nb)+0.75)*dc, (mod(i-1,Nb)+0.5)*db-0.5*b];
  for j = 1:N, % loop over vortices
      xj = [(floor((j-1)/Nb)+0.25)*dc, (mod(j-1,Nb)+0.5)*db-0.5*b];
      dx = xi(1)-xj(1); dy = xi(2)-xj(2);
      r1 = sqrt(dx^2 + (dy+0.5*db)^2);
      r2 = sqrt(dx^2 + (dy-0.5*db)^2);
      A(i,j) = ((1+dx/r2)/(dy-db/2) - (1+dx/r1)/(dy+db/2) - ...
               ((dy+db/2)/r1 - (dy-db/2)/r2)/dx)/(4*pi);
  end
  F(i) = -U*alpha;
end

G = A\F;  % solve for vortex strengths
L = rho*U*db*sum(G); % lift
CL = L/(0.5*rho*U^2*b*c); % lift coefficient
```

Code notes:

- The geometry and discretization are hard-coded.
- Certain parameters are arbitrary: e.g. density and speed do not affect the lift coefficient.
- The indices i and j loop over all unknowns, so that the `floor` and `mod` functions are used to identify the x and y positions in the lattice.
- The control point (i) is at the 3/4 chord panel location, whereas the bound portion of the vortex (j) is at the 1/4 chord.
- r_1, r_2, Δx, and Δy are defined in Figure 7.46.

By the Kutta–Joukowsky theorem, the lift is the sum of the lift forces on the horseshoe vortices, each of which has a span of Δb, so that

$$L = \rho U_\infty \Delta b \sum_j \Gamma_j \quad \Rightarrow \quad C_L = \frac{L}{\frac{1}{2}\rho U_\infty^2 bc}. \tag{7.117}$$

To obtain the drag, we can also use the Kutta–Joukowsky theorem, which requires a separate evaluation of the downwash at every vortex midpoint.

Example 7.10 Vortex lattice versus lifting line
Prandtl's lifting-line and the vortex-lattice method use the same potential flow assumption and small angle approximation for the angle of attack. For high aspect-ratio wings, they should provide similar results. As the aspect ratio decreases, the vortex-lattice method will be more accurate because it accounts for the chordwise variation of the lift distribution.

We run the codes in Listings 7.3 and 7.6 for several aspect ratios and compare their lift curves, C_L versus α. Figure 7.47 shows the comparison. We see that the difference between the two analyses diminishes as the aspect ratio increases. Note that relative to vortex lattice, lifting line overpredicts the lift at low aspect ratios.

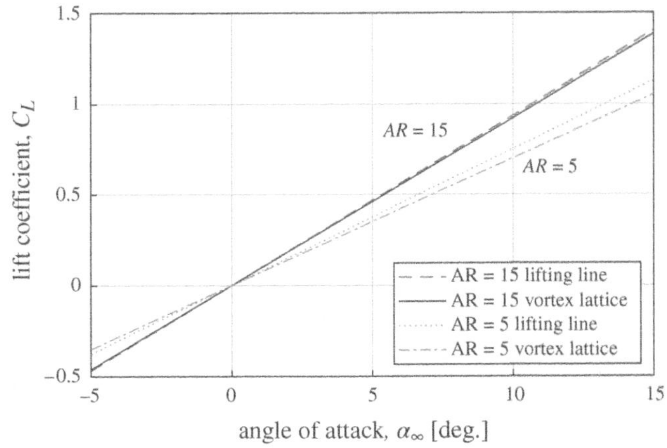

Figure 7.47 Lift-curve slope comparison for rectangular wings of aspect ratios 5 and 15, analyzed using the lifting-line and vortex-lattice methods. Lifting line overpredicts the lift compared to vortex lattice.

The vortex-lattice method provides much more information than just the lift coefficient. Aside from other integrated quantities, like lift and induced drag, we can also compute distributions across the span and chord. On panel j, the vortex of strength Γ_j and span Δb experiences a lift force of $L_j = \rho U_\infty \Delta b \Gamma_j$. Dividing this force by the panel area, $\Delta b \Delta c$, gives the stress, which is the pressure difference Δp between the lower and upper surfaces. Normalizing Δp by the dynamic pressure, $\frac{1}{2}\rho U_\infty^2$, gives the pressure coefficient difference, Δc_p. Figure 7.48 shows a surface plot of this pressure difference distribution for an $AR = 5$ wing at $\alpha_\infty = 5°$. The pressure difference is largest at the leading edge and decreases along the chord. Also, the pressure difference is larger at the root compared to the wing tips.

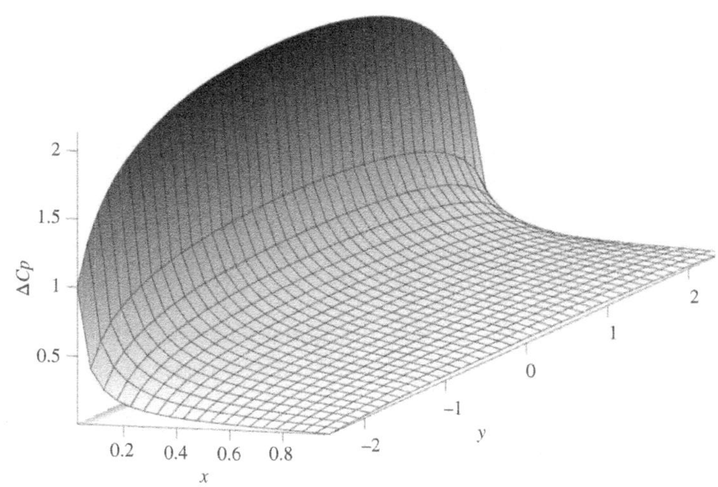

Figure 7.48 Pressure difference between the lower and upper surfaces for an $AR = 5$ rectangular wing analyzed using the vortex-lattice method.

7.7 Trefftz-Plane Analysis

So far our discussion of finite wings has focused on what happens at the wing, in terms of downwash and forces. In this section, we will instead look at the flow far downstream of the wing. We will see that considering the effect of a wing on the flow far downstream provides another interpretation of induced drag.

7.7.1 Crossflow Kinetic Energy

Consider potential flow around a wing, and a box-shaped control volume drawn around the wing,* as shown in Figure 7.49. The back of this control volume, cutting through the wake, is known as the *Trefftz plane* (McLean 2012; Anderson 2016).

To obtain the force on the wing, we apply the integral momentum conservation equation,†

$$\int_{S_{\text{wing}}+S_\infty} (\rho\vec{v})\vec{v}\cdot\vec{n}\,dS = -\int_{S_{\text{wing}}+S_\infty} p\vec{n}\,dS. \tag{7.118}$$

On the wing, S_{wing}, zero flow-through requires $\vec{v}\cdot\vec{n}=0$. Also, the pressure integral over the wing is the force of the fluid on the wing,‡

* Although the figure depicts a wake profile typical of a wing, the following analysis applies to general shapes.

† We assume steady flow without body forces.

‡ The normal vector $\vec{n}$ points out of the control volume, and hence into the wing on S_{wing}.

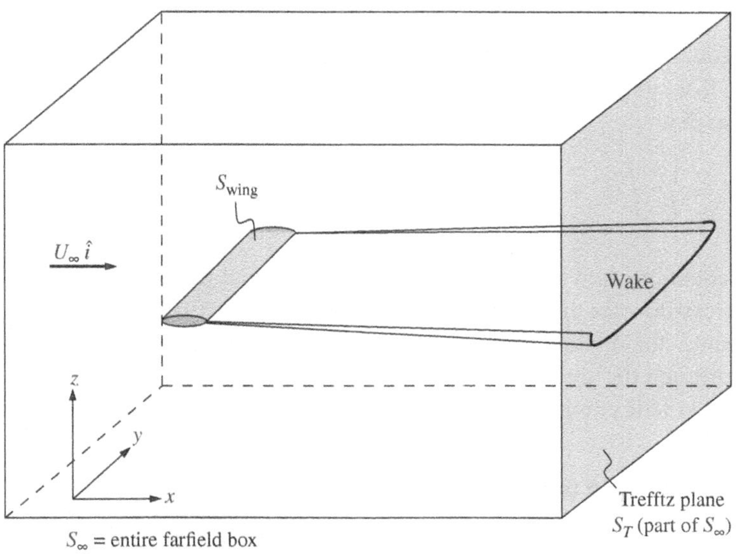

Figure 7.49 Control volume around a wing, showing the Trefftz plane.

$$\vec{F}_{\text{on-wing}} = \int_{S_{\text{wing}}} p\vec{n}\,dS = D_i\hat{i} + Y\hat{j} + L\hat{k}, \tag{7.119}$$

where D_i is the induced drag, Y is the side force, and L is the lift. Therefore, the conservation of momentum equation becomes

$$D_i\hat{i} + Y\hat{j} + L\hat{k} = -\int_{S_\infty} p\vec{n}\,dS - \int_{S_\infty} (\rho\vec{v})\vec{v}\cdot\vec{n}\,dS. \tag{7.120}$$

The drag is the horizontal component of the right-hand side, which is, using $\vec{v} = u\hat{i} + v\hat{j} + w\hat{k}$,

$$D_i = -\int_{S_\infty} p\vec{n}\cdot\hat{i}\,dS - \int_{S_\infty} (\rho u)\vec{v}\cdot\vec{n}\,dS. \tag{7.121}$$

Bernoulli's equation§ is used to eliminate the pressure,¶

$$D_i = \int_{S_\infty} \frac{1}{2}\rho(u^2+v^2+w^2)\vec{n}\cdot\hat{i}\,dS - \int_{S_\infty} (\rho u)\vec{v}\cdot\vec{n}\,dS. \tag{7.122}$$

§ We ignore gravitational potential, so Bernoulli's equation becomes

$$p_\infty + \frac{1}{2}\rho U_\infty^2 = p + \frac{1}{2}\rho|\vec{v}|^2.$$

¶ Since S_∞ is a closed curve, any integrals of a constant involving the normal evaluate to zero, e.g.

$$\int_{S_\infty} C\vec{n}\cdot\vec{i} = 0$$

Separating the velocity into the freestream, $U_\infty\hat{\imath}$, and a perturbation, $\hat{u}\hat{\imath} + \hat{v}\hat{\jmath} + \hat{w}\hat{k}$, which is not assumed to be small yet, and applying conservation of mass,|| the drag expression becomes

$$D_i = \frac{1}{2}\rho\int_{S_\infty} (2U_\infty\hat{u} + \hat{u}^2 + \hat{v}^2 + \hat{w}^2)\,\vec{n}\cdot\hat{\imath}\,dS - \rho\int_{S_\infty} (\hat{u})\vec{v}\cdot\vec{n}\,dS. \tag{7.123}$$

|| The integral form of conservation of mass for an incompressible fluid states that

$$\int_{S_\infty} \vec{v}\cdot\vec{n} = 0.$$

If we put most of control volume boundary, except S_T, far away from the wing, then the velocity perturbations go to zero except downstream, where the presence of trailing vortices will create non-zero perturbations. Thus, $\hat{u}, \hat{v}, \hat{w} \to 0$ except on the Trefftz plane, S_T, where $\vec{n}\cdot\hat{\imath} = 1$, so that Equation 7.123 becomes

$$D_i = \frac{1}{2}\rho\int_{S_T} (\hat{v}^2 + \hat{w}^2 - \hat{u}^2)\,dS. \tag{7.124}$$

This equation holds when S_∞ is far from the wing, whereas S_T may still be close. Finally, if we move S_T far downstream, the x-velocity perturbation must go to zero,** so that

** The reason why these perturbations go to zero is that the trailing vortices far downstream must be in the x direction and cannot induce an x component of velocity.

$$D_i = \frac{1}{2}\rho\int_{S_T} (\hat{v}^2 + \hat{w}^2)\,dS. \tag{7.125}$$

This equation shows that the induced drag is equal to the kinetic energy left in the cross-flow far downstream (Landau and Lifshitz 1959). This should make sense, since for inviscid flow, the work per unit time done by the wing as it moves through the air ($D_i U_\infty$) cannot be dissipated as heat. It instead remains as deposited kinetic energy that we can measure far downstream in the crossflow.

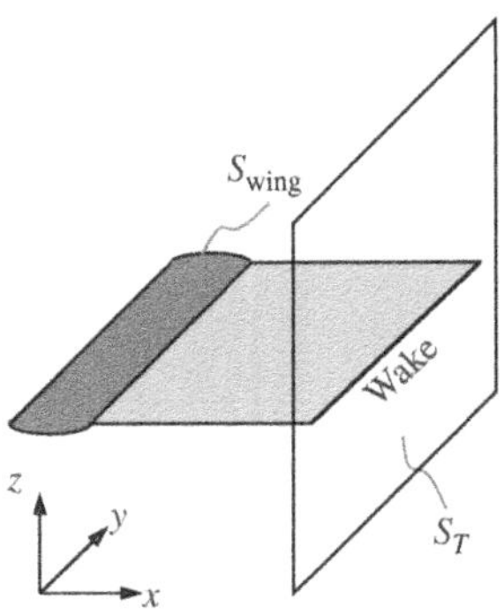

Figure 7.50 A horizontal, flat wake.

7.7.2 Relationship to Downwash

The drag in Equation 7.125 can be related to the downwash interpretation of Prandtl's lifting-line model. To do this, we model the wake as a thin, planar horizontal vortex sheet, as shown in Figure 7.50. Recall that although in reality the vortex sheet rolls up and is "force free" (as in Figure 7.49), modeling it as horizontal and flat will still yield the correct drag value, since the force on the wake required to keep it flat has no x component.

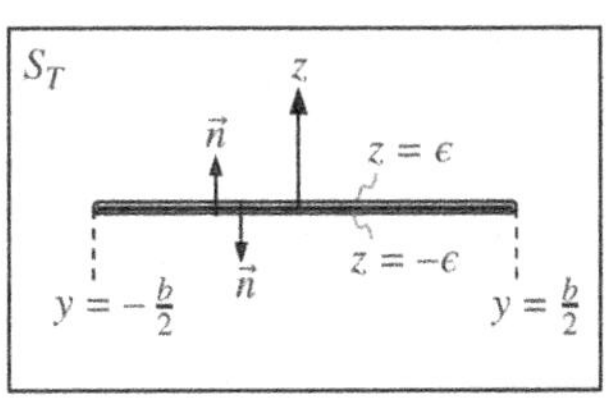

Figure 7.51 Trefftz-plane wake.

By the potential flow assumption, let $\vec{v} = \nabla\phi$. Note, $\vec{v} = U_\infty\hat{\imath} + \hat{v}\hat{\jmath} + \hat{w}\hat{k}$ in the Trefftz plane. Thus, Equation 7.125 becomes,

$$D_i = \frac{1}{2}\rho\int_{S_T} \left[\left(\frac{\partial\phi}{\partial y}\right)^2 + \left(\frac{\partial\phi}{\partial z}\right)^2\right] dS. \tag{7.126}$$

A general result of integration by parts on S_T, where the two coordinates are y and z, is that*

$$\int_{S_T} (\tilde{\nabla}\phi\cdot\tilde{\nabla}\phi + \phi\tilde{\nabla}^2\phi)\,dS = \int_{\partial S_T} \phi\tilde{\nabla}\phi\cdot\vec{n}\,dl, \tag{7.127}$$

* $\tilde{\nabla}$ is the two-dimensional gradient operator in y and z,

$$\tilde{\nabla}\phi = \frac{\partial\phi}{\partial y}\hat{\jmath} + \frac{\partial\phi}{\partial z}\hat{k}.$$

where ∂S_T is the boundary of S_T and $\vec{n}$ is a normal vector that points outward from S_T. As indicated in Figure 7.51, ∂S_T consists of the wake line and the farfield boundary. Continuity requires $\tilde{\nabla}^2\phi = 0$,† so the induced drag is

$$D_i = \frac{1}{2}\rho\int_{S_T} \nabla\phi\cdot\nabla\phi\,dS = \frac{1}{2}\rho\int_{\partial S_T} \phi\nabla\phi\cdot\vec{n}\,dl. \tag{7.128}$$

† The continuity equation in potential flow is

$$\nabla^2\phi = \frac{\partial^2\phi}{\partial x^2} + \tilde{\nabla}^2\phi = 0.$$

Since far downstream, $\frac{\partial\phi}{\partial x} = U_\infty =$ const., $\frac{\partial^2\phi}{\partial x^2} = 0$, and so $\tilde{\nabla}^2\phi = 0$.

Since far away from the wake (in y and z), the velocity perturbations go to zero, the integral over ∂S_T reduces to a contour integral along the wake, which we split into two pieces: one over the top, and one over the bottom,

$$D_i = \frac{1}{2}\rho \int_{-b/2}^{b/2} \left\{ \left[\phi(\nabla\phi \cdot \vec{n})\right]_{z=\epsilon} + \left[\phi(\nabla\phi \cdot \vec{n})\right]_{z=-\epsilon} \right\} dy. \tag{7.129}$$

On the horizontal wake, the outward-pointing normal vector $\vec{n}$ points down along the top portion of the integral, and up along the bottom portion. Taking $\epsilon \to 0$, by continuity we must have

$$[\nabla\phi \cdot \vec{n}]_{z=\epsilon} = -[\nabla\phi \cdot \vec{n}]_{z=-\epsilon} = -w(y) = \text{downwash on wake},$$

so Equation 7.129 becomes

$$D_i = -\frac{1}{2}\rho \int_{-b/2}^{b/2} \left[\phi(y)_{z=\epsilon} - \phi(y)_{z=-\epsilon}\right] w(y) dy. \tag{7.130}$$

We can relate the potential jump across the wake to $\Gamma(y)$, the vortex strength in the lifting-line model, by noting that the strength of the vortex sheet is $\gamma(y) = \frac{d\Gamma}{dy}$. From Stokes's theorem,‡ the vortex-sheet strength is the tangential velocity jump across the sheet,

‡ We obtained this result in thin-airfoil theory, by applying a thin rectangular control volume around the vortex sheet.

$$\frac{d\Gamma}{dy} = v_{z=\epsilon} - v_{z=-\epsilon} = \left.\frac{\partial\phi}{\partial y}\right|_{z=\epsilon} - \left.\frac{\partial\phi}{\partial y}\right|_{z=-\epsilon}. \tag{7.131}$$

Integrating from $-b/2$ to y and noting that at $y = -b/2$, $\Gamma = 0$ and $\phi_{z=\epsilon} = \phi_{z=-\epsilon}$, yields $\phi(y)_{z=\epsilon} - \phi(y)_{z=-\epsilon} = \Gamma(y)$, so that Equation 7.130 becomes

$$D_i = -\frac{1}{2}\rho \int_{-b/2}^{b/2} w(y)\Gamma(y) dy. \tag{7.132}$$

This equation holds in the wake for $x \to \infty$. From Prandtl's lifting-line theory, Equation 7.46,

$$D_i = \int_{-b/2}^{b/2} D_i'(y)\, dy = -\rho \int_{-b/2}^{b/2} w(y)\Gamma(y)\, dy. \tag{7.133}$$

This equation looks almost like Equation 7.132, except for the missing factor of $\frac{1}{2}$. The resolution to this difference is that in Equation 7.133, the downwash is measured at the wing, where the trailing vortices are semi-infinite, while in Equation 7.132, the downwash is measured far downstream, where the trailing vortices appear infinite in both directions. Since a semi-infinite vortex induces a velocity of half that of an infinite vortex, $w(x \to \infty, y) = 2w(x=0, y)$. Therefore, the Trefftz-plane drag calculation agrees with the lifting-line result.

7.8 Summary and Concepts

This chapter extended potential flow to the analysis of finite wings. Whereas airfoils can be thought of as wings of infinite span, the three-dimensional nature of real wings introduces new challenges in their analysis, and a new force: induced drag. A single vortex used to model sectional flow turning becomes a vortex line along the span, and any changes in lift require changes in the strength of that line, leading to shed vorticity. A horseshoe vortex is a crude model of this effect, one that can be extended numerically by using a collection of horseshoe vortices to model the wing. This collection can be in a line for high aspect ratios, or in a lattice for general wings. In the former case, a continuous model of the vortex strength along the span is also accurate, leading to Prandtl's lifting-line equation, which ultimately requires a numerical solution. Finally, a control volume analysis of the wing can take advantage of the signature of the shed vorticity in the wake to calculate the induced drag.

Some of the key concepts covered in this chapter are as follows:

- A vortex line models potential-flow lift generation along a wing, but it cannot end at the tips, leading to a horseshoe-vortex model with shed vorticity.
- Most wings have a non-constant lift distribution across the span, which can be modeled with multiple horseshoe vortices or a continuous function of vortex strength across the span and a vortex sheet wake.
- To model chordwise variation of the lift distribution, important for low aspect ratio wings, a numerical vortex-lattice approach can be used.
- The wing models presented have flat wakes, but a real wake curls up due to its self-induced velocities. Fortunately, the flat-wake assumption does not affect the lift and induced drag calculations.
- Means to reduce induced drag at given flight conditions include decreasing the span loading or increasing the span efficiency, by modifying the lift distribution.
- An elliptical distribution of lift is the most efficient in terms of the lowest induced drag for a given lift, in the lifting-line model.
- Geometrical twist and zero-lift angle of attack have the same effect in lifting-line theory: they can modify the lift distribution but not the lift-curve slope of the wing.
- Induced drag appears on three-dimensional wings due to the downwash caused by the trailing vortices, which affects the velocity at the wing and tilts the force vector.
- Another way to interpret the induced drag is that the power required to overcome it is the rate at which crossflow kinetic energy is deposited into the wake through the trailing vortices.
- No viscous models were considered in this chapter, which means that skin friction drag, pressure drag, separation, and stall were not taken into account. Prior to stall, boundary-layer analyses can often be coupled to the potential-flow solver.

7.9 Problems

1. Justify each answer to the below questions with an explanation.
   a True or false: according to the horseshoe vortex wing model, as an aircraft in steady-level flight increases speed, the induced drag decreases.
   b True or false: in Prandtl's lifting-line theory, the induced drag is zero whenever the lift is zero.
   c True or false: a wing with an elliptical chord distribution always produces an elliptical lift distribution.
   d True or false: vertical winglets can be modeled with Prandtl's lifting-line theory.

e True or false: In Prandtl's lifting-line theory, deploying flaps does not change a wing's lift-curve slope.

f True or false: in Prandtl's lifting-line theory, the span efficiency factor can change with the angle of attack.

g Explain how a flap deflection affects the matrix and right-hand-side vector in the vortex-lattice method.

2. An aircraft of span b is flying a distance of $d = b/2$ away from a cliff wall as shown in Figure 7.52. Using a horseshoe vortex model, calculate the percent savings in induced drag obtained by flying next to the cliff, compared to flying away from any walls.

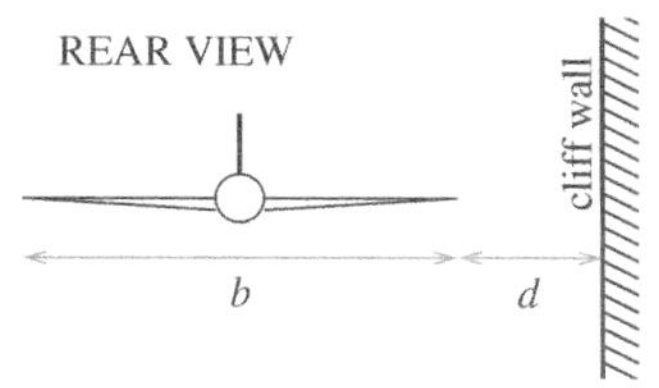

Figure 7.52 Aircraft next to a cliff.

3. A wing has the following sectional lift distribution:

$$L'(y) = \frac{1}{50}\rho b U_\infty^2 \left[1 - \left(\frac{2y}{b}\right)^4\right],$$

where ρ is the density, b is the span, and U_∞ is the freestream speed. Using Prandtl's lifting-line theory, determine the span efficiency factor, e.

4. Assuming Prandtl's lifting-line theory, prove that for a twisted wing with an elliptical chord distribution, $dA_n/d\alpha_\infty = 0$ for $n > 1$. Assume that a_0 is constant.

5. A tapered, left-right symmetrical wing, shown in Figure 7.53, has a NACA 2408 section, a geometric twist that varies linearly from the root to the tip, and a trailing-edge flap hinged in the middle of the airfoil thickness. For $AR = 8$, $\sigma = 0.5$, $g = -3°$ down, $f = 10°$ down, and $\alpha_\infty = 3°$, determine C_L, C_{Di}, and e, and plot both c_ℓ and $c_\ell c/\bar{c}$ versus y/b. Use Prandtl's lifting-line theory and an airfoil panel solver (e.g. mfoil) for sectional data.

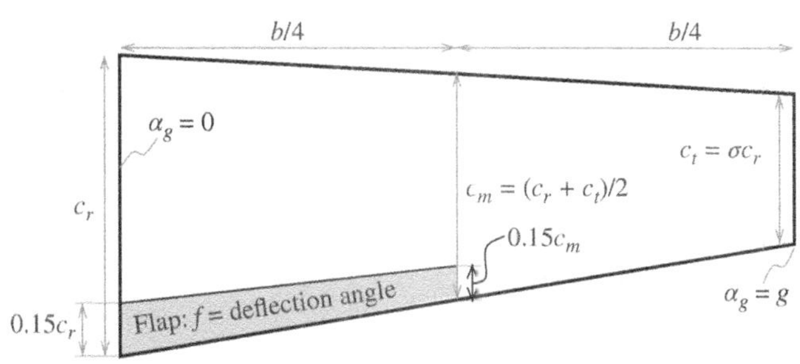

Figure 7.53 Wing with flap, taper, and twist for Problems 5 and 6.

6. For the same wing as in Problem 5, determine the flap deflection f and tip twist g that minimize drag at $C_L = 0.5$. Calculate α_∞, C_{Di}, and e, and plot $c_\ell c/\bar{c}$ with the elliptical distribution overlaid.

7. Modify the vortex-lattice code developed in Section 7.6.2 to analyze a tapered wing with a given leading-edge sweep angle. A key step will be velocity calculation for a horseshoe vortex with non-right angles, for which you can use Equation 7.13. The general Kutta–Joukowsky formula $\vec{F} = \rho\Gamma\vec{v}_0 \times \vec{b}$ gives the force when the bound segment $\vec{b}$ of the horseshoe vortex is in an arbitrary direction. Compare linear versus cosine spanwise spacings for the vortices and present lift and drag coefficient polars for a tapered wing with a taper ratio of 0.5, a leading-edge sweep angle of 15°, and an aspect ratio of 12.

8. Implement a Trefftz-plane induced-drag calculation, using Equation 7.132, into your code from the previous problem. Note that the downwash calculation is simpler because it uses infinite vortex lines. For the same tapered wing at $\alpha_\infty = 5°$, compare the Kutta–Joukowsky and Trefftz-plane induced-drag coefficients at different panel resolutions, e.g. $8 \times 2, 16 \times 4, \ldots$.

8

Compressible Flow

In previous chapters, the assumption of incompressibility was often made. This is a good assumption for low-speed flight, but it breaks down as the speed of the vehicle analyzed increases relative to the speed of sound. When flying faster than the speed of sound, the governing equations undergo a fundamental change and require a different solution process. The behavior of bodies in supersonic flow also changes significantly, as does their design. Even in the subsonic regime, compressibility effects change the performance of airfoils and wings, albeit less drastically than in supersonic flow. This chapter discusses qualitative aspects of compressibility, presents quantitative analysis tools that are valid under certain enabling assumptions, and shows how these tools can be used for airfoil and wing design.

8.1 Compressibility Effects

For engineering purposes, flows are generally considered *incompressible* if the density changes are negligible. Isentropic relations tell us that the density changes are proportional to the square of the *Mach number*, defined as

$$M \equiv \frac{|\vec{v}|}{c}, \tag{8.1}$$

where $|\vec{v}|$ is the magnitude of the velocity and c is the speed of sound. Therefore, if $M \lesssim 0.3$, density changes will be on the order of 10%, under isentropic assumptions, and compressibility effects will be small. However, when we talk about flows over wings and airfoils, it is not enough to just consider the freestream Mach number, M_∞. This is because the flow is slowed down in some regions and sped up in others. For example, over the upper surface of a wing producing lift, the flow is typically faster than the freestream, which translates to a higher Mach number. Therefore, even if an airfoil is flying at $M_\infty < 1$, the local Mach number somewhere in the vicinity of the airfoil may be greater than 1, which is supersonic flow. Some approximate generalizations of airfoil flow regimes based on M_∞ are given in Figure 8.1. Note that the precise numbers depend on the airfoil geometry and angle of attack.

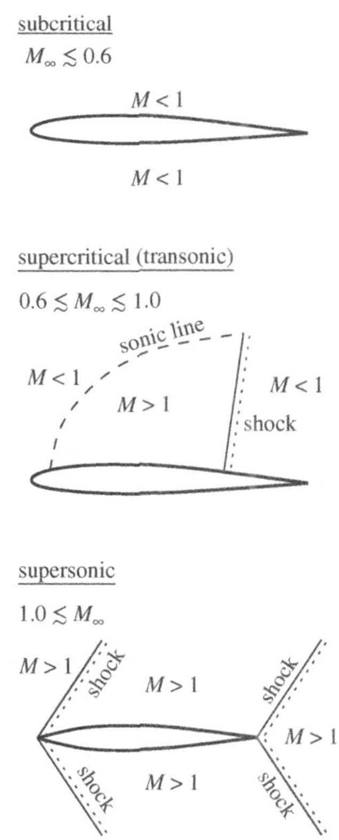

Figure 8.1 General characterization of flow regimes based on the freestream Mach number, M_∞.

The general behavior of the lift-curve slope, $dc_\ell/d\alpha$, and the drag coefficient, c_d, as functions of M_∞ is shown in Figure 8.2. Note that the lift-curve slope first increases with M_∞ in the subsonic regime, $M_\infty < 1$. For $M_\infty > 1$, however, the lift-curve slope decreases with M_∞. Also, the drag coefficient remains relatively constant in the subcritical regime: if there were no viscosity, we would expect c_d to remain zero in this two-dimensional flow. The positive value shown is due to viscous effects. As the Mach number increases closer to $M_\infty = 1$, the drag coefficient

Aerodynamics: A Computational Introduction, First Edition. Krzysztof J. Fidkowski.
© 2026 John Wiley & Sons, Inc. Published 2026 by John Wiley & Sons, Inc.

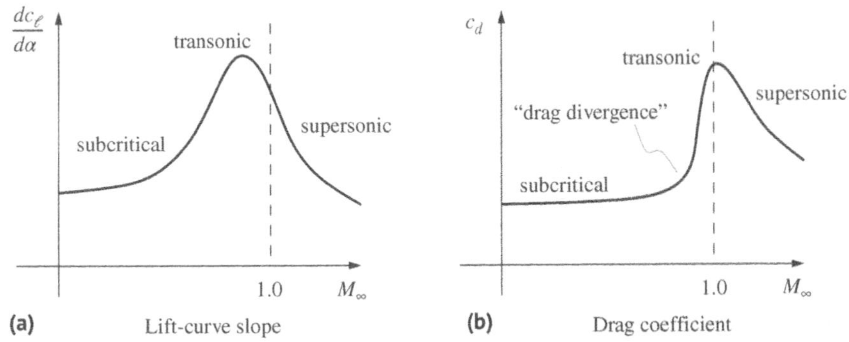

Figure 8.2 Behavior of the lift-curve slope and the drag coefficient as a function of M_∞.

begins to rise. This rise in c_d is called *drag divergence*, and it is due to the appearance of local supersonic flow (pockets of $M > 1$) on the airfoil, and the associated shocks. Shocks are irreversible processes in which flow energy is dissipated as heat, and this effect manifests itself as drag on the airfoil. In the following sections, we will analyze subcritical and supersonic flows quantitatively under some assumptions. Quantitative analysis of transonic flows is more difficult, and we will not discuss this regime at great length.

8.2 Compressible Potential Flow

Compressible flow can be analyzed using the Navier–Stokes equations, or if assuming inviscid flow, the Euler equations. These equations are nonlinear, and their solution requires computationally intensive methods, such as ones that use meshes of the flow domain. More tractable and still quite useful are linearized approaches based on compressible potential flow, as these lend themselves to efficient solution by superposition.

In this section, we show how the governing equations of compressible flow reduce to a single velocity potential equation under some assumptions. These assumptions are generally quite good, in that they are consistent with the design characteristics of efficient high-speed vehicles. More details about small-disturbance theory and supersonic flow in general can be found in the work of Liepmann and Roshko (2002).

8.2.1 The Full-Potential Equation

To model the flow, we assume potential flow, which is valid as long as the flow is irrotational.* In practice, this means that we assume inviscid flow, since viscosity and no-slip boundaries cause fluid elements to rotate.

The gradient of the velocity potential, ϕ, gives the velocity. For example, in two dimensions, we write

$$\vec{v} = \nabla\phi = \phi_x\hat{\imath} + \phi_y\hat{\jmath}, \tag{8.2}$$

where the subscripts on ϕ denote differentiation with respect to x and y. Conservation of mass, assuming steady-state flow, requires†

$$0 = \nabla\cdot(\rho\vec{v}) = \nabla\rho\cdot\vec{v} + \rho\nabla\cdot\vec{v}. \tag{8.3}$$

* Previously, we also assumed incompressibility as part of potential flow, but only the irrotational assumption is required for ϕ to exist.

† Note, when not assuming incompressible flow, the density is not constant and must stay inside the divergence.

To obtain a single closed equation, we need to express the density in terms of the unknown velocity potential, and possibly freestream constants. To do this, we assume isentropic flow,‡ so that, by the isentropic relation for density,

‡ This is a valid assumption if there are no shocks, or if any shocks present are weak.

$$\rho = \rho_0\left(1 + \frac{\gamma - 1}{2}M^2\right)^{\frac{-1}{\gamma-1}}, \tag{8.4}$$

where ρ_0 is the stagnation density, γ is the ratio of specific heats, and M is the local Mach number. Taking the gradient,

$$\nabla\rho = -\frac{\rho}{2}\left(1 + \frac{\gamma - 1}{2}M^2\right)^{-1} \nabla(M^2). \tag{8.5}$$

Substituting $\nabla\rho$ into Equation 8.3 and dividing by ρ yield

$$-\frac{1}{2}\left(1 + \frac{\gamma - 1}{2}M^2\right)^{-1} \nabla(M^2)\cdot\vec{v} + \nabla\cdot\vec{v} = 0. \tag{8.6}$$

To close this equation, we need to express M^2 in terms of the velocity and freestream quantities. This is possible by using the speed of sound and isentropic relation for temperature,

$$\frac{c_0^2}{(q/M)^2} = \frac{c_0^2}{c^2} = \frac{\gamma R T_0}{\gamma R T} = \frac{T_0}{T} = 1 + \frac{\gamma - 1}{2}M^2, \tag{8.7}$$

where $q = |\vec{v}|$ is the velocity magnitude, and c, c_0 are the local and stagnation speeds of sounds, respectively. Solving this equation for M^2 yields

$$M^2 = \left(\frac{c_0^2}{q^2} - \frac{\gamma - 1}{2}\right)^{-1}. \tag{8.8}$$

This expression involves only constants and the speed, which is in terms of the potential, since $q = |\nabla\phi|$. Thus, substituting Equation 8.8 into Equation 8.6 gives a single closed equation for the velocity potential. The substitution requires the gradient of the squared Mach number, which is

$$\nabla(M^2) = -M^4(-1)\frac{c_0^2}{q^4}\nabla(q^2) = \frac{c_0^2}{c^4}\nabla(q^2). \tag{8.9}$$

Substituting this expression into Equation 8.6 gives§

§ We also use Equation 8.7 here to substitute

$$1 + \frac{\gamma - 1}{2}M^2 = \frac{c_0^2}{c^2}.$$

Note that this expression can be inverted to give c^2 in terms of the Mach number, which is expressed solely in terms of ϕ through Equation 8.8.

$$-\frac{1}{2c^2}\nabla(q^2)\cdot\vec{v} + \nabla\cdot\vec{v} = 0. \tag{8.10}$$

This is called the *full-potential equation*. It is nonlinear in the potential because of the appearance of $q^2 = |\nabla\phi|^2$ and in the nonlinear dependence of c^2 on ϕ. Solving this equation is therefore not trivial, and doing so numerically requires a computational mesh, as the equations do not admit solution by superposition.

8.2.2 Small-Disturbance Approximation

To linearize the full-potential equation, we assume small velocity disturbances,

$$u = U_\infty + \hat{u}, \quad v = \hat{v}, \tag{8.11}$$

where $\hat{u}$ and $\hat{v}$ satisfy

$$\left|\frac{\hat{u}}{U_\infty}\right| \ll 1 \quad \text{and} \quad \left|\frac{\hat{v}}{U_\infty}\right| \ll 1. \tag{8.12}$$

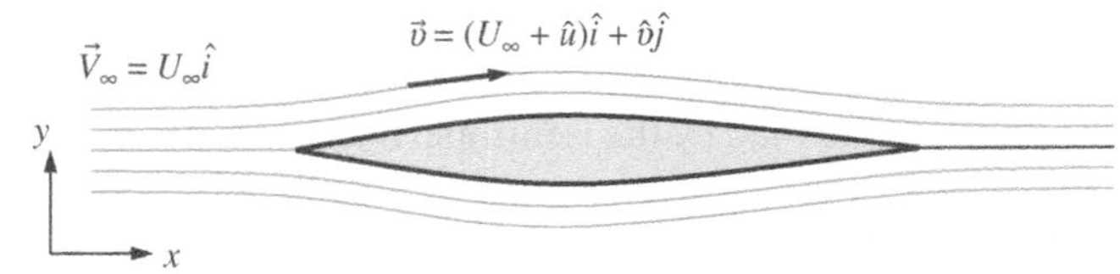

Figure 8.3 A slender aerodynamic body in high-speed flow creates small velocity perturbations, $\hat{u}$ and $\hat{v}$.

Note, we assume that the freestream is purely in the $\hat{i}$ direction. The small-disturbance assumption is often a good approximation for slender aerodynamic bodies at high speeds and low angles of attack, as illustrated in Figure 8.3.

The squared velocity magnitude becomes, in two dimensions,

$$q^2 = (U_\infty + \hat{u})^2 + (\hat{v})^2 = U_\infty^2 + 2U_\infty \hat{u} + \hat{u}^2 + \hat{v}^2. \tag{8.13}$$

Dropping the small quadratic perturbation terms, the gradient is

$$\nabla(q^2) \approx 2U_\infty \nabla \hat{u}. \tag{8.14}$$

Also, by definition of the speed of sound,*

$$\frac{1}{c^2} = \frac{M_\infty^2}{U_\infty^2}\frac{c_\infty^2}{c^2} \approx \frac{M_\infty^2}{U_\infty^2}. \tag{8.15}$$

* From the temperature isentropic relation, we can show that

$$\frac{c_\infty^2}{c^2} = 1 + \mathcal{O}(\hat{u}/U_\infty).$$

Substituting these expressions into Equation 8.10, we obtain

$$-\frac{M_\infty^2}{2U_\infty^2} 2U_\infty \nabla \hat{u} \cdot \vec{v} + \nabla \cdot \vec{v} = 0. \tag{8.16}$$

Now, $\nabla \hat{u} \cdot \vec{v} \approx U_\infty \hat{u}_x$,† so the above equation simplifies to

$$-M_\infty^2 \hat{u}_x + \hat{u}_x + \hat{v}_y = 0. \tag{8.17}$$

† This equation follows after dropping second-order perturbation terms,

$$\nabla \hat{u} \cdot \vec{v} = \hat{u}_x(U_\infty + \hat{u}) + \hat{u}_y \hat{v} \approx \hat{u}_x U_\infty.$$

Finally, substituting in the velocity potential, which we decompose as $\phi = \phi_\infty + \hat{\phi}$, gives

$$(1 - M_\infty^2)\hat{\phi}_{xx} + \hat{\phi}_{yy} = 0. \tag{8.18}$$

This is the *linearized compressible potential-flow equation*. It is valid for small disturbances, i.e. slender bodies at small angles of attack, and for both subsonic and supersonic flows. It is not valid for transonic or hypersonic flows, or for flows with large disturbances. In particular, in the transonic regime, $M_\infty \approx 1$, and $1 - M_\infty^2 \approx 0$. This means that the second-order terms dropped in front of $\hat{\phi}_{xx}$ should not have been dropped and need to be put back in. Doing so gives the *transonic small-disturbance equation* (TSDE), which is nonlinear. In the hypersonic regime, M_∞^2 becomes large, and its product with the dropped second-order velocity perturbation terms may not be small. These terms would have to be left in, again resulting in a nonlinear equation.

The mathematical *character* of Equation 8.18 changes between subsonic and supersonic flows. First, for $M_\infty < 1$, the subsonic regime, both coefficients in the equation are positive, and the partial differential equation is *elliptic*,

$$\underbrace{(1 - M_\infty^2)}_{\text{positive for } M_\infty < 1} \hat{\phi}_{xx} + \hat{\phi}_{yy} = 0 \quad \Rightarrow \text{elliptic PDE.} \tag{8.19}$$

In flows governed by an elliptic PDE, perturbations at a point propagate throughout the entire flowfield, as illustrated in Figure 8.4. On the other hand, for

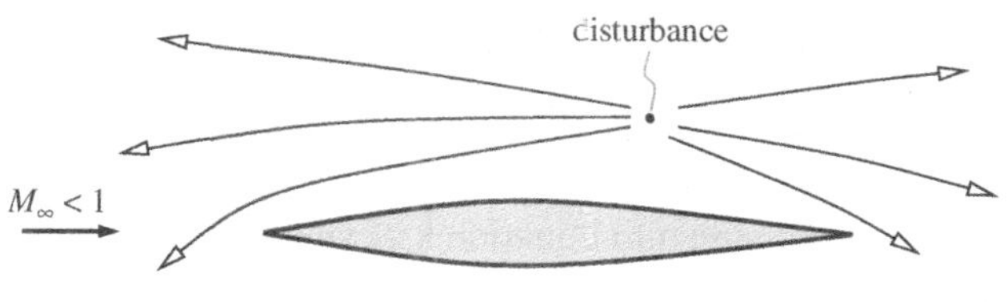

Figure 8.4 When the governing equation is elliptic, perturbations propagate throughout the entire flow domain.

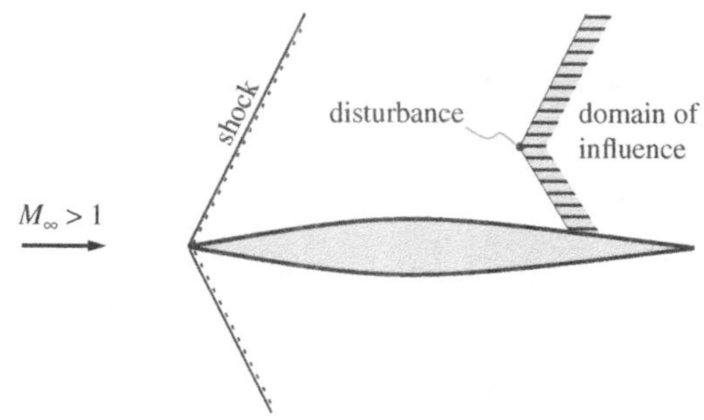

Figure 8.5 When the governing equation is hyperbolic, perturbations only affect the flow in a certain domain of influence.

$M_\infty > 1$, the supersonic regime, the signs on the two terms in Equation 8.18 are opposite, and we have a *hyperbolic* PDE,

$$\underbrace{\left(M_\infty^2 - 1\right)}_{\text{positive for } M_\infty > 1} \hat{\phi}_{xx} - \hat{\phi}_{yy} = 0 \quad \Rightarrow \text{hyperbolic PDE.}$$

In flows governed by a hyperbolic PDE, perturbations affect the flow only in a certain domain of influence, as shown in Figure 8.5. These differences in character lead to different solution approaches between the subsonic and supersonic regimes. Before delving into these, we present two formulas that are common to both regimes.

8.2.3 Pressure Coefficient

The pressure coefficient formula simplifies in compressible flow under the small-disturbance assumptions. Starting from the definition of the pressure coefficient, we have*

$$c_p = \frac{p - p_\infty}{\frac{1}{2}\rho_\infty U_\infty^2} = \frac{2}{\gamma M_\infty^2}\left(\frac{p}{p_\infty} - 1\right). \tag{8.20}$$

The pressure ratio, p/p_∞, can be written in terms of the local and freestream Mach numbers by the isentropic assumption,†

$$\frac{p}{p_\infty} = \left(\frac{1 + \frac{\gamma-1}{2}M_\infty^2}{1 + \frac{\gamma-1}{2}M^2}\right)^{\frac{\gamma}{\gamma-1}} = \left(\frac{M_\infty^2 c_0^2/U_\infty^2}{M^2 c_0^2/q^2}\right)^{\frac{\gamma}{\gamma-1}} = \left(\frac{c^2}{c_\infty^2}\right)^{\frac{\gamma}{\gamma-1}}. \tag{8.21}$$

We have not assumed small disturbances yet. Using the isentropic relation for the temperature,

$$\begin{aligned} c_0^2 = c^2\left(1 + \frac{\gamma-1}{2}M^2\right) &= c_\infty^2\left(1 + \frac{\gamma-1}{2}M_\infty^2\right) \\ c^2 + \frac{\gamma-1}{2}q^2 &= c_\infty^2 + \frac{\gamma-1}{2}U_\infty^2 \\ \frac{c^2}{c_\infty^2} &= 1 + \frac{\gamma-1}{2}\left(M_\infty^2 - \frac{q^2}{c_\infty^2}\right). \end{aligned} \tag{8.22}$$

* We use $c_\infty^2 = \gamma p_\infty/\rho_\infty$ to write $\rho_\infty U_\infty^2 = \gamma p_\infty U_\infty^2/c_\infty^2 = \gamma p_\infty M_\infty^2$.

† In the second step, we used Equation 8.8, which states that

$$\frac{1}{M^2} + \frac{\gamma-1}{2} = \frac{c_0^2}{q^2}.$$

By the small-disturbance approximation, $q^2 \approx U_\infty^2 + 2U_\infty \hat{u}$, so that

$$\frac{c^2}{c_\infty^2} \approx 1 + \frac{\gamma - 1}{2} M_\infty^2 \left(1 - 1 - \frac{2\hat{u}}{U_\infty}\right) = 1 - (\gamma - 1) M_\infty^2 \frac{\hat{u}}{U_\infty}. \tag{8.23}$$

Substituting this expression into Equation 8.21 yields‡

‡ Here, we use $(1 - \epsilon)^a \approx 1 - a\epsilon$.

$$\frac{p}{p_\infty} \approx 1 - \frac{\gamma}{\gamma - 1}(\gamma - 1) M_\infty^2 \frac{\hat{u}}{U_\infty} = 1 - \gamma M_\infty^2 \frac{\hat{u}}{U_\infty}. \tag{8.24}$$

So the pressure coefficient under small-disturbance assumptions is

$$c_p \approx \frac{2}{\gamma M_\infty^2}\left(-\gamma M_\infty^2 \frac{\hat{u}}{U_\infty}\right) = -\frac{2\hat{u}}{U_\infty}. \tag{8.25}$$

This equation holds for both subsonic and supersonic flows.§

§ It also holds in the incompressible potential-flow case, since by Equation 4.81,

$$\begin{aligned} c_p &= 1 - \frac{u^2 + v^2}{U_\infty^2} \\ &= 1 - \frac{(U_\infty + \hat{u})^2 + \hat{v}^2}{U_\infty^2} \\ &= -\frac{2\hat{u}}{U_\infty} - \frac{\hat{u}^2 + \hat{v}^2}{U_\infty^2} \\ &\approx -\frac{2\hat{u}}{U_\infty}. \end{aligned}$$

8.2.4 Wall Boundary Conditions

The flow is assumed inviscid, and hence near a wall, we only need to satisfy flow tangency. As shown in Figure 8.6, this means that the velocity vector at the wall must be tangent to the surface. Let θ be the angle of the wall, measured in radians. Assuming a small angle θ, which is consistent with the slender-body assumption, the tangency boundary condition is

$$\frac{v}{u} = \tan\theta \approx \theta. \tag{8.26}$$

Solving for v we have $v = u\theta = (U_\infty + \hat{u})\theta \approx U_\infty \theta$, where in the last step we used the fact that $\hat{u}$ is small compared to U_∞. Therefore, the *flow-tangency boundary condition* for small-disturbance flow is

$$v = \hat{v} = U_\infty \theta. \tag{8.27}$$

Again, this equation holds for both subsonic and supersonic flows.

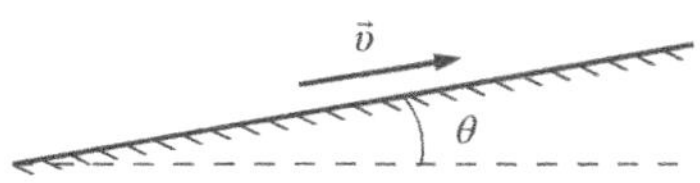

Figure 8.6 Flow-tangency boundary condition for the small-disturbance equations.

8.3 Subsonic Flow

We first consider the small-disturbance potential flow equation in the subsonic case, for which $M_\infty < 1$. The governing equation for the subsonic case can be written, using Equation 8.18, as

$$\underbrace{\left(1 - M_\infty^2\right)}_{\beta^2} \hat{\phi}_{xx} + \hat{\phi}_{yy} = 0, \tag{8.28}$$

where β is a real number since, for $M_\infty < 1$, $1 - M_\infty^2 > 0$. Our solution approach will be to make this equation look like Laplace's equation, which can be solved by the techniques described in Chapter 4.

8.3.1 Coordinate Transformation

To make Equation 8.28 look like Laplace's equation, we use the *Prandtl–Glauert transformation*, introduced by Glauert and Taylor (1928). This transformation is a coordinate scaling that can take a couple forms, the simpler one of which changes only the x coordinate.*

* The alternative is to multiply y and $\hat{\phi}$ by β.

$$\bar{x} \equiv \frac{x}{\beta}, \quad \bar{y} \equiv y. \tag{8.29}$$

With this transformation, the derivatives in Equation 8.28 are

$$\hat{\phi}_{xx} = \hat{\phi}_{\bar{x}\bar{x}}(\bar{x}_x)^2 = \hat{\phi}_{\bar{x}\bar{x}}\frac{1}{\beta^2}, \tag{8.30}$$

$$\hat{\phi}_{yy} = \hat{\phi}_{\bar{y}\bar{y}}. \tag{8.31}$$

Substituting these expressions into Equation 8.28, we obtain

$$\hat{\phi}_{\bar{x}\bar{x}} + \hat{\phi}_{\bar{y}\bar{y}} = 0. \tag{8.32}$$

This is Laplace's equation, which governs incompressible flow. The potential is in terms of the new coordinates, $\bar{x}$ and $\bar{y}$.

We now consider how the boundary condition transforms. At a wall of angle θ, we must have

$$U_\infty\theta = \hat{v} = \hat{\phi}_y = \hat{\phi}_{\bar{y}}.$$

Therefore, the *wall boundary condition does not change* under the transformation. The implication of this transformation is that subsonic compressible flow around a slender body can be related to incompressible flow about the same geometry stretched along the x axis.

Suppose we use an incompressible flow solver, such as a panel method, to solve the transformed equation, Equation 8.32. The result of the solver is a set of velocity perturbations,†

$$\bar{u} = \hat{\phi}_{\bar{x}}, \quad \bar{v} = \hat{\phi}_{\bar{y}}.$$

These velocities represent the incompressible, i.e. $M_\infty = 0$, solution. To translate back to the compressible solution, we use the Prandtl–Glauert transformation,

$$\hat{u} = \hat{\phi}_x = \frac{1}{\beta}\hat{\phi}_{\bar{x}} = \frac{1}{\beta}\bar{u}. \tag{8.33}$$

By Equation 8.25, the pressure coefficient is‡

$$c_p = -\frac{2\hat{u}}{U_\infty} = -\frac{1}{\beta}\frac{2\bar{u}}{U_\infty} = \frac{c_{p0}}{\beta}. \tag{8.34}$$

Since $\beta = \sqrt{1 - M_\infty^2} < 1$, this relation tells us that c_p for subsonic, compressible flow is *larger in magnitude* than for incompressible flow over the same geometry. Furthermore, since forces in potential flow are obtained by integrating the pressure, we have that the lift and moment coefficients transform as

$$c_\ell = \frac{c_{l0}}{\beta}, \quad c_m = \frac{c_{m0}}{\beta}, \tag{8.35}$$

where c_{l0} and c_{m0} are the lift and moment coefficients in incompressible flow.

For example, extending the incompressible thin-airfoil theory result to $M_\infty > 0$ gives

$$c_\ell(M_\infty) = \frac{c_{l0}}{\beta} = \frac{2\pi(\alpha - \alpha_{L=0})}{\sqrt{1 - M_\infty^2}}. \tag{8.36}$$

The lift-curve slope is then

$$\frac{dc_\ell}{d\alpha} = \frac{2\pi}{\beta} = \frac{2\pi}{\sqrt{1 - M_\infty^2}}. \tag{8.37}$$

† Equation 8.32 is for the disturbance potential, $\hat{\phi}$, the derivatives of which are the velocity disturbances.

‡ We define the incompressible pressure coefficient as

$$c_{p0} = -\frac{2\bar{u}}{U_\infty}.$$

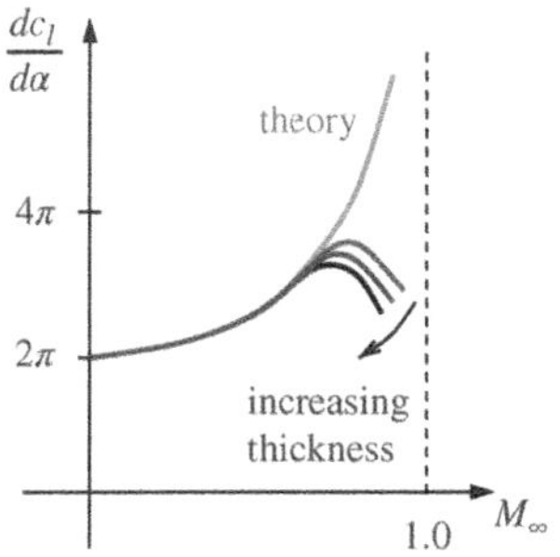

Figure 8.7 Theoretical and realistic results for the lift-curve slope obtained using the Prandtl–Glauert transformation.

This slope *increases* with increasing M_∞, consistent with the qualitative depiction in Figure 8.2, and reproduced in Figure 8.7. In practice, however, we expect to see plots more like the other curves shown in this figure, where the difference between theory and reality becomes more prominent for thicker airfoils. The reason for this departure from theory is that as the Mach number increases, the flow eventually goes supercritical. Strong shocks form, and small-disturbance assumptions no longer apply.

8.3.2 Critical Mach Number

As an airfoil flies in faster freestream Mach numbers, M_∞, the Mach number distribution around the airfoil also generally increases. Of interest in design is the freestream Mach number at which the flow first goes supercritical; that is, when the flow first accelerates above $M > 1$ locally, somewhere on the airfoil. From the Prandtl–Glauert transformation, we know that

$$c_p(M_\infty) = \frac{c_{p0}}{\sqrt{1 - M_\infty^2}}. \tag{8.38}$$

If we have the incompressible pressure coefficient distribution, c_{p0}, e.g. from a panel method, we can use the above equation to calculate c_p at $M_\infty > 0$. From the isentropic relations,

$$c_p = \frac{2}{\gamma M_\infty^2}\left(\frac{p}{p_\infty} - 1\right) = \frac{2}{\gamma M_\infty^2}\left[\left(\frac{1 + \frac{1}{2}(\gamma - 1)M_\infty^2}{1 + \frac{1}{2}(\gamma - 1)M^2}\right)^{\frac{\gamma}{\gamma - 1}} - 1\right]. \tag{8.39}$$

The *critical pressure coefficient*, defined as the pressure coefficient at sonic conditions, is a function only of M_∞ and corresponds to the above formula evaluated at $M = 1$,

$$c_{p,\text{crit}} = \frac{2}{\gamma M_\infty^2}\left[\left(\frac{1 + \frac{1}{2}(\gamma - 1)M_\infty^2}{1 + \frac{1}{2}(\gamma - 1)}\right)^{\frac{\gamma}{\gamma - 1}} - 1\right]. \tag{8.40}$$

The critical freestream Mach number, $M_{\infty,\text{crit}}$ is the smallest one at which $c_p(M_\infty)$ first drops below $c_{p,\text{crit}}$ somewhere on the airfoil.* The steps for finding $M_{\infty,\text{crit}}$ are then as follows:

1. Find the minimum c_p at $M_\infty = 0$. Call this $c_{p0,\text{min}}$. Note, we are interested in the minimum pressure coefficient since this will be the location where the flow speed is highest.
2. Calculate the corresponding transformed c_p at M_∞,

$$c_{p,\text{min}}(M_\infty) = \frac{c_{p0,\text{min}}}{\sqrt{1 - M_\infty^2}}, \tag{8.41}$$

and plot this quantity versus M_∞.

3. Also plot $c_{p,\text{crit}}(M_\infty)$ from the isentropic relations, Equation 8.40. The critical freestream Mach number, $M_{\infty,\text{crit}}$, is the intersection of the two curves, as shown in Figure 8.8.

* When $c_p < c_{p,\text{crit}}$, the flow has gone from M_∞ to supersonic, assuming isentropic changes.

A thick airfoil will have more negative $c_{p0,\min}$ and hence more negative $c_{p,\min}(M_\infty)$ compared to a thin airfoil. Therefore, as shown in Figure 8.8, the critical Mach number is expected to decrease with increasing thickness.

Incompressible analysis codes that employ the Prandtl–Glauert transformation will often include a $c_{p,\text{crit}}$ line for a given M_∞, as shown in Figure 8.9. If the c_p calculated from the code drops below this $c_{p,\text{crit}}$, i.e. if the c_p curve is above the $c_{p,\text{crit}}$ line on an inverted c_p axis, then the result of the Prandtl–Glauert transformation should not be trusted.

A conventional airfoil operating in supercritical flow will result in a strong shock and associated loss of lift and increased drag. This is illustrated in Figure 8.10. One can use thinner airfoils to mitigate this effect, but there are practical limits (e.g. structural, fuel volume, etc.) to how thin an airfoil can be.

Reasonably thick *supercritical airfoils* have been designed with "flattened" c_p curves that do not drop much below $c_{p,\text{crit}}$ and instead result in a larger supersonic region terminated by a weak shock. An example of such an airfoil is illustrated qualitatively in Figure 8.11.

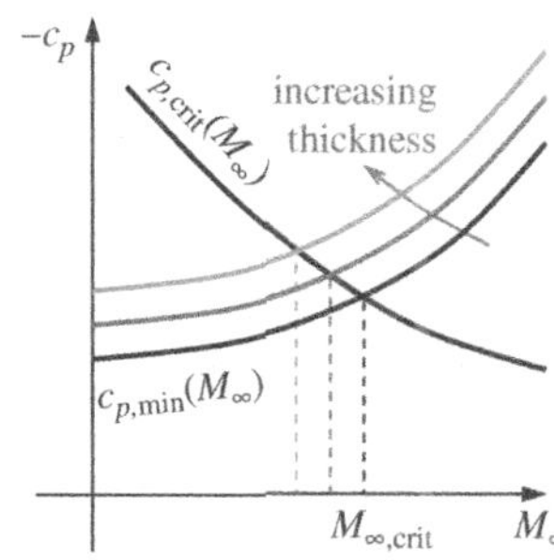

Figure 8.8 Critical freestream Mach number determination as the intersection between the critical pressure coefficient and the transformed minimum pressure coefficient.

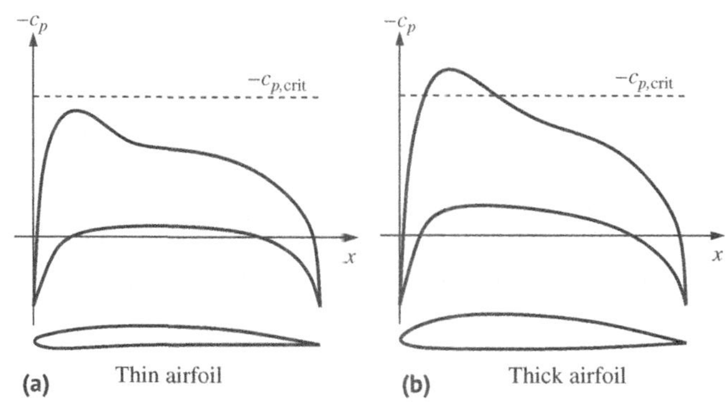

Figure 8.9 Typical c_p plots and the associated $c_{p,\text{crit}}$ lines for incompressible solutions scaled with the Prandtl–Glauert transformation. Shown are plots for a thin airfoil and a thick airfoil at the same M_∞ and the same lift coefficient. The thick airfoil minimum c_p is below the critical c_p which means that the flow is supercritical, and the result of the Prandtl–Glauert transformation is suspect.

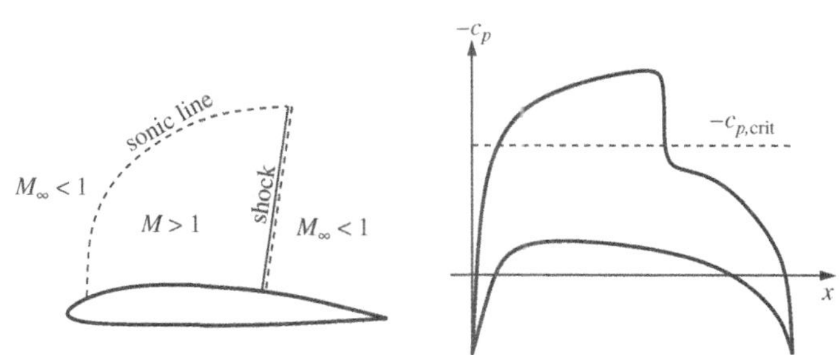

Figure 8.10 Conventional airfoil operating in supercritical flow, and the resulting pressure coefficient distribution (not obtained using a Prandtl–Glauert transformation).

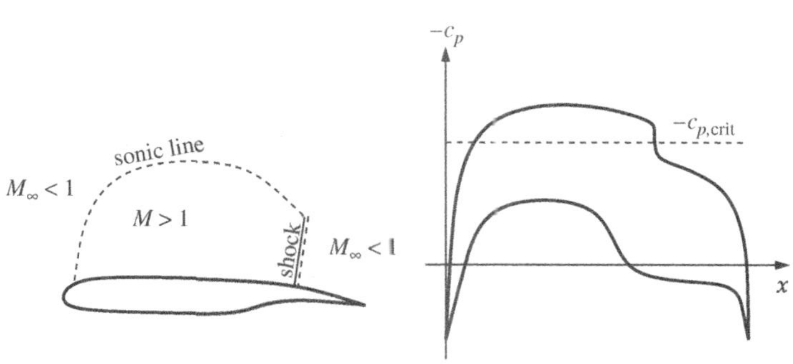

Figure 8.11 A supercritical airfoil with a "flattened" c_p curve. The supersonic region terminates with a weaker shock, resulting in more lift at lower drag.

Example 8.1 Critical Mach number for a cylinder
Using the Prandtl–Glauert transformation, we can calculate the critical Mach number, $M_{\infty,\text{crit}}$, of a lifting cylinder, as a function of the non-dimensional circulation, $\Lambda = \Gamma/(4\pi a U_\infty)$. Recall from Chapter 4 that Γ is the circulation, a is the cylinder radius, and U_∞ is the freestream speed. We assume that $\Lambda \geq 0$.

In the incompressible case, for positive Λ, the lowest pressure occurs on the top of the cylinder. This can be seen from Equation 4.87, where $\theta = \pi/2$ makes $\sin\theta = 1$, which minimizes c_p. Therefore, we have

$$c_{p0,\text{min}} = 1 - 4(\Lambda + 1)^2. \tag{8.42}$$

In the compressible case, this pressure coefficient is corrected via the Prandtl–Glauert transformation,

$$c_{p,\text{min}}(M_\infty) = \frac{1 - 4(\Lambda + 1)^2}{\sqrt{1 - M_\infty^2}}. \tag{8.43}$$

The critical pressure coefficient is given in Equation 8.40. Equating the two gives us an equation for $M_{\infty,\text{crit}}$:

$$\frac{1 - 4(\Lambda + 1)^2}{\sqrt{1 - M_{\infty,\text{crit}}^2}} = \frac{2}{\gamma M_{\infty,\text{crit}}^2}\left[\left(\frac{1 + \frac{1}{2}(\gamma - 1)M_{\infty,\text{crit}}^2}{1 + \frac{1}{2}(\gamma - 1)}\right)^{\frac{\gamma}{\gamma-1}} - 1\right]. \tag{8.44}$$

We solve this equation using the `fzero` function in MATLAB, which finds $M_{\infty,\text{crit}}$ that makes the difference between the left- and right-hand sides of Equation 8.44 zero. Listing 8.1 shows the code that does this.

Code notes:

- `Lv` is a vector of the non-dimensional circulation values that are tested.
- `Mv` is the vector of critical Mach numbers.
- The starting guess of 0.5 for the `fzero` function helps it to quickly converge to the correct critical Mach number.

Listing 8.1 Cylinder critical Mach number calculation.

```
function mcritcyl
% calculates the critical Mach number for a cylinder

Lv = linspace(0,4,41); % vector of Lambda values
for iL = 1:length(Lv),
  Mv(iL) = fzero(@(M) myfun(M,Lv(iL)), 0.5);
end

figure(1); clf;
plot(Lv, Mv, 'k-', 'linewidth', 2);
set(gcf,'position', [400, 400, 400, 400]); grid on;
writeplot('$\Lambda = \Gamma/(4\pi a U_{\infty})$', ...
          '$M_{\infty,\textrm{crit}}$', '../figs/mcritcyl.pdf', 4);

%-----------------------------
function f = myfun(M, L);
g = 1.4;
LHS = (1 - 4*(L+1)^2)/sqrt(1-M^2);
RHS = 2/(g*M^2)*( ((1+.5*(g-1)*M^2)/(1+.5*(g-1)))^(g/(g-1)) - 1);
f = LHS-RHS;
```

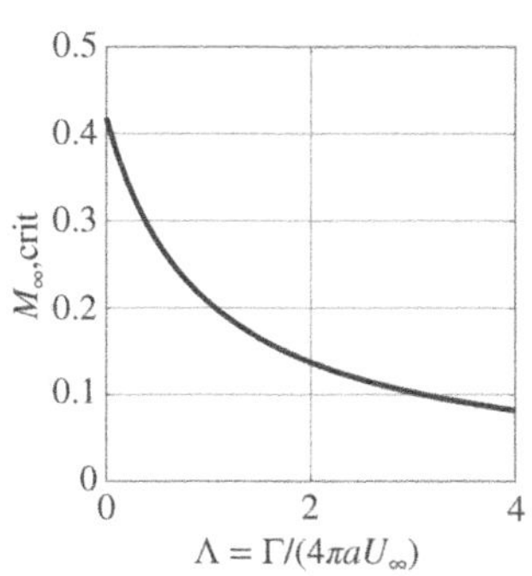

Figure 8.12 Dependence of the critical Mach number on the circulation of the flow over a cylinder.

Figure 8.12 shows the plot produced by the code. As expected, the critical Mach number decreases as the circulation increases. This is because the minimum pressure on the top surface of the cylinder decreases with increasing circulation.

8.4 Supersonic Flow

We now consider the small-disturbance potential flow in the supersonic case, for which $M_\infty > 1$. The governing equation for the *supersonic* case can be written, using Equation 8.18, as

$$\underbrace{(M_\infty^2 - 1)}_{\lambda^2} \hat{\phi}_{xx} - \hat{\phi}_{yy} = 0, \tag{8.45}$$

where λ is a real number since, for $M_\infty > 1$, $M_\infty^2 - 1 > 0$. It turns out that this equation is somewhat easier to solve than its elliptic counterpart, and the behavior of the solution is much different.

8.4.1 Characteristics

The solution to Equation 8.45 can be written quite simply as*

$$\hat{\phi}(x, y) = F(x - \lambda y) + G(x + \lambda y), \tag{8.46}$$

where $F(\cdot)$ and $G(\cdot)$ are scalar functions that depend on the boundary conditions.

To verify that this is indeed the solution, we write

$$\begin{aligned}\hat{\phi}_{xx} &= F''(x - \lambda y) + G''(x + \lambda y)\\ \hat{\phi}_{yy} &= F''(x - \lambda y)\,(-\lambda)^2 + G''(x + \lambda y)\,\lambda^2\\ &= \lambda^2\left[F''(x - \lambda y) + G''(x + \lambda y)\right],\end{aligned}$$

so that

$$\begin{aligned}\lambda^2\hat{\phi}_{xx} - \hat{\phi}_{yy} = \lambda^2 &\left[F''(x - \lambda y) + G''(x + \lambda y)\right]\\ &-\lambda^2\left[F''(x - \lambda y) + G''(x + \lambda y)\right] = 0.\end{aligned}$$

Since Equation 8.45 is identically satisfied, Equation 8.46 is indeed the solution.† What does this solution look like? Considering the individual functions F and G separately, we have

- $F(x - \lambda y) =$ constant along lines for which $x - \lambda y =$ constant, i.e. on lines with slope $dy/dx = 1/\lambda$.
- $G(x + \lambda y) =$ constant on lines with slope $dy/dx = -1/\lambda$.

Such lines are called *characteristics*.‡ They carry information about the solution, in that one part of the solution remains constant along the lines.

In supersonic flow, information cannot propagate upstream, and any disturbances come from changes to the flow as it interacts with solid walls, through boundary conditions on the surface of an object like an airfoil. If an airfoil is flying in freestream conditions, where the flow upstream is uniform and at $M_\infty > 1$, the non-freestream characteristics emanating from the airfoil boundary will appear as the solid lines shown in Figure 8.13.

Consider a point P above the airfoil's surface. By Equation 8.46, the perturbation potential there is given by the sum of F and G. The values of F and G at the point of interest come from the two characteristics that intersect the point. First, the downward-sloping characteristic that carries information on G comes from the freestream, so we write $G = G_\infty$. In fact, for all points above this airfoil, the value of G will be G_∞. However, the value of F comes from a characteristic that

* Both $F(\cdot)$ and $G(\cdot)$ are scalar *functions* of one variable. So, for example, $F(x - \lambda y)$ means the quantity $(x - \lambda y)$ is the input into the function $F(\cdot)$, not that F multiplies $(x - \lambda y)$.

† Until we specify boundary conditions, arbitrary $F(\cdot)$ and $G(\cdot)$ are valid solutions!

‡ The slope of the characteristics should be familiar. Recall that the *Mach angle*, μ is defined by

$$\tan\mu = \frac{1}{\sqrt{M_\infty^2 - 1}} = \frac{1}{\lambda}.$$

This is the angle at which infinitesimal disturbances in the flow propagate. It can be derived by considering a beacon moving at speed $U_\infty = M_\infty c_\infty$, sending out a continuous signal that propagates outward from each point at the speed of sound, c_∞. In our case, we see that *the slope of the characteristics is the tangent of the Mach angle.* It should not be surprising that our analysis with small-disturbance assumptions has led us to the result for an infinitesimal source.

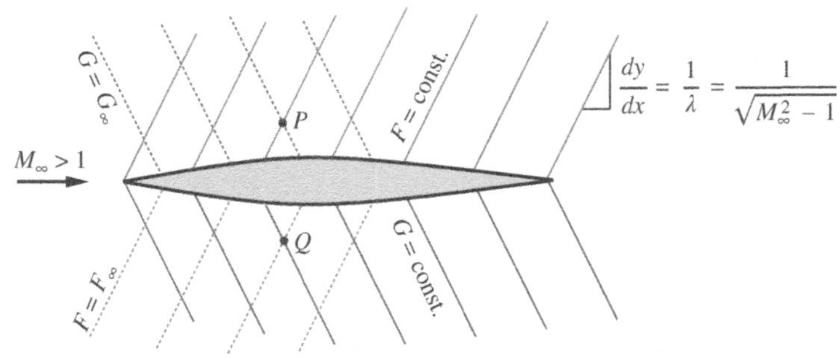

Figure 8.13 Characteristics emanating from an airfoil boundary in small-disturbance supersonic flow.

intersects the geometry. It will depend on the boundary condition at the intersection of the characteristic and the geometry, so that in general, different points above the airfoil will have different F values.§

§ All upward-sloping F characteristics have the same slope, $1/\lambda$, so they will never intersect. Similarly, all G characteristics are parallel and will never intersect.

Conversely, consider a point Q below the airfoil surface. The value of F at such a point comes from the freestream, so we write $F = F_\infty$. For all points below this airfoil, F will be constant, since it will always come from a characteristic that connects to the freestream. On the other hand, G will now depend on position, since the characteristic that determines G intersects the airfoil.

Differentiating Equation 8.46 with respect to x and y gives the perturbation velocities¶

¶ F' and G' are constant, respectively, along the same characteristic lines along which F and G are constant.

$$\hat{u} = \hat{\phi}_x = F' + G', \qquad \hat{v} = \hat{\phi}_y = \lambda(-F' + G'). \tag{8.47}$$

These equations will be generally more useful than those involving $\hat{\phi}$ for analyzing supersonic flow problems, since velocity perturbations relate directly to the pressure coefficient and the flow angle.

8.4.2 Pressure Coefficient in External Flow

As discussed in the previous section, perturbations in external flow* over an object originate on the boundary. We now look at how these perturbations affect the pressure coefficient on the upper and lower surfaces of an airfoil.

* By external flow, we mean uniform freestream flow over a single object.

First, consider a point on the upper surface, shown in Figure 8.14. The small perturbation boundary condition at this point is

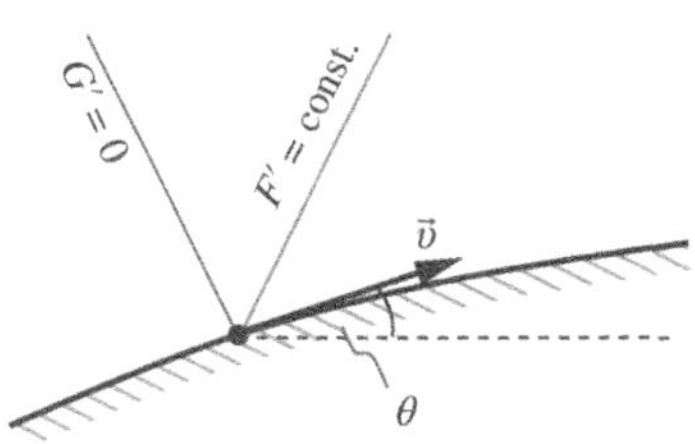

Figure 8.14 Boundary condition on the upper surface of an airfoil.

$$\hat{v} = U_\infty \theta. \tag{8.48}$$

From the previous section, we know that G on the upper surface is everywhere (not just along a line) constant, which means that its derivative is zero, $G' = 0$. This derivative tells us how G changes as we move from one characteristic to another, and on the top surface, $G = G_\infty$ for all characteristics. So, by Equation 8.47, the velocity perturbations at the point are

$$\hat{u} = F' + G' = F', \qquad \hat{v} = \lambda(-F' + G') = -\lambda F'. \tag{8.49}$$

Using $\hat{v} = U_\infty \theta$ from the boundary condition requirement, Equation 8.48, we obtain $F' = -U_\infty \theta / \lambda$. So the horizontal velocity perturbation and pressure coefficient (from Equation 8.25) are

$$\hat{u} = F' = -U_\infty \frac{\theta}{\lambda}, \qquad c_p = -\frac{2\hat{u}}{U_\infty} = \frac{2\theta}{\lambda}. \tag{8.50}$$

This equation, which is valid for the upper surface, states that the pressure coefficient depends only on the deflection angle θ, where θ is measured from the freestream direction (x axis in our case). On the upper surface, a positive θ corresponds to a compression, and hence $c_p > 0$, while $\theta < 0$ corresponds to an expansion, and hence $c_p < 0$.

Second, for a point on the lower surface, illustrated in Figure 8.15, the small perturbation boundary condition is the same as in Equation 8.48, $\hat{v} = U_\infty \theta$. However, instead of $G' = 0$, we now have that $F' = 0$. This is because on the lower surface, all F characteristics have the same value, $F = F_\infty$, as shown in Figure 8.13. So, by Equation 8.47, the velocity perturbations on the lower surface are

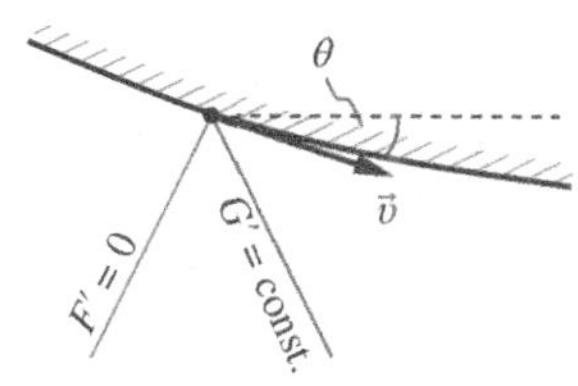

Figure 8.15 Boundary condition on the lower surface of an airfoil.

$$\hat{u} = F' + G' = G', \qquad \hat{v} = \lambda(-F' + G') = \lambda G'. \tag{8.51}$$

Using the boundary condition $\hat{v} = U_\infty \theta$, we have $G' = U_\infty \theta / \lambda$, so that $\hat{u}$ and c_p are

$$\hat{u} = G' = U_\infty \frac{\theta}{\lambda}, \qquad c_p = -\frac{2\hat{u}}{U_\infty} = \frac{-2\theta}{\lambda}. \tag{8.52}$$

Again, the pressure coefficient depends only on the deflection angle, although the result is negative of that over the upper surface. This is because a positive angle on the lower surface corresponds to an expansion (negative c_p), while a negative angle corresponds to a compression.†

Figure 8.16 shows that the same physical principle holds on both the lower and upper surfaces: the pressure coefficient increases in a compression and decreases in an expansion. Knowing the pressure coefficient as a function of the geometry, we can integrate it to obtain the force and moment coefficients on an airfoil.

† In summary, for either the upper or lower surface,

- c_p increases if the flow turns into itself (compression).
- c_p decreases if the flow turns away from itself (expansion).

The magnitude of c_p is the same in both cases and is given by

$$|c_p| = \frac{2|\theta|}{\sqrt{M_\infty^2 - 1}}. \tag{8.53}$$

The sign depends on whether the flow is in compression ($c_p > 0$) or expansion ($c_p < 0$).

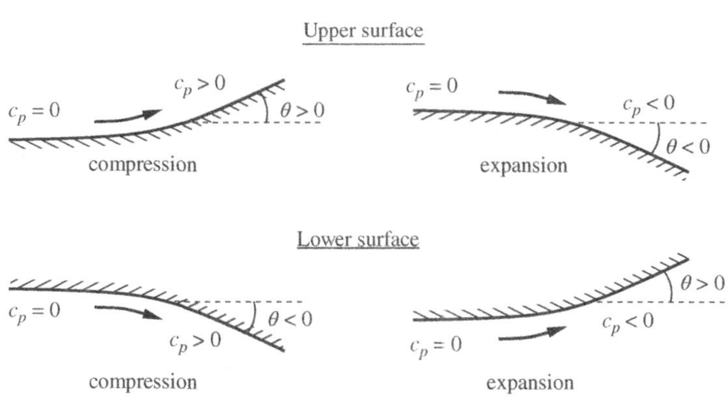

Figure 8.16 Geometries leading to compressions and expansions in small-disturbance potential flow.

Example 8.2 Flat plate in supersonic flow

In this example, we will use the pressure coefficient result in Equation 8.53 to calculate the force coefficients on a simple geometry: a flat plate. We consider a flat plate of chord length c at an angle of attack α in supersonic flow at $M_\infty > 1$, as shown in Figure 8.17.

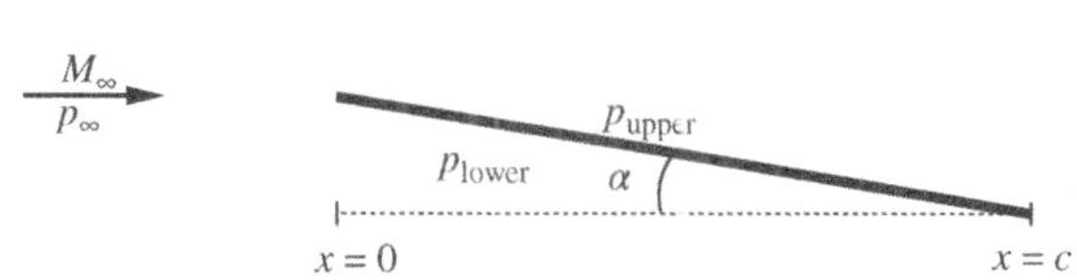

Figure 8.17 A flat plate in supersonic flow.

Assuming a small angle α, the two-dimensional force on the flat plate is

$$\vec{F}' = \int_0^c (p_{\text{lower}} - p_{\text{upper}})\vec{n}\, dx, \tag{8.54}$$

where, by the small-angle approximation for α in radians,

$$\vec{n} = \sin\alpha\hat{\imath} + \cos\alpha\hat{\jmath} \approx \alpha\hat{\imath} + \hat{\jmath}. \tag{8.55}$$

Therefore, the two-dimensional lift is

$$L' = \vec{F}' \cdot \hat{\jmath} = \int_0^c (p_{\text{lower}} - p_{\text{upper}})\, dx, \tag{8.56}$$

which means that the lift coefficient is

$$c_\ell = \frac{L'}{\frac{1}{2}\rho_\infty U_\infty^2 c} = \frac{1}{c}\int_0^c (c_{p,\text{lower}} - c_{p,\text{upper}})\, dx. \tag{8.57}$$

Using the results of the previous section, the pressure coefficients on the upper and lower surface are

$$c_{p,\text{upper}} = \frac{-2\alpha}{\lambda}, \quad c_{p,\text{lower}} = \frac{2\alpha}{\lambda}. \tag{8.58}$$

Substituting into our expression for the lift coefficient, we have[‡]

‡ The integral is straightforward to evaluate because the integrand is constant.

$$c_\ell = \frac{1}{c}\int_0^c \left(\frac{2\alpha}{\lambda} - \frac{(-2\alpha)}{\lambda}\right) dx = \frac{4\alpha}{\lambda}. \tag{8.59}$$

This equation tells us that higher M_∞ will result in a lower lift-curve slope, because $\lambda = \sqrt{M_\infty^2 - 1}$. The drag is the $\hat{\imath}$ component of the force, $D' = \vec{F}' \cdot \hat{\imath}$, so the drag coefficient is the same as the lift but with an extra α, which comes from the $\hat{\imath}$ component of the normal vector in Equation 8.55,[§]

§ The dependence on α^2 shows that the drag coefficient is non-negative regardless of the sign of α.

$$c_d = \frac{1}{c}\int_0^c (c_{p,\text{lower}} - c_{p,\text{upper}})\alpha\, dx = \frac{4\alpha^2}{\lambda}. \tag{8.60}$$

Example 8.3 Linearized supersonic flow over a ramp
We can analyze supersonic flow over the ramp pictured in Figure 8.18 using linearized supersonic theory. We will be able to calculate the velocity perturbations, $\hat{u}$ and $\hat{v}$, and the pressure coefficient, c_p, in each distinct region of the flow.

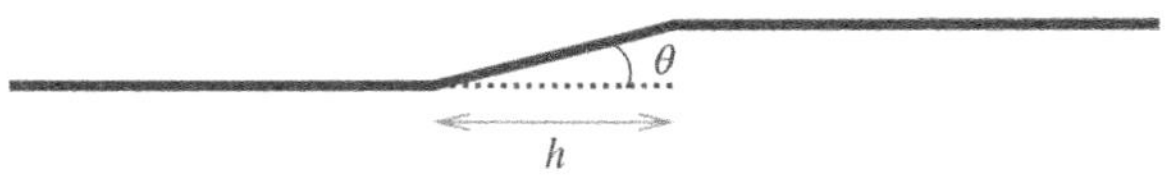

Figure 8.18 Ramp in supersonic flow.

To simplify the calculations, suppose that $\lambda = \sqrt{M_\infty^2 - 1} = 1$, which means that $M_\infty = \sqrt{2}$. We now define three distinct flow regions, A, B, and C, as shown in Figure 8.19.

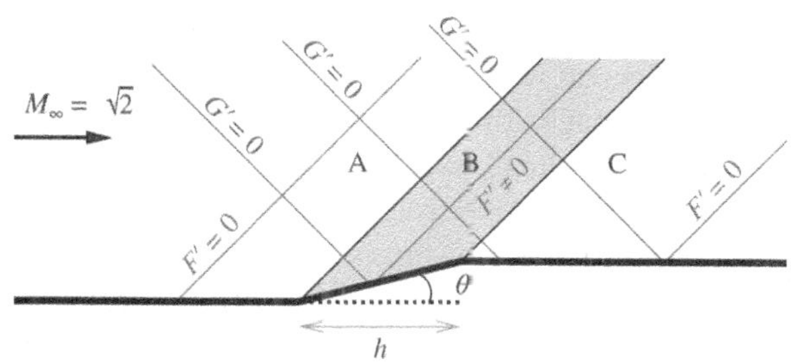

Figure 8.19 Region definitions for supersonic flow over a ramp.

A This region is still in the freestream, so $\hat{u} = \hat{v} = 0$, $c_p = 0$, and $F' = G' = 0$.

B In this region, the flow is perturbed upward by the ramp. The wall boundary condition is $\hat{v} = U_\infty \theta$, and $G' = 0$ because these characteristics come from the freestream (region A). So

$$\hat{v} = U_\infty \theta = \lambda(-F' + G') = -\lambda F'. \tag{8.61}$$

Solving for F' gives $F' = -U_\infty \theta/\lambda$. The horizontal velocity perturbation and pressure coefficient are then

$$\hat{u} = F' + G' = F' = \frac{-U_\infty \theta}{\lambda}, \quad c_p = \frac{-2\hat{u}}{U_\infty} = \frac{2\theta}{\lambda}. \tag{8.62}$$

C In this region, the flow angle is zero because of the horizontal wall, so $\hat{v} = 0$. The G' characteristic is still zero due to its connection to the freestream. Using the equation $\hat{v} = \lambda(-F' + G')$, we see that $F' = 0$, and so this region is like the freestream: no perturbations and $c_p = 0$.

Example 8.4 Linearized supersonic flow over a ramp in a duct
Extending Example 8.3, assume now that there is a wall enclosing the ramp, making a duct as shown in Figure 8.20.

Figure 8.20 Supersonic flow over a ramp in a duct.

Again, we are interested in the c_p and velocity perturbations in each distinct region of the flow. This situation is more complicated than the previous one due to the presence of the upper wall. We have to account for reflections of characteristics from the walls in our analysis. To start, the boundaries between distinct regions are shown in Figure 8.21.

In identifying the regions, we also already determined which lines are compressions and which are expansions. At the start of the ramp, the flow turns into itself and is compressed, so the line dividing the first regions A and B is a compression wave. This compression wave reflects off the top wall as a compression wave: we deduce this from the fact that the pressure increases when the flow crosses from region B to region C, since the flow again turns into itself. So compressions reflect off walls as compressions. At the end of the ramp, the flow turns away from itself, so the line dividing the first region B and the second region A is an expansion – our c_p calculation also tells us this, since c_p decreases when crossing from region B to the second region A. The expansion reflects off the top wall as an expansion: it now forms the boundary between regions D and the third region A, over which the pressure decreases. The pair of compression and expansion waves continues to reflect off the bottom and top walls for the remainder of the duct.

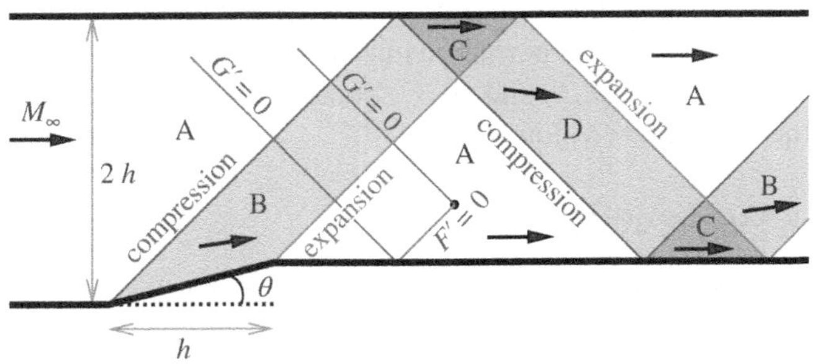

Figure 8.21 Identification of regions, flow directions, and characteristics in supersonic flow over a ramp in a duct.

Using Figure 8.21, we can analyze the state of the flow in each region.

A This region is always the freestream, so $\hat{u} = \hat{v} = 0$, $c_p = 0$. Note that this region keeps appearing downstream of the ramp, as the reflected disturbances only affect the flow in between the lines originating from the beginning/end of the ramp. A graphical analysis with characteristics shown above confirms that $F' = 0$ and $G' = 0$ in the first *A* region after the ramp, and this pattern continues downstream.

B In this region, the flow is perturbed upward by the ramp. As in the previous example, we have that $F' = -U_\infty\theta/\lambda$, $G' = 0$, $\hat{u} = F' = -U_\infty\theta/\lambda$, and $c_p = 2\theta/\lambda$.

C In this region, the flow has just reflected off the upper wall but is still influenced by the ramp on the bottom wall. The F' characteristic in this region is the same as in region *B*, but the G' characteristic is no longer zero due to a lack of connection to the freestream. Physically, something has to change from region *B*, where the flow is going upward, to region *C*, where the flow must be horizontal again to not go through the wall. The wall boundary condition is precisely the extra piece of information we need: $\hat{v} = 0$ combined with $F' = -U_\infty\theta/\lambda$ (from region B) gives

$$0 = \hat{v} = \lambda(-F' + G') \quad \Rightarrow \quad G' = F' = \frac{-U_\infty\theta}{\lambda}. \tag{8.63}$$

Using the $\hat{u}$ equation, we have

$$\hat{u} = F' + G' = \frac{-2U_\infty\theta}{\lambda}, \quad c_p = \frac{-2\hat{u}}{U_\infty} = \frac{4\theta}{\lambda}. \tag{8.64}$$

D In this region, the flow is only influenced by the reflection off the top wall, so that $G' = -U_\infty\theta/\lambda$. The F' characteristic, however, comes from region *A* (the second one) where the flow is freestream, so $F' = 0$. Therefore,

$$\hat{v} = \lambda(-F' + G') = \lambda G' = -U_\infty\theta. \tag{8.65}$$

and $\hat{u}$ and c_p are

$$\hat{u} = F' + G' = \frac{-U_\infty \theta}{\lambda}, \quad c_p = \frac{-2\hat{u}}{U_\infty} = \frac{2\theta}{\lambda}. \tag{8.66}$$

8.4.3 Lift and Drag Calculation

So far we have derived formulas for the pressure coefficient on either side of an airfoil in external flow, and we have analyzed a flat plate airfoil at a zero angle of attack. Consider now an arbitrary slender airfoil in supersonic flow, as illustrated in Figure 8.22. In this analysis, we use z for the vertical coordinate, to be consistent with our airfoil theory notation from Chapter 5. We also use a wind-aligned coordinate system in which the x axis points in the freestream direction, $\vec{V}_\infty = U_\infty \hat{i}$. The angle of attack α is assumed small, so that the chord length c is also the projected length of the airfoil along the x axis.* The lower and upper surface functions are given by $z_{\text{lower}}(x)$ and $z_{\text{upper}}(x)$, respectively.

* This equality is due to $\cos\alpha \approx 1$ for a small angle α.

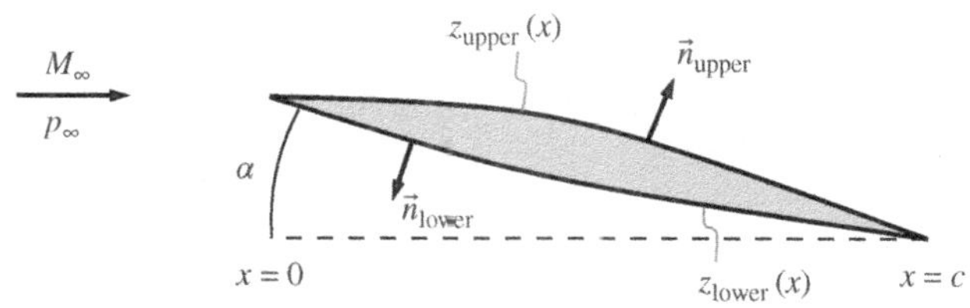

Figure 8.22 A slender airfoil at an angle of attack in supersonic flow.

Let $\theta_{\text{upper}}(x)$ be the flow angle on the upper surface, and $\theta_{\text{lower}}(x)$ the flow angle on the lower surface. Assuming that these angles are small, we have that†

$$\theta_{\text{upper}}(x) \approx \frac{dz_{\text{upper}}}{dx}, \qquad \theta_{\text{lower}}(x) \approx \frac{dz_{\text{lower}}}{dx}. \tag{8.67}$$

† This equality is due to $\tan\theta \approx \theta$ for small angles θ measured in radians.

The surface normals pointing out of the geometry and into the fluid are, using small-angle approximations,‡

$$\vec{n}_{\text{upper}}(x) = -\sin(\theta_{\text{upper}})\hat{i} + \cos(\theta_{\text{upper}})\hat{k} \approx -\theta_{\text{upper}}\hat{i} + \hat{k}, \tag{8.68}$$

$$\vec{n}_{\text{lower}}(x) = \sin(\theta_{\text{lower}})\hat{i} - \cos(\theta_{\text{lower}})\hat{k} \approx \theta_{\text{lower}}\hat{i} - \hat{k}. \tag{8.69}$$

‡ Note the sign switch in the formula for the lower normal versus the upper normal, due to the fact that they point in opposite directions for a given θ.

The total sectional force on the airfoil is the sum of the upper and lower forces, $\vec{F}' = \vec{F}'_{\text{upper}} + \vec{F}'_{\text{lower}}$, where

$$\vec{F}'_{\text{upper}} = \int_0^c p_{\text{upper}}(x)(-\vec{n}_{\text{upper}}(x))\, dx, \tag{8.70}$$

$$\vec{F}'_{\text{lower}} = \int_0^c p_{\text{lower}}(x)(-\vec{n}_{\text{lower}}(x))\, dx \tag{8.71}$$

The direction of this force depends on the normal vectors, $\vec{n}_{\text{upper}}$ and $\vec{n}_{\text{lower}}$. The sectional lift is the $\hat{k}$ component of $\vec{F}'$,§

§ The $\hat{k}$ component of the normal is ± 1.

$$L' = -\int_0^c p_{\text{upper}}(x)\, dx + \int_0^c p_{\text{lower}}(x)\, dx.$$

The lift coefficient is the sectional lift force normalized by the product of the dynamic pressure and the chord length,

$$
\begin{aligned}
c_\ell = \frac{L'}{\frac{1}{2}\rho U_\infty^2 c} &= -\frac{1}{c}\int_0^c c_{pu}(x)\,dx + \frac{1}{c}\int_0^c c_{pl}(x)\,dx \\
&= -\frac{1}{c}\int_0^c \frac{2\theta_{\text{upper}}(x)}{\lambda}\,dx + \frac{1}{c}\int_0^c \frac{-2\theta_{\text{lower}}(x)}{\lambda}\,dx \\
&= -\frac{2}{\lambda c}\left[\int_0^c \frac{dz_{\text{upper}}}{dx}\,dx + \int_0^c \frac{dz_{\text{lower}}}{dx}\,dx\right] \\
&= -\frac{2}{\lambda c}\left[z(c) - z(0) + z(c) - z(0)\right] \\
&= \frac{4}{\lambda}\underbrace{\left[\frac{z(0) - z(c)}{c}\right]}_{\tan\alpha \approx \alpha}.
\end{aligned}
$$

In the last step, we used the definition of the angle of attack as the angle between the chord line, which connects $z(0)$ to $z(c)$, and the freestream. Therefore, the airfoil's lift coefficient is

$$
c_\ell = \frac{4\alpha}{\lambda}. \tag{8.72}
$$

This result is the same as for a flat plate, in Equation 8.59, and it does not depend on the airfoil shape (i.e. camber and thickness) as long as the airfoil is slender and at a small angle of attack (small angles $\theta_{\text{upper}}(x)$ and $\theta_{\text{lower}}(x)$).¶ The lift-curve slope is $4/\lambda = 4/\sqrt{M_\infty^2 - 1}$, which is consistent with the qualitative depiction in Figure 8.2.

¶ The angle of attack is contained in the definitions of the upper and lower surface angles, due to the wind-aligned coordinate system.

The sectional drag is the $\hat{\imath}$ component of $\vec{F}'$,||

$$
D' \approx \int_0^c p_{\text{upper}}(x)\theta_{\text{upper}}(x)\,dx - \int_0^c p_{\text{lower}}(x)\theta_{\text{lower}}(x)\,dx.
$$

|| The $\hat{\imath}$ component of the normal vectors is $-\theta_{\text{upper}}$ on the upper surface and θ_{lower} on the lower surface.

Dividing by $\frac{1}{2}\rho U_\infty^2 c$, the drag coefficient is

$$
\begin{aligned}
c_d &= \frac{1}{c}\int_0^c c_{pu}(x)\theta_{\text{upper}}(x)\,dx - \frac{1}{c}\int_0^c c_{pl}(x)\theta_{\text{lower}}(x)\,dx \\
&= \frac{2}{c\lambda}\left[\int_0^c \theta_{\text{upper}}(x)^2\,dx + \int_0^c \theta_{\text{lower}}(x)^2\,dx\right] \\
&= \frac{2}{\lambda c}\int_0^c \left[\theta_{\text{upper}}(x)^2 + \theta_{\text{lower}}(x)^2\right]dx.
\end{aligned} \tag{8.73}
$$

This equation shows that the drag coefficient depends on the airfoil geometry and angle of attack, which is in the definition of the angles. The lowest drag coefficient occurs when the integrals of the squares of $\theta_{\text{upper}} \approx dz_{\text{upper}}/dx$ and $\theta_{\text{lower}} \approx dz_{\text{lower}}/dx$ are as small as possible, which happens for a flat plate. Real airfoils have thickness, however, and this thickness increases the integrand and hence the drag. A design goal is to make the thickness distribution as smooth as possible, to avoid unnecessary slopes that increase the squares of the angles in the integral in Equation 8.73. Therefore, to minimize drag on an airfoil in linearized supersonic flow, the waviness of the geometry, illustrated in Figure 8.23, should be kept to a minimum.

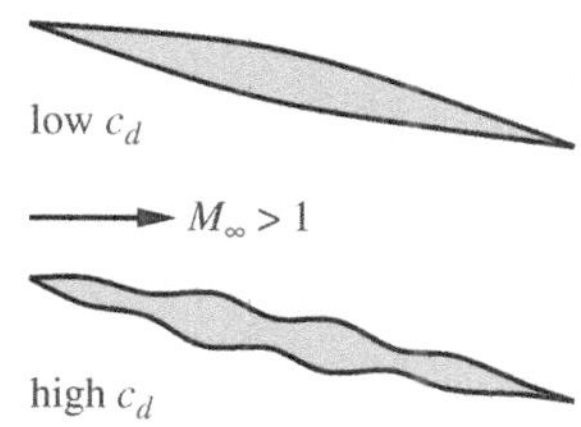

Figure 8.23 In linearized supersonic flow, wavy geometry variations add to drag without affecting the lift.

The above lift and drag coefficient analysis is valid for thin airfoils in external supersonic flow. When calculating the force on a thin airfoil in general supersonic flow conditions, e.g. close to another airfoil or to a wall, we cannot always use the simple form for c_ℓ in Equation 8.72, nor for c_d in Equation 8.73, since the characteristics hitting the airfoil may not be coming from the freestream. For example, F' may not be zero on the lower surface, or G' may not be zero on the upper surface. In such cases, the more general forms, without substituting for c_p, are valid,**

$$c_l = \frac{1}{c}\int_0^c \left[c_{p,\text{lower}}(x) - c_{p,\text{upper}}(x)\right]\, dx, \tag{8.74}$$

$$c_d = \frac{1}{c}\int_0^c \left[c_{p,\text{upper}}(x)\theta_{\text{upper}} - c_{p,\text{lower}}(x)\theta_{\text{lower}}\right]\, dx. \tag{8.75}$$

** Note that θ in these formulas is defined as positive for upward-sloping geometry, and negative for downward-sloping geometry. This definition is the same for lower and upper surfaces.

Example 8.5 Accuracy of small-disturbance potential flow
Consider again a flat-plate airfoil, as in Example 8.2. Suppose that this flat plate is flying at $M_\infty = 2$, $\alpha = 3°$. The results from linear theory are

$$c_\ell = \frac{4\alpha}{\sqrt{M_\infty^2 - 1}} = 0.12092, \quad c_d = \frac{4\alpha^2}{\sqrt{M_\infty^2 - 1}} = 0.00633. \tag{8.76}$$

In reality, without small-disturbance assumptions, we will have a system of oblique shocks and expansion fans, as shown in Figure 8.24.

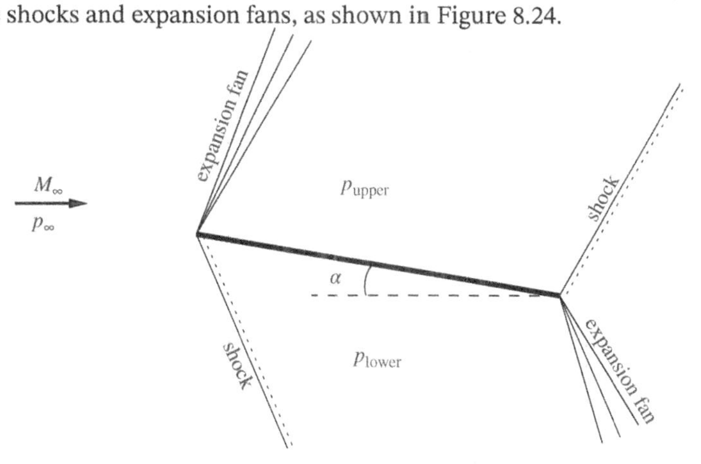

Figure 8.24 Oblique shocks and expansion fans for a flat plate at an angle of attack in supersonic flow, analyzed in Example 8.5.

Using shock–expansion theory, we find that the Mach number after the expansion fan, that is over the upper surface of the flat plate, is

$$M_{\text{upper}} = \nu^{-1}\left[\nu(M_\infty) + \alpha\right] = 2.1106, \tag{8.77}$$

where $\nu(M)$ is the Prandtl–Meyer function. The pressure on the upper surface is then given by the isentropic relations,

$$p_{\text{upper}} = \left[\frac{1 + \frac{\gamma-1}{2} M_{\infty}^2}{1 + \frac{\gamma-1}{2} M_{\text{upper}}^2} \right]^{\frac{\gamma}{\gamma-1}} p_{\infty} = 0.8416 \, p_{\infty}. \tag{8.78}$$

On the lower surface, the flow is post oblique shock. Using the oblique shock "$\theta - \beta - M$" relationship, we find

$$p_{\text{lower}} = 1.18097 p_{\infty}. \tag{8.79}$$

Finally, using the pressures over the upper and lower surfaces, we can calculate the lift and drag coefficients (note, we do not assume small angles),

$$\begin{aligned} c_{\ell} = \frac{L'}{\frac{1}{2}\rho_{\infty} U_{\infty}^2 c} &= \frac{(p_{\text{lower}} - p_{\text{upper}}) c \cos \alpha}{\frac{\gamma}{2} p_{\infty} M_{\infty}^2 c} \\ &= \frac{2}{\gamma M_{\infty}^2} [1.18097 - 0.8416] \cos(3^{\circ}) \\ &= 0.121036, \\ c_d = \frac{D'}{\frac{1}{2}\rho_{\infty} U_{\infty}^2 c} &= \frac{(p_{\text{lower}} - p_{\text{upper}}) c \sin \alpha}{\frac{\gamma}{2} p_{\infty} M_{\infty}^2 c} \\ &= 0.00634. \end{aligned}$$

In this case, the difference from linear theory is very small: only 0.1% for c_{ℓ} and 0.18% for c_d. We expect the deviation to increase as the angle of attack increases, as thickness and camber are introduced, and as M_{∞} approaches 1 (transonic flow) or M_{∞} becomes large (hypersonic flow).

8.5 Airfoil Design

This section synthesizes some of the compressible-flow analysis tools developed in the previous section and applies them to the design of airfoils. A typical design objective for an airfoil is the amount of drag it experiences at a given lift. Sources of drag include skin friction, pressure drag due to separation and a non-closed effective displaced body, and wave drag due to formation of shocks. As the Mach number increases, wave drag becomes a sizable contribution to the total drag, and airfoils intended to fly at high speeds are designed to reduce it.

It is difficult to find a single airfoil shape that is optimal in drag minimization across a range of Mach numbers. Airfoil shapes will thus vary depending on the intended flow regime. Figure 8.25 shows a qualitative illustration of optimal airfoil shapes across flow regimes.

- In the incompressible limit, there is no wave drag, and an airfoil can be thick to accommodate structural considerations. Such airfoils often have a large radius of curvature at the leading edge to prevent suction peaks.
- At higher speeds but still in the subcritical regime, a goal becomes to not create shocks. A critical Mach number calculation becomes the relevant analysis, and airfoils tend to be thinner since increasing thickness reduces the critical Mach number.
- In transonic flight, which is in the supercritical regime but below Mach 1, shocks will be present, and the goal is to minimize their strength. Supercritical airfoils with an aft-shifted camber are designed to do this.

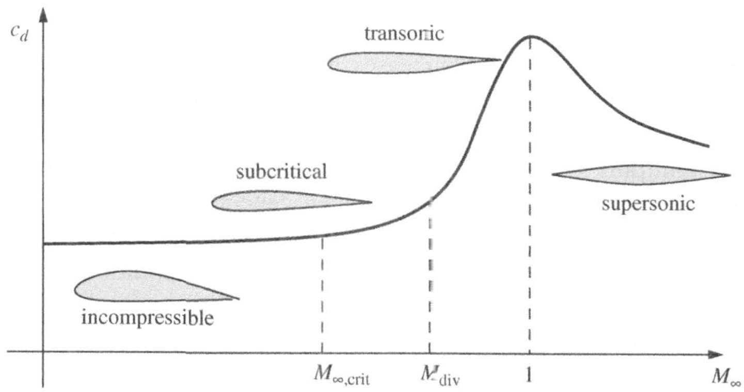

Figure 8.25 Variation of optimal airfoil shapes across flow regimes.

- Above Mach 1, in supersonic flight, airfoil design changes drastically. These airfoils are generally thin to minimize drag, and they have a sharp leading edge to not create large disturbances.

8.5.1 Subcritical Airfoils

A subcritical airfoil flies below the critical Mach number, so that no shocks are present. In order to be able to fly fast, we would like to make $M_{\infty,\text{crit}}$ as high as possible to fly fast. From Figure 8.8, decreasing the magnitude of the minimum incompressible c_p shifts the corrected $c_{p,\min}$ curve down and increases $M_{\infty,\text{crit}}$. Hence, the design goal becomes to minimize the magnitude of $c_{p0,\min}$ (typically negative), while still producing a prescribed amount of lift.*

A simple way to affect $c_{p0,\min}$ is to vary the camber. The lift distribution of a symmetric airfoil is biased toward the front.† This distribution results in the lowest c_p values occurring toward the front. A similar result holds for airfoils with maximum camber in the front or middle of the airfoil.

For example, Figure 8.26 shows a critical Mach number analysis for a NACA 2410, which has its maximum camber at the 40% chord location. At a lift coefficient of $c_\ell = 0.5$, the critical Mach number is 0.675, as predicted by the Kármán–Tsien (Tsien 1939; Kármán 1941) compressibility correction in mfoil. This critical Mach number is limited by the lowest pressure, which occurs on the upper surface at approximately 20% chord.

Figure 8.27 shows the effect of moving the camber back to 80% chord while keeping the thickness and maximum camber the same. This airfoil, the NACA 2810, has a higher critical Mach number of 0.725 at the same lift coefficient. The reason for this increase is the change in the pressure distribution: larger pressure differences now occur on the aft portion of the airfoil. That is, the NACA 2810 generates more lift toward the back of the airfoil compared to the front or middle, resulting in upper-surface pressures that are overall not as low compared to the NACA 2410. The shift in the lift distribution is also evident in the moment coefficient, which becomes more negative, pitch-down, for the NACA 2810 airfoil.

In summary, the general design principle for subcritical airfoils is to make their critical Mach number high: to fly as fast as possible without creating shocks. This can be done with reduced thickness and an aft-shifted camber, which also shifts the lift distribution toward the back. The upper-surface c_p should remain relatively flat and not have low-pressure spikes or bumps, which would be the first locations of shock formation.

* This is an incompressible analysis goal, so we can use potential flow tools, such as thin-airfoil theory or a vortex-panel method.

† Recall from thin-airfoil theory that the center of pressure is at the quarter-chord for a symmetric airfoil.

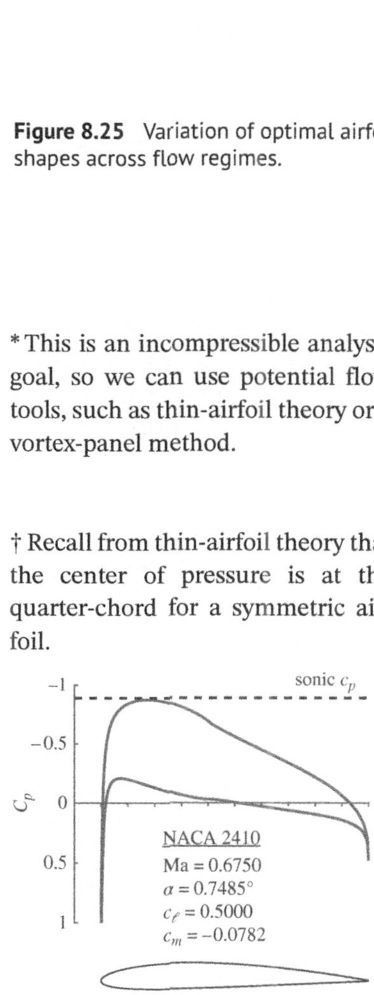

Figure 8.26 40% camber position.

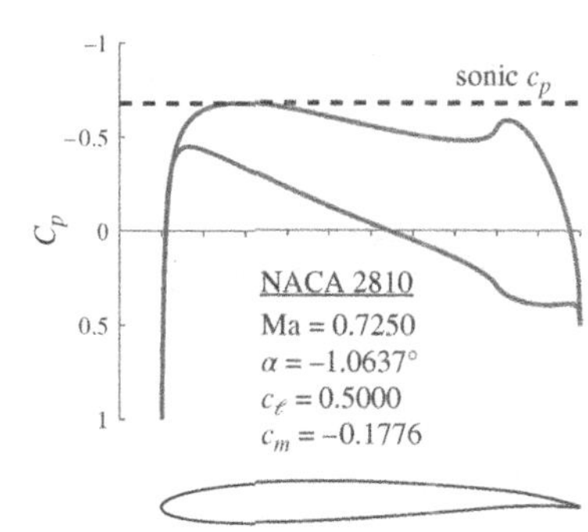

Figure 8.27 80% camber position.

8.5.2 Supercritical Airfoils

As the Mach number of an airfoil increases, maintaining subcritical flow becomes no longer feasible at a practical thickness and lift coefficient. Supercritical airfoils have, however, been designed to fly at $M_\infty > M_{\infty,\mathrm{crit}}$. Shocks are now present, and the goal is to minimize their strength. Stronger shocks lead to more drag because of the increased generation of entropy, which depends on the upstream Mach number. Entropy production is directly related to drag, for example through a formula due to Oswatitsch (1956).* Hence, the design goal is to minimize the size and strength of any shocks present.

* This formula gives the wave drag as an integral over the farfield boundary of a quantity that depends on the jump in entropy across the shock.

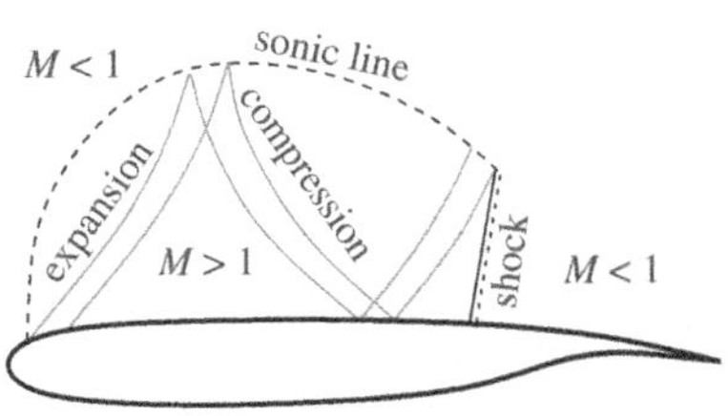

Figure 8.28 A supercritical airfoil.

Figure 8.28 shows a generic supercritical airfoil. Such an airfoil is characterized by a large leading-edge radius, a relatively flat upper surface, and a high camber on the aft portion due to deformation of the lower surface. The rounded leading edge accelerates the flow over a short distance to supersonic speeds. Once supersonic, further acceleration of the flow near the leading edge sends out expansion waves that propagate within the supersonic bubble, as shown in Figure 8.28.

When these waves hit the sonic line, a free surface, they reflect as compression waves that return to the airfoil surface and reflect again as compression waves. These compressions help slow the flow down isentropically, although a complete isentropic deceleration is generally not possible due to the compressions coalescing into a shock at the end of the supersonic bubble.

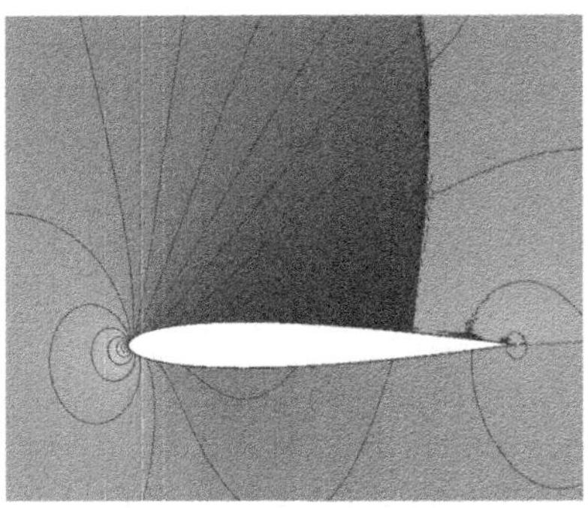

Figure 8.29 Mach contours for a NACA 0012 airfoil in transonic flow.

Computational fluid dynamics can provide a more comprehensive view of the flow over a supercritical airfoil, and the differences between a conventional one. An example is the pair of flowfields shown in Figures 8.29 and 8.30, which were obtained using a high-order finite-element method (Fidkowski 2013). Figure 8.29 shows the Mach number distribution for a NACA 0012 airfoil at freestream Mach number $M_\infty = 0.8$ and a lift coefficient of $c_\ell = 0.6$. A large region of supersonic flow is evident on the upper surface, and it is terminated by a strong and large shock. The resulting wave drag coefficient is $c_{dw} = 0.0537$, or 537 counts.

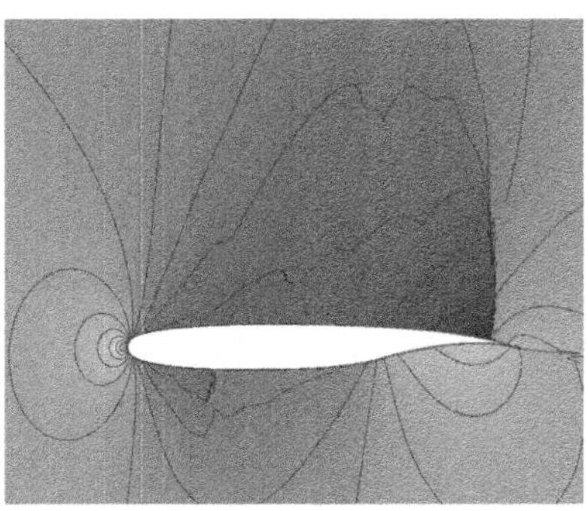

Figure 8.30 Mach contours for a supercritical airfoil in transonic flow.

A supercritical airfoil can achieve the same lift with less drag, through lower flow speeds on the upper surface and a smaller shock, as shown in Figure 8.30. This airfoil generates more of its lift on the aft portion, where on the lower surface, the Mach numbers are smaller due to the high camber and large amount of flow turning. For the same lift coefficient of $c_\ell = 0.6$, the supercritical airfoil achieves a wave-drag coefficient of $c_{dw} = 0.0142$, or 142 counts, over three times lower than the NACA 0012.

Additional design considerations for supercritical airfoils include off-design performance and moment coefficient constraints. A supercritical airfoil that is designed for minimum drag at one Mach number and c_ℓ combination may not perform well as these change. A multi-point optimization is often performed to find the best airfoil over a range of operating conditions. In addition, the aft loading of a supercritical airfoil changes its moment, and the new location of center of pressure and its variation with respect to operating conditions has to be taken into account in the design.

8.5.3 Airfoils in Supersonic Flow

At supersonic speeds, the flow physics fundamentally change. Mathematically, the equations become hyperbolic instead of elliptic. This means that the flow does not see the airfoil until it hits it, at which point the flow must turn and create a shock or an expansion. In contrast, at subsonic speeds, the flow turns before hitting the airfoil, due to pressure gradients established ahead of the airfoil. Figure 8.31 illustrates this difference.

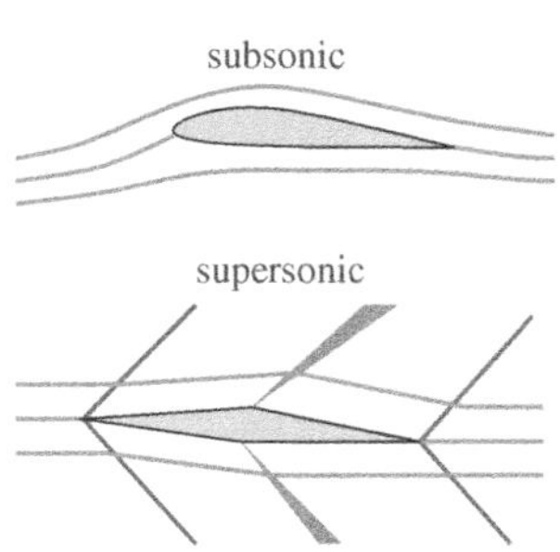

Figure 8.31 Comparison of subsonic and supersonic flow streamlines.

Supersonic airfoil design changes as drastically as the flow physics. In supersonic flight, disturbing the freestream flow creates shocks and wave drag. Thus, subsonic airfoils with rounded leading edges are not a great choice because a strong bow shock will appear in supersonic flow, as shown in Figure 8.32.

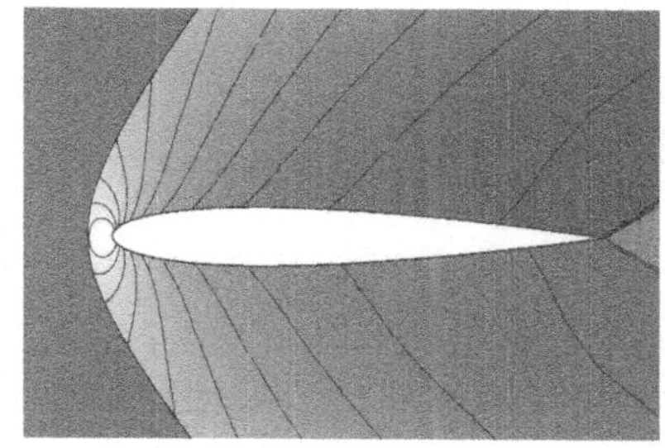

Figure 8.32 NACA 0012, $c_{dw} = 0.098$.

Instead, supersonic airfoils usually have sharp leading edges, which create oblique shocks that do not decelerate the flow as much and hence also do not generate as much entropy. Figure 8.33 shows a prototypical "diamond" supersonic airfoil with 12% thickness, like the NACA 0012. Compared to the NACA 0012, at the same freestream Mach number and lift coefficient, the diamond airfoil has less than half the drag coefficient. This drag reduction comes from keeping the disturbances small, in particular by avoiding normal shocks.

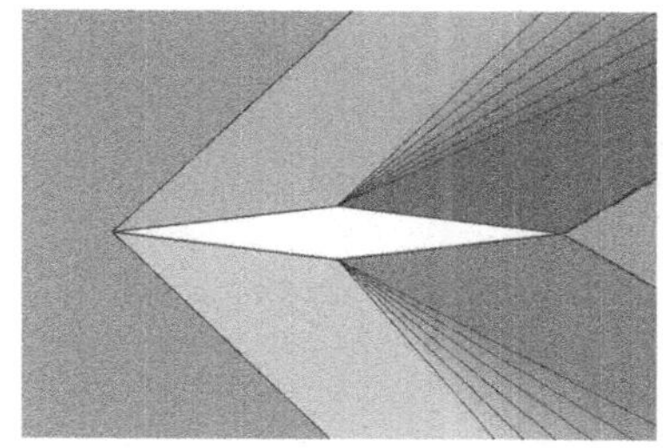

Figure 8.33 Diamond airfoil, $c_{dw} = 0.046$.

Assuming small disturbances, we obtain solutions to Equation 8.45 using the method of characteristics. By Equation 8.53, the pressure coefficient on the airfoil's surface has magnitude $2|\theta|/\lambda$, where θ is the local flow angle relative to the freestream – this angle includes the angle of attack. The sign of c_p depends on whether the flow is being compressed or expanded. The pressure distribution for an airfoil in supersonic flow is therefore straightforward to calculate under the small-disturbance potential flow assumptions.

Figure 8.34 shows a qualitative c_p distribution for a biconvex airfoil. If the upper and lower surfaces are quadratic functions of x, then $\theta(x)$ will be approximately linear. Note the discontinuity in pressure at the leading and trailing edges: the flow at these points goes through different shocks or expansions depending on if it is on the upper or lower surface.

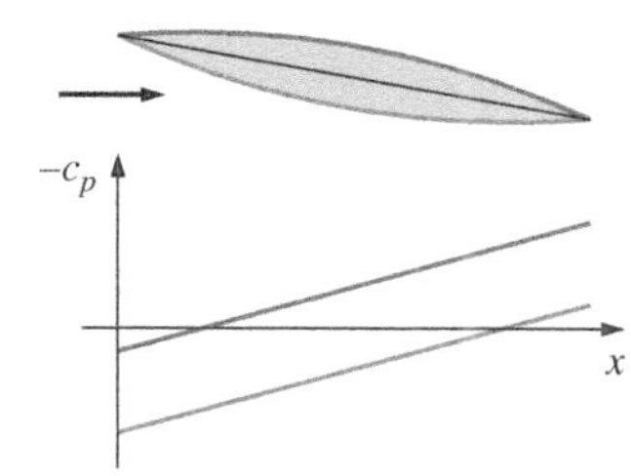

Figure 8.34 Pressure distribution for a biconvex airfoil in supersonic flow.

For comparison, Figure 8.35 shows a qualitative c_p distribution for a diamond airfoil. The straight segments of the airfoil surface make $\theta(x)$ piecewise constant. The lift coefficient generated by the two airfoils is the area (divided by c) between the lower and upper c_p distributions. This area is the same for both airfoils if they are flying at the same angle of attack. In addition, in both cases, the center of pressure is at the half-chord location. This contrasts with the approximate quarter-chord center of pressure of a (symmetric) airfoil in subsonic flow.

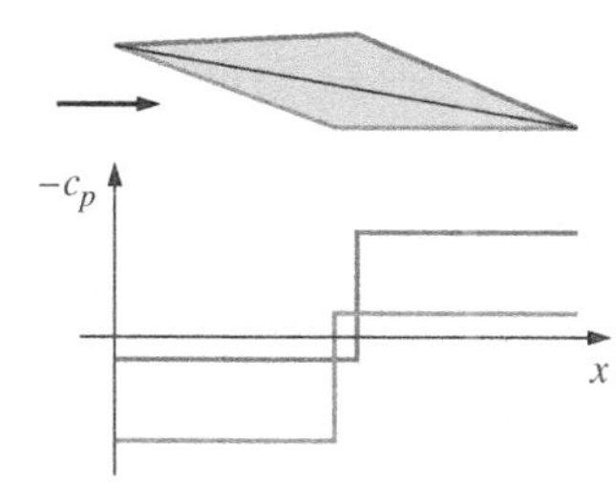

Figure 8.35 Pressure distribution for a diamond airfoil in supersonic flow.

Wave drag occurs due to the formation of shocks with their associated entropy increase. Linearized potential flow theory assumed isentropic changes to the flow, which means that, for small disturbances, the wave drag will be second order in the disturbances. We see this in Equation 8.73, where the drag coefficient depends on θ^2 on the lower and upper surfaces of an airfoil.

Example 8.6 Wave drag on a triangular airfoil
Consider a triangular airfoil formed by an isosceles triangle of base angles β, shown in Figure 8.36.

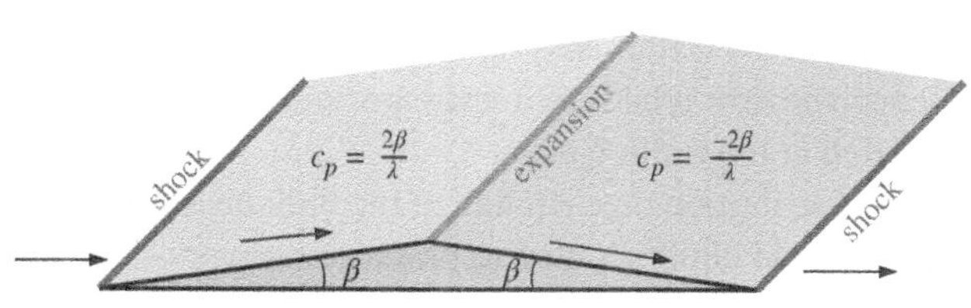

Figure 8.36 Supersonic flow over a triangular airfoil at zero angle of attack.

At zero angle of attack, the pressure coefficient distribution is as shown in the Figure. On the upper surface, it is positive after the compression of the leading-edge shock, negative after the expansion from the midpoint vertex, and back to zero after the trailing-edge shock. On the lower surface, it is zero because the geometry is aligned with the flow. Integrating the pressure coefficient or using Equation 8.73 shows that the drag on this airfoil is

$$c_d = \frac{2\beta^2}{\lambda}. \tag{8.80}$$

The previous example showed that in inviscid supersonic flow, an airfoil with thickness* experiences drag even when no lift is being generated. Indeed, Equation 8.73 confirms this: the only way to achieve zero drag is to have $\theta = 0$ on both surfaces, which means no thickness and zero angle of attack. However, this result was derived for a single airfoil in external flow. When multiple airfoils are present in proximity, their characteristics can interact, sometimes in beneficial ways.

* A flat plate at zero angle of attack would produce no wave drag.

Example 8.7 The Busemann biplane
A classic example of beneficial shock–expansion interaction is the Busemann biplane, conceived by Adolf Busemann around 1935 (Sears and Tan 1951). This biplane consists of two triangular airfoils stacked one above the other, with the triangle vertices pointing toward each other, as shown in Figure 8.37.

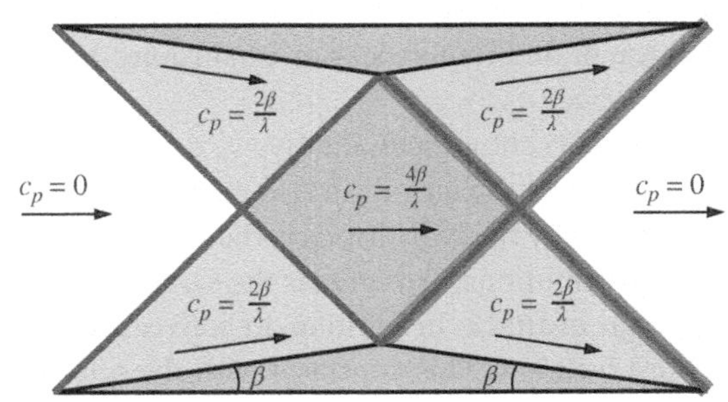

Figure 8.37 The Busemann biplane at zero angle of attack. Red lines indicate shocks, and green lines are expansions. The shocks and expansions overlap when the biplane is flying at its design Mach number, which is the one for which the leading shocks impinge on the triangle vertices.

This figure shows the distribution of pressure coefficient in the various regions formed at zero angle of attack when the biplane is in its design condition: the one in which the leading-edge shocks impinge exactly on the triangle vertices. Since the equations of small-disturbance potential flow are linear, the solutions for the flows over the two triangle airfoils superimpose.† This superposition results in a pattern of crossing shocks and coincident shocks/expansions. Every time the flow passes through a shock, the pressure coefficient increases by $2\beta/\lambda$. The expansions are twice as strong (double the flow turning), and so flow crossing an expansion experiences a c_p decrease of $4\beta/\lambda$. This information is sufficient to make the c_p sketch shown in Figure 8.37, with the result that $c_p = 0$ after passing through the last shock–expansion combination. Hence, the flow is back to freestream, and no further shocks or expansions are required at the trailing edge.

† A new addition consists of the reflected shocks and expansions.

The pressure distribution on each half of the biplane is fore-aft symmetrical, which means that we obtain complete pressure recovery. Therefore, the drag on each triangle, and hence the biplane, is zero. Of course, the biplane also produces no lift at the zero angle of attack, and once the angle of attack is increased, it generates both lift and drag like a flat plate. In this case, it still has a benefit compared to a single airfoil in an external flow because the drag associated with the thickness is eliminated.‡

‡ Or at least mitigated – the zero drag result only holds under the assumption of our theory, in particular infinitesimal disturbances and no viscosity. In addition, the biplane is assumed to fly at exactly on-design conditions, whereas deviations in angle or Mach number will prevent the beneficial cancellation.

8.6 Wing Design

Just as with airfoils, the designs of wings also vary across flow regimes. Figure 8.38 shows representative sketches of how wing shapes change with Mach number. At low speeds, high aspect-ratio, unswept wings are desirable for their low induced drag. Sailplanes are a good example of such wings, and their aspect ratios can reach 20 or 30. As the flight Mach number increases, the aspect ratio often decreases due to structural considerations. Approaching the critical Mach number, sweep is often introduced to prevent the formation of shocks, by reducing the Mach number normal to the wing. For the same reason, sweep continues to be important in the transonic and supersonic regimes. At high supersonic speeds, the wing shape often changes further into a "delta" wing for improved structural rigidity as aerodynamic loads increase.

The analysis of wings in compressible flow also changes with flow regime. Wings in low-speed flight can be analyzed with lifting-line or vortex-lattice methods, as long as they satisfy the necessary assumptions. A compressibility correction, such as Prandtl–Glauert, can be applied as the Mach number increases, if the flow remains subcritical. As with airfoils, the transonic regime is harder to analyze due to its nonlinearity, but semi-quantitative design rules can still be formulated. In the supersonic case, if disturbances are small, linearized potential flow applies and superposition makes quasi-analytical and numerical panel methods possible again. The next sections review these methods and present some quantitative solution techniques.

8.6.1 Prandtl–Glauert Correction

The three-dimensional subsonic small-disturbance potential flow equation is*

$$\underbrace{(1 - M_\infty^2)}_{\beta^2}\,\hat{\phi}_{xx} + \hat{\phi}_{yy} + \hat{\phi}_{zz} = 0. \tag{8.81}$$

* Recall that $\hat{\phi}$ is the disturbance potential, the gradient of which gives the velocity perturbations. In addition, the subscripts denote derivatives.

Just like in two dimensions, we can scale the x coordinate to obtain Laplace's equation for incompressible flow,

$$x = \beta\bar{x}, \quad \Rightarrow \quad \hat{\phi}_{\bar{x}\bar{x}} + \hat{\phi}_{yy} + \hat{\phi}_{zz} = 0.$$

The effect of compressibility on the pressure coefficient is the same as in two dimensions: it increases by $1/\beta$. However, the scaling of only one coordinate affects the shape of the wing.† Figure 8.39 shows this effect applied to a swept wing. All of the sectional chords of the original wing are multiplied by $1/\beta$, which is greater

† In two dimensions, the shape also changes by the scaling of the chord, but we are not concerned with variations in relative thickness under the assumptions of small disturbances. In three dimensions, the entire wing planform changes.

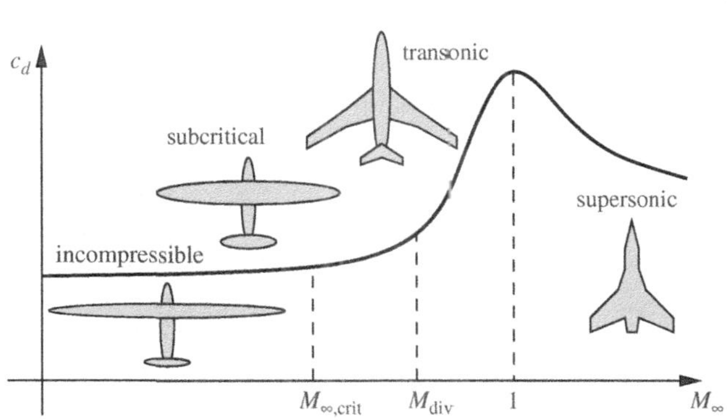

Figure 8.38 Variation of optimal wing shapes across flow regimes.

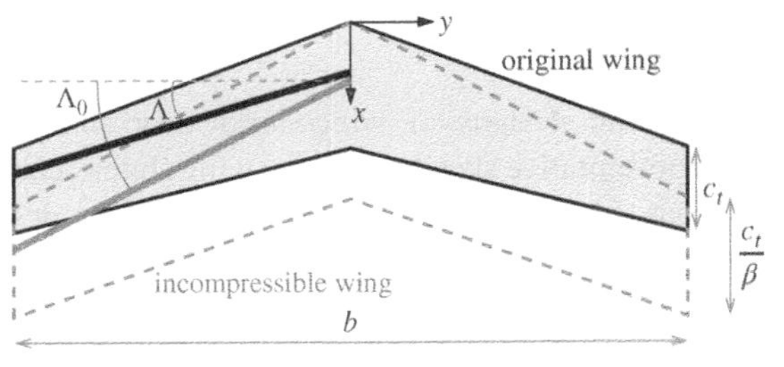

Figure 8.39 Wing planform stretching under the Prandtl–Glauert transformation.

than 1, to obtain the incompressible wing. From the geometry, the sweep angle and aspect ratio are also affected,

$$\frac{\tan \Lambda_0}{\tan \Lambda} = \frac{1}{\beta}, \quad AR = \frac{AR_0}{\beta}. \tag{8.82}$$

The sectional and total lift coefficients of the original wing are $1/\beta$ times those of the incompressible wing. However, since the area of the original wing is β times that of the incompressible one, the total and sectional lifts are the same for both wings.

8.6.2 Wing Sweep

Many high-speed aircraft have swept wings. Sweep is defined by the angle Λ between the quarter-chord line and the y axis. Various swept configurations are possible, as shown in Figure 8.40, with the most common being aft-swept wings. In many cases, a wing may not be defined by a single value of Λ. It may have variable sweep, in which Λ is not constant but changes with spanwise position. Or, for very low aspect ratio wings, such as delta wings, an alternate definition of sweep, based on the leading edge, may be more relevant than the quarter-chord sweep. For high aspect-ratio wings, sweep is generally well defined and has a specific role.

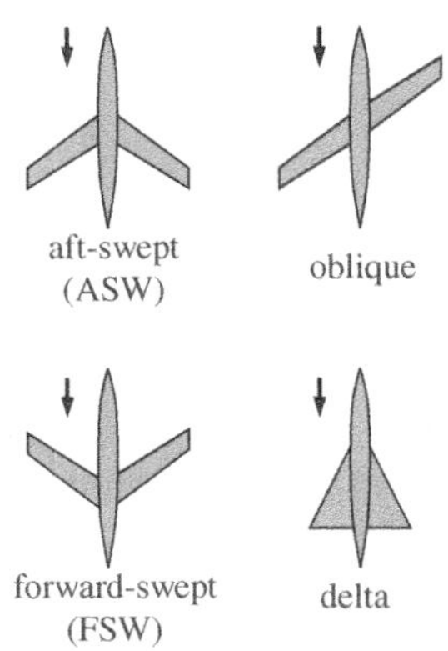

Figure 8.40 Different swept-wing configurations.

The overall goal of sweep is to reduce wave drag. At subsonic speeds, sweep increases the critical Mach number of the aircraft, while at transonic and supersonic speeds, it reduces the shock strength.

Sweep is based on the independence principle derived from the consideration of infinite wings. First, consider two infinite *unswept* wings in a uniform freestream of speed $U_\perp$, as shown in Figure 8.41. Whereas wing 1 is stationary in one external observer's frame, wing 2 is moving spanwise, perpendicular to the flow, at speed $U_\parallel$ in the same frame. If the flow is inviscid, the motion of wing 2 does not affect the airflow over it, so it has the same pressure distribution and force as wing 1.

Now imagine yourself as an observer standing on the second wing, as shown in Figure 8.42. You see a combination of normal and spanwise flow, but only the normal velocity affects the force, by Galilean invariance, discussed in Section 2.1.2. The combination of chordwise and spanwise flow in the observer's frame is what a swept wing sees. Hence, we conclude that an infinite swept wing in inviscid flow

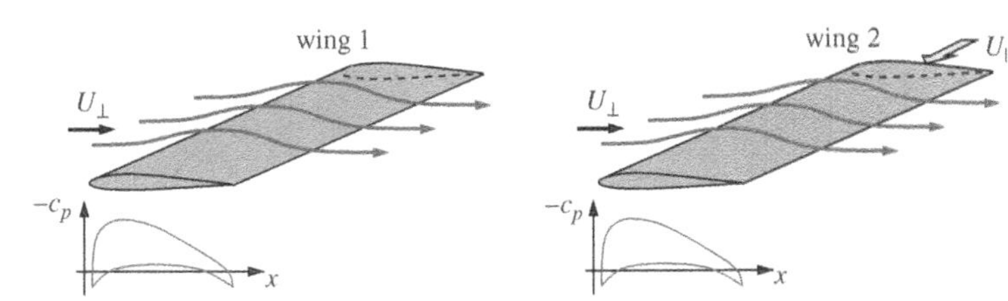

Figure 8.41 Two infinite wings in a uniform freestream flow.

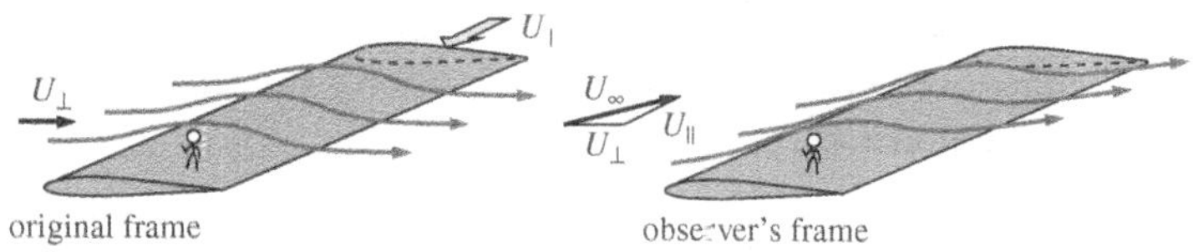

Figure 8.42 An observer on a moving wing sees crossflow, as if the wing were swept.

behaves the same way as an infinite unswept wing in a freestream of speed $U_\perp$, the velocity component perpendicular to the swept wing.*

As shown in Figure 8.43, in a swept wing, the wing sections perpendicular to the wing see an incoming Mach number $M_\perp$, given by

$$M_\perp = M_\infty \cos \Lambda. \tag{8.83}$$

Increasing the sweep angle Λ decreases $M_\perp$. For example, a 25° sweep makes $M_\perp$ 10% lower than M_∞. This decrease in turn increases $M_{\infty,\text{crit}}$ of the aircraft, since the aircraft can now fly at a Mach number that is higher than the critical Mach number of the airfoil sections, without having shocks form. Sweep does have its downsides, however. It leads to increased crossflow for finite wings and extra structural weight. The optimum sweep angle is therefore generally set by multidisciplinary design considerations.

*Viscosity and finite-wing effects make this analogy imperfect by introducing boundary layers and crossflow.

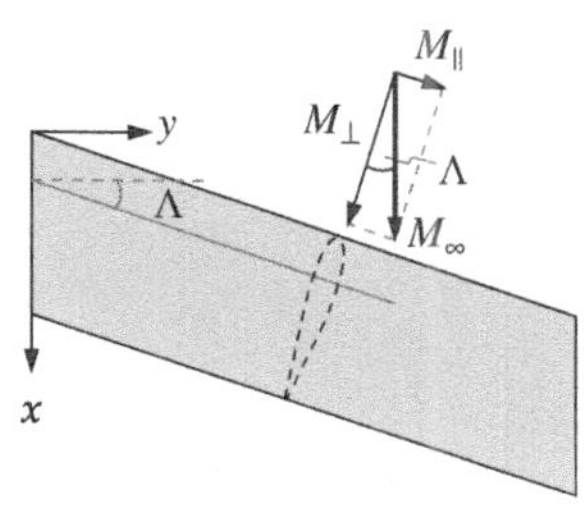

Figure 8.43 Perpendicular Mach number on a swept wing.

8.6.3 Slender-Body Theory

For aircraft in compressible flow, particularly in the transonic and supersonic regime, it is not enough to analyze the wing alone. The body of the aircraft, i.e. the fuselage, also creates disturbances, and these interact with the wing. Transonic and supersonic analysis therefore applies to wing-body configurations. The fuselage contributes to drag via skin friction and pressure drag, interference drag at the wing-body juncture, and wave drag due to shock formation and interaction with the wing. In this section, we consider how to design the wing and fuselage so as to reduce the wave drag.

Aircraft flying at high speeds are usually long and slender. A useful analysis technique is to approximate the wing-body combination as a single body of revolution, as shown in Figure 8.44. The cross-sectional area, $S(x)$, includes the fuselage, wings, and appendages at each streamwise location x. It is the blockage that the flow must go around, so changes in $S(x)$ lead to shocks and hence wave drag.

We can model supersonic flow over an axisymmetric body of revolution at zero angle of attack using a line of sources of strength $q(x)$, as shown in Figure 8.46.

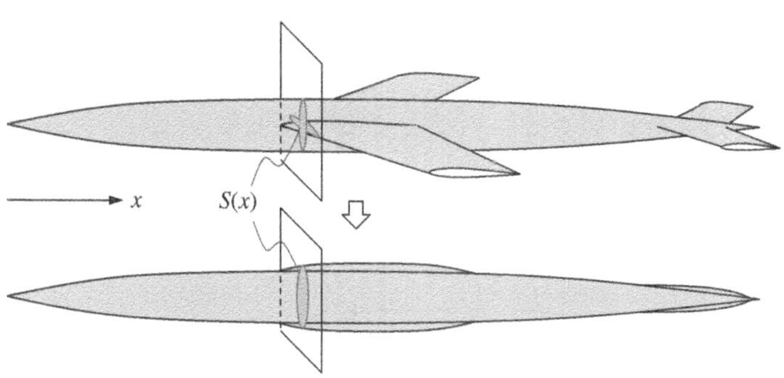

Figure 8.44 Slender-body approximation of an aircraft.

These sources are not the same as in incompressible flow. Instead, they are elementary solutions to the three-dimensional version of the *supersonic* potential flow equation, Equation 8.45,

$$\lambda^2\phi_{xx} - \phi_{yy} - \phi_{zz} = 0, \quad \lambda = \sqrt{M_\infty^2 - 1}.$$

A point source of strength Q at the origin has the velocity potential

$$\phi = -\frac{Q}{2\pi h}, \quad h = \sqrt{x^2 - \lambda^2(y^2 + z^2)}.$$

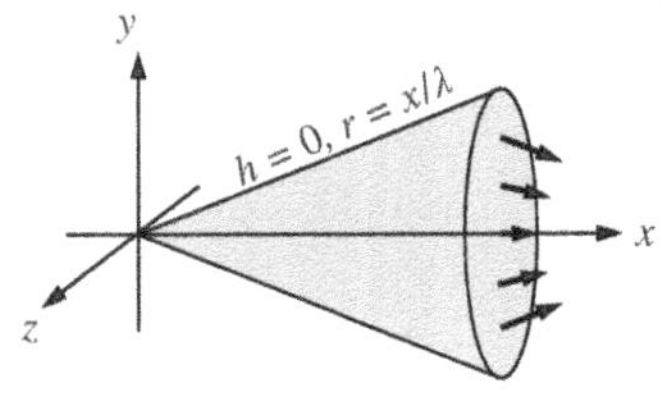

Figure 8.45 A point source in supersonic flow.

As shown in Figure 8.45, this velocity potential is undefined outside of a cone with apex at the origin and opening up along the downstream x axis. This form is therefore consistent with the physics of supersonic flow: point disturbances cannot affect the upstream flow and are only felt in a downstream Mach cone. The strength of these sources depends on the area distribution: at every location where the area changes, we need to place a source to account for the change. The continuous source strength is therefore directly related to the rate of area change according to $q(x) = U_\infty dS/dx$ (Kuethe and Chow 1997).

8.6.4 Slender-Body Wave Drag

To calculate the wave drag on the slender body, we can use a control-volume approach. Placing a cylindrical control volume around the slender body, as shown in Figure 8.46, conservation of x momentum gives the drag exerted by the fluid on the body as*

$$D_w = -\int_{S_{cyl}} \rho u(\vec{v} \cdot \vec{n})\, dS = -\rho_\infty \int_{S_{cyl}} \frac{d\phi}{dx}\frac{d\phi}{dr}\, dS. \tag{8.84}$$

*This calculation assumes that the inflow and outflow of momentum through the cylinder front and back balance, which is valid if the cylinder is long, since the states there will be freestream. The horizontal velocity is $u = \frac{\partial \phi}{\partial x}$, the radial velocity is $\vec{v} \cdot \vec{n} = \frac{\partial \phi}{\partial r}$, and by small disturbances, the density is approximately ρ_∞.

Note that in this expression, ϕ is the total potential, including freestream, not just the perturbations. Axisymmetry reduces the integral to $2\pi R$ times a line integral, and using the equation for the source potential from Figure 8.46 yields, after a few steps,

$$D_w = -\frac{\rho_\infty}{4\pi}\int_0^L\int_0^L q'(x)q'(\xi)\ln|x-\xi|\, d\xi dx. \tag{8.85}$$

Substituting the potential-flow source model $q(x) = U_\infty S'(x)$, this drag expression becomes (Kármán 1935; Jones 1953)

$$D_w = -\frac{\rho_\infty U_\infty^2}{4\pi}\int_0^L\int_0^L S''(x)S''(\xi)\ln|x-\xi|\, d\xi dx. \tag{8.86}$$

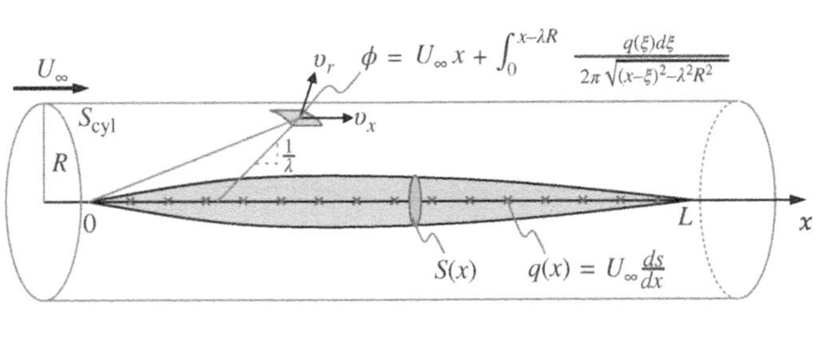

Figure 8.46 Drag calculation on a slender body in supersonic flow.

This equation can be used to calculate the wave drag from knowledge of the cross-sectional area of a slender body. In designing a body shape, a natural question is what should the shape be to minimize the drag, under some constraint such as a fixed body volume. Such a minimization is simplified by a variable transformation combined with a Fourier expansion of the source strength,†

$$x = \frac{L}{2}(1 - \cos\theta), \quad q = U_\infty L \sum_{n=1}^{\infty} A_n \sin n\theta. \tag{8.87}$$

Substituting these equations into Equation 8.85, gives, after some manipulation (Jones 1953),

$$D_w = \frac{\pi \rho U_\infty^2 L^2}{8} \sum_{n=1}^{\infty} n A_n^2. \tag{8.88}$$

Only A_1 contributes to the base area, which is $S(L)$,‡ and only A_1 and A_2 contribute to the body volume,

$$V = \frac{\pi L^3}{8}\left(A_1 + \frac{1}{2}A_2\right). \tag{8.89}$$

However, from Equation 8.88, all A_n contribute to the drag. So we can find the minimum-drag body by setting $A_n = 0$ for all $n > 2$.

Keeping only $A_1 \neq 0$ gives the minimum-drag shape for a non-zero base area, shown in Figure 8.47. This is the *von Kármán ogive*, presented by Kármán (1935), with area distribution and wave-drag coefficient based on the base area, $S(L)$,

$$S(\theta) = \frac{S(L)}{\pi}\left(\theta - \frac{1}{2}\sin 2\theta\right), \quad C_{Dw} = \frac{4}{\pi}\frac{S(L)}{L^2}. \tag{8.90}$$

Keeping only $A_2 \neq 0$ gives the minimum-drag shape with zero base area, which is pointed at both ends, as shown in Figure 8.47. This is the *Sears–Haack body* (Haack 1941; Sears 1947; Jones 1953), for which the area distribution and drag coefficient based on the maximum cross-sectional area at $x = L/2$ are

$$S(\theta) = \frac{4V}{\pi L}\left(\sin\theta - \frac{1}{3}\sin 3\theta\right), \quad C_{Dw} = \frac{24V}{L^3}. \tag{8.91}$$

† The Fourier expansion coefficients A_n are of course not the same as the ones in finite-wing theory, but they serve a similar purpose to simplify the analysis.

‡ The relationship between A_1 and $S(L)$ is left to Problem 7.

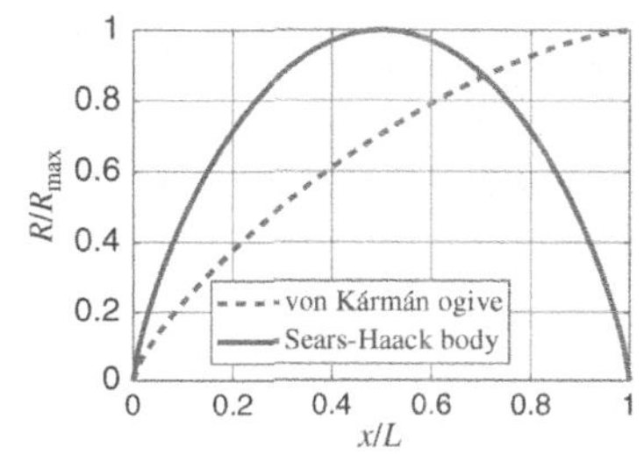

Figure 8.47 Optimal slender-body shapes for a given base area and volume.

Example 8.8 Sears–Haack A_n calculation

To obtain the A_n coefficients given $S(x)$ we use Equation 8.87, with $q = U_\infty dS/dx$, so that

$$\begin{aligned} U_\infty L \sum_{n=1}^{\infty} A_n \sin n\theta &= U_\infty \frac{dS}{dx}, \\ L \sum_{n=1}^{\infty} A_n \sin n\theta &= \frac{dS}{dx}. \end{aligned} \tag{8.92}$$

Multiplying both sides by $\sin m\theta$ and integrating over the body, from $\theta = 0$ to $\theta = \pi$, give§

$$A_n = \frac{2}{\pi L}\int_0^{\pi} \frac{dS}{dx} \sin n\theta \, d\theta. \tag{8.93}$$

We can verify that this formula returns the correct A_n values for the Sears–Haack body. From Equation 8.91, we calculate dS/dx as

$$\frac{dS}{dx} = \frac{dS}{d\theta}\frac{d\theta}{dx} = \frac{4V}{\pi L}(\cos\theta - \cos 3\theta)\frac{2}{L\sin\theta}. \tag{8.94}$$

§ The left-hand side becomes

$$L\frac{\pi}{2}\delta_{mn}A_n$$

due to orthogonality of the sine functions over the range $\theta \in [0, \pi]$. Since $\delta_{mn}A_n = A_m$, we can revert back to the index n instead of m.

Substituting into the A_n formula,

$$A_n = \frac{16V}{\pi^2 L^3} \int_0^{\pi} (\cos\theta - \cos 3\theta) \frac{\sin n\theta}{\sin\theta}\, d\theta. \tag{8.95}$$

The only nonzero A_n occurs for $n = 2$,¶ since $\sin 2\theta = 2\sin\theta\cos\theta$, so that

¶ For all other n, the integration yields 0 due to orthogonality of sine and cosine functions.

$$A_2 = \frac{16V}{\pi^2 L^3} \int_0^{\pi} (\cos\theta - \cos 3\theta)\, 2\cos\theta\, d\theta = \frac{16V}{\pi L^3}. \tag{8.96}$$

This value of A_2 is consistent with the volume formula in Equation 8.89 for $A_1 = 0$.

Example 8.9 Slender-body drag calculation

Suppose that we are given the shape of a slender body as a discrete set of sampled areas, S_i. Using Equation 8.86 to compute the drag requires the calculation of second derivatives of $S(x)$, which we can do with finite differences. However, it also requires integrating through a singularity, $\ln|x - \xi|$, which can pose challenges numerically.

An alternative approach to calculating the wave drag uses Equation 8.88. To obtain the A_n coefficients, we use Equation 8.93. We assume that we have S_i samples at equally spaced points in x and use a central difference to calculate dS/dx at all points.|| We then integrate over θ by trapezoidal integration, as shown in Listing 8.2.

|| A central difference works at all points except at the ends, where we do not need a value since $\sin n\theta = 0$ at $\theta = 0$ and π.

Code notes:

- The physical parameters are arbitrary, as A_n and C_{Dw} are non-dimensional.
- The code uses the Sears–Haack body for verification, but any other shape could be substituted in the S formula.
- Only a few (4) A_n coefficients are calculated for brevity in the output, but this can easily be increased.
- A_2 and C_{Dw} are compared to the analytical values for the Sears–Haack body.

Listing 8.2 Drag calculation for a slender body.

```
function slenderdrag
% calculates the drag of the Sears-Haack body

rho = 1; U = 1; L = 10; V = 1;   % arbitrary
N = 500; dx = L/N; % number of points and dx
x = linspace(0,L, N+1); % equal spacing in x
theta = acos(1-2*x/L); % corresponding theta values

% S for the Sears-Haack body
S = 4*V/(pi*L) * (sin(theta) - (1/3)*sin(3*theta));

% central difference for dS/dx
S_x = [0, (S(3:end)-S(1:end-2))/(2*dx), 0];

% An and Dw calculation
Dw = 0; A = zeros(4,1);
for n = 1:4,
  A(n) = 2/(pi*L)*trapz(theta, S_x.*sin(n*theta));
  Dw = Dw + pi*rho*(U*L)^2/8 *n*A(n)^2;
end
A
A2true = 16*V/(pi*L^3) % Sears-Haack body
CDw = Dw/(0.5*rho*U^2*max(S))
CDWtrue = 24*V/L^3
```

Listing 8.3 Drag calculation output.

```
A =
    0.0000
    0.0051
    0.0000
    0.0000
A2true =
    0.0051
CDw =
    0.0240
CDWtrue =
    0.0240
```

Listing 8.3 shows the output of the code, which consists of the first four A_n coefficients and the wave drag coefficient, C_{Dw}, normalized by the maximum cross-sectional area. From this output, we see that only A_2 is nonzero, and that it matches the analytical result for the Sears–Haack body. The same holds for the wave drag coefficient.

8.6.5 The Area Rule

The previous section presented the optimal shape of a slender body of revolution in supersonic flow, which is the Sears–Haack body if it is pointed at both ends. However, an aircraft is clearly not an axisymmetric slender body, since it has wings, a tail, and other asymmetries. Furthermore, the optimality of the Sears–Haack body is Mach number independent, whereas we observe different aircraft drag coefficients at different Mach numbers. The question is therefore how to apply the slender-body theory results to the analysis of a full aircraft. Ideally, we could map the aircraft to a slender body by consolidating all of the non-symmetric features into an equivalent area, as shown in Figure 8.48, but this only works for Mach numbers close to 1, as we shall see shortly. For general Mach numbers, we can map an aircraft to a body of revolution by looking at Mach planes, which are tangent to Mach cones, as illustrated in Figure 8.49. The underlying principle is that disturbances propagate along Mach cones. If a Mach cone intersects a lot of area, a lot of flow turning occurs, since the disturbances superimpose in our linear theory. An optimal supersonic design at a given Mach number will therefore minimize the amount of area in any given Mach cone.

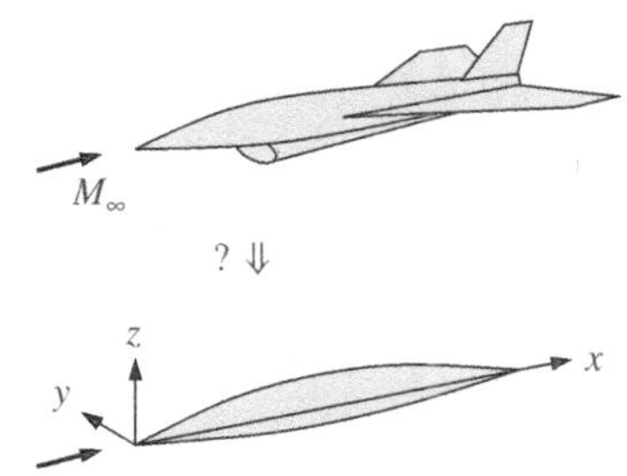

Figure 8.48 Applying slender-body theory to aircraft requires more than just an area calculation.

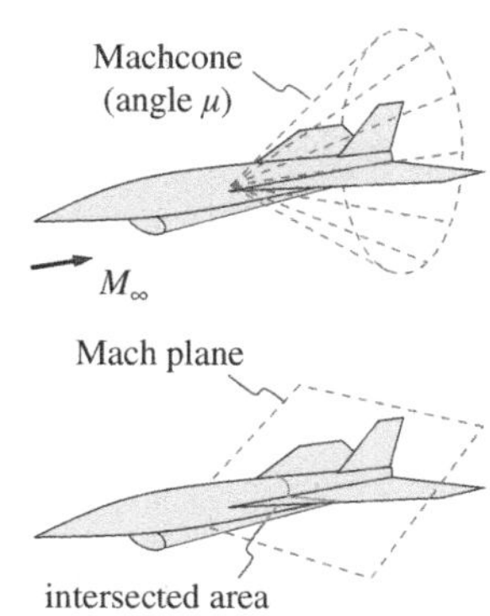

Figure 8.49 A Mach plane cutting through an aircraft yields an intersected area.

The drag calculation for a general aircraft shape in supersonic flow becomes more complicated due to the fact that Mach planes can intersect the aircraft at different angles. For a given x location, we consider a Mach plane at an azimuthal angle φ shown in figure 8.50, and we compute the intersected area, $s(\varphi, x)$. Using the Mach angle, μ, we then project this area onto the y–z plane to obtain the projected area distribution,

$$S(\varphi, x) = \sin(\mu)s(\varphi, x). \tag{8.97}$$

Slender-body theory applies to this distribution, which means that we can compute the drag using Equation 8.86, which is an integral over x.* The result of the slender-body calculation using planes at a single azimuthal angle, φ, is $D_w(\varphi)$. The final wave drag is then the average of these drag values,

$$D_w = \frac{1}{2\pi} \int_0^{2\pi} D_w(\varphi)\, d\varphi. \tag{8.98}$$

*The length of the slender body, L, will generally depend on φ, since some parts of the aircraft may intersect a particular Mach plane at, for example, $x < 0$.

It is important to remember the assumptions that underlie this equation, which are small-disturbance supersonic potential flow. Strong normal shocks, viscous effects, and large velocity deviations from the freestream cannot be captured in this theory.

In transonic flow, for $M_\infty \approx 1$, the drag calculation simplifies because $\mu \approx 90°$, which means that the Mach planes are nearly perpendicular to the aircraft centerline. In this case, the wave drag is approximated by the slender-body formula, Equation 8.86, using the aircraft cross-sectional area. This is known as Hayes's

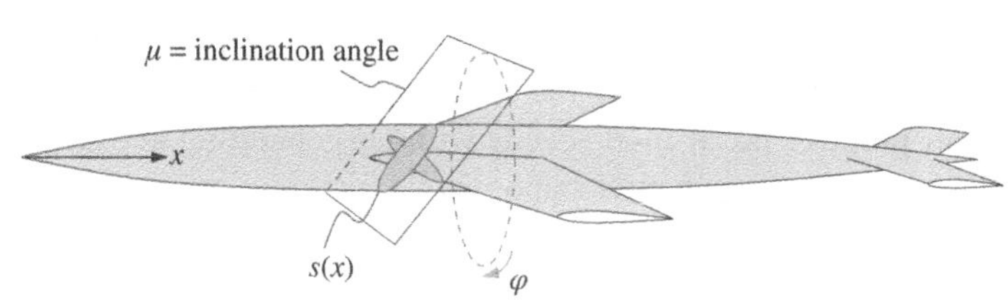

Figure 8.50 The supersonic area rule for calculating wave drag.

area rule (Hayes 1947; Lomax and Heaslet 1956), which was tested by Whitcomb (1956).

The fact that changes in the cross-sectional area yield increased wave drag leads to the "coke-bottle" fuselage design, shown in Figure 8.51, which narrows at locations of the wing to keep the sum of fuselage and wing cross-sectional areas nearly constant. This design applies to Mach numbers in the transonic regime. Higher-speed flight requires analysis using the supersonic area rule with Mach plane intersections.

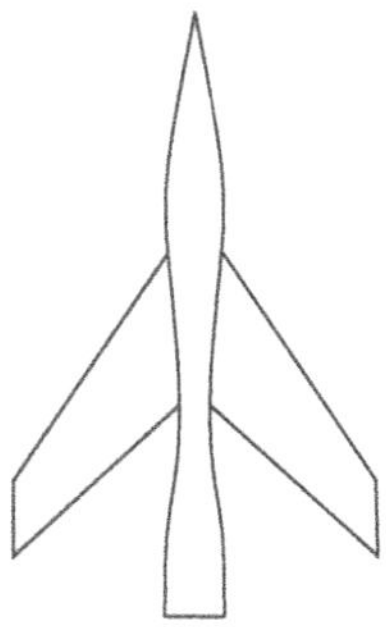

Figure 8.51 A coke-bottle fuselage based on the transonic area rule.

8.7 Summary and Concepts

Whereas the previous chapters considered almost exclusively incompressible flow, this chapter introduced compressibility and studied its effects on both subsonic and supersonic flight. Most of the quantitative results came from assuming small disturbances, which is usually a good assumption for vehicles designed for efficient high-speed flight. This assumption effectively linearized the nonlinear compressible flow equations, simplifying their application. In the subsonic regime, corrections to incompressible flow were derived, and these were used for predicting the critical Mach number at which shocks appear. In the supersonic regime, small-disturbance theory yielded a fundamentally different solution approach based on characteristics, with simplified lift and drag expressions. The chapter concluded by presenting the application of these tools to the design of both airfoils and wings in compressible flow.

The key concepts covered in this chapter are as follows:

- All fluids are compressible, and a measure of compressibility is the Mach number, which is the ratio of the fluid speed to the speed of sound.
- Increasing the Mach number changes the flow regime from subcritical to transonic to supersonic.
- The appearance of shocks in transonic and supersonic flow creates wave drag, caused by the irreversible entropy generation within a shock.
- Assuming small disturbances lets us linearize the governing equations to simple compressible potential flow, which is valid in both the subsonic and supersonic regimes. Transonic and hypersonic flow cannot be analyzed with this theory.
- The same equation governs linearized subsonic and supersonic flow, but a sign switch on one of the terms when M_∞ crosses 1 fundamentally changes the character from elliptic to hyperbolic, which also changes the solution process.
- In the subsonic regime, prior to the onset of shocks, pressure is amplified by $1/\sqrt{1 - M_\infty^2}$.
- The lift coefficient of supersonic airfoils depends only on the angle of attack, $c_\ell = 4\pi/\sqrt{M_\infty^2 - 1}$.
- Wave drag in supersonic flow is hard to avoid, and in particular, it is present any time lift is generated.
- Supercritical airfoils maximize the critical Mach number, which means that they can produce lift at high speeds without generating shocks, and their shocks are weak when they first appear.

- Wings are swept to reduce the perpendicular Mach number and delay the onset of shocks at high speeds.
- To minimize wave drag in transonic and low-supersonic flow, wing-body shapes should avoid large changes in the cross-sectional area.

8.8 Problems

1. Justify each answer to the below questions with an explanation.
   a True or false: if an aircraft creates a shock, it must be flying faster than the speed of sound.
   b True or false: given two airfoils, the one producing a higher lift coefficient always has the lower critical Mach number.
   c True or false: in subsonic small-disturbance flow, the center of pressure does not change with Mach number.
   d True or false: in supersonic small-disturbance flow, increasing M_∞ increases the drag coefficient.
   e True or false: an airfoil producing zero lift in supersonic flow also produces zero lift in subsonic flow.
   f What is the strength of an F' characteristic resulting from a G' characteristic hitting a horizontal wall as shown in figure 8.52?
   g True or false: when supersonic horizontal flow encounters a compression wave of positive slope, it deflects upward.

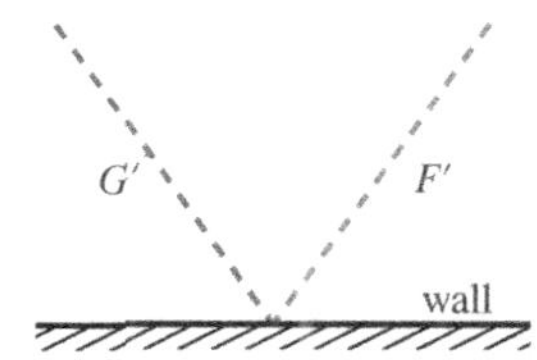

Figure 8.52 Characteristic reflecting off a horizontal wall.

2. Consider two airfoils, a NACA 4312 and a NACA 4812, designed to operate in the compressible lift coefficient range of $0.6 \leq c_\ell \leq 1.4$. On one Figure, plot the critical Mach number corresponding to each compressible lift coefficient for both airfoils. Use the Prandtl–Glauert correction and a panel solver to obtain the incompressible c_p. Discuss the differences, using c_p diagrams to demonstrate your points.

3. A thin flat-plate airfoil flying at angle of attack α deflects a flap down at an angle η as shown in figure 8.53. Assuming small-disturbance supersonic flow, sketch the compressions and expansions that define the various regions in the flowfield, and calculate the lift coefficient, drag coefficient, center of pressure, and the aerodynamic center.

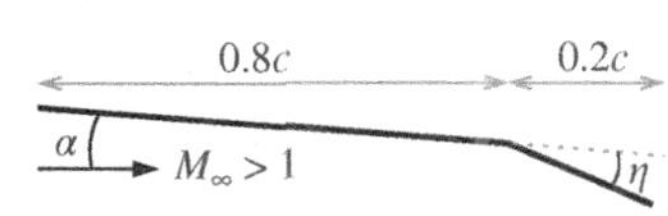

Figure 8.53 Flat plate with a flap in supersonic flow.

4. Using characteristic analysis, verify that when a compression wave reflects off a wall, it remains a compression wave. Draw a flow diagram and analyze the flow direction and pressure coefficient in each region.

5. An airfoil of 10% thickness as shown in figure 8.54 has a quadratic upper surface and a flat lower surface. At an angle of attack α, determine its drag coefficient and center of pressure in supersonic flow, using small-disturbance theory.

Figure 8.54 Airfoil with quadratic upper surface.

6. A symmetric diamond airfoil of chord c is flying at a height h above the ground at $M_\infty = \sqrt{2}$ and zero angle of attack as shown in figure 8.55. Assuming small disturbances, sketch lines separating different regions of the flow for $h = c$, and calculate the flow properties in each region. At what h (in terms of c) does the airfoil first feel the effects of the ground?

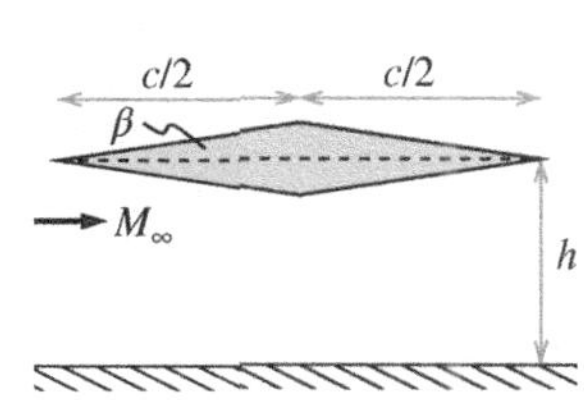

Figure 8.55 Diamond airfoil near the ground.

7. In slender-body theory, starting with the source-strength expansion in Equation 8.87, express the base area, $S(L)$ in terms of A_1 and verify that Equation 8.89 for the volume holds.

8. An aircraft flying at $M_\infty \approx 1$ is approximated a slender body with the lumped cross-sectional area distributions shown in Figure 8.56. Determine the wave drag at $\rho_\infty = 0.2\,\text{kg/m}^3$, $U_\infty = 300\,\text{m/s}$.*

* Hint: use a large number of integration points and calculate the drag using various numbers of A_n coefficients.

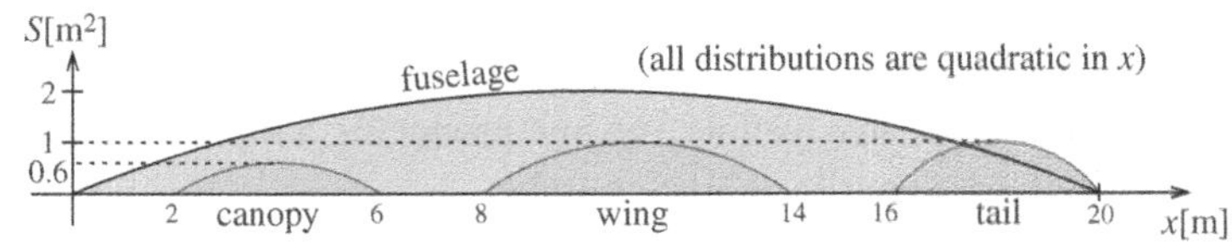

Figure 8.56 Cross-sectional area distribution for an aircraft.

References

Acheson, David J. (1990). *Elementary Fluid Dynamics.* Oxford Applied Mathematics and Computing Science Series. Oxford: Clarendon Press (cited on pages 1, 76, 83).

Anderson, John D. (2012). *Introduction to Flight.* Seventh. McGraw Hill (cited on page 98).

Anderson, John D. (2016). *Fundamentals of Aerodynamics.* McGraw Hill (cited on pages 60, 61, 152, 153, 208, 275, 302).

Batchelor, George K. (2000). *An Introduction to Fluid Dynamics.* Cambridge: Cambridge University Press (cited on pages 1, 75, 77, 190).

Bertin, John J. and Russell M. Cummings (2001). *Aerodynamics for Engineers.* Cambridge University Press (cited on page 275).

Blasius, Paul Richard Heinrich (1950). *The Boundary Layers in Fluids with Little Friction.* NACA Technical Memorandum. 1256 (cited on page 213).

Cebeci, Tuncer. and Apollo M.O. Smith (1974). *Analysis of Turbulent Boundary Layers.* New York: Academic Press (cited on page 208).

d'Alembert, Jean Le Rond (2012). *Essai d'une Nouvell Théorie de la Résistance des Fluides.* Hachette Livre BNF (cited on page 110).

Drela, Mark (1989). "XFOIL: An Analysis and Design System for Low Reynolds Number Airfoils." In: *Low Reynolds Number Aerodynamics, Lecture Notes in Engineering.* Ed. by T.J. Mueller. Vol. 54. Springer Berlin Heidelberg, pp. 1–12. doi: 10.1007/978-3-642-84010-4_1 (cited on pages 169, 246).

Drela, Mark (2014). *Flight Vehicle Aerodynamics.* Cambridge, MA: MIT Press (cited on pages 235, 246).

Fidkowski, Krzysztof J. (2013). "High-Order Output-Based Adaptive Methods for Steady and Unsteady Aerodynamics." In: *37th Advanced CFD Lectures series; Von Karman Institute for Fluid Dynamics (December 9–12 2013).* Ed. by Herman Deconinck and Rémi Abgrall. von Karman Institute for Fluid Dynamics (cited on page 336).

Fidkowski, Krzysztof J. (2022). "A Coupled Inviscid-Viscous Airfoil Analysis Solver, Revisited." In: *AIAA Journal* 60.5, pp. 2961–2971. doi: 10.2514/1.J061341 (cited on pages 167, 169).

Gad-el-Hak, Mohamed (1995). "Stokes' Hypothesis for a Newtonian, Isotropic Fluid." In: *Journal of Fluids Engineering* 117.1, pp. 3–5 (cited on page 78).

Glauert, Hermann (1923). *The Elements of Aerofoil and Airscrew Theory.* London: Cambridge University Press (cited on page 270).

Glauert, Hermann and Geoffrey Ingram, Taylor (1928). "The Effect of Compressibility on the Lift of an Aerofoil." In: *Proceedings of the Royal Society of London. Series A, Containing Papers of a Mathematical and Physical Character* 118.779, pp. 113–119. doi: 10.1098/rspa.1928.0039 (cited on page 318).

Haack, Wolfgang (1941). "Geshossformen Kleinsten Wellenwiderstandes." In: *Lilienthal-Gesellschaft für Lüftfahrtforschung* 139, pp. 14–28 (cited on page 345).

Hayes, Wallace D. (1947). "Linearized Supersonic Flow." PhD thesis. California Institute of Technology (cited on page 348).

Helmholtz, Hermann von (1867). "LXIII. On Integrals of the Hydrodynamical Equations, Which Express Vortex-Motion." In: *The London, Edinburgh, and Dublin Philosophical Magazine and Journal of Science* 33.226, pp. 485–512. doi: 10.1080/14786446708639824 (cited on pages 260, 261).

Houghton, Edward L. et al. (2016). *Aerodynamics for Engineering Students.* New York: Butterworth-Heinemann (cited on page 264).

Aerodynamics: A Computational Introduction, First Edition. Krzysztof J. Fidkowski.
© 2026 John Wiley & Sons, Inc. Published 2026 by John Wiley & Sons, Inc.

Jones, Robert T. (1953). *Theory of Wing-Body Drag at Supersonic Speeds.* NACA Technical Report 1284 (cited on pages 344, 345).

Joukowsky, Nikolai E. (1910). "Über die Konturen der Tragflächen der Drachenflieger." In: *Zeitschrift für Flugtechnik und Motorluftschiffahrt* 1, pp. 281–284 (cited on page 191).

Kármán, Theodore von (1935). "The Problem of Resistance in Compressible Fluids." In: *Volta Congress.* Rome, pp. 222–276 (cited on pages 344, 345).

Kármán, Theodore von (1941). "Compressibility Effects in Aerodynamics." In: *Journal of the Aeronautical Sciences* 8.9, pp. 337–356. doi: 10.2514/8.10737 (cited on page 335).

Katz, Joseph and Allen Plotkin (2001). *Low-Speed Aerodynamics.* Cambridge Aerospace Series 13. Cambridge University Press (cited on pages 167, 275).

Kroo, Ilan and Stephen Smith (1991). *The Computation of Induced Drag with Nonplanar and Deformed Wakes.* NASA Technical Report N91-24106 (cited on page 262).

Kuethe, Arnold M. and Chuen-Yen, Chow (1997). *Foundations of Aerodynamics: Bases of Aerodynamic Design.* Wiley (cited on pages 249, 343).

Lamb, Horace (1932). *Hydrodynamics.* London: Cambridge University Press (cited on page 264).

Landau, Lev D. and Evgeny M. Lifshitz (1959). *Fluid Mechanics.* Pergamon Press (cited on pages 1, 304).

Liepmann, Hans W. and Anatol Roshko (2002). *Elements of Gasdynamics.* New York: Dover Publications (cited on page 312).

Lomax, Harvard and Max A. Heaslet (1956). "Recent Developments in the Theory of Wing-Body Wave Drag." In: *Journal of the Aeronautical Sciences* 23.12, pp. 1061–1074. doi: 10.2514/8.3731 (cited on page 348).

McBain, Geordie D. (2012). *Theory of Lift: Introductory Computational Aerodynamics in MATLAB/Octave.* Wiley (cited on page 110).

McLean, Doug (2012). *Understanding Aerodynamics: Arguing from the Real Physics.* Wiley (cited on page 302).

Munk, Max M. (1921). *The Minimum Induced Drag of Aerofoils.* NACA Technical Report 121 (cited on page 270).

Oswatitsch, Klaus (1956). *Gas Dynamics.* New York: Academic, pp. 177–210 (cited on page 336).

Pope, Stephen B. (2000). *Turbulent Flows.* Cambridge: Cambridge University Press (cited on pages 33, 246).

Prandtl, Ludwig and Oskar G. Tietjens (1934). *Fundamentals of Hydro- and Aeromechanics.* New York: McGraw-Hill Book Company, Inc. (cited on pages 34, 38, 260, 261, 264).

Reynolds, William C. (1987). *Fundamentals of Turbulence for Turbulence Modeling and Simulation.* Tech. rep. Advisory Group for Aerospace Research and Development, North Atlantic Treaty Organization, pp. 1–66 (cited on page 246).

Reynolds, William C. (2004). "The Potential and Limitations of Direct and Large Eddy Simulations." In: *Whither Turbulence? Turbulence at the Crossroads.* Vol. 357. Lecture Notes in Physics. Springer Berlin Heidelberg, pp. 313–343. doi: 10.1007/3-540-52535-1_52 (cited on page 246).

Saffman, Philip G. (1995). *Vortex Dynamics.* Cambridge Monographs on Mechanics. Cambridge University Press (cited on pages 260, 261).

Schlichting, Hermann and Klaus Gersten (2016). *Boundary-Layer Theory.* Springer Berlin Heidelberg (cited on page 226).

Sears, William R. (1947). "On Projectiles of Minimum Wave Drag." In: *Quarterly of Applied Mathematics* 4.4, pp. 361–366 (cited on page 345).

Sears, William R. and Hao Sung Tan (1951). "The Aerodynamics of Supersonic Biplanes." In: *Quarterly of Applied Mathematics* 9.1, pp. 67–76 (cited on page 338).

Stokes, George Gabriel (1845). "On the theories of the internal friction of fluids in motion, and of the equilibrium and motion of elastic solids". In: *Transactions of the Cambridge Philosophical Society* (cited on page 83).

The MathWorks Inc. (2024). *MATLAB.* Natick, MA: The MathWorks Inc. (cited on page 19).

Thwaites, Bryan (1949). "Approximate Calculation of the Laminar Boundary Layer." In: *Aeronautical Quarterly* 1.3, pp. 245–280. doi: 10.1017/S0001925900000184 (cited on page 241).

Trefethen, Lloyd N. (2000). *Spectral Methods in MATLAB.* Society for Industrial and Applied Mathematics (cited on page 282).

Tsien, Hsue-Shen. (1939). "Two-Dimensional Subsonic Flow of Compressible Fluids." In: *Journal of the Aeronautical Sciences* 6.10, pp. 399–407. doi: 10.2514/8.916 (cited on page 335).

Whitcomb, Richard T. (1956). *A Study of the Zero-Lift Drag-Rise Characteristics of Wing-Body Combinations Near the Speed of Sound.* NACA Technical Report 1273 (cited on page 348).

White, Frank M. (1974). *Viscous Fluid Flow.* McGraw Hill (cited on pages 244, 245).

Wilcox, David C. (2006). *Turbulence Modeling for CFD.* La Cañada, CA: DCW Industries, Inc. (cited on page 248).

Index

Aerodynamics: A Computational Introduction, First Edition. Krzysztof J. Fidkowski.
© 2026 John Wiley & Sons, Inc. Published 2025 by John Wiley & Sons, Inc.